D0574016

Piloting Basics Handbook

Other McGraw-Hill Books of Interest

Aviation Weather Handbook By Terry T. Lankford
Instrument Flying Handbook By Thomas P. Turner

Piloting Basics Handbook

Lewis Bjork, Editor

McGraw-Hill

New York   San Francisco   Washington, D.C.   Auckland   Bogotá
Caracas   Lisbon   London   Madrid   Mexico City   Milan
Montreal   New Delhi   San Juan   Singapore
Sydney   Tokyo   Toronto

Library of Congress Cataloging-in-Publication Data

Bjork, Lewis.
    Piloting basics handbook / Lewis Bjork.
        p.   cm.
    Includes index.
    ISBN 0-07-136104-9
    1. Airplanes—Piloting—Handbooks, manuals, etc.   I. Title.

TL710.B553   2000
629.132'52—dc21   00-060577

McGraw-Hill

A Division of The McGraw-Hill Companies

  2 3 4 5 6 7 8 9 0 DOC/DOC   0 9 8 7 6 5 4 3 2 1

ISBN 0-07-136104-9

The sponsoring editor for this book was Shelley Ingram Carr, the editing supervisor was Sally Glover, and the production supervisor was Pamela Pelton. It was set in Garamond by Kim Sheran and Joanne Morbit of McGraw-Hill's desktop publishing department, Hightstown, N.J.

Printed and bound by R. R. Donnelley & Sons Company.

McGraw-Hill books are available at special quantity discounts to use as premiums and sales promotions, or for use in corporate training programs. For more information, please write to the Director of Special Sales, McGraw-Hill, Two Penn Plaza, New York, NY 10121-2298. Or contact your local bookstore.

This book is printed on recycled, acid-free paper containing a minimum of 50% recycled, de-inked fiber.

Contents

17 A closer look at the airspace system   *423*

18 Special-use airspace   *463*

Part four: INTEGRATE   *655*

Introduction

No book can exactly duplicate a seat-of-the-pants feel of the airplane, a smooth touch on the flight controls, or even the sometimes rapidly changing conditions that are part of any normal flight. As with riding a bicycle, the only way to learn flying is to fly. This book is therefore intended only as a supplement to your flight training.

In spite of the foregoing, however, there is a great deal offered here that would stretch the talents of even the best flight instructors. You hold in your hands the collected guidance of several experts in various aspects of flying. There are appropriate comments in basic flight techniques that will enhance your knowledge of why the airplane behaves as it does. These come from two talented flight instructors; one, the director of a large flight school (David Frazier) with over 10,000 hours of instruction given, and the other a bonafide "stick-and-rudder" aerobatic enthusiast (Michael Love). Their comments are mixed throughout the discussions of how to fly maneuvers, each interjected to suit the greater purpose. You could treat this book very much like sitting down with a panel of experts as they explain the details of sometimes complicated aspects of flying.

If you listen well, it soon becomes clear that some careful ground work and preparation may greatly improve your flying—for aviation, even with its lofty image, has an immense printed component. From regulations to charts, to simple weather reports, pilots are required to absorb a huge amount of information for even the simplest flights. At this, the book excels any other learning medium. Terry Lankford (an aviation weather specialist) takes you on a thorough tour of aeronautical charts and their related publications, carefully integrating the more practical aspects of chart use. Then Paul Illman leads a careful discussion of radio procedures and the air traffic control system. These discussions are bolstered by numerous related appendices.

The book organization generally follows a famous avaiation dictum: aviate, navigate and communicate—a basic hierarchy of pilot tasks. For your sake, however, I have taken the liberty of adding a fourth section, Integrate, in an attempt to develop your understanding of how separate air operations must work as one.

All of this has been carefully edited so that you may get the benefit of extensively comprehensive information in a single resource. The editor is an airline captain with a fondness for experimental air-plaines. So you hold a wealth of knowledge in your hands. Read it carefully and your flying ability will be richly enhanced.

Happy flying.

Lewis Bjork

Part One: Aviate

Begin with the machine itself. Study it closely, pay careful attention, and it may become an extension of your own body, where flying becomes as natural and easy as walking...

1

The total preflight

The preflight begins and ends with the one who is responsible for the safe conduct of each flight—the pilot. Depending on the type of flight to be undertaken, the pilot has many plans to make. The flight could be anything from a nice Sunday afternoon hop around the patch to a transcontinental flight. The preparations will vary in details, but not much in the basics, such as pilot, equipment, route, and weather. Of these, the most critical player is the pilot, so the preflight preparations begin on a rather personal level.

Fatigue

The complexities of operating an aircraft are more fatiguing than many people realize. The constant attention to heading, altitude, airspeed, and communications can tire a person rather quickly. This is especially true if the pilot is inexperienced.

For student pilots, the physical duties, when combined with the mental stress of learning to fly, can tire even the most robust person in a hurry. This means you must learn to pace yourself. If you notice you are making random errors that you normally would not make, you are probably reaching the saturation point and should end the lesson. If your instructor does not realize that you are tired, speak up. Very little learning can occur if a person is fatigued, and the dangers of flying in a fatigued state cannot be overemphasized.

Physical or mental fatigue can be very deceptive. It can sneak up on anyone at anytime and it is frequently deadly. A number of years ago, a student graduated with his Commercial, Instrument, Multi, and CFI certificates. This young man decided he would enter the aviation field as an air taxi/charter pilot. He was an average pilot, or maybe slightly better, and found a job flying charter for a firm in the north. I saw him a few months later and inquired as to how everything was going. He said he was "flying his butt off," building a lot of twin time but was a little tired of going places at all hours of the day and night.

The next time I heard his name mentioned was after he crashed and died. It seems he had been flying parts to some factory in Canada and had flown several trips in succession starting in the evening and continuing through until daybreak. On the last trip, just as a new day was dawning, he fell asleep. He slumped over the yoke and went nearly straight in. It was a needless tragedy induced by fatigue. He needn't have taken that last load. He could have told someone to forget it. But he didn't. He probably thought he could make it. Or maybe he chose not to recognize the signs of fatigue. Maybe he was afraid of getting fired for refusing that one more flight. Which is worse—fired or dead? I realize this case is extreme, but it happened—and it doesn't always happen to the other guy. Be aware of the fatigue factor.

Every case of pilot fatigue does not necessarily end in disaster. Fatigue happens at some degree to all pilots as they fly. Not too many years ago, my boss and I flew a Cessna 310 from our home base at Lawrenceville, Illinois, well up into Michigan, to purchase a new Decathlon. We stopped only long enough to sign the paperwork and have a Coke. We then proceeded across Lake Michigan to a place in northern Illinois where the plane was actually located. It was nearly noon by this time, and the new plane was not yet ready.

Finally, at about 2 p.m., I departed for home in the Decathlon. I was fresh in the knowledge that I had at least two hours to play with the new plane during the 225-mile flight home—and play I did. After some inverted flight, I broke into the full envelope of aerobatic maneuvers for nearly an hour. Settling down at last, I was a little off course but knew the country well enough and proceeded home. Having flown most of the day already, and as the exhilaration of the new purchase began to fade, I noticed a peculiar feeling beginning to overtake me. My arms felt as though they were made of lead. I couldn't think quite right. The VOR needle kept wandering around (I knew it was just an unreliable radio). Nothing seemed to go right and the last 100 miles seemed to last forever. Nearing home, a sudden burst of adrenaline hit me and I was overtaken with a case of the "look at me's." I made an inverted pass at a safe altitude and landed.

I taxied the beautiful red, white, and blue machine toward the ramp, eagerly awaiting the accolades from the crowd of people who wanted to get a look at the new plane. As I neared the hangar, I swung the Decathlon around, pulling the mixture as I neared a stop, jumped out of the plane, and found I could not stand up well. My head was swirling and my legs would not support me. My whole body was writhing in agony. Not wanting anyone to know of my plight, I knelt down and pretended to examine the belly of the plane. I knew that sooner or later I would have to get up or call for help. After a few minutes, I slowly made it to my feet and leaned against the plane. I did my best to answer the questions about the trip, how the plane flew, and so on. My answers were very short.

I recalled that I still had to fly our Stearman over to an adjacent field in preparation for an airshow the next day. After an hour or so, I did. I then went home, took two aspirins, and sat down to think.

The things I thought were not very pleasant. To put it mildly, I was fatigued. I was so tired that I unknowingly put myself in a dangerous situation. Students are taught early in flight training never to exceed their or the aircraft's capabilities. I certainly didn't exceed the Decathlon's but I exceeded my own physical capabilities about as far as I ever care to. No one is immune from fatigue.

Illness

Most diseases of a serious nature, that is, ones that could cause a total disruption of the pilot's functions, should be caught at the first flight physical and would not allow the prospective pilot to hold a medical. Pilots already flying, however, may develop such an illness, such as the everyday cold, the flu, and headaches. Everyone experiences these maladies once in a while. They are not generally incapacitating and people tend to suffer silently until they recover. The smart pilots stay home and take care of themselves. Pilots who fly when they are sick put themselves in a potentially hazardous situation.

Suppose you have a mild case of stomach flu, and maybe you "just have" to make a flight. It should be not big deal, right? Wrong. Did you ever try to do anything constructive, like land an airplane, while throwing up? Stay home or have someone else fly if you really must go. Suffering from *get-there-itis* has caused more grief than any other form of aviation-related illness.

Medication

Amphetamines, barbiturates, tranquilizers, laxatives, and some antibiotics are but a few of the drugs the FAA says should be taken only during a nonflying period. They can cause sleepiness, lack of coordination, and reduced reflex action and should be taken only as directed. The effects of these drugs might last for hours or days after they are taken, so pilots should follow the advice of a doctor or pharmacist regarding any drug in question. (Fig. 1-1.)

Here is an example from NTSB File #3-3883. On December 4, 1976, a 57-year-old private pilot with 2,204 total hours departed the Depere, Wisconsin, airport in a Cessna 150E. The flight was listed as a local solo, noncommercial, pleasure flight. The pilot was ob-

Fig. 1-1 *Even most over-the-counter drugs are forbidden for consumption while flying. If in doubt, check with your Aviation Medical Examiner.*

served performing maneuvers at low altitude before spinning into the ground.

An autopsy revealed .92 Placidyl/100 and salicylate in the pilot's urine. (Placidyl is a barbiturate and can cause extreme drowsiness. Salicylate is more like an aspirin and is often used as a painkiller.) The probable causes listed by the NTSB are physical impairment, failure to obtain/maintain flying speed, and unwarranted low flying. It is unknown exactly how much of an effect the drugs had on the pilot, except that he is dead, and drugs are listed as a factor in the cause.

Alcohol

One FAA official, when questioned off the record, said that alcohol is probably a contributing factor in more aircraft accidents than anyone cares to believe. How many nonfatal accidents actually involve alcohol is anybody's guess. If the accident is nonfatal, the FAA might not arrive for hours or days, if they arrive at all. Sometimes they mail out an accident report for the pilot to fill out when it's convenient. This allows time for a pilot to sober up and invent a story devoid of the presence of alcohol. Therefore, the numbers regarding alcohol as a factor in accidents are probably skewed to the low end of the spectrum.

Here is another example, from NTSB File #3-3644. On November 6, 1976, a Piper PA-24 Departed Evanston, Wyoming, for Salt Lake City, Utah. It carried a pilot and a passenger. The pilot had 488 total hours, 241 in type, and he was not instrument rated. No information is given on the copilot. Near Grantsville, Utah (20 miles past their destination), the plane was observed flying low to the ground. The terrain is listed as containing high obstructions.

The aircraft collided with wires and poles and crashed. Both occupants died. The NTSB lists the probable causes as exercising poor judgment, failing to see and avoid objects or obstructions, and physical impairment due to alcohol. A subsequent autopsy revealed the pilot had a blood alcohol level of .212 percent and the copilot had .200 percent.

The FAA allows no more than .04 percent alcohol by weight. Anything over this the FAA considers a pilot legally (or illegally?) drunk. These pilots were five times that level.

Alcohol is a depressant. It slows the reflex action, dulls the senses, and perhaps worst of all, can create overconfidence. How many pilots, fortified by the remnants of a six-pack, have flown on to their deaths?

Mental preparedness

Apathy, anxiety, or any other state of mind that could hinder judgment should be laid to rest before a flight is undertaken. A pilot whose mind is occupied with something other than flying is apt to be heading for trouble.

We all have our good days and our bad, our peaks and valleys of mental readiness. People have problems and pilots are not immune to these. It could be trouble with a spouse, money, job, the IRS, or whatever. The point is that these problems have to be addressed before or after the flight—not during it.

Apathy is a lack of emotion, feeling, or passion. It is a mental state of indifference, lethargy, and sluggishness. What it really is, in plain language, is a state of just not giving a damn. You could care less if the flight is made, the sun comes up, or anything else. If you attempt a flight in this condition, you could be setting yourself up for a bad experience. You must have your mind clear and ready for the sometimes-difficult tasks of flight. If possible, clear up the problem prior to flight time. Try to put it *completely* from your mind. If you don't, you might be brought back to reality by doing something like landing with the gear up, or running out of gas. That is guaranteed to take your mind off your old troubles and put it to work on a new set.

Anxiety

Anxiety is a state of mental uneasiness arising from fear or apprehension. The causes of anxiety are hard to detect and the cure is usually more difficult than it is with apathy.

A person suffering from anxiety feels that no matter what he or she does, it probably won't be right. It is a self-consuming, defeatist attitude and must be overcome before safe flight can take place. Sometimes you must look deeply inward to find the cause of the anxiety. Maybe an open, honest talk with a good friend, flight instructor, or professional psychologist will bring the problem to the fore. Then it can be met head on and conquered. Whatever it takes, it has to be solved before you can perform proficiently as a pilot.

Vertigo

Vertigo, or spatial disorientation, as it is sometimes called, inhibits a person's ability to perceive attitude with respect to the horizon. A pilot experiencing vertigo is unable to tell whether the airplane is climbing, descending, or turning without referencing the instruments or a good look at the horizon.

When on the ground, you perceive attitude with respect to the earth by seeing fixed objects around you, by feeling the weight of your body on your feet, and by the vestibular organs in your inner ear. You may orient yourself by any one of these means for short periods of time.

While in flight, however, all three of these normal means of orientation may become obscured or confused. The pilot might only be able to see objects that are in or attached to the aircraft when ground references are obscured by clouds or darkness. Your ability to sense the direction of the earth's gravity, by the weight on your body and through your vestibular organs, can be confused by accelerations in different directions caused by centrifugal force and turbulence. For example, the senses are unable to distinguish between the force of gravity and the horizontal force resulting from a steep turn.

Because of these facts, gyroscopic instruments are necessary to fly for more than a few minutes without visual access to outside reference points. The use of such instruments does not ensure freedom from vertigo, for no one is immune; they do enable a pilot to overcome its effects, however. Skilled pilots learn to disregard the psychological discomfort that occurs when the instrument indications appear to disagree with perceived senses, and follow the instruments anyway.

With respect to the pilot preflight, the possibility of vertigo and other aerial challenges requires a certain mental focus and discipline. If the pilot is not mentally prepared, and adequately skilled for all conditions expected on a particular flight, the risk increases substantially.

Pilot licenses

Aside from the physical and mental preparation, there are a few more things a good pilot must check before each flight. Are your medical certificate and license in your possession? Are they current? Are you current in the aircraft you are going to fly this day? If these and all the other things in the previous pages are in good shape, then you are probably physically, mentally, and emotionally ready for the flight. Remember, the pilot-in-command is responsible for the safe conduct of each and every flight. Prepare well for it and it will probably go as planned. (Fig. 1-2.)

The aircraft

Many pilots believe a total preflight of an aircraft consists of a thorough walk-around looking for any loose nuts and bolts, but this is only part of it.

Fig. 1-2 *To fly legally, you must have on your person a current medical certificate and pilot's license.*

Logbooks

The preflight really starts with the paperwork. In addition to the certificates required of the pilot, the airplane has a plethora of its own. I start with the logbooks. The FAA requires that every aircraft owner maintains a current set of airframe and engine logbooks. These logbooks do not have to be carried in the airplane, but must be readily available. What should be evident in these logbooks? What do you look for in order to prove your aircraft was airworthy, at least at the time of the logbook entry? It varies as to whether the aircraft is used privately, or for commercial purposes.

If the aircraft is your own and you do not rent, lease, or otherwise make money directly from it, then the first item to check is whether the aircraft is within the time constraints of its last annual inspection. Every aircraft must have an annual inspection, and the evidence of the inspection is found in the aircraft airframe and engine logbooks. An annual inspection is valid for 12 calendar months, after which if you were to fly this aircraft, you would be technically illegal. Your aircraft would be considered unairworthy.

If the logbooks state that your aircraft has last had an annual on June 18, 1999, then it would be considered airworthy until 11.59 p.m. of June 30, 2000. After that, the aircraft would become illegal for flight. In short, the annual inspection is valid from the date of inspection for one year *plus* the number of days left until 11:59:59 on the last day of the month in which it was inspected.

All aircraft must have an annual inspection unless they are on a progressive maintenance program. In addition to the annual inspection, aircraft used for hire must have an inspection every 100 flight hours that should be entered in the airframe and engine logbooks, as well.

While looking through the logbooks, it doesn't hurt to scan the pages for any sign of substantial damage, how and when it was repaired, and any other pertinent information you can find. You might gain a clue as to where to look more closely in the preflight. Required inspections listed in the logbooks only mean the airplane is *legally* airworthy at the time the inspections were completed—a lot could have happened to the aircraft since.

Many airlines and rental companies make use of a *squawk sheet*. A squawk sheet is a form used to indicate any malfunctions, gripes, or complaints made by prior pilots after a flight. This should be fascinating preflight reading. Note the date and type of comments made by the previous pilots about the aircraft you are about to fly. Additionally, you might ask other pilots who have recently flown the particular aircraft if they noticed anything unusual.

General overall look

As you approach the aircraft, give it a good general overall look (Fig. 1-3). Sometimes you can spot a wrinkle in the skin or some other problem better if you are a short distance away and not right on top of the aircraft.

Sometimes a defect that should be obvious can be overlooked if it is unexpected. We once had a Cessna 150 whose prop had been removed for some reason or another. Since the plane was taking up room in the shop, we decided to take it out and tie it down on the line with all the other aircraft. As a test, one instructor sent his student out to preflight the propless 150. Off he went. He got to the aircraft, checked the paperwork, drained the sumps, checked the oil, ran the flaps down, checked all the controls, and returned to report the aircraft ready for flight. His instructor nearly died. I didn't relate this little tale to make the student look foolish. There is a message

Fig. 1-3 *Approaching the aircraft, look for any obvious problems that might render it unairworthy.*

here: DO NOT TAKE ANYONE'S WORD THAT YOUR PLANE IS READY FOR FLIGHT. CHECK IT YOURSELF! The pilot-in-command is solely responsible for the safe operation of the aircraft, and that most certainly includes the preflight.

While methods and specifics of a preflight may vary from plane to plane and from pilot to pilot, the main thing to remember is to do it the same way every time. It is probably best to use a written checklist so the procedure never varies.

I teach my students to arrive at the aircraft with their eyes wide open. If he or she has given the aircraft a thorough overview while approaching the plane, the pilot should recognize any major problem (including missing aircraft parts) by the time they arrive at the tie-down. Additionally, it is wise to leave the aircraft tied down during the preflight until you're ready to climb inside. This precaution helps avoid some interesting square dancing between you and the aircraft in case of unexpected, sudden winds caused either by nature or other aircraft.

Aircraft documents

Begin by opening the door to make a visual sweep around the interior, checking for any irregularities or problems. If all seems to be in order, the next step is to check the paperwork. Make sure the airworthiness certificate, registration, operation limitations, radio station license, weight and balance data, and all placards are in order. Remember, the FAA requires that the airworthiness certificate be in full view at all times. Keep it in front of the little plastic window that most aircraft have for document storage.

A note on the airworthiness certificate: its mere presence isn't enough; it needs to be valid. Inspect the airframe and engine logbooks to make sure the aircraft has had all pertinent AD notes, inspections, and damage repair endorsed by the proper authority. Then, providing you have checked the logs prior to arriving at the aircraft and you don't discover anything during the preflight inspection to prove otherwise, the airworthiness certificate is indeed valid. The date of issuance is attached to the airworthiness certificate, indicating simply that on the date of issuance, the aircraft was airworthy. After a detailed examination of all required aircraft documents, return them to their proper place and continue with the preflight (Fig. 1-4).

Fig. 1-4 *The interior preflight consists of checking the paperwork, fuel quantity, fuel selector, mixture, flaps down, throttle idle, master and mags OFF.*

Interior preflight

Next, position the fuel selector so that the engine would draw fuel from a full tank. For aircraft equipped with multiple positions, this may be the left, right, or both tanks or for aircraft that has only ON/OFF positions, merely switch to the ON position. The fuel is turned on at this point in the preflight so that the sumps will drain from the tank rather than just the line, thus keeping any water or sediment from remaining settled on the bottom to cause problems later.

With the fuel selector in a supply position (ON), make sure the mixture is pulled full lean, throttle idle, and carburetor heat or alternate air in the cold, or off, position. Make sure the magnetos are OFF. If your aircraft is equipped with a key, don't insert it until you are ready to start the aircraft. This method is the most effective way to prevent an accidental hot prop during the exterior preflight. Even with the key out, always treat a prop as if it were going to bite you, because it can.

Next, turn on the master switch and check the fuel gauges to get an indication of available fuel. Later, during the exterior preflight, verify the readings given by the interior gauges by actually opening the fuel tanks and taking a look inside because gauges can sometimes give false information.

If your aircraft is equipped with electric flaps, lower them while the master switch is still on. They will then be in a better position to be checked during the exterior preflight.

While checking the fuel indications and lowering the flaps, listen for the whirling sound of the electric gyro instruments as they begin to wind up. These will be checked in more detail later, during the pre-takeoff run-up.

Next, turn off the master switch in order to save battery power, and set all trim tabs for takeoff. If possible, watch for the movement of the trim tabs as you turn the trim tab in the cockpit. This will tell you that they are indeed in good working order and should function properly in flight.

Unless there is a strong wind blowing, remove the control lock to accurately verify all control surfaces are functioning properly.

Now, go back through and double-check to see that the fuel is on, the mixture lean, carburetor heat cold, throttle idle, trim tabs set, magnetos off (leave the key out), and control lock removed. Stow any baggage, maps, etc., and you will be ready to begin the exterior preflight.

Exterior preflight

There are many ways to complete the preflight. Choose your favorite method then follow the same procedures every time. A written checklist followed step-by-step will ensure checking everything in the proper order.

For preflighting multiengine aircraft, I prefer to start at the door and continue around until reaching the door again, now ready to get in and go. The exterior preflight of a single-engine trainer, a Cessna 152, for example, might go as follows: Start at the engine access door, open it up, and take a good look inside. Look for any obvious discrepancies, such as a magneto lying in the bottom of the cowling. Look for security of nuts, bolts, wires, and harnesses. If a wire is disconnected, point it out to a mechanic. If the aircraft hasn't been flown for a few days, check for any signs of bird nests or the presence of mice. Once I found a nest complete with mother bird and five little blue eggs...it happens. If a nest is not removed, it can ignite when the engine is heated up (usually in the air).

Check the oil level and add as necessary. Some circumstances may cause false indications and may need to be retested. For example, cold oil has a tendency to climb up the stick; wipe the dipstick and recheck for accuracy. Also, if the engine has run recently some of the oil will remain up in the engine for a time; wait to test until everything's settled. Never lay the dipstick down; hold it in your hand until you replace it securely. This helps prevent getting it dirty, ruining clothes, or forgetting it entirely.

Next drain the engine fuel sump. Give it time to drain out any unwanted water or debris. Three to five seconds drain time should be enough for most situations. If possible, it is best to try to catch the drained fuel in a clear container. This way you can see any debris or water that might be contained in the fuel. Continue draining the sump until the fuel is clear. Remember, water is heavier than fuel, so it will quickly settle to the bottom.

If you don't have a glass container to catch the fuel, draining it out onto a hard surface (not grass) and then looking for any bubbles of water also works, only not as well. Be sure the valve is completely off and not leaking.

Close the engine access door and move toward the front of the aircraft. Examine the cowling for any sign of loose rivets, etc., as you make your way to the prop area. Once in front of the aircraft, look inside the cowl on top of the cylinders for any signs of birds' nests or other material that might have collected on them. Be sure all visible wires and harnesses are in place and secure.

Next grasp the propeller about half way out and gently push and pull it to make sure it is secure. There should be a small amount of movement in and out, but nothing monumental. Check the leading edges of the propeller for any signs of nicks or cracks. As a propeller rotates, it creates a tornadolike suction that can pull up rocks, dirt, or other foreign material. These particles can strike the leading edge of the prop and cause nicks (Fig. 1-5). If you find a nick on the leading edge of the prop, have a mechanic file it down. Remember, the prop is a very delicately balanced airfoil; even a small nick is capable of posing a potential problem. If a prop has many nicks and blemishes it would be wise to remove the prop and send it to a prop repair shop for rebalancing or possible replacement.

Next, squat down directly in front of the aircraft and check the air filter that is located directly beneath the prop and landing light. It should be clean and secure. Check for security of the exhaust pipes with the toe of your shoe (this prevents burnt fingertips if the aircraft has run recently).

Check the nose gear for security and proper inflation, tread wear, and any signs of cords showing through. If the cord is evident, it's time for a tire change. Also check the nose gear for stability and any sign of missing cotter pins, washers, and nuts. I tell my students that while it is important to check the things that are there, it

Fig. 1-5 *Always check the prop for any signs of nicks or cracks.*

is even more important to find anything that is not there during the entire preflight.

With the front of the aircraft secure, move on to the cowl on the opposite side from the engine access door. Check for any loose or missing rivets and be sure the static port, which is located just in front of the pilot-side door, is clean and free of any wax, dirt, or insects. For some reason, insects like to nest in the tiny static ports of aircraft.

Now, to check the fuel, grasp the handhold located on the fuselage and place your left foot on the step provided along the lower fuselage. Step up and place your right foot gently to the step provided in the middle of the wing strut. It is important that you keep the majority of your weight on the fuselage step because the strut is not formed to take much weight at a right angle. If your aircraft is not equipped with steps, it is best to use a ladder. Look in to see the fuel level. If you cannot see into the tank, you can try to feel the fuel level with your fingers, or use a clean stick to measure the fuel level. However inconvenient, it is important to verify fuel levels; don't trust electric fuel gauges because, as mentioned previously, sometimes they lie.

Secure the fuel cap and begin the check of the leading edge of the wing (Fig. 1-6). As you move along the wing, check for any dents, loose rivets, or tears if your aircraft is fabric covered. Also be sure

Fig. 1-6 *The wing should be checked for any obvious problems such as dents, loose rivets, blocked pitot tube, etc...*

the strut is secure and free from any bends, dents, etc. As you look over the wing and strut, lean down and take an overall look at the wing bottom. Be sure it is free from any problems that might hinder a safe flight.

Moving down the wing, you come to the pitot tube at about the point where the strut joins the spar. It must be free of any foreign matter if it is to supply adequate ram air pressure to the pitot instruments. A reminder: Do not blow into the pitot tube to free any debris. This usually results in ruining the airspeed indicator. The force of the human lung has such power that a short blast of air expelled from the mouth can have more pressure than the pitot tube experiences in flight. If the pitot tube has any foreign debris, try to pick it out with a small, pointed object such as a paperclip. If this fails, you might try blowing *gently* into the pitot tube while someone else watches the airspeed indicator to make sure you don't blow it away. If you aren't sure of what to do, take it to a mechanic and let him solve the problem.

Finish the preflight of the leading edge of the wing and arrive at the wingtip and examine the navigation light to be sure it is secure. While at the wingtip, check the general condition to be sure it has suffered no *hangar rash*. (Hangar rash is a common disease caused by the careless movement of aircraft in or around other aircraft that can result in wrinkled wingtips.)

Moving to the aft section of the wing, pivot the aileron up and down, checking for freedom of travel. At the same time, watch the aileron on the opposite wing to make sure it is moving in the reverse direction. Besides freedom of travel, you want to check the aileron hinges for security and smooth operation, and to be sure that all cotter pins are in place. When examining the aileron hinges and actuating rods, hold the aileron firmly in the up position so that any wind gust will not crush your fingers (Fig. 1-7).

Since the flaps are already in the down position, examine them and their tracks, operating rods, and hinges very carefully. Most flaps will have a slight amount of play in the down position, but not much— maybe a half inch. If there is much more play than that, have them checked by a mechanic before attempting flight.

Now is a perfect time to drain the wing fuel sump. While on the subject of checking the fuel, always be certain the fuel you drain out is

Fig. 1-7 *When you check the actuating rods of the ailerons, hold the aileron in the up position so a sudden gust of wind doesn't shorten your finger.*

the correct octane for your particular aircraft. You can tell the octane rating of the fuel by the color. 80-Octane fuel is red. 100-Octane fuel is green. 100 low-lead fuel is blue, and jet fuel is clear.

If two different octane grades are mixed in your tank, they will turn clear like water or jet fuel. This chemical process is formulated at the fuel refinery as a safety measure to alert the pilot of the problem. Since a mixture of two different fuel octanes may or may not be healthy to feed the aircraft, some precautions should be taken if the drained fuel comes out clear.

First, check the possibility that you have a tank of water or jet fuel. Jet fuel has a very distinctive smell like kerosene. If this is the case, drain the tank and replace it with the proper octane. If you have a load of water, continue draining through the sump until you get the proper color fuel for your particular aircraft.

After eliminating the water or jet fuel possibilities, it is likely that the fuel in the tank is a mixture of two differing grades of fuel. Now the problem lies in researching and discovering just which two are in the tank. If the octane mixture in your tank is of a higher number than your aircraft requires, then it should be okay to fly. But if the octane is of a lower number than you require, you may want to drain and replace the entire tank with fuel of the desired octane.

Directly below the fuel sump, check the landing gear, tire, and brake assembly. Watch for any sign of moisture on the ground beneath the inner portion of the wheel. If your brake line has developed a leak, this is one of the primary spots to find an accumulation of fluid. Also check the brake and tire for any signs of unusual wear.

As you move back along the side of the aircraft toward the tail section, check the side and underneath for any loose rivets, tears, dents, etc. Be sure the surface has not buckled from any undue stress or strain.

Reaching the tail section, check the leading edge of the horizontal stabilizer as you did the leading edge of the wing. It should be free of any large dents (although you are apt to find quite a few rock chips and nicks) and fairly solid as you try to move it gently up and down.

The elevator should move very freely and should travel all the way to the stops that are found just beneath the rudder. Now is a good time to check the external rudder cables for proper tension and security. Watch for any loose or missing cotter pins, nuts, or bolts. Move the rudder from side to side and feel for smooth travel all the way to its stops. Look up and check your VOR antenna and anticollision light (strove or rotating beacon). If you can reach them, feel for the proper security.

Now complete the right side of the aircraft as per the methods utilized for the left side. With the completion of the walk-around preflight, you have the peace of mind that comes from knowing you have done everything in your power to determine that the aircraft is ready for flight.

Mistakes

Many needless accidents are directly attributable to a careless or nonexistent preflight. The following example, which I did not make up, demonstrates just how dangerous it can be when a pilot decides for some reason or other that he or she is the exception to the rules.

The pilot, who held an Airline Transport Certificate, skipped the preflight altogether and hopped into his friend's Cessna 182 intending to simply fly away, but a dead battery delayed his plans. He decided to try a prop start and on the second pull on the prop the engine roared to life, and to his surprise and horror rushed past him off on its own. Still on the ground, the aircraft gathered speed as it crossed the active runway, lost its tail as it passed under (almost) an irrigation system, and crashed a half a mile away in a gravel pit. The pilot, who had quit chasing the aircraft after the first 100 yards, could only watch with frazzled nerves as a $150,000 aircraft was smashed into pieces, knowing his carelessness was to blame and now it was too late to do anything about it. Oops, there goes the plane and your reputation and career went with it and you know you're going to sound stupid trying to explain this one.

Proper planning and preflight would have certainly avoided this accident. Even a student pilot should know to set the parking brakes; chock the aircraft; tie it down; turn off the mags; turn off the fuel; or leave the mixture full lean. After all, wasn't the plan to start the plane? So having it start shouldn't come as a complete surprise.

There are more than enough problems associated with flight without going out of your way to tempt fate. A safe flight begins with proper preplanning and continues from there. Don't shift the odds by a poor or nonexistent preflight of yourself or your aircraft.

The weather

An integral part of preflight planning is checking weather conditions to be encountered during the flight. Improper respect for weather conditions can transform a pilot into a statistic. The NTSB and FAA files are full of accident reports of pilots who continued flying beyond experience/capability limits into known adverse weather conditions.

The word *limitations,* so often found in accident reports, is the key to much of the problem. Each and every aircraft has its limitations. Each has its own gross weight, Vx, Vy, etc. Each pilot also has his or her own set of limitations, whether they are known or unknown. You should find your own limitations for a given condition and then never try to push yourself beyond them.

These limitations include such items as ceiling and visibility minimums, wind velocity and direction, and duration of flight. Of course, these will vary depending on whether the flight is to be VFR or IFR, local or cross-country, dual or solo. My own personal minimum preference is a 500-foot ceiling, 1-mile visibility, about 30 knots of wind at the surface, and not more than 2 hours per leg if on a cross-country. Also, I'll do just about anything to avoid flying into icing conditions. I have developed a very healthy respect for the forces of weather. Weather is undoubtedly the most powerful force in aviation (contrary to popular opinion at the FAA); weather has brought down everything humans have ever put into the air at one time or another.

If your proposed flight is going to be local, you can usually look out the window and tell if the conditions are favorable for the safe completion of the flight. However, the weather can change for the worse in a short period of time, so rather than take a chance, why not place a call to the nearest Flight Service Station or National Weather Service? The personnel at the FSS and NWS can inform you of present conditions, weather at nearby areas, and a forecast for the next 24 hours. For a cross-country flight, this service becomes invaluable to ensure you won't encounter weather conditions beyond your limitations later on during the flight.

Weather information can be obtained via personal computer as well. A pilot can dial up the Weather Channel or AccuWeather and gather weather forecasts, winds aloft, etc., for the entire nation and most of the world. Another source that requires a computer and telephone modem is the government-sponsored DUAT (direct user access terminal). DUAT will bring you up-to-the-minute weather to the screen. One drawback of DUAT is that you have to decode it and are solely responsible for your interpretation of the weather briefing. I prefer to call FSS and let them read it to me; as a bonus, they keep a log of all briefings given.

Of course, everyone knows that the prediction of weather is not an exact science. Sometimes they are wrong, so you must exercise common sense and compare your own limitations to the forecast. Speaking of common sense, it doesn't make much sense to utilize a weather briefing service and then attempt a flight regardless of the forecast. NTSB File # 3-3803 is an example:

On December 31,1976, a 59-year-old, non-instrument-rated private pilot took off from the Auburn Municipal Airport in Auburn, Washington. The intended destination was Royal City, Washington. Witnesses reported seeing the aircraft take off, enter an overcast layer in a climbing turn, dive steeply, pull up into the clouds again, and then dive into the ground. The weather at the site was 150-foot ceiling with one mile or less visibility, with fog. The pilot had been briefed by FSS personnel. The type of weather was IFR. Type of flight plan—none. Moral of the story: If the weather conditions are IFR and you are only rated for VFR flight conditions—don't go. The situation exceeds the pilot's limitations.

Here are some guidelines that will help both you and the FSS personnel when obtaining a weather briefing. You must be specific and clear and know what you want. Open the conversation by telling them who you are, where you are, and where you are going. Then, tell them when you propose to leave and if the flight is to be VFR or IFR. Also include your proposed route of flight and aircraft type. Now they have the appropriate information and will inform you of all pertinent weather data, such as general weather along the route of flight, winds aloft, terminal forecasts, area forecast, NOTAMs, etc. Be sure to write down the information. If you are unclear on any point, or if they happen to leave out something you need, ask for it. Ask for any pertinent NOTAMs. It beats getting ready, flying to an airport, and finding it closed for repairs.

Pilots, whether a student or more advanced, are going to have to deal with some form of weather from the time they untie the aircraft until it is safely tied back down. Get to know the weather and the people who predict it. Set your own limitations and stick to them. You will be glad you did.

Drug testing

At the present time, all ATC personnel, airline cockpit crews, flight attendants, dispatchers, maintenance employees, security personnel, air-taxi pilots, crop dusters, banner towers, bird chasers, hot air balloonists who give rides to sightseers, and just about everyone else who makes a living in aviation is subject to drug testing. If you are going to fly for hire, you will be tested for drugs.

The aviation industry, as well as the nation, is deeply divided over the drug testing issue. Most pilots don't have a problem with testing any pilot after an accident or incident as an addition to the fact-finding process. The problem many pilots have is that they object to unilaterally giving up their rights as Americans, protected, up until now, by our Fourth Amendment right to privacy.

Under pressure from Congress, the FAA rushed into the drug testing business without a plan. What ensued was a folly of the fifth magnitude. All pilots were summarily issued a summons to drug testing without evidence of any real problem existing in the first place. For instance, as of this writing, there has been one airline accident attributable to drugs since the dawn of aviation. One! And there has never been an accident during an instructional flight attributable to drugs. Sort of like killing a rabbit with a nuclear missile, isn't it?

While no one will argue with the need for a clear-headed, drug-free pilot, surgeon, dentist, congressman, etc., the real problem comes when a false positive test can ruin your career. Unfortunately, until more intelligent thinking comes to the fore, if you are going to fly for hire, you will be subject to drug testing.

The decision and flight plan

The decision and flight plan are the culmination of the preflight process. After taking into consideration the readiness of yourself, the aircraft, the weather, and any other pertinent data, you have to make the decision, "Do I go or not?" Remember, it's your decision. The pilot-in-command has the responsibility for the safe conduct of each and every flight. Don't let anything (ego, itinerary, boss, flight instructor, etc.) force you into a flight situation you feel is unsafe.

If your decision is to go, then you should file a flight plan (Fig. 1-8). Many pilots only file a flight plan if they intend to go on a cross-country. It's really not a bad idea to file even if you are planning to remain in the local area. A flight plan is *mandatory* only if you intend to go IFR. You can go just about anywhere you want VFR without having to file a flight plan. However, filing even when flying VFR has its advantages. With a flight plan, someone knows where you are and approximately when you are to return. If you should run into any trouble, someone will be looking for you about 30 minutes after you are scheduled to close your flight plan. This is cheap insurance since the most it can cost is one short phone call. You can even file your flight plan on your computer by using the DUAT mentioned previously.

Flight training

Whether you are thinking about learning how to fly or have 20 years and 10,000 flight hours logged, there is always the need to get good flight training. When you receive flight instruction, it should be from a competent, qualified fight instructor that you feel comfortable flying

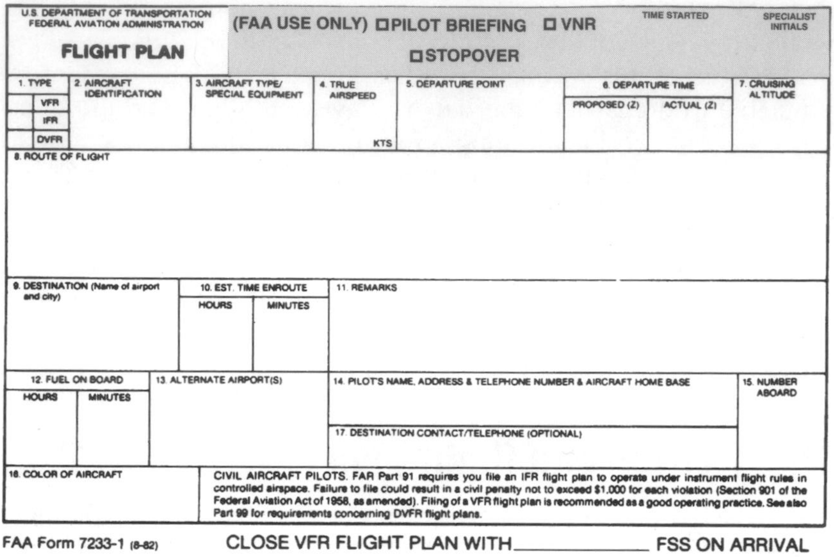

U.S. DEPARTMENT OF TRANSPORTATION FEDERAL AVIATION ADMINISTRATION	(FAA USE ONLY) □PILOT BRIEFING □ VNR	TIME STARTED	SPECIALIST INITIALS
FLIGHT PLAN	□STOPOVER		

1. TYPE	2. AIRCRAFT IDENTIFICATION	3. AIRCRAFT TYPE/ SPECIAL EQUIPMENT	4. TRUE AIRSPEED	5. DEPARTURE POINT	6. DEPARTURE TIME		7. CRUISING ALTITUDE
VFR					PROPOSED (Z)	ACTUAL (Z)	
IFR							
DVFR			KTS				

8. ROUTE OF FLIGHT

9. DESTINATION (Name of airport and city)	10. EST. TIME ENROUTE		11. REMARKS
	HOURS	MINUTES	

12. FUEL ON BOARD		13. ALTERNATE AIRPORT(S)	14. PILOT'S NAME, ADDRESS & TELEPHONE NUMBER & AIRCRAFT HOME BASE	15. NUMBER ABOARD
HOURS	MINUTES			
			17. DESTINATION CONTACT/TELEPHONE (OPTIONAL)	

16. COLOR OF AIRCRAFT	CIVIL AIRCRAFT PILOTS. FAR Part 91 requires you file an IFR flight plan to operate under instrument flight rules in controlled airspace. Failure to file could result in a civil penalty not to exceed $1,000 for each violation (Section 901 of the Federal Aviation Act of 1958, as amended). Filing of a VFR flight plan is recommended as a good operating practice. See also Part 99 for requirements concerning DVFR flight plans.

FAA Form 7233-1 (8-82)   **CLOSE VFR FLIGHT PLAN WITH_____ FSS ON ARRIVAL**

Fig. 1-8 *A flight plan is cheap insurance.*

with. There needs to be a channel of communication between you and your flight instructor that flows in both directions. We have become a service-oriented society, and you should expect nothing less from your instructors than someone who thoroughly understands your goals and helps you achieve them.

Your flight instructors should act professionally. They must communicate well, in a manner that is clear to you. Any flight, whether a first lesson or a biennial flight review, should involve pre- and post-flight discussions as needed until your understanding of the subject matter is complete. If they are not training you in a way that seems beneficial to you, find other instructors—there are many excellent ones out there.

When you are in the air, you should be doing the majority of the flying, with the instructor demonstrating a maneuver or technique then letting you practice. If you have an instructor that does as much flying as you do during a lesson, you may want to look elsewhere. During a lesson the instructor should critique your progress in clear, concise terms that help you become more proficient at each maneuver. If you do not understand what you are supposed to do, or how something needs to be changed to get a maneuver right, ask. Your instructor should be able to answer your question immediately or get the answer for you once you are on the ground. The flying techniques you learn from your flight instructors, good or bad, may stick with you for your entire flying career.

Don't feel limited to just the local flight school. If you are serious about learning to fly, it is worth checking out the schools within a reasonable drive of your home. Finally, ask questions as you learn to fly. Flight instruction is a two-way communication process, and your instructor may not be able to anticipate all of your needs. If you enter flight instruction with a desire to learn and the time to be able to fly on a regular basis, you will find that the time spent is one of the most enjoyable of your life.

Aircraft safety

As a pilot you shoulder tremendous responsibility that you should not take lightly. Your passengers depend on your judgment and skills to safely get them to their destination. On several occasions during my flying career, I have made the decision to stay on the ground when the weather appeared to be a potential problem. In each of these instances the passengers were somewhat put out, but respected the fact that I put their safety first. Other pilots depend on you to smoothly fit into the airspace that you use and follow standard procedures.

When you fly, try to anticipate contingencies, plan ahead, and be prepared. Do not let time pressures, stress, or other factors override your decision-making processes. If you approach flying with an open mind and are willing to commit the time it takes to be a proficient, safe pilot, you will spend a lifetime of enjoyment with the activity. The magic of flight is as fresh and wonderful today as it was with the Wright Brothers' first successful attempt.

2

Flying

The myriad forces acting upon an airplane in flight may be simply depicted as the basic four: lift, gravity, thrust, and drag (Fig. 2-1).

In order for a plane to get off the ground, the lift it is generating must be greater than the weight of the plane, or the pull of gravity. Once in the air, gravity is still trying to pull the plane back to the ground, so lift must be maintained for it to stay at a given altitude. When lift is reduced, such as in a descent, the plane loses altitude as a result of the force of gravity. When the plane is not accelerating up or down, the forces of lift and weight are equal.

Thrust moves the plane forward. Whether thrust is generated through a propeller, jet engine, rocket, or even the gentle glide of powerless flight, the ever-present need for it in sustaining flight is the same. Thrust opposes drag and the two forces balance when the airplane sustains a constant speed. Acceleration or deceleration depend upon a variance of thrust with respect to drag.

Drag

Drag is the resistance of the air to a body moving through it, and it comes in several types. The two most basic types of drag are parasite drag and induced drag. *Parasite drag* is the combined effects of form drag and skin friction. *Form drag* is the result of the air displacement due to the physical shape of the airplane. It can be minimized through careful design and *streamlining,* giving most airplanes, especially gliders, their characteristically beautiful shape. Skin friction drag is caused by the resistance of airflow across the surface of a body and can be reduced by making the skin surfaces as smooth as possible, and especially by making the airplane as small as possible, which reduces total surface, or wetted area. Parasite drag increases with the square of the increase in airspeed. For

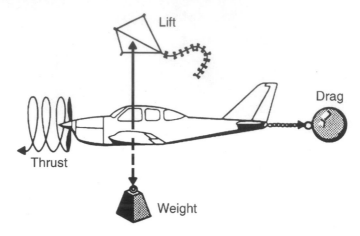

Fig. 2-1 *The basic forces acting on an airplane.*

example, if you double your airspeed, the parasite drag of the plane will increase to four times the amount that was present at the original airspeed.

Induced drag is the other half of the pie. Induced drag is present whenever the airplane is generating lift, indeed, the drag is *lift induced*. This drag is due to a characteristic rearward inclination of the lift vector, a natural by-product of the air displacement created by the wing as it flies. (See Fig. 2-2.)

Unlike parasite drag, which increases as airspeed increases, induced drag varies inversely as the square of the plane's airspeed. In this case, the slower an airplane is flying, the greater the induced drag it generates. This means that induced drag is highest when the plane is flying slowest, such as during landing, and much less at normal cruise speeds.

Lift

Lift may be defined by the same Newtonian laws that refer to gravity. That is, a force is the product of a mass and acceleration. In this case, the air is the mass, and the wing provides the acceleration. The wing basically displaces air downward, and the resulting force is upward. The amount of lift is strongly affected by changing either part of the formula, that is mass, or acceleration. The mass of air that may be affected by the wing is increased or decreased by corresponding

Induced Drag

Fig. 2-2 *Induced drag—drag that is produced as a by-product of lift, or lift induced.*

changes in the aircraft's velocity. A faster plane encounters more air, enlarging the mass part of the formula, and therefore may produce more lift. An airplane flying relatively slow may have to compensate for the low air mass by imparting a correspondingly large acceleration to the air, which may be directly accomplished by increasing the *angle of attack.*

Figure 2-3 depicts the angle of attack of a wing. By definition, the angle of attack is the angle created between the chord line of the wing and the relative wind. The *chord line* is an imaginary straight line that extends from the trailing edge of the wing through the leading edge. The *relative wind* is the direction of the airflow caused by the forward velocity of the airplane.

As the angle of attack increases, the lift generated by the wing also increases, up to a point. As the angle of attack increases, the smooth airflow present at lower angles of attack begins to become turbulent. Beyond a certain angle of attack, known as the critical angle of attack, there is so much turbulence created in the airflow over the wing that the amount of lift generated may be sharply reduced.

Figure 2-4 depicts a wing at progressively higher angles of attack, up to the critical angle of attack, and subsequent disruption of smooth airflow. Although highly specific to a particular airplane and configuration, for most general-aviation airplanes, the critical angle of attack is approximately 17 degrees.

The disruption of an otherwise smooth airflow over the wing due to excessive angle of attack is called a *stall.* Wings generally do not stall all at once. Instead, the turbulent flow of air works its way across the

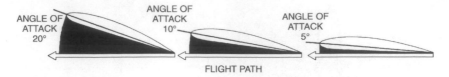

Fig. 2-3 *Angle of attack has a direct and immediate effect on lift production.*

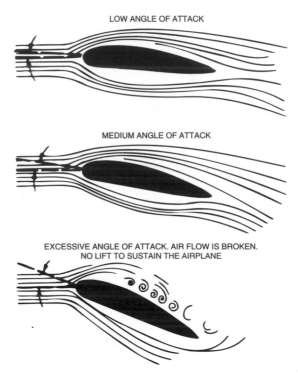

Fig. 2-4 *If angle of attack exceeds a critical point, the wing stalls and lift is diminished.*

wing, in patterns specific to a given wing shape. Aircraft engineers pick a wing shape, or planform, that results in characteristics best suited for the plane.

Figure 2-5 shows six different wing planforms and the pattern of stall progression on each of them at critical angles of attack. Notice that the stall normally begins on the aft portion of the wing and moves forward toward the wing's leading edge. On some planforms, the stall occurs first near the wingtips and moves inward.

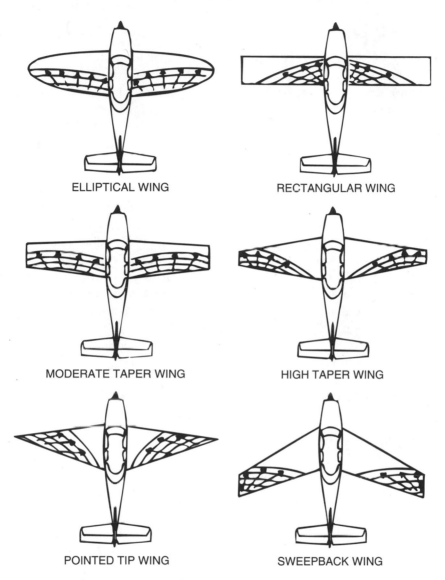

ELLIPTICAL WING   RECTANGULAR WING

MODERATE TAPER WING   HIGH TAPER WING

POINTED TIP WING   SWEEPBACK WING

Fig. 2-5 *Wing planform may affect stall patterns.*

Manufacturers of light general-aviation aircraft normally try to design the wing so that the stall begins near the fuselage and progresses outwards toward the wing-tips, ensuring that the ailerons remain effective as the stall develops. A wing planform that possesses undesirable stalling characteristics may be altered by twisting or other aerodynamic modifications until its stall behavior is acceptable.

Airfoil types

Figure 2-6 depicts a very common general-aviation airfoil. Notice the flat underside of the shape, with the relatively large curvature along the airfoil's upper surface. Visible among the various labels are leading edge, trailing edge, the chord, and the camber of the upper and lower surfaces of the wing. The leading and trailing edge definitions are self-explanatory. Chord line has already been discussed. *Camber* refers to the curvature of the upper and lower surfaces of the airfoil shape.

Figure 2-7 shows six airfoil designs. As you can see, the shape of these airfoils is very different, depending on the application the aircraft will be used for. Early airfoils had a great deal of camber on both the upper and lower surfaces of the wing, allowing flight at slow airspeeds, which was the best they could muster at the time. However, as aircraft performance increased, airfoil shapes underwent fairly radical design changes to best suit the airplane's operating envelope. Aircraft engineers must pick airfoil designs and wing planforms that best suit the design criteria they need to meet. The result is usually some kind of practical compromise. Airplane wings optimized for high speeds usually have poor slow-flight characteristics, and vice-versa. The aid of mechanical devices, which change the airfoil's shape, can do much to expand a wings' possibilities, however.

Lift enhancement devices

Aircraft designers would like to build a plane that cruises at very high speeds and lands at very low speeds, but these are difficult

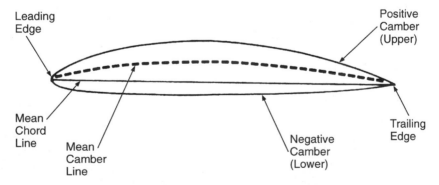

Fig. 2-6 *Airfoil components.*

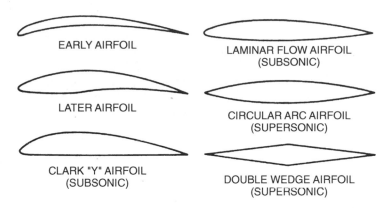

Fig. 2-7 *Some different airfoil shapes.*

characteristics to build into the same plane. They are able to achieve this goal to some degree with the use of lift enhancement devices such as flaps, slats, and leading-edge slots. Each of these devices allows the wing to fly at slower airspeeds without stalling, yet still fly at acceptable cruise airspeeds. We will look at each of these devices in this section.

Flaps

Within limits, greater camber produces greater lift. However, a less cambered wing allows for higher airspeeds at cruise. With the invention of retractable flaps we can have both, to a point. Adding flaps increases the camber of the airfoil. As a result, the airplane can fly at slower airspeeds without stalling. Then with the flaps retracted the camber is reduced and the airplane can achieve higher airspeeds at cruise.

Different designs have a wide range of effects on stall speeds. Some planes may only get two- to five-knot decreases in stall speed when flaps are used, while others allow them to fly fifteen to twenty knots slower. Specialized slow-flight aircraft can seemingly hang suspended motionless when their flaps are deployed and land in less than a few hundred feet of runway.

There are many different flap designs, but we will focus on three: the plain flap, the split flap, and the fowler flap (Fig. 2-8). The plain flap is hinged at the rear of the wing and basically pivots down around its hinge point.

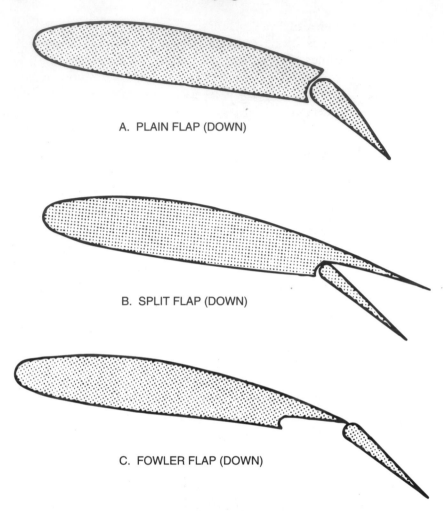

A. PLAIN FLAP (DOWN)

B. SPLIT FLAP (DOWN)

C. FOWLER FLAP (DOWN)

Fig. 2-8 *Plain, split, and Fowler flap types.*

The second flap design, split flaps, also increases the camber of the underside of the airfoil, but is located on the aft, bottom portion of the wing. Unlike the plain and fowler flaps, it does not change the camber of the upper surface of the wing.

The fowler flaps are the most efficient of the three shown in Fig. 2-8. In addition to moving down, they also move back from the trailing edge of the wing. This not only increases the camber of the airfoil on the upper and lower surface, but also increases wing area. Additionally, depending on how they are designed, slots between the wing and sections of flap can improve the flow of air over the wing at high angles of attack, further reducing the speed at which turbulence becomes strong enough to stall the wing. The next time you ride the airlines, notice how the flaps extend down and away from the trailing edge of the wing during takeoff or landing. This is part of the system that allows a jet that cruises at Mach .8 to land at relatively slow airspeeds.

Leading-edge devices

Flaps usually describe high-lift devices located on the wing's trailing edge, but similar effects can be achieved at the wings leading edge. The following mechanisms are but a few among many fitting a broad category of *leading-edge devices,* which generally serve to accomplish the same goal as trailing-edge flaps.

Slots

Slots are a cleverly designed gap in the wing or flap and are used primarily to enable the wing to fly at higher angles of attack. Many of the more complicated fowler flaps employ double, and occasionally triple, rows of slots as required to meet specific runway performance requirements. Flaps of this nature are referred to as *slotted flaps.* Slots can do much to enhance the effectiveness of a wing flap, and also work well at the forward edge of the wing.

A typical *leading-edge slot* is comprised of an open gap aft of the leading edge, running spanwise, parallel to the spar.

Figure 2-9 illustrates the difference between the airflow over a slotted airfoil at high angle of attack, as compared to that of a wing without slots. As you can see, air is able to flow through the

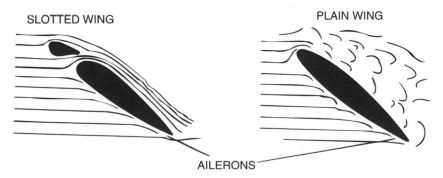

Fig. 2-9 *Leading-edge slots increase a wing's critical angle of attack.*

slot, then along the upper surface of the wing in a manner preventing an undesirable stall. When the wing is at lower angles of
attack, the slot has less effect on the flow of air. Slots are often
placed in front of the ailerons, allowing better aileron control at
high angles of attack.

Slats

Slats work by extending the leading edge downward, and forward,
much like flaps work on the trailing edge of the wing. Figure 2-10
depicts a leading-edge slat. Slats normally work on slides and rollers,
allowing the leading edge of the wing to move down and out from
the wing.

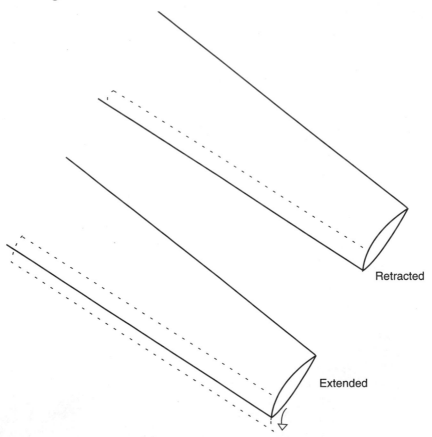

Retracted

Extended

Fig. 2-10 *Leading-edge slats are retractable and usually function
automatically with changes in angle of attack.*

Slats are frequently designed to create a slot between the slat and the wing when the slat is extended. This provides additional ability for the wing to produce lift at higher angles of attack. Leading-edge slats can be extended mechanically with motors or hydraulics, or by natural aerodynamic loads. In the case of the latter, the force of air pushing against the slat holds it in a streamlined position at low angles of attack/high speeds. But when the plane slows sufficiently and angle of attack increases, the slat begins developing lift and flies itself into the extended position. When combined with fowler flaps, leading-edge slats can dramatically increase the camber of the airfoil.

Leading-edge flaps

A simple increase in camber can be achieved by simply drooping the leading edge of the wing. The effects are similar to trailing-edge flaps, and when combined with everything else, can dramatically increase an airplane's ability to fly at slow speeds.

Figure 2-11 depicts a leading-edge flap in both retracted and extended positions. Notice that the extended position greatly increases the airfoil camber. This increase in camber also increases both lift and drag. Interestingly, an airliner with both leading- and trailing-edge devices fully extended requires an enormous amount of thrust to maintain level flight, and substantial amounts just to make a landing approach. Some airports restrict full-flap landings for large jets whenever possible for noise reasons.

Flight controls

There are three primary flight control surfaces: the ailerons, rudder, and elevators. These control movement of the airplane around three corresponding axes: the vertical, longitudinal, and lateral. Figure 2-12 illustrates the three axes of movement. These axes intersect at the airplane's center of gravity.

Yaw is movement about the vertical axis—the nose appears to swing to the pilot's left and right. Yaw is be controlled with the rudder via the rudder pedals. *Roll* takes place around the longitudinal axis and is controlled by the ailerons. Finally, *pitch,* or the movement of the nose up or down, as perceived by the pilot, takes place around the lateral axis, as governed through the elevators. As you will come to understand as you read through this book, understanding the correct

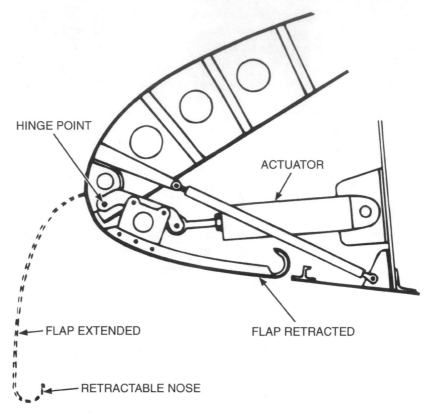

HINGE POINT

ACTUATOR

FLAP EXTENDED   FLAP RETRACTED

RETRACTABLE NOSE

Fig. 2-11 *Leading-edge flap.*

use of flight controls—and how control inputs will affect the attitude, airspeed, and control of the aircraft—is one of the basic fundamentals to becoming a safe, competent pilot.

Figure 2-13 depicts the ailerons, rudder, and elevator. This is the
standard arrangement for most general-aviation airplanes. There are
variations, such as the Beechcraft Bonanza, which has a V-tail, and
some planes incorporate spoilers on the wings as opposed to
ailerons, the Mitsubishi MU-2 being an example, and there are even
a few with the elevators at the front, as in the Beechraft Starship, in
a canard arrangement, but these represent the minority.

Each of these flight surfaces is controlled by the pilot from inside the
cockpit. Ailerons are controlled by the control wheel, or stick in
some aircraft, through turning the control wheel or moving the stick
from left to right. To roll the plane to the left, you turn the control

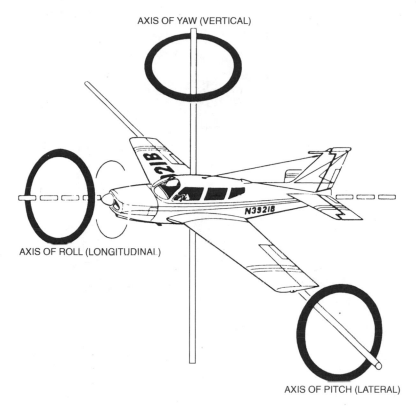

AXIS OF YAW (VERTICAL)

AXIS OF ROLL (LONGITUDINAL)

AXIS OF PITCH (LATERAL)

Fig. 2-12 *Three primary axes that pass through the airplane's center of gravity.*

wheel to the left; to roll right, you turn the control wheel to the right. If the plane you are flying has a control stick, then lateral movement of the stick to the left or right rolls the plane to the left or right.

Figure 2-14 illustrates how ailerons cause the plane to roll. As you can see, in a roll to the left the right aileron moves down, while the left aileron moves up. We learned earlier in the chapter that greater camber on the airfoil generates more lift. The ailerons essentially increase the camber of the wing that has the lowered aileron, which increases lift for that wing. The wing with the raised aileron suffers from a reduction in lift, the end result in this example being that the right wing is generating more lift than the left wing. This imbalance in lift generated by the wings causes the right wing to rise and the left wing to drop. As a result, the airplane rolls to the left about the longitudinal axis. The same holds true for rolls in the opposite direction.

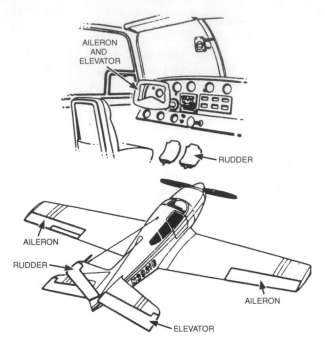

Fig.2-13 *Basic flight controls.*

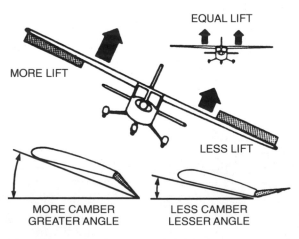

Fig. 2-14 *Ailerons manipulate wing lift by changing angle of attack.*

The elevator is also manipulated through the control wheel with forward and backward movement. Pulling back on the control wheel, or *yoke,* raises the nose of the airplane; moving the yoke forward lowers the nose—as depicted in Fig. 2-15.

As with the ailerons, pitch control through the elevators results from lift variation at the control surface.

Figure 2-16 shows movement of the rudder to the left and right generating changes in camber, and lift, just as with the other flight control. Like the other motions, this movement takes place about the plane's center of gravity.

Lift vectors

When an airplane is in level flight, i.e, constant altitude, lift equals gravity. Figure 2-17 shows a plane in both straight and level flight, and in a turn. With wings level, all lift is exactly opposite the effect of gravity. This is not true when the airplane flies in a banked turn.

Notice how the banked wings direct the lift in the direction the plane is banked. If this total lift were divided into vertical and horizontal

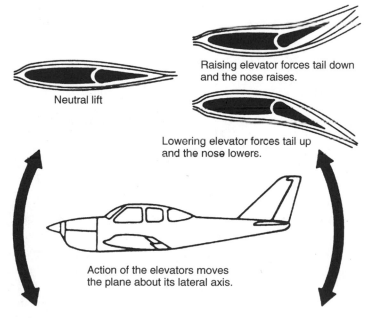

Fig. 2-15 *Elevator controls pitch attitude.*

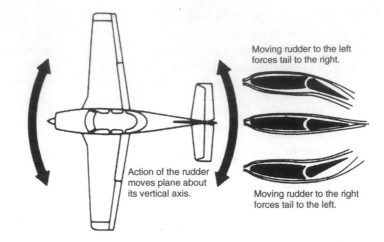

Fig. 2-16 *Rudder movement governs movement about the yaw axis.*

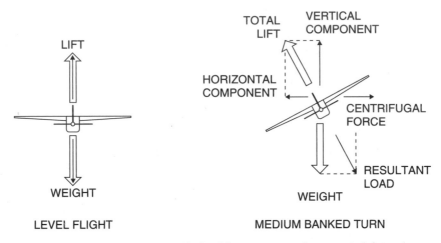

Fig. 2-17 *Turns are accomplished by pointing the wing's lift in the direction the pilot desires to go. Technically, it is the horizontal component of the total lift (in a bank) that produces a turn.*

components, for ease of discussion, the vertical portion would continue to oppose gravity, while the horizontal component would turn the airplane. The total of these two would necessarily be greater than when the airplane had the wings level. For pilots, this means that while turning, the plane must produce more lift, depending upon the angle of bank, which is usually accomplished by an increase in angle of attack and some increase in power. Simply stated, the pilot has to hold back pressure on the yoke while the airplane turns.

The complexity does not cease there, however, as the mechanics of banking the wings may generate an imbalance of drag at the wingtips, which may work against the direction of the desired turn. The drag occurs because the lowered aileron generates more lift, and correspondingly more drag than the raised aileron. This asymmetrical drag causes the nose of the plane to yaw in the direction of the lowered aileron. For example, in a left turn the right aileron moves down, causing greater drag on the right wing. This drag causes the nose of the plane to yaw right; this is adverse yaw. Coordinated rudder input compensates for adverse yaw. For a left turn, left rudder is used; in a right turn, right rudder is used. When done correctly, the amount of rudder exactly cancels out adverse yaw from the aileron.

An instrument known as the inclinometer is found on the instrument panel of most aircraft and is generally referred to among pilots as "the ball."

Figure 2-18 shows the *ball,* a small black bead in a curved tube filled with liquid. In a coordinated turn the ball stays centered in the glass tube, but in an uncoordinated turn the ball slips to the inside of the turn or skids to the outside of the turn. How much rudder will be required to keep the ball centered depends on the airplane, its airspeed, the angle of the bank, and a number of other factors. Early in flight training, students will often stare at the ball during a turn, trying to get the correct amount of rudder input. With practice, though, they develop a feel for whether a turn is coordinated or not.

Trim tabs

A trim tab acts in the same relationship as an aileron does to the airfoil, using the change in camber to affect the position of a flight

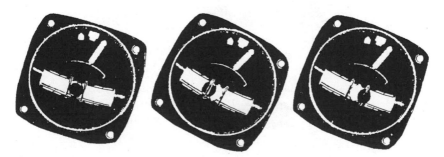

Fig. 2-18 *The inclinometer, or "ball."*

control surface. Figure 2-19 shows an elevator trim tab, but the same principles apply to rudder or aileron trim tabs.

In this case, the elevator trim is hinged at the trailing edge of the elevator. The trim tab positions may be directly controlled by the pilot to alleviate the need for constant elevator control pressure during various phases of flight, such as in a turn. Trim adjustments will also be necessary as fuel loads and airspeeds change.

Figure 2-20 depicts both aileron and rudder trim tabs. Most single-engine general-aviation planes are equipped with elevator trim, but a smaller percentage have aileron or rudder trim. Like the elevator trim tabs, aileron and rudder trim cause the control surface to be deflected, alleviating the need for the pilot to maintain constant control pressure. Aileron trim may be useful to eliminate any rolling tendency a plane may exhibit. This could be due to unequal weight distribution, such as more fuel in one wing tank than another, or more passengers on one side of the plane. Sometimes how a plane is rigged, or the twist that is built into the wings, can also result in a rolling tendency. Rudder trim is used to overcome any yawing tendency a plane may have, such as during a climb, when imbalanced propeller thrust may induce a tendency to turn, or when op-

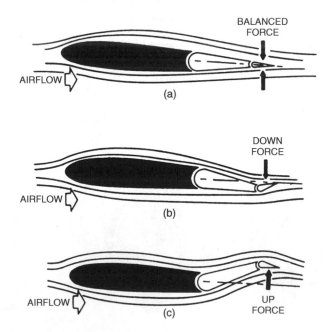

Fig. 2-19 *Elevator trim tab.*

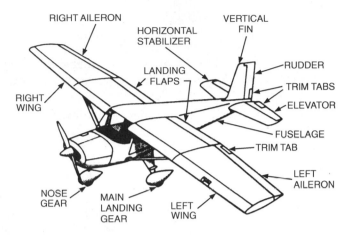

Fig. 2-20 *Aileron and rudder trim.*

erating a multiengine aircraft after one engine has failed. Trim, when properly used, goes a long way toward easing a pilot's workload in the cockpit.

Center of gravity

There are several methods used to compute the center of gravity, or weight and balance as it is sometimes referred to. The operations manual for the plane you are flying will have a weight and balance section, which should detail computational methods and C.G. (Center of Gravity) limits for that aircraft. Loading the airplane such that its balance point is outside of its proscribed limits invites serious control difficulties that may render the airplane uncontrollable in some flight regimes. Individual planes are different in their specific weight and balance, so information specific to that plane (not type) must be available to the pilot.

V-speeds

V-speeds, or the various critical velocities needed for the airplane you fly, are an important part of flight maneuvers. Stall speeds should be uppermost in your mind as you take off and land. Maneuvering and climb speeds play a crucial part in extracting maximum performance and safety from the airplane. V-speed abbreviations provide a notation standard common to the aviation industry. Here are a few to be concerned with.

Stall speeds

Depending on the plane, there may be a number of different V-speeds associated with different aircraft configurations, from flaps up, to partial flap extension, to full flap extension, gear extended, gear retracted, and anything else that might be useful to the pilot. Here are two modifiers to stall speed that should be of interest.

V_{so} (Stall speed landing configuration)

V_{so} is minimum flight speed in the landing configuration. This configuration can vary from plane to plane, but generally is gear down, flaps down, and power off at maximum landing weight. Pilots will often refer to this configuration as "dirty" because of the additional drag caused by having the gear and flaps extended into the airstream. Some pilots use a memory aid here; V_{so} = stall speed with all the "stuff out."

V_s (Stall speed clean configuration)

V_s is the stall speed for the clean configuration. This means the plane has the flaps (and landing gear, if the plane is a retractable) retracted. Planes normally stall at a higher airspeed in the clean configuration due to the loss of lift resulting from the flaps being retracted. Keep in mind that V_{so} and V_s stall speeds are for when the plane is in level flight.

V_{fe} (Flap extension speed)

V_{fe} refers to the maximum airspeed at which flaps should be in the extended position. If you fly the airplane at speeds that are greater than V_{fe} with the flaps extended, it is possible to cause physical damage to aircraft structures. The flap connecting rods, rollers, tracks, or the flaps themselves could be bent or ripped from the plane if you exceed V_{fe} with the flaps extended. The same damage could take place if you extend the flaps above V_{fe}. The airspeed indicator commonly depicts safe operating airspeeds for the flaps with a white arc.

V_{ne} (Velocity never exceed)

V_{ne}, or the airspeed above which the airplane should never be flown, is a very important number to keep in mind. Indicated by the red line on the airspeed indicator, V_{ne} is the maximum speed an airplane can safely be flown at in smooth air. V_{ne} is determined by the aircraft manufacturer based on engineering design and aircraft testing. Faster speeds than V_{ne} are not recommended as they will eventually result in premature structural failure.

V_a (Maneuvering speed)

Maneuvering speed, known as V_a, is the airspeed at or below which rapid movements of the controls or turbulence will not cause damage to the aircraft at gross weight. V_a is not normally marked on the airspeed indicator because it may change dramatically with lighter weights—typically slower; but may be placarded on the instrument panel. If it is not placarded, you should look it up in the operations manual for the plane you fly. If you should encounter turbulence, you should reduce the airspeed you are flying at to at or below V_a. If you fly at these airspeeds, the plane will theoretically stall before it generates loads that are sufficient to damage the aircraft. At speeds above V_a, turbulence or rapid control movements may generate g-loads that exceed the structural design of the plane and may damage the plane to the point of structural failure.

V_x (Best angle of climb speed)

The best angle of climb speed for an airplane allows you to gain the greatest amount of altitude within a given distance. Best angle of climb is the airspeed intended by aircraft manufacturers for climbs after a short-field takeoff or when there are obstacles you must clear at the end of the runway.

Due to the relatively high angle of attack that is achieved during a climb at V_x, the engine may have a tendency to overheat; for this reason you should avoid prolonged climbs at best angle of climb.

V_y (Best rate of climb)

The best rate of climb will allow the plane to gain the largest amount of altitude over a given period of time. V_y is often recommended by aircraft manufacturers for extended climbs to cruise altitudes. The angle of attack used during a climb at V_y airspeeds is lower than V_x, allowing better airflow through the engine compartment and lower engine operating temperatures. You will notice substantial differences in engine oil temperatures during the summer between flying at V_x or V_y.

3

Mastering the maneuvers

All airplanes have the same basic characteristics. They all pitch, roll, and yaw. The same basic forces apply, only the magnitudes of those forces seem to change. The fact that one airplane is 500 times heavier than another is of little significance when thinking in terms of why they fly.

The size of a 747, for example, is large by any standards. But does it take a person of superhuman abilities to fly one? Of course not. The pilots flying this large airplane operate under the same principles that make the simplest trainer take wing. These fundamental principles of flight are so basic, and so pervasive, that they must be completely understood by every pilot—no matter what equipment they fly.

Straight-and-level flight, climbs, descents, and turns are the foundations upon which all normal flight is built. Every single maneuver has its origin in one or more of these fundamentals. Sounds simple, doesn't it?

Straight and level is one of the first maneuvers a student pilot learns. It is the ability to hold a constant heading, altitude, and airspeed. A good definition for straight and level would be slight corrections from any climb, turn, or dive—correcting any deviation from a straight, level, line of flight. It is the most fundamental of all the fundamentals. At the same time, it may also be one of the most challenging. Since conditions around the airplane are rarely constant, the task of *maintaining* constant flight parameters in turbulent air, for example, may require intense concentration.

There are several ways to get the feel of straight-and-level flight through the use of outside visual references in combination with instrument interpretation, a technique known as *integrated flight*. First, set up straight-and-level flight using mostly instruments.

Then, look out over the nose and get a good picture of what you see. Try to take a picture with your mind of what you want to see every time you are straight and level. In other words, you want to see how much ground and horizon are visible and where the horizon line crosses the windshield. You might also note where the cowling of your aircraft is situated with respect to all the other points. If anything changes, then so do the flight parameters. This careful positioning of the aircraft with respect to the horizon is known as flight *attitude*. Generally speaking, once correct flight parameters are achieved, maintaining the flight attitude constant at that point will preserve those parameters until the attitude is changed.

In the level flight example, if the nose of the airplane has appeared to rise, even slightly, you are probably climbing. Conversely, if the nose appears a little low on the horizon, you are probably diving. It's that simple. Pilots must become highly sensitized to their aircraft's flight attitude—that is the key to all flight maneuvers.

Sometimes pilots become so accustomed to a particular "look" out the cockpit windows that a simple seat adjustment can throw off their landings for days. Bring the pilot of the 747 back to a simple Cessna trainer, for example, and there will be comedy afoot. The jumbo pilots will be delightfully out of their element, until they become accustomed to the "view" out of the smaller plane.

Using the flight instruments

Figure 3-1 shows a basic panel layout. Let's begin with the center of the panel. At the top center is the artificial horizon, or attitude indicator. This is a gyroscopic instrument. A vacuum pump attached to the engine causes a flow of air through the instrument, which spins its gyroscope up to speed. When powered, the gyroscope will attempt to maintain a constant orientation as the plane moves, giving it the ability to mimic the horizon's constant position.

Figure 3-2 shows the artificial horizon when the plane is in level flight. The small airplane in the instrument's foreground represents the plane you are flying, while the background represents the horizon, earth, and sky, outside the airplane. As the plane moves, the background of the artificial horizon moves. When viewed in relation

Fig. 3-1 *Basic instrument panel layout—sometimes referred to as the "sacred six."*

to the small airplane in the instrument, you can maintain a reference to the horizon, even if the outside horizon is obscured.

The instrument located directly below the artificial horizon is the directional gyro, or DG. Like the artificial horizon, it is a gyroscopic instrument powered by suction on most single-engine aircraft. Figure 3-3 shows the directional gyro. The instrument acts much like a compass but does not use the magnetic field of the earth for direction-finding purposes. Instead, as part of the run-up prior to takeoff, you set the DG to the same heading indicated on the compass. The gyro in the instrument then maintains this alignment and indicates the plane's heading with more stability than the floating compass.

A word of warning concerning the DG, though. The directional gyro in most airplanes has some tendency to drift off the correct heading as time passes. This precession is easily corrected by periodic comparisons with the DG and magnetic compass while in steady flight. For planes with older instruments, or those with failing suction pumps, you may find that the DG will not hold the correct heading for any period of time. When it comes to low vacuum pressure, the same is true of the artificial horizon. I have had flights where low vacuum pressure from the pump caused the artificial horizon to give erroneous readings, indicating a bank when the plane was actually flying straight and level. In heavy instrument conditions this can cause the pilot some distress, but with a proper instrument technique you should be able to quickly determine that an instrument is

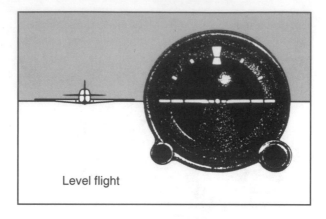

Level flight

Climbing turn to the left

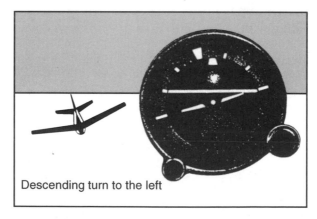

Descending turn to the left

Fig. 3-2 *Attitude indicator or "artificial horizon" showing level and turns both right and left.*

Fig. 3-3 *Directional gyro.*

failing, which instrument it is, and then continue to fly using alternative instruments.

Figure 3-4 shows the airspeed indicator. The needle moves clockwise to indicate higher airspeed. The airspeed indicator measures pressure differentials between air entering a probe called a pitot tube and static air pressure from the static vent. There are typically several color bands on the airspeed indicator: white, green, yellow, and red. The white band denotes the recommended speeds for flap operation. The high end is the maximum airspeed the flaps may be extended, while the lower end of it is the stall speed of the plane with flaps fully extended. If flaps are kept out above the airspeed indicated at the top of the white arc, damage to the flaps and other structures of the wing could potentially result.

The bottom of the green arc is the stalling speed of the plane with flaps retracted, or in the clean configuration. Please note the stalling speed is higher when flaps are not used, which fits with our discussion of flaps earlier. At the top of the green arc is the airspeed that should be used for normal cruise, although some planes do not have sufficient power at altitude to actually reach this indicated airspeed.

A yellow arc begins at the end of the green arc and depicts the range of airspeeds wherein the plane should be flown only with caution. At these airspeeds it is possible to damage the airplane if severe turbulence is encountered or rapid control inputs are made.

At the top of the yellow arc is a red line, the maximum airspeed. Above this airspeed the airplane can be structurally damaged due to the dynamic load.

Fig. 3-4 *Airspeed indicator.*

Fig. 3-5 *Altimeter.*

Figure 3-5 shows an altimeter, which operates by measuring changes in barometric pressure. Altimeters normally have three needles: one showing feet, one showing hundreds of feet, and finally one showing thousands of feet. Most altimeters also have a small window that is used to set the reference barometric pressure for your location.

Like the altimeter, the vertical speed indicator (lower right in Fig. 3-1) is a pressure-sensing device that is used to indicate the rate of change in your airplane's altitude. This is usually expressed in hundreds of feet per minute.

Straight-and-level flight

Look out each side of the aircraft and see where the wings are in relation to the horizon. If one wing is up, the airplane is in a banked attitude and will be turning. Straight flight can also be maintained quite readily by monitoring the view out the front windshield. Simply watch the horizon. If the airplane is wings-level with respect to the horizon, it will fly straight.

Sometimes a pilot subconsciously allows the weight of a hand to pull the yoke down to one side, causing the aircraft to bank in that direction. Holding the yoke lightly alleviates this problem.

The pilot must also hold his or her head upright to see everything in proper perspective. If the aircraft is in a 5-degree bank to the left and you subconsciously tilt your head 5 degrees to the right, the horizon will appear to be in its proper alignment, when in reality, things are not as they should be. The sooner you learn to ride with your aircraft and not to lean against the turn, the sooner you will master attitude flight control.

Use trim. Set up the desired attitude and adjust the trim tab to lighten any pressure you feel on the controls. When you no longer feel any pressure on the yoke, let go of everything and the aircraft should remain in that attitude. If it doesn't, put the aircraft back where you want it and trim it a little more. The most important thing to remember when using the trim tab is that it is not designed for use as a primary flight control, so you must first establish the desired attitude, maintain the control pressures, and then adjust trim delicately until the controls seem to maintain themselves.

Have an idea of where the horizon will be in relation to the nose in the level flight attitude. For most aircraft, the nose will be below the line of the horizon during level flight. After you gain experience with a plane, you can normally lower the nose to just about the right position in relation to the horizon just from experience. Once you have the nose at the right position, you should check the altimeter and vertical speed indicator to determine if you are climbing, descending, or maintaining the correct altitude. The vertical speed indicator can be very useful in figuring out if there are small corrections that need to be made.

If you are transitioning to straight and level from a climb, you will probably need to pull back the engine power from its climb setting to the cruise setting you want to use. As you reduce power it is likely the nose of the plane will want to drop slightly, and you may need to readjust the control yoke's position to maintain the correct altitude. Power settings can have a tremendous effect on how the airplane will fly. For instance, if you are flying straight and level and you increase the power without changing any control inputs, in most cases the plane picks up more airspeed. This greater airspeed results in more lift, and the airplane has a tendency to climb. To maintain the same altitude after a power increase, you will need to push forward on the control yoke, which keeps the airplane from climbing. The amount you will push forward depends on how much power was added and the characteristics of the plane you are flying. You should also retrim the elevator, setting it to reduce the pressure you are holding on the control yoke to neutral.

Once you have the airplane trimmed for level flight you can begin to verify that you are holding straight and level. At this point you have scanned outside the plane to make sure your attitude is correct, and verified that with the artificial horizon. At the same time you should verify that the wings are level. If the plane is banked left or right, the horizon will appear banked as you look outside of the airplane. The artificial horizon will also show that you are banked. You can then use the compass and directional gyro to confirm your heading.

As you fly along in straight-and-level flight, the plane will be occasionally influenced by the air you are flying through. Wind gusts, turbulence, updrafts, and downdrafts all affect the airplane as it flies. You will find that you need to continuously make minor inputs to correct for these outside disturbances to straight-and-level flight. If the wings are rocked to the left or right, you can use aileron to get them back to level. The same holds true for nose up or down. You can be flying along and encounter an updraft, which wants to push the airplane upward. To counteract the effects of the updraft you will need to push the nose down. As you fly out of the updraft you will need to release the forward pressure to avoid losing altitude.

Once you gain experience in holding a plane at straight and level, you will find it is not nearly so mechanical as the previous description might indicate. Flying the airplane is a fluid activity, with you making constant, small inputs to hold the plane's wings level and the nose at the correct attitude. You should also continuously scan both the instruments inside and outside the plane.

Climbs

Many of the same reference points used in straight-and-level flight can be used for climbs. Reference to wingtips may be used to complement instrument indications for laterally level flight and can also be used for pitch information.

In some aircraft, the forward field of view is limited while climbing, making frequent reference to the flight instruments rather important for directional control. However, lateral references may still be used for directional orientation after a bit of practice.

Let's say we're cruising along at 3000 feet above ground level (AGL), and we are asked by air traffic control to climb to 5000 feet. To initiate the climb you will do two things: First, add power as recommended by the aircraft's manufacturer. This will vary from plane to plane, but for many aircraft, climb settings normally are full power. Then increase backpressure on the control yoke, raising the nose of the plane. There are two airspeeds that you will be concerned with during climbs: the best angle of climb airspeed and the best rate of climb airspeed.

The best rate of climb, expressed as V_y, gives you the greatest gain in altitude over a period of time. The best angle of climb airspeed, known as V_x, results in the greatest altitude gain in a given distance. Generally, for takeoffs you will initially climb at V_x airspeeds for that

Fig. 3-6 *The view outside is a primary attitude reference.*

plane. For climbs from a given altitude, such as in our example, many aircraft manufacturers recommend using the V_y airspeed. For most airplanes, V_x is a lower airspeed than V_y. As a result, the angle of attack is greater at V_x than it is at V_y, resulting in the nose of the plane being higher.

As you raise the nose, the airspeed will begin to fall. You are now taking some of the additional power and lift and converting it to an altitude gain. Normally you will want the airspeed to stabilize at the correct climb airspeed, whether that is V_x, V_y, or another airspeed recommended by the manufacturer. Once the airspeed has stabilized, you can retrim the elevator to reduce backpressure on the control yoke to neutral.

As you approach the altitude you are climbing to, you will need to gradually reduce the nose-up angle. Correctly done, you will level off at the altitude you are supposed to attain. This is where the vertical speed indicator (VSI) comes in handy in helping you judge when to begin the level off. A rule of thumb is to begin leveling off the plane when you are 10 percent of the VSI's reading below the desire altitude. Let's say we're climbing at 500 feet per minute and want to level off at 3000 feet. Ten percent of 500 feet per minute is 50. In this case, at 2950 feet we would start to level off, hopefully ending up at 3000 feet as the final altitude. If we were climbing at 1000 feet per minute, we would start to level off at 2900 feet.

When releasing backpressure on the control yoke, the input should be gradual. You don't want to force your passengers up against their seatbelts by shoving the control yoke forward too forcefully, giving them the feeling of weightlessness. Normally, a nice, steady release of the backpressure on the control yoke gives your passengers a pleasant transition from the climb to level flight. Once you are in level flight, you will need to reset your engine power, then retrim the elevators for level flight.

Left-turning tendencies

It may seem odd to discuss a left turn in itself, but many airplanes turn left of their own accord during some phases of flight. The reasons for this stem from a clockwise-turning propeller (when viewed from the cockpit). The propeller exerts some natural turning forces on the aircraft, among which are torque, gyroscopic precession, spiral slipstream effects, and P-factor. The sum of all of these forces results in a left-turning tendency that must be countered by the pilot. If the propeller turns in the opposite direction, as is common with some European engines, and power plants employing a speed reduction mechanism, the natural tendency would be a turn to the right, for the same reasons.

Torque force is present in all propeller-driven aircraft. Simply stated, the airplane reacts to the engine torque on the propeller by rolling in the opposite direction. The amount of torque felt depends entirely upon the amount of power applied to the propeller. The amount of pilot effort required to combat this roll depends upon the effectiveness of the flight controls, and indirectly on the speed of the airplane.

Gyroscopic precession results from the spinning mass of the propeller itself. The propeller may be rather heavy and is a wonderful gyroscope. That is to say, it exhibits some resistance to movement, such as an aircraft turn, climb, or descent. This resistance is felt by the pilot as a turning tendency that is prevalent mostly when attitude adjustments are made.

Naturally, the spinning propeller creates a great deal of spinning airflow behind it. This is the notorious spiral slipstream, which engenders visions of the propeller blast resembling a tornado that shrouds the airplane. A clockwise-turning propeller creates a clockwise-turning airflow, which spirals back around the airplane and strikes the rudder from the left side, causing a left turn. The strength of the turn is also related to propeller thrust and aircraft speed. The problem could be solved by placing the rudder on the bottom of the fuselage, but that would complicate the landing gear system.

P-factor tends to yaw the aircraft to the left most powerfully in a climbing attitude. P-factor yaws the aircraft due to unequal thrust across the propeller disc. As a two-bladed propeller rotates in straight-and-level flight, both blades get an equal "bite" of air, which produces an equal amount of thrust from each blade. However, in a climbing attitude, the descending blade (the one on the right when viewed from the cockpit) gets a much larger bite of air and creates more thrust than the ascending blade. Because of this unequal thrust, the aircraft is yawed to the left. The proper corrective measure to offset the effects of both torque and P-factor is the application of right rudder as needed to maintain heading.

Torque

The airplane must resist powerful torque effects from the rotating propeller. Left alone, the engines' effort at the propeller would be mirrored when the airplane attempts to roll the opposite direction. Since most aircraft have clockwise-turning propellers, when viewed from the cockpit, the natural response of the airplane to increased power (torque) is a roll to the left. (Fig. 3-7.)

Slipstream

Thrust created through the spinning propeller is imparted a spin of its own—like a tornado spiraling back around the fuselage or nacelles of the airplane. This spiral motion of the slipstream causes the air to strike the rudder at an angle that produces a natural yaw (Fig. 3-8). Air rotating clockwise around the fuselage would strike the vertical stabilizer from the left and cause a left turn.

Gyroscopic effect

The propeller itself develops rather powerful gyroscopic effects. Indeed, at high power settings the propeller may offer powerful resistance

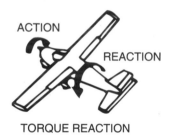

Fig. 3-7 *Torque.*

Fig. 3-8 *Spiral slipstream creates lift across the vertical stabilizer, which swings the nose to the left.*

to changes in aircraft attitude. This is mostly unfelt at the flight controls, except for a dynamic response that also pries the airplane into a left turn (Fig. 3-9). Other effects are felt through the engine bearings and are responsible for high wear in the crankshaft of many aerobatic airplanes.

P-factor

The propeller thrust may be unbalanced by the effects of angle of attack. As the angle of attack increases, the thrust becomes more powerful on the right side of the propeller arc, which pulls the airplane to the left. The unbalancing of thrust results from the propeller blade on the downward side of its arc (right side, when viewed from the cockpit) getting a more efficient bite of air—greater angle of attack—than the opposite side (Fig. 3-10). This unbalancing of thrust increases as the airplane's angle of attack is increased.

Descents

Descents require a little advanced planning. The following discussion begins at cruise altitude and integrates a landing approach.

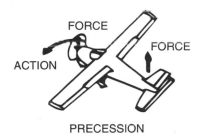

Fig. 3-9 *Gyroscopic precession translates pitch inputs at the tail into a left yaw at the nose—especially at low airspeeds.*

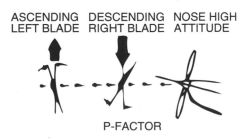

Fig. 3-10 *P-factor, or* propeller *factor, produces a left turn at high angles of attack.*

For a cruise descent, begin at normal cruise speed at a given altitude, and slowly reduce power as you trim the aircraft to maintain the cruise airspeed. Watch the vertical-speed indicator as you begin to descend. When it settles on the desired rate, trim the aircraft to maintain the airspeed and use your power to control the rate of descent. It's that simple. Sit back and monitor other traffic arriving at or departing from the airport.

Let's assume you are arriving in the vicinity of your destination airport after a cross-country and you have to let down from 5000 feet to an airport at sea level. If a comfortable descent rate for your aircraft is 500 feet per minute (FPM), then it will take you about 10 minutes to come down. How far out will you have to begin the cruise descent in order to arrive at the destination airport at the traffic pattern altitude? If the destination airport has a 1000-foot traffic pattern, you will have to lose 4000 feet before beginning the approach. At 500 FPM, it will take about eight minutes. For simplicity, let's say you have a ground speed of 120 MPH. Remember, it's ground speed, not indicated airspeed, that will give you the most accurate information. However, if you do not know your ground speed, indicated airspeed will probably get you close enough unless you have a very strong wind.

You have eight minutes to lose 4000 feet at a ground speed of 120 MPH. Since 120 MPH equates to two miles per minute and you have eight minutes to get down, merely multiply the miles per minute by the minutes of descent in order to find the distance out to initiate your descent. In this case, it would be $2 \times 8 = 16$. You will start your 500-FPM descent 16 miles out in order to arrive at the destination airport at a 1000-foot traffic pattern altitude.

Once in the pattern, the airplane must be slowed, and the descent continues while configuring for landing. This is the basis for the landing approach. The approach descent can be accomplished in any one of many different configurations. For simplicity, consider the technique used for a Cessna 152. With only minor modifications, this method can be used for most general-aviation aircraft.

Starting at a given heading, altitude, and airspeed, slowly reduce the power to about 1500 RPM, and hold enough backpressure on the yoke to maintain level flight attitude. As the aircraft slows, begin trimming to relieve the pressure on the stick as you put down whatever flap setting you desire to practice for this particular configuration.

As the airspeed slows to approach speed, keep the power constant and trim for hands-off flight at that speed. At the reduced power setting, the nose will be lower and you will be in an approach descent. Rate of descent is usually controlled with power and airspeed with pitch. In other words, if you see you are descending too fast, you can bring the pitch up a little, add power, or both. You will have to work a little to find the right combinations, especially in rough air.

Select an appropriate airspeed for descent. In smooth air, a high indicated airspeed may be suitable. If the air is rough, you may consider slowing down, perhaps even to maneuvering speed, or V_a. Begin by adjusting pitch attitude and reducing power to achieve a desired rate of descent. If our initial airspeed in the descent is too high, we need to raise the nose of the plane slightly, which will reduce the airspeed; if the airspeed is too low, increase the airspeed by dropping the nose slightly. With airspeed established, you can then set your rate of descent using the VSI as an indicator. For most descents you should aim for approximately 500 feet per minute. This descent rate prevents discomfort for your passengers. While you may need to increase the rate of descent due to various factors, a higher rate of descent causes your passenger's ears to pop more frequently, and if they have any head congestion it may make the situation painful.

As with climbs, begin to level off about 10 percent of the descent rate before your target altitude. Ease back on the control yoke, slowly raising the nose. You can also begin to add power to help prevent the airspeed from falling off. Correctly executed, you can smoothly transition from the descent to level flight and hit your target altitude. Once you have leveled off and set your engine power, retrim the plane for level flight.

Turns

Although somewhat different from a car, steering an airplane is really not very difficult. An aircraft turns because of a change in the direction of lift—a change from completely vertical lift to lift at a horizontal angle relative to the horizon. This change is implemented when the ailerons are manipulated to place the airplane in a banked attitude—aiming the wings lift in the direction of the desired turn. It is the *bank* that causes the turn—not the ailerons themselves. As long as the airplane remains in a banked attitude, the turn will continue.

A properly coordinated level turn begins when the aileron and rudder are added simultaneously to cause the aircraft to bank to the desired position. A small amount of backpressure is added to the elevator to keep the nose up and maintain altitude. When the desired amount of bank is reached, everything except the elevator is returned to neutral. The reason the aileron and rudder are returned to neutral is to stop the aircraft in the desired bank. If you were to hold the aileron and rudder in, the aircraft would continue to roll, and neither you nor your trainer aircraft is probably ready for descending rolls.

The elevator should be kept slightly aft of neutral to maintain altitude because an aircraft loses vertical lift any time it is in a banked configuration. To make up for this loss, a small amount of up elevator must be held through the turn.

If the ailerons produce the banking motion, then what is the rudder good for? Why is it on the aircraft at all? It is there for two very good reasons. One is to allow the pilot to have control over the yaw, or side-to-side movement, of the aircraft. The other is to overcome adverse yaw produced by the ailerons during the turn entry.

Try this little experiment the next time you fly and you should gain some insight into turns. Fly straight and level and put your feet on the floor. Pick a prominent object or reference point over the nose, and roll into a turn without the use of the rudder. The nose will actually seem to go in the opposite direction for an instant, and then it will resume its correct flight path. This phenomenon is called adverse yaw and is caused by the induced drag of the downward aileron.

Now, level the wings and start another turn using the same reference point. This time purposely lead the turn with the rudder and then apply aileron. The result will be the plane yawing in the direction of the turn and is known as a skid. With too much rudder, the plane moves laterally to the outside of the turn, as shown by the ball on the right side of the turn-and-slip indicator.

In a properly coordinated level turn using outside visual references, the nose of the aircraft will roll about a point and then continue in the direction of the turn as the bank is established. There should be no undesired yaw motion as ailerons are applied.

Let's begin with entry into the turn. Figure 3-11 shows the forward view looking from the pilot's seat over the nose of the plane in straight and level flight. The dot in the figure is an imaginary reference point

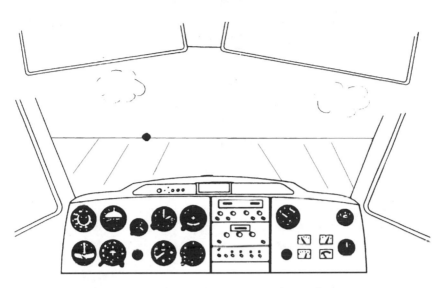

Fig. 3-11 *Forward view from pilot's seat with a reference point on the horizon.*

on the windscreen that is directly in front of the pilot. As you enter the turn you want to maintain level flight and not climb or descend. The best visual reference you can have is the point on the horizon on the windscreen directly in front of you, which is represented by the dot. During turns, that point will remain at almost the same relation to the horizon above the control panel as it did in level flight.

Figure 3-12 illustrates when the plane is banked in a left turn. The dot shows where the horizon was during level flight. In this case the horizon has stayed on the dot, and the plane does not climb or descend during the turn. Figure 3-13 shows the horizon has moved below the dot, which means the nose of the plane has risen and the plane is climbing during entry into the turn. Figure 3-14 shows that the horizon has moved up in relation to the dot, meaning the nose of the plane has dropped and the plane is losing altitude. This visual reference can be verified by using the artificial horizon, altimeter, and VSI, which will also show when the plane is climbing or descending. A word of caution, though. Many student pilots become so focused on the instruments during turns that they stop looking outside the plane and stare intently at the instrument panel. When flying in VFR conditions, these instruments should verify what you are seeing by looking outside the plane, not act as your primary reference during turns.

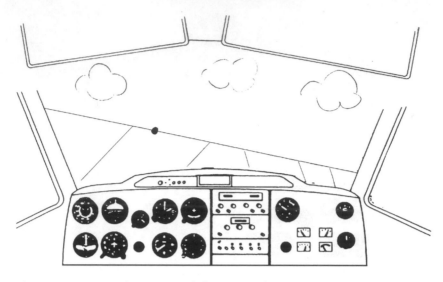

Fig. 3-12 *Forward view in a left turn. Reference point will be near the horizon if the plane is maintaining altitude.*

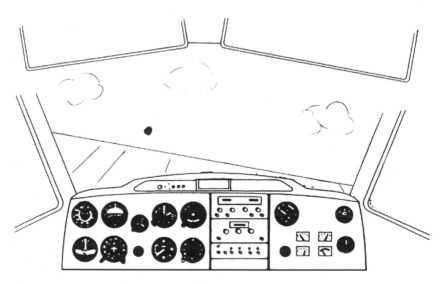

Fig. 3-13 *Forward view in a climbing left turn.*

Before you begin a turn, always look in the direction you are turning toward to make sure there are no other aircraft or obstacles in that direction of flight. In a high-wing airplane this may require you to raise the wing slightly to clear the area. Once the turn area is cleared, you can look forward again and set your horizon reference

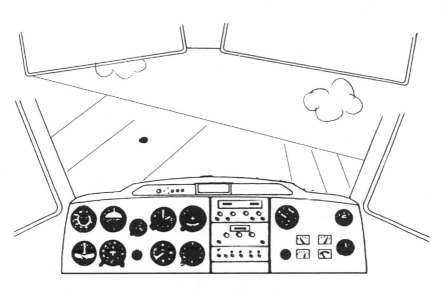

Fig. 3-14 *Forward view in a descending left turn.*

point for attitude. Apply aileron in the direction of the turn, at the same time applying just enough rudder to prevent the nose from yawing and to keep the ball centered in the turn and bank indicator. You will also need to apply just enough up elevator to maintain your reference point on the horizon. It takes practice to become used to how much of each control should be used during a turn, and you will find that this changes not only from plane to plane, but also is affected by the airspeed.

To exit the turn you will roll the wings back to level flight, using rudder and aileron in a coordinated movement, just as when you entered the turn. As you reduce the bank, the lift vectors will again be used to keep the plane at altitude, and you will need to reduce backpressure on the control yoke.

There's also another important factor you need to take into consideration during turns. In a shallow banked turn, from 0 to 20 degrees bank, the dihedral of the wings will attempt to restore the plane to wings-level flight, so you will need to keep ailerons input throughout the turn. In a medium banked turn, 20 to 45 degrees, the plane will have a tendency to remain in the bank, and you will be able to neutralize the ailerons to some extent once the bank has been established. In a steep banked turn, greater than 45 degrees, the plane will have a tendency to roll steeper, and you may need to feed in some opposite aileron to counteract this tendency. Remember, as

you make these varying aileron inputs you will also need to adjust
your rudder inputs to maintain coordinated flight.

Steep turns

(Fig. 3-15.) A steep turn requires an advanced sense of coordination
and timing. Maintaining altitude and orientation are only two of the
problems this maneuver presents. Steep turns take the aircraft to the
opposite end of the performance spectrum from slow flight. However, the possibility of an accelerated stall still exists as the aircraft
reaches a 50- to 60-degree bank and large doses of backpressure
must be applied to help maintain altitude.

After clearing the area, set up straight-and-level flight below maneuvering speed. (Steep turns should be done at or below maneuvering
speed because of the possibility of momentarily high G-loads caused
by the pitch changes required to maintain altitude.) Pick a heading
and altitude at which to begin the steep turn. After completing a 360-
degree or 720-degree steep turn, this will be the heading and altitude at which you want to wind up. Then, begin the roll into the
steep turn at a higher-than-normal rate. As the bank reaches about
half the desired maximum, 25 to 30 degrees, smooth application of
backpressure is initiated as the roll continues to the desired amount
of bank. If your backpressure input versus bank angle is relatively
close to the desired amount, altitude can then be controlled by slight

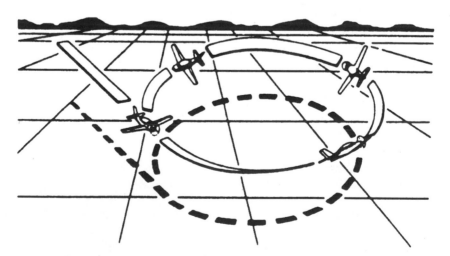

Fig. 3-15 *Steep power turn.*

changes of bank to either increase or decrease the vertical lift. At the same time, small pitch changes might be necessary to complement the bank and aid in proper coordination.

Unless you're a master at steep turns or enjoy high G-loads and feeling centrifugal force to the point that your fillings are pulled from your upper teeth, the vertical lift component can be taken care of through small changes in bank instead of large changes in pitch. Try it.

Because a steep turn usually produces about two Gs, the airspeed will decay due to the higher load factor. Power should be added as needed to help keep the airspeed up. Remember, as the bank increases, so does your stall speed. Therefore, power application during the steep turn should help decrease the chances of an accelerated stall.

The most common problem associated with steep turns is uncoordinated entry and recovery technique. Pilots entering with too little backpressure for a given bank lose effective lift due to the steep bank and enter a descending spiral. Pilots entering with too much backpressure for a steep bank find themselves in a tight, climbing turn. Proper aileron and rudder input and the correct amount of backpressure help assure a good entry into a steep turn.

To keep a steep turn coordinated, a slight amount of opposite aileron must be used to overcome the overbanking tendency as the bank reaches the 50- to 60-degree range. Do not use top rudder to correct the overbanking tendency.

Recovery from a steep turn is as with all turn recoveries, only more control pressure is required to make it come out in a coordinated fashion. Since you turn at a high rate, it takes more rudder application than normal to overcome the turn and also the adverse yaw effect of the ailerons as they are applied to reverse the bank. Also, as you proceed through the rollout, relax the backpressure to prevent a climb after the turn is completed. In other words, as you roll back to level flight, if you don't release the backpressure you held through the maneuver, you will increase your altitude at the end and ruin a maneuver that might have been otherwise alright.

In steep turns you will find that it requires more effort to maintain altitude. Because so much of the lift is vectored to turn the aircraft, the increase in the angle of attack must be greater as compared to shallower banks. You will find in some planes that not only do the elevator inputs become fairly heavy, but also the plane may not be able to maintain altitude unless additional power is added during entry to the steep banked turn. With some planes, no matter how much power you add you will find they have a tendency to lose altitude once the bank exceeds a certain angle. You do not normally trim the airplane for turns, since they are normally done over a short time frame and elevator pressures are held for only a short period of time.

A common error during steep turns is to become uncoordinated as the bank steepens. When in a steep turn, pilots will hold rudder opposite the direction of the turn, which tends to yaw the nose of the plane up slightly. For example, in a 60-degree left bank turn, some pilots hold right rudder to help hold the nose of the airplane above the horizon. This is often a response to the heavy elevator pressure they are holding and is often an unconscious act on their part, but this places them in an uncoordinated turn.

Level flight in a steep turn induces a load factor greater than one. The wings are producing lift that exceeds that required for straight-and-level flight. The wings are flying at a higher angle of attack and behave as though the airplane is heavier. Translated, this means the airplane will stall at a higher indicated airspeed during a turn; the steeper the turn, the faster the stall will be. If the pilot has the plane in an uncoordinated steep turn when it stalls, it is quite possible the plane will enter a spin. Some pilots enjoy spins and practice them on a regular basis. But an unplanned spin from a steep turn can make for an exciting day for you and your passengers. If you are uncomfortable with the elevator pressures you must hold to maintain altitude during a steep banked turn, reduce the angle of bank to make things more controllable.

As you roll out of a steep turn you will need to remember to reduce the engine's power setting and to reduce the backpressure on the control yoke. There is a common tendency to erroneously gain a great deal of altitude while exiting a steep turn as a result of the additional power and increased elevator backpressure.

Slow flight

Slow flight may be defined as maneuvering the airplane at airspeeds close to the stall speed of the plane. For the private pilot flight test, the examiner will be looking for speeds 5 knots above stall speed during several maneuvers centered on slow flight.

During flight at cruise airspeeds, the controls of most general-aviation aircraft have a solid, authoritative feel to them. When you roll in aileron, or use the elevator, the plane reacts in a crisp, responsive manner. As you slow a plane down, though, the airflow over the controls is reduced. As a result, the controls become less effective and acquire a "mushy" feeling. You will find that they not only feel softer due to the lower airspeed, but also the plane is much less responsive to control inputs and it takes a greater amount of control input to get the plane to execute the maneuver you are attempting to fly. When you are in slow flight mode, make the control inputs relatively gently, keeping in mind how the feedback on the controls is and how the plane reacts to the inputs. Depending on your familiarity with the aircraft, you may need to build a "feel" for control inputs at different airspeeds, from cruise down to just above stall speed. This will allow you to be able to anticipate how the airplane is going to react when you are flying at different speeds. Go out and practice flying at slow airspeeds until you can make smooth, coordinated inputs on the controls. As you become more practiced at slow flight you will find you can hold the airplane at just the desired attitude with smooth, coordinated control inputs.

Slow flight entry

Any time you are practicing slow flight, be sure you have sufficient altitude to recover from any unplanned stalls or spins.

Essentially, if you start in cruise configuration, begin slowing the plane by reducing power slowly. In airplanes like a Cessna 152 or a Piper Cherokee, start by reducing the engine's power to about 1700 RPM, then adjust as necessary to maintain altitude when the desired airspeed has been achieved. As the plane begins to slow after the initial power reduction, case back on the control yoke to slow the plane and establish a pitch altitude appropriate for the new speed. How much and how quickly you will need to input backpressure on the control yoke will depend on the plane and how much power has been reduced. As the plane slows and the pitch of the nose is raised, you will need to input right rudder to keep the ball centered as the left-turning tendencies begin to come into play. Many students forget to make the rudder correction, and the nose of the plane begins to drift slowly to the left. It is also common for students to lose altitude while the plane slows. Be sure to increase the pitch of the plane enough to help slow it down and to maintain altitude as it slows.

Work pitch attitude and power setting simultaneously to maintain the selected slow flight parameters. With a little practice you will be able to anticipate the correct power and pitch inputs and slow the plane to the correct airspeed while maintaining altitude very consistently. If you want to use flaps as part of the slow flight mode, add them once the plane has slowed to the white arc of the airspeed, then apply flaps in increments. You can use any flap settings, from no flaps to full flaps during slow fight practice.

When you are adding flaps, be prepared for the pitch change that will accompany their deployment. Each airplane will react differently, so learn the plane you are flying and become familiar with its flight characteristics.

If the plane is descending and you have the proper airspeed dialed in, you will need to increase your power setting slowly. Just as with the power reduction, you will also need to make a corresponding pitch change to maintain a constant airspeed. With an increase in power setting, you will need to increase the plane's pitch to maintain the desired airspeed. By making power changes slowly, you will avoid overcompensating with too much or too little power.

The one thing you should notice immediately is the softness of the controls. Ailerons are mushy, with relatively large inputs needed to hold the wings straight and level as compared to cruise airspeed. The elevators are also less authoritative, and you will find that you will need a fairly large amount of right rudder to overcome the left-turning tendencies we discussed in the last section. The noise of the plane has also changed. The engine is producing less power, and the sound of airflow over the surface of the plane is also quieter at slower airspeeds. Get the feel for all of these sensations because you can detect changes in airspeed from these sensory inputs without ever looking at the airspeed indicator.

Figure 3-16 depicts the load factor placed on the plane in relation to the angle of bank of the plane. As you can see, as the bank of the plane increases, the load factor also increases. This increase in load as a result of a bank results in an increase in the stall speed of the airplane. If we are flying just five knots above stall speed, and we attempt to make a turn at a 45-degree bank, we are quite likely to end up in a stall due to the increase in stall speed as a result of the bank. The slower we fly, the more likely we are to stall during a steep turn as a result of this relationship between load factor increase and a corresponding increase in stall speed.

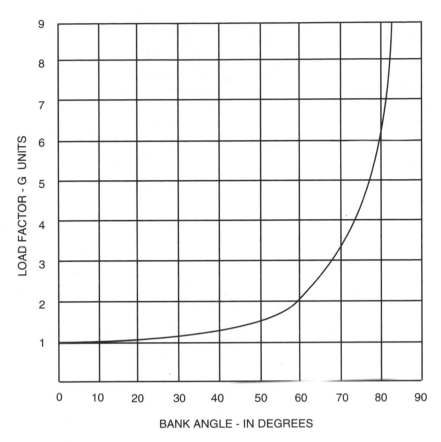

Fig. 3-16 *Load factor chart.*

Turns are flown just as we would when making a turn in cruise flight, but we must realize the controls will be less effective, and you will need to keep your banks shallow to avoid a stall. You may need to keep right rudder inputs applied, even in turns to the left, to keep the ball centered. In slow flight, you can feel a certain amount of control buffet. The buffet may increase during the turn as a result of airflow disruptions and the increased load as a result of the turn. Monitor your airspeed and altitude during turns. The plane has a tendency to lose altitude and airspeed during a turn, and it may be necessary to add power and adjust the pitch to keep your airspeed and altitude locked in. The amount of the power and pitch change will depend on the plane and how steep the bank is, but you should practice at different bank angles to become familiar with these changing settings. As you roll out of the turn you will need to reset power and pitch to avoid gaining altitude or airspeed as the load on the wings is reduced. A final note about turns in slow flight. In turns to the right you will find that a large amount of right rudder is necessary to keep the ball centered. Not only are you trying to overcome the left-turning tendencies of the plane, but you are also overcoming adverse yaw with a rudder greatly diminished in control effectiveness.

Once you become comfortable with turns in slow flight, you may practice banking the airplane to the point of causing it to stall, to help you learn how the plane feels and reacts at this limit. As with any maneuver, you should have a qualified flight instructor along until you become familiar with the maneuvers. The instructor can help you learn and critique your performance and help prevent you from getting into trouble.

Slow flight recovery

Recovery is transition back to cruise power settings without gaining or losing altitude and maintaining heading. First, add engine power slowly, back to cruise power settings. As you add power you must reduce backpressure on the control yoke to avoid gaining altitude. If flaps were extended during entry into slow flight, remove them in increments. The reason for the incremental approach is to avoid a sudden loss of lift, with the associated risk of stall or aggressive pitch movement from the plane. After each notch of flaps is retracted, give the plane a few seconds to gain additional airspeed before retracting to the next setting. If you have extended landing gear, you will also want to retract them during this time. Do not exceed either the maximum flap extension speed, V_f, or the maximum speed the landing gear can be down or operated. If these airspeeds are exceeded while the flaps or landing gear are extended, you may cause damage to the airplane or the components themselves. As configuration changes, the plane accelerates to cruise airspeeds fairly quickly and you must adjust pitch attitude accordingly, back to cruise flight. With increasing speeds, you will also need to retrim the airplane.

Stalls

Simply stated, an aircraft cannot stall unless it exceeds its critical angle of attack. Remember that angle of attack is the angle between the chord line of the wing and the relative wind. Most airfoils become critical and stall somewhere between 15 and 20 degrees of pitch.

Don't exceed your critical angle of attack and you won't find your-self in a stall. It's that simple, and yet people continue to stall, spin, and crash.

A plane can stall at any airspeed. A normal stall is usually entered by slowing the airplane to the point that it stalls at a slow airspeed. But if you have flown on a hot summer day, when the turbulence is banging against the plane, and heard the stall warning horn go off as a result of turbulence, you have encountered a situation that could result in a stall. If you have put the airplane into a turn and pulled back hard on the yoke and felt the plane shudder, this is also an indicator of a high-speed, or accelerated, stall. Figure 3-17 illus-trates a plane flying along at cruise speeds in level flight, the hori-zontal arrow indicating the direction of flight. The second arrow represents a gust of wind from turbulence. This wind gust can result in an angle of attack that briefly exceeds the critical angle of attack, causing part or all of the wing to stall.

A normal stall is one that does not require abrupt control inputs to initiate. For example, a normal stall would be entered in the manner discussed for slow flight entry, but without subsequent addition of power to stabilize the speed. The airplane continues to decelerate in level flight until the increasing angle of attack causes turbulent air-flow—a stall.

In contrast, an accelerated stall normally takes place at a higher air-speed, with rapid or excessive control inputs initiating the stall. The airplane is normally under higher g-loads and stalls more abruptly.

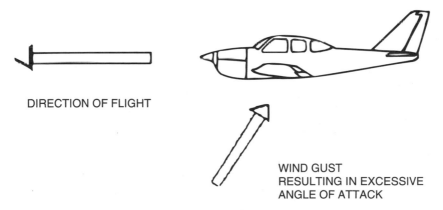

DIRECTION OF FLIGHT

WIND GUST
RESULTING IN EXCESSIVE
ANGLE OF ATTACK

Fig. 3-17 *Gusts may stall the wing in addition to causing a turbulent ride.*

Accelerated stalls are usually the result of a need to maneuver quickly. As we discuss normal and accelerated stalls, keep this difference in mind.

Normal, power-off stall

Before you practice any flight maneuver, be sure to execute clearing turns to ensure that the airspace you are flying in is clear. You will not only want to check for traffic at your altitude, but also above and below your altitude. You may have a tendency to gain or lose altitude as you practice, and it is important to verify that no other airplanes are in your vicinity. After reaching a safe altitude and clearing the area, slowly ease the throttle to idle. The reason we do this slowly is to prevent rapid cooling of the engine. After a long climb to altitude, the engine has heated up as a result of generating more power and the reduced airflow around it due to the increased angle of climb. If the engine is cooled too rapidly, this can result in excessive engine wear or damage to engine components.

You will start the maneuver at a specific altitude. Maintain that altitude as the plane slows by easing back on the control yoke, just as you did during slow flight practice. But rather than maintain a given airspeed, you will want to bleed off speed until the airplane stalls. When the stall takes place, you should have the control yoke at or near its aft stop. As the stall approaches, you should notice the same sensations we discussed in the slow flight section: soft control feel, reduced engine noise, and the sound of airflow around the plane will become quieter.

Most planes experience a certain amount of buffet as the plane nears stall speed, but the amount will vary for each plane. As the plane stalls it will normally experience a nose-down movement known as the "break." The break signals that the stall has occurred and is a pitch-down movement. The amount of break will also be different from plane to plane. In some cases the break can be sharp and very noticeable, while in others the plane may wallow along without a significant pitch down.

While the plane is approaching the stall, or as it stalls, it may have a tendency to drop off on a wing. This can be the result of how the airplane is rigged, or minor differences in the plane's wings' angles of attack. For many pilots the first response is to use the ailerons to

Fig. 3-18 *Departure stall, forward view.*

maintain wings level during the stall. This should be avoided, instead using the rudder to keep the wings level. Ailerons generate their roll ability by changing the shape of the airfoil. When a plane stalls, this can result in a deeper stall of a wing and a more aggravated stall situation. In some cases this can cause the plane to enter a spin. The use of rudders to hold the wings level as the plane stalls helps prevent this from taking place. If the left wing of the plane drops, you will want to use right rudder to help lift it back up. If the right wing drops, use left rudder to keep it level. Make sure you neutralize the rudder once the wings are level again, and don't overcontrol the amount of rudder you input. Practice with a plane will help you get used to how much rudder should be used.

As the plane stalls, you will initiate stall recovery procedures. For normal stalls this will include relaxing backpressure on the control yoke to reduce the angle of attack and adding full power. Reducing elevator backpressure will result in a nose-down attitude, which puts the angle of attack less than the critical angle of attack. In most cases you only need to drop the nose of the plane slightly below the horizon. Figures 3-18 and 3-19 illustrate the position of the nose just prior to stall entry and as the plane recovers from the stall. In this case the nose of the plane has been lowered only a small distance below the horizon.

Fig. 3-19 *Departure stall recovery, forward and side views.*

The use of engine power will help the plane recover more quickly from the stall and reduce the amount of altitude the plane loses. Make sure you use full power when you apply the throttle. Some pilots have a habit of easing in only a small amount of power, which results in greater altitude loss. As the plane achieves flying speed, ease the elevator back to produce a positive rate of climb.

Avoid pulling the nose up too quickly; this could result in a secondary stall and the need to recover from the stall again with an associated altitude loss.

One of the most common errors is not causing the plane to stall completely. This is normally the result of a timid use of the elevator and not getting it back quickly enough, or all the way to the back stop. The results will vary, but the plane may wallow, not getting a clean break. The use of ailerons to hold the wings level during the stall is also common as pilots begin to learn this maneuver. They have been trained to use the ailerons to control the roll of the plane, and now they must learn to use the rudder in place of the ailerons.

As the plane stalls, some pilots will shove forward on the control yoke too aggressively, resulting in a significant nose-down attitude. Not only does this pin you against the seat belt and cause all the dirt to float up from the floor, but it also results in excessive altitude loss. In most cases gently releasing backpressure and lowering the nose to just below the horizon is sufficient to break the stall and let the wings start to fly again. Regarding the use of the engine during power-off stall recovery, pilots may forget to add engine power and only use partial power as they recover from the stall. Always add full power in a smooth, constant application. Finally, some pilots will then use too much elevator or apply it too rapidly as they attempt to establish a positive rate of climb. As we discussed, this can cause the airplane to stall again and result in even more altitude loss. Smoothly ease the nose of the plane up as flying speed is increased until the descent is stopped.

Normal, power-on stall

Power-on stalls are very similar to the power-off stalls just covered. Flight control use is the same to enter and recover from both stalls, the major difference being that full power is being produced by the engine during entry into a power-on stall.

There are a number of different ways to enter a power-on stall, but this discussion concerns entry from cruise power settings. First ease the nose of the plane up, letting airspeed bleed off to the best angle of climb speed. At that point advance the throttle smoothly to full power. Keep pulling back on the control yoke in a firm, constant motion to keep the nose of the plane coming up and bleeding off airspeed. As with the power-off stall, you will notice that the flight controls become less effective. The sound of the engine will seem to change as the plane slows.

The plane will begin to buffet, as with the power-off stall. The additional airflow generated by the propeller tends to increase the amount of air flowing around the wings and fuselage, and this causes a stronger buffet. The strength of the buffet will be different in each plane and may be more pronounced in some models than in others.

As the plane stalls it will once again have a tendency to pitch nose down. As the plane stalls, release backpressure on the elevators and allow the nose to drop just below the horizon to reduce the angle of attack. The amount of nose-down attitude you will need to achieve to break the stall will differ for each plane, and some planes may require a greater pitch-down attitude than others.

Since power should already be set at full power, you will not need to advance the throttle during the stall recovery. It is a good idea to keep your free hand on the throttle to make sure it doesn't creep back during the stall. Once your airspeed has increased sufficiently, ease the nose up and establish a positive rate of climb.

Pilots seem to feel less comfortable with the pitch angles the plane achieves during power-on stalls and how much elevator it takes to get many airplanes to stall when the throttle is at full power. Figure 3-20 shows the pitch angles just before a power-on stall, then after the plane stalls. Compare these to the pitch angles in the power-off stall, and you can see how much steeper the angles are. Because the engine is producing full power, it also takes a considerable amount of right rudder to keep the ball centered as the plane's airspeed drops. Because the nose rises so much, it is often difficult to see the horizon over the nose of the plane, adding to the disorientation that is common for pilots learning the maneuver.

Most of the errors discussed in power-off stalls apply to power-on stalls as well. Remember to use the rudder to hold the wings level during a power-on stall. If a plane has a tendency to drop a wing during power-off stalls, you may find that it is more exaggerated in the power-on stall configuration.

The nose of the plane can drop more rapidly due to the sharper break during the stall and may have a tendency to sink lower below the horizon. Try not to aggravate the situation by adding too much forward elevator input, which can drop the nose even further below

Fig. 3-20 *Departure stall, left view.*

the horizon. Every plane will react differently during a stall, but stall recovery is a task that requires that you are light and smooth with flight control inputs. You will also want to make sure you do not set yourself up for a secondary stall by raising the nose too abruptly after your initial stall recovery. The airspeed will have a tendency to build more quickly since the power was on during the entire stall and there can be greater altitude loss as a result of this.

The accelerated stall

The accelerated stall occurs when the critical angle of attack is exceeded by abrupt control movement, causing the aircraft to stall at a higher-than-normal indicated airspeed. Just because your aircraft has a V_{so} of 40 doesn't mean that there is no way it can be stalled at a higher speed. You don't necessarily have to be nose high to stall it, either. If you exceed the critical angle of attack, you can stall your aircraft going straight up, straight down, or straight and level. Remember, in an accelerated stall, the stall occurs at a higher-than-normal airspeed and usually requires some very rapid control movement to induce the stall. But it will stall.

Everything has its limits and so does your aircraft. What is the highest airspeed at which you can stall your aircraft safely? The answer is the maneuvering speed (V_a)—the speed at which you can apply full, abrupt control travel without causing structural damage to your aircraft. Anytime you are at or below maneuvering speed, you can stall your aircraft without worrying about whether you or your wings are going to get to the ground first. It is a built-in safety factor tested by the aircraft manufacturer and certified in your aircraft's operations limitations.

To practice accelerated stalls, first climb to a minimum stall recovery altitude of 1500 feet above the ground, and then climb another couple of thousand feet for mistakes. Clear the area by doing a 90-degree turn in each direction (or a 180, whichever makes you happy). Look for other aircraft while doing the turns. Too many people go through the motions of clearing turns and use the time to unwrap a piece of gum or adjust the ventilation in the cabin.

The entry speed for the accelerated stall should be no more than 1.25 times the unaccelerated stall speed in a clean configuration: V_{sl} times 1.25. If your aircraft has a V_{sl} of 60 knots, this would work out to an entry speed of 75 knots ($60 \times 1.25 = 75$). This rather low speed provides a margin for safety because the load factor at the time of the stall will be lower if it begins at 1.25 V_{sl} rather than up near maneuvering speed. The stall is less violent and causes less stress on you and your aircraft.

After you arrive at altitude and clear the area, slow your aircraft down to entry speed, and while maintaining constant altitude and a low power setting, roll into a coordinated 45-degree bank. Rapidly increase the pitch as you hold your altitude and the aircraft should stall at a higher-than-normal airspeed. If properly executed, it will stall quickly and break rather sharply. If you turn more than 90 degrees during the execution of the accelerated stall, you aren't pulling back fast enough on the yoke and probably aren't getting a true accelerated stall. You don't have time to eat a hamburger while doing an accelerated stall properly. It happens in a hurry.

Recovery from the accelerated stall is as with all stalls:

- Reduce the pitch to break the stall.
- Add power, if available.
- Level the wings and return to normal flight attitude.

Complete these procedures in that order and quickly. The stall recovery can be initiated at any of the normal recognition cues peculiar to stalls (or impending stalls). These cues are common to all and include the warning horn and/or light, decreasing control effectiveness, buffeting, and finally, the break. The most sinister thing about the accelerated stall is that these cues happen much more rapidly than is the case with most stalls. But if you practice recovery at all the different cues, you will be ready should you ever get into a situation that calls for prompt action.

As far as leveling the wings is concerned, the ailerons will be nearly useless during stalled flight, because smooth airflow over the wings is disrupted. Since the airplane may roll off sharply during the stall break, pilots should counter this rolling tendency, at least initially, with judicial amounts of rudder. The ailerons will become effective again during recovery.

Takeoff and departure stall

In order to practice the recognition of and recovery from this stall, take your aircraft to at least 1500 feet above ground level. Clear the area, as always, before practicing stalls. This particular stall is initiated at liftoff speed, so retard the throttle well below cruise RPM and maintain a constant altitude as the aircraft slows to liftoff speed. As you reach liftoff speed, advance the throttle to full takeoff or climb power and increase the pitch simultaneously, not allowing any acceleration of the aircraft. After the desired pitch is established, all controls are returned to neutral except for the rudder, which must be used to overcome torque and P-factor. Recovery can be initiated at the first indication of the impending stall or after the full break.

The takeoff and departure stall should be practiced from a straight climb as well as climbing turns in both directions. When you practice it while turning, use a moderate bank of 15 to 20 degrees. Notice how the cues in this stall seem to come one at a time. You should be able to recover at any one of the cues. In fact, to the competent pilot, these cues almost scream for action on the part of the pilot. Recovery will be normal except for the power. Because you already have a high-power setting, reduce the pitch to break the stall and level the wings using coordinated control forces. Then fly the aircraft out of the stall with as little altitude loss as possible, and set up a normal climb.

Approach-to-landing stall

The approach-to-landing stall often occurs on the turn from base leg to final approach, but it has been known to happen when a pilot attempts to stretch the glide on final without sufficient power to maintain altitude. The turn from base leg to final approach is the prime area for this stall because pilots often start their turn to final too late and are tempted to add a little inside rudder and some backpressure to tighten the turn. Seeing that this helps a little but that they are still going to overshoot the runway centerline, they sneak in a little more rudder and backpressure to tighten the turn even more. Often they use opposite aileron to try to keep the turn from overbanking due to the effect of the rudder and backpressure. They are so intent on making the runway on the first try, rather than going around and flying a better pattern, that they fail to notice the cues the aircraft is giving them. It is the perfect setup for an approach-to-landing stall.

Approach-to-landing stalls are executed from normal approach speed and should be practiced from straight glides as well as gliding turns (the latter simulates the turn from base to final). Clean and full-flap configurations should be practiced because you might be called on to make various approaches in various configurations during the conduct of actual flight. Also, these stalls should be practiced at an altitude higher than the minimum recovery altitude of 1500 feet above the ground because you will be in a descent during the maneuver.

After carefully clearing the area, smoothly begin to retard the throttle to the approach power setting. Maintain your altitude as you slow to the approach speed. During this time, the flaps should be lowered to landing position (assuming the aircraft is so equipped). When normal approach speed is reached, initiate a descent. After descending about 200 feet, begin to smoothly increase the pitch as the power is reduced to near idle. Continue to increase the angle of attack until the stall occurs. The stall usually occurs with very little nose-high attitude. Pulling the nose up too sharply ruins the intent of the maneuver. Because this stall almost always happens slowly and gradually, you should be able to recognize and feel the cues as they come one at a time.

Recovery from an approach-to-landing stall can be initiated at any point, up to and including a full break. Recovery is much like a full-flap go-around from a landing approach. Relax backpressure and smoothly add power as you level the wings and transition to best-angle-of-climb or best-rate-of-climb airspeed. While you are doing this, the flaps should be brought up to the manufacturer's recommended setting for a go-around. In most aircraft, the flaps are not brought up all at once because that can cause a momentary sink, which could prove problematic at a very low altitude.

Stall avoidance

To learn to avoid stalls and be able to recognize them instinctively, you must first be very familiar with stalls. That is precisely the reason we have spent so much time discussing the various types of stalls you might encounter: normal, accelerated, approach, and departure are all stalls that you should be very comfortable with. Once you can enter and recover from these stalls, and as you practice each of these maneuvers, you gain a feel for them. This feel, the ability to know what the stalls make the plane's control surfaces feel like, the attitudes and sounds of the engine, and slipstream are the core of being able to recognize the onset of a potential stall. Through practice you can not only get better at entry and recovery from stalls, but you also gain the insight to know when the airplane is about to stall, without looking at the instruments. This knowledge is crucial to avoiding stalls when situations become critical. So go out and practice stalls as often as you can; it will help you become more proficient at stall avoidance as well.

As you are practicing, what should you be paying attention to that can help in learning stall avoidance? There are a number of warning signs, many of which we have already covered. The flight instruments can provide a great deal of information via the airspeed indicator and the turn and bank indicator. If you notice the airspeed indicator dropping toward the bottom of the green arc, you know the airplane is becoming a little too slow. If the artificial horizon shows an abnormally large pitch, it might be a good idea to check the airspeed and verify that you are not bleeding off too much airspeed.

Most stalls are not caused when the pilot is closely monitoring the instruments, though. It is not possible to document each potential accidental stall scenario, but some common factors should be considered important. Often the pilot is distracted in one way or another and does not pay attention to the plane. This distraction can come in the form of trying to force the plane into a steep bank as the pilot attempts to avoid overshooting the turn from base to final. The pilot can become so focused on the turn that he or she loses focus on how steep the bank is in relation to the speed of the plane. An accelerated stall can have very rapid onset, with little time between prestall buffet and the actual stall, making it more difficult to recover once the stalls become imminent. For these reasons it is very important to avoid accelerated stall situations, especially when flying in the pattern. Do not focus so heavily on one aspect of the flight that you ignore other pieces. Keep a broad view of what is going on; get into the habit of being "situationally aware" as you fly.

Avoid distractions while maneuvering at low speed. Some of these may include radio chatter or Air Traffic Control, looking at an approach chart or sectional, a passenger asking you questions as you get ready to land, or any number of other things that can pop up during a flight. Emergencies are a great way to become distracted. An engine that suddenly begins to run rough and loses power can cause you to focus almost exclusively on the oil and fuel pressure gauges if you don't make yourself stay aware of the entire situation. Gear lights not showing green as you get ready to land can also be a situation that might have you looking at the lights, head buried in the cockpit. In a case like this you are already slow, head looking down at the gear indicator lights, and if you lean forward to tap on them, you may unknowingly pull back on the control yoke as a brace.

Pay attention to the feel of the controls. If you have your head in the cockpit, you should still be able to notice when the controls become softer due to dropping airspeeds. This is a definite indicator that you are getting slower than you want to, especially if you are in the pattern or at low altitudes.

Sounds are another major sensory input that can help you maintain situational awareness. As the plane slows, the sounds generated by the slipstream will become quieter. Like the feel of the controls, you do not need to be looking outside the plane to notice this change in sound levels. Conversely, if the plane is picking up speed, the noise of the slipstream will increase. Engine noises will also change with changes in airspeed, and any difference in sounds from the engine can also be an indicator that your airspeed is changing either up or down. With practice you can become fairly accurate at "guesstimating" the airspeed based on the sounds of the plane.

On takeoff you should be familiar with the normal pitch attitude used to achieve airspeeds such as V_x and V_y. If you notice that the nose is higher than normal during takeoff, you should verify your airspeed. In some cases this pitch angle/airspeed difference can be caused by improper use of flaps, the gear being down/up when it should be in the opposite position, or improper propeller or engine power settings. Obviously you will need to be looking outside the plane to notice excessive or shallow pitch attitude, but this can be a clue that other factors are affecting the performance of the plane. High angles of attack can lead to low-altitude stalls, so be very careful.

Spins

Before you begin spin training, talk with the flight instructor and get a feel for his or her spin training experience. Then check out the airplane the instructor wants you to use. Many aircraft are not certified for spins; the certification is based on passing specific tests laid down by the FAA. If you don't feel comfortable with what you see and hear, find another instructor that will meet your needs.

Spin definition

I have heard spins described in various ways, often with expletives scattered amongst the description. There is no mystery; for a plane to spin, first both wings must have stalled. Second, you must have a situation where one wing has a greater angle of attack than the other. The spin usually swings around the wing with the highest angle of attack. Just a simple yaw motion at the time of stall is enough to cause an apparent difference in angle of attack between the two wings. Since the outside wing is moving at a higher speed due to its rotation, the angle of attack at the tip will be much less than the tip of the inside wing. Poor rigging, propeller torque, and even sloppy footwork from the pilot may be sufficient to induce the initial spin entry.

Whatever the reason, the stalled plane then begins to yaw in the direction of the wing that is more stalled and the plane begins to slip in the direction it is beginning to yaw. As the plane yaws, the situation becomes worse as the side of the fuselage, the vertical stabilizer, and other vertical surfaces weathervane into the wind. The asymmetrical lift also results in a rolling moment, causing the plane to roll in the direction of the stalled wing. Due to the yawing, rolling, drag, and centrifugal forces at work, the plane continues to pitch nose down, rolling and yawing, until the spin is fully developed. Each plane reacts differently during the course of a spin; the same plane can even exhibit different characteristics, depending on how it is loaded and other factors. The pitch angles and rate of rotation during a spin will also vary as the spin progresses.

Once a plane enters a spin, it will usually remain there until recovery is initiated. If you are flying a plane certified for spins, multiple-turn spins are safe as long as you remain within the limits specified by the manufacturer. A note of warning, though—never purposely spin a plane that is not certified by the manufacturer for spins. For those aircraft certified in the normal category, the manufacturer is only required to demonstrate that the plane is able to recover from a one-turn spin, or three second spin, whichever takes longer. If you spin a plane in this category, once the plane passes one turn, you have become a test pilot, because the flight characteristics are unknown. In some cases the plane may enter a spin mode that you may not be able to recover from. Practice spins only in spin-certified aircraft.

Stall/spin relationship

To spin the airplane you must first stall it. Stalls can take place at any airspeed, so it is conceivably possible to spin an airplane at any airspeed. In fact, an aerobatic maneuver known as the snap roll is a high-speed stall/spin, resulting in the plane spinning in the direction the plane is moving. This can be going horizontally, vertically, or straight down and is a very rapid, quick maneuver.

You may reason that if a plane is difficult to get into a spin, it may also be difficult to get out of the spin. While this is not always the case, be certain you know the correct spin entry and recovery procedures for your airplane. The following discussion will be generally applicable, but the plane you are flying may have a unique set of techniques that are documented in the aircraft operations manual.

Spin versus spiral

From the cockpit these two maneuvers can appear very similar; the plane is in a nose-down attitude rotating to the left or right. There is, however, a very large difference between spins and spirals. During a spin the plane is stalled. A spiral is not.

A spiral can be entered from a botched attempt to spin the plane, but a spiral is significantly different in what is aerodynamically taking place. Unlike the spin, the wings are not stalled in a spiral; they are both producing lift. The rolling motion found in a spiral is normally due to aileron and/or rudder control inputs. Another major difference is that the airspeed will be rapidly increasing in a spiral. This can become a dangerous situation. If the plane accelerates to speeds beyond V_{ne}, structural damage could occur.

Recovery from spins and spirals also differ. For a spin you must reduce power on the engine to idle, input full rudder in the direction opposite to the spin, and ease forward on the elevator control to break the stall. To recover from a spiral, you also must reduce engine power to prevent excessive speed buildup, but you will not need full rudder opposite the direction of the spiral's rotation and you will not want to push the nose of the plane down further through forward elevator control inputs. This would actually result in the plane entering a greater nose-down attitude and gaining even more airspeed. The use of full rudder opposite the direction of the spiral could actually cause excessive yawing and additional controllability problems. Rather, reduce engine power to keep the plane from gaining speed too quickly, level the wings, and ease the nose back to a level flight attitude. Avoid pulling back too sharply on the elevator control; too much elevator could inflict excessive g-forces and result in damage to the aircraft.

Spiral entry will not happen the same way for every airplane. You may accidentally roll the plane into a spiral or find some other unique manner for putting the plane into a spiral. It may be the result of attempting to spin a plane that is reluctant to do so, or simple pilot disorientation. Accordingly, experience is the best teacher for getting a feel for what a spin feels like, as compared to a spiral.

Spin phases

A spin is not a static maneuver; instead the plane moves through several distinct phases with the characteristics of the spin changing in each phase. The four phases are spin entry, incipient spin, developed spin, and spin recovery. Other books may use other names, but the basic concepts documented during each phase of the spin are still the same.

Spin entry phase

The spin begins as the plane stalls and the rolling/yawing motions begin. The nose of the plane normally drops well below the horizon as the spin begins to take effect. To enter a spin, first slow the airplane, just as with a stall. As the plane stalls, make sure the elevator control is fully aft, assuring a complete stall, then push full rudder in the direction you want to spin. For example, if you want to spin to the right, push full right rudder and hold it there. This will cause the plane to enter the second phase of the spin, the incipient phase.

Normal spin entry is made from a power-off, flaps-retracted stall. Before beginning the maneuver, fly clearing turns to the left and right, paying particular attention to the airspace below. You could lose several thousand feet of altitude, so it is important to make sure there are no aircraft below that could pose a safety concern. A photographic sequence of events follows:

Figure 3-21 shows the view looking out the left window from the pilot's seat. The angle of attack is very apparent in this view. As the plane stalls, the ailerons should be in the neutral position and maintained in that position throughout the course of the spin. If the ailerons are turned into the spin, or away from the spin, they could cause a significant change in its characteristics. The rudder should be briskly applied, and the elevator should be moved to the full aft position if it is not already there.

Figure 3-22 depicts the forward and left views of a spin as it progresses. As you can see, the change in pitch as the plane enters the spin is dramatic. The plane used in this photo sequence has a ten-

Fig. 3-21 *Left view, just prior to stall. The airplane is moving parallel with the horizon—note the angle of attack.*

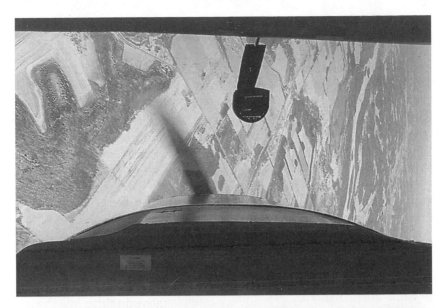

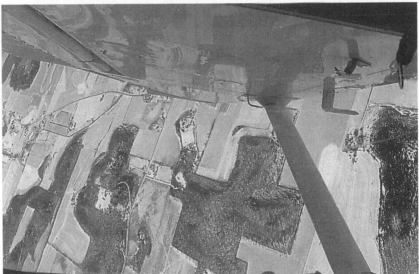

Fig. 3-22 *Forward and left views of a spin entry—just after the stall.*

dency to roll slightly inverted as it first enters the spin, then it settles down into a steep nose-down attitude as the spin develops. The steep nose-down angles and rapid rate of rotation can be disorienting to many pilots new to spins. Once the spin begins and the nose is pointed at the ground, the section lines, trees, or roads are the best ground reference that the pilot has to maintain orientation. Keep the ailerons in a neutral position, the stick or control yoke fully back, and full rudder toward the direction of spin rotation in. Changing these control inputs will also change the spin.

Insufficient backpressure on the elevator control can prevent the plane from stalling cleanly and getting a good break as the stall takes place. This can make it more difficult to get the plane to enter a spin.

Using ailerons relates to the insufficient elevator backpressure situation. As students attempt to force the plane into the spin while it mushes along, they frequently roll aileron in, thinking their use will help things along. This can cause the plane to spiral, which looks similar to the spin but is a completely different maneuver.

Use of less than full rudder, briskly applied as the plane stalls, may prevent the plane from cleanly entering the spin. A common reaction from the plane is that it yaws in the direction of the rudder input but never really breaks cleanly and enters the spin. Do not be overly forceful in the rudder input, but be authoritative in its application and you will get a nice, clean entry into the spin.

If power is carried into a spin, it can cause the nose of the plane to rise, putting the plane into a flat spin mode—this can be very dangerous. Always be certain that power is at idle when you stall the airplane and plan to enter a spin. If you should happen to inadvertently carry power as you enter the spin, immediately reduce it to idle.

Like unplanned stalls, the purpose of learning to do spins is to be able to recognize the potential for a spin and how to avoid it. Then, if that should fail, how to recover from it as rapidly as possible. Almost without exception, pilots who are exposed to spins for the first time have a strong tendency to pull back on the control yoke, trying to raise the nose—exactly opposite of what is required. Although most pilots could read this and assume they are well informed for spin recovery, there is no substitute for experience and practice.

If you should happen to accidentally enter a spin, determine which direction the plane is spinning before you make rudder input. Directly over the nose of the plane is the best location to look to figure out whether the plane is spinning to the left or right. If you are so disoriented that you can't tell which way the plane is spinning, step on the rudder pedal that has more resistance. The difference in airflow over the rudder can make the rudder pedal that would be used for spin recovery more resistant to movement. And if using that rudder pedal fails, try the other one.

Incipient spin phase

The incipient phase of the spin takes place as the plane transitions from forward flight to the nose-down, rolling, yawing descent present in a spin. Inertia, lift, drag, and yaw all affect the plane during the incipient phase. In fact, NASA has expended a great deal of time and resources studying spins and what factors affect them.

It generally takes approximately two turns for the plane to make the transition from horizontal to vertical flight path. Due to the tendency of a plane to want to continue in its original motion, in many cases it is easier to recover from a spin while the plane is in the incipient phase. In fact, while not an approved method of recovery for most aircraft, letting go of the controls while the plane is in the incipient phase can often result in the plane recovering from the spin on its own. DO NOT assume this is the case for the plane you are flying and go off and practice spins without competent instruction. Self-taught spins are not recommended for anyone. At the end of the incipient phase, the plane's momentum has transitioned into a fully developed spin, and it is much less influenced by the forces originally acting on it during spin entry.

Developed spin phase

The developed phase of a spin is not an unchanging situation. The plane actually goes through cyclical periods as part of the developed phase. The changes here include differing pitch attitudes and rates of spin rotation—oscillations in nearly every axis. For example, the Pitts Special S-2B has a tendency to be nose low at the 180-degree point of the spin, and the nose is higher at the 360-degree point.

Every plane is different in the way it reacts during the developed spin phase. Some airplanes will have a more nose-down tendency during the spin, while others may have faster rotation rates. Depending on how you enter the spin and apply controls, the same plane may behave differently from spin to spin. The center of gravity can also affect the spin characteristics of a plane. Generally, the more aft the center of gravity is, the flatter the spin will be (Fig. 3-23).

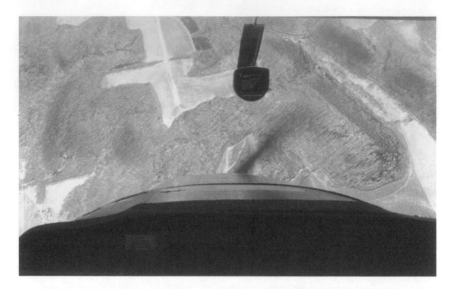

Fig. 3-23 *Forward and left side views of a developed spin.*

Spin recovery phase

The recovery phase of the spin begins when you apply correct flight control inputs that stop spin. It may take several turns from the point the spin recovery begins before the spin is finally over. It is very important that you know whether your plane is reacting correctly.

According to the FAA, the proper spin recovery technique is as follows:

1. Retard power.
2. Apply opposite rudder to slow rotation.
3. Apply positive forward elevator movement to break stall.
4. Neutralize rudder as spinning stops.
5. Return to level flight.

If the operations manual for your plane lists a different set of procedures to recover from the spin, use those rather than those listed here. Generally, the plane reacts immediately, recovering from the spin in less than one turn if the control inputs are correctly applied.

Figure 3-24 shows a pilot taking these steps to recover from a spin. We have already indicated that a plane can take several turns to recover from a spin AFTER the correct control inputs have been applied. Many pilots become concerned if the recovery control inputs do not immediately produce a recovery from the spin, and they attempt other inputs, prolonging the time and altitude lost before they actually stop the spin.

The rudder input opposite the rotation of the spin should be brisk and to the full limit of the rudder travel. This input stops the yawing motion of the plane and ends the rotation of the spin. The forward application of elevator reduces the angle of attack and breaks the stall. The amount of elevator input necessary and the briskness of the input will vary from plane to plane. For many aircraft, a small forward movement is enough to break the stall, while with others it will be necessary to get the elevator control forward of neutral before the angle of attack is reduced to below the critical angle of attack. Keep the ailerons neutral as you recover; it is not uncommon for pilots to roll in ailerons in the same direction as the counter spin rudder input. This can aggravate the stall, making it more difficult to recover from the spin.

Once the spin is stopped and the rudder is neutral, ease the nose of the plane back to level flight. Avoid being overaggressive on the nose-up elevator input; like stall recovery, this can result in a secondary stall. At higher airspeeds, excessive use of the elevator can result in higher g-loads on the plane and damage the airframe, not to mention making for an uncomfortable recovery.

Spins can be disorienting to pilots who are new to them, especially the recovery phase of the spin. They have just had the plane pitch

CLOSE THROTTLE
FULL OPPOSITE RUDDER
BRISK FORWARD ELEVATOR

HOLD ELEVATOR FORWARD
NEUTRALIZE RUDDER

EASE ELEVATOR BACK
TOWARD NEUTRAL

Fig. 3-24 *Spin recovery procedure.*

nose down and begin to rotate as they watch the terrain below spin crazily about the nose. When it comes time to begin the recovery, the pilot is often more than ready to exit the spin, and he or she applies the recovery controls in an aggressive, and sometimes counterproductive, manner. After a little practice, most pilots begin to feel more comfortable practicing spins and settle down as they make control inputs during recovery. But during the first few spins, a number of common errors seem to come to the surface.

The first error is incorrect order of application of the controls. One pilot knew spin recovery required power-off, full opposite rudder and forward elevator control input, but he figured the order didn't matter. The order he wanted to use reversed the rudder and elevator steps, which can result in an accelerated spin. Improper order of the steps could prevent recovery or even make the spin worse.

Partial rudder input opposite the direction of the spin, or slow input of the rudder, can slow the recovery from the spin. Make sure you take the rudder pedal right to the stop. Once rotation has halted, you must then remember to neutralize the rudder pedals. Many pilots will forget to center the rudder after the spin stops, and the plane begins to yaw in the opposite direction as a result of the continued rudder input. This can result in the plane entering a spin in the opposite direction of the one it just recovered from.

Many pilots use an excessive amount of forward elevator as they break the stall. Use the minimum forward elevator needed to break the stall. Additional elevator actually increases the rate of descent and the loss of altitude. Figure 3-25 shows the altitude lost during one, two, and three-turn spins, comparing the loss between a Cessna 152 and Pitts Special S-2B. These are for comparison only and should not be considered indicative of the altitude those model aircraft will lose in a spin. Notice that a one-turn spin results in a 600- to 800-foot loss of altitude. If in the pattern, you don't have much room to recover if you actually get into a spin. Excessive forward elevator pressure as you recover could mean the difference between having or not having a successful recovery.

Once pilots have stopped the spin and broken the stall, they often pull back on the elevator control too rapidly or with too much force as they attempt to get back to straight-and-level flight. Use the elevator with authority, but do not overreact and cause an accelerated stall during spin recovery. At the other end of the spectrum, some

NUMBER OF TURNS	CESSNA 152	PITTS SPECIAL S-2B
1	600 ft.	800 ft.
2	1,100 ft.	1,500 ft.
3	1,500 ft.	2,000 ft.

Fig. 3-25 *Spin altitude loss will vary greatly with different airplane types.*

students do not pull out of the dive after spin recovery quickly enough. Not only does this cause additional altitude loss, but also it allows the airspeed to rapidly build. If left unchecked, V_{ne} can be reached very quickly, again with the potential for structural damage.

Exotic spins

The basic spin is only the tip of the iceberg. Properly designed and well-flown aerobatic aircraft commonly perform variations on at least six different kinds of spins. In addition to the normal spin, there are accelerated spins, and flat spins flown upright and inverted—six in all. Recently, some daring pilots have perfected a maneuver first developed by radio-control fliers, the knife-edge spin. Some radio-control airplanes are capable of *climbing* in a flat spin. Presumably, given enough time, creative pilots and capable airplanes will some-day duplicate that maneuver as well.

There are several aerobatic books which cover exotic spins as a matter of course. As such, this book focuses only on the normal, upright variety, and how to recover.

Emergency spin recovery

In addition to the normal, FAA-approved spin recovery procedures we have discussed, there is another recovery procedure. First publicized by Mr. Eric Muller and now promoted by Mr. Gene Beggs, this procedure is based on findings that planes are able to recover from spins if a simplified spin recovery procedure is used. This procedure reduces the pilot's need to correctly position the elevator and ailerons for spin recovery, or to need to know what type of spin they are in. To summarize, the procedure is:

1. Cut that throttle!
2. Take your hands off the stick!

3. Kick full rudder opposite until the spin stops!
4. Neutralize rudder and pull out of the dive! (*Sport Aerobatics*, p. 31, April 1994)

You can see that this procedure differs from the recommended version by the FAA. One of the largest differences is releasing the control stick, which allows the airflow to position the elevator and ailerons in spin recovery positions. Once rudder opposite the direction of the spin is applied, the plane recovers from the spin.

Mr. Beggs has successfully tested the recovery procedure in several aircraft with success, as long as the plane was loaded within proper weight and balance ranges. The aircraft tested by Mr. Beggs included several Pitts Special models, the Christen Eagle II, the Cessna 150, Cessna 172, and the Beechcraft Skipper trainer (*Sport Aerobatics*, p. 31, April 1994).

I must mention certain cautions regarding this spin recovery technique, though. NASA studies have found that the procedure does not work with all aircraft and should not be depended on in all cases. Some authors have suggested that the Muller/ Beggs recovery procedure may work at certain points in a spin while not working at others.

4

Ground reference maneuvers

You are at the mercy of the wind from the time you lift off until you land. Take it even one step further—you're contending with the wind from the time you untie your aircraft until you tie it safely down again. Although relatively unnoticeable in flight, wind plays havoc with course and speed over the ground. For this reason, it is imperative that you master ground track procedures very early in your flight experience.

Turns about a point

Turns about a point, as well as all ground track maneuvers, are entered downwind (Fig. 4-1). The primary reason for the downwind entry is safety. Because the angle of bank is proportional to the ground speed, the faster the ground speed, the steeper the bank. The slower the ground speed, the shallower the bank. Therefore, if the turn about a point is entered downwind, the initial bank will be the steepest encountered in the maneuver. It prevents a pilot from making the mistake of entering the maneuver upwind (into the wind) at a steep bank, only to find that to hold the desired radius, the bank must be increased to a point that might exceed the capabilities of the pilot or the aircraft.

All ground track maneuvers incorporate an infinite number of bank angles, but the three most used reference banks are shallow, medium, and steep. Once you have mastered the basic rules of bank as they apply to ground track, you should have no difficulty with any ground track maneuver. The basic rules of bank are:

- Steep bank downwind (wind behind you).
- Shallow bank upwind (wind in front of you).
- Medium bank crosswind (wind from your side).

Remember that the wingtips point at the reference point only when you are directly upwind or downwind. Everywhere else your wingtips will be either ahead of or behind the reference point. Don't try to keep the wingtips on the point or you will make a very egg-shaped circle. Crabbing is important at all times, not just directly up-wind or downwind.

The turn about a point can be a very large or very small circle. It all depends on how close you are to the reference point. The closer the aircraft is to the reference point, the steeper the banks will be, and the smaller the circle. The farther away from the reference point the aircraft is, the shallower the bank, and the larger the circle. And re-member, banks are relative. Certainly a 45-degree bank is steeper than one of 15 degrees, but it is only three times as steep, whereas a bank of five degrees is five times as steep as a bank of one degree. So think of your angle of bank as one relates to the other, not in terms of 45, 15, 10 degrees, etc. Think of them as shallow, medium, and steep.

To properly execute a turn about a point, you first need to attain the proper altitude. The FAA recommends you be from 600 feet to 1000 feet above any obstacles as you choose a point about which to pivot. Make sure the point is stationary. I once had a student try to do a turn about a point around a car. And the car was going about 60

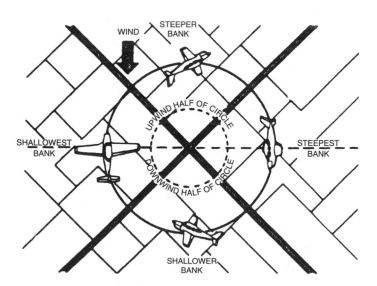

Fig. 4-1 *Turns about a point.*

MPH down the road—truly an advanced maneuver for which the student was not ready.

As you enter downwind and cross abeam of the point, the bank will be initiated. It should be a coordinated turn resulting in a steep bank. The steep bank will be made progressively shallower through the first half of the circle. As the crosswind point is reached, the bank will be medium, and it will also be the point of maximum crab angle since you will be directly crosswind. As the aircraft proceeds from the crosswind to the upwind position, the bank will get progressively shallower until you arrive at the upwind position with the shallowest bank of the entire maneuver. You now have turned 180 degrees. At this point, the entire banking procedure reverses and gets progressively steeper as you go from upwind to crosswind. Proceeding from crosswind, the bank continues to steepen until you arrive at the downwind starting point and again have the steepest bank of the maneuver.

If there is no wind, theoretically, the bank will be constant throughout the 360-degree turn. Therefore, it is probably best to practice this maneuver in a steady breeze. A gusty wind condition can cause you some difficulty in acquiring a feel for changes in ground speed and resultant banks required to form a constant radius about the point. Remember to strive to hold a constant altitude throughout the maneuver.

And do not become so engrossed in your maneuver that you forget to look out for other aircraft. I remember flying at 6000 feet giving some aerobatic training and looking down to see what I at first thought was two students dogfighting. They were far below us, in unison, going around a circle in perfect symmetry. As I watched them for a minute, it began to dawn on me that they were, in fact, both doing turns-around-a-point around the same point! And they never saw each other! Imagine the potential for tragedy. I quickly called one of the aircraft on the radio and instructed them to break to the right since they were in effect flying in formation without the benefit of seeing the other aircraft.

S-turns

The S-turn is a fine training maneuver, requiring the same bank techniques incorporated in the turn about a point. The S-turn differs from the turn about a point in that instead of flying a complete

360-degree circle, you make a series of 180-degree turns along a road or other straight landmark. Once again, crab angle, along with shallow, medium, and steep banks, are your primary guides in obtaining the desired track over the ground.

Begin at about 600 feet above obstructions and over a road that lies perpendicular to the wind (Fig. 4-2). Enter downwind for the same reasons mentioned for the turn about a point. As you approach the road, decide how large you intend the half circles to be. Remember, the points of maximum distance from the road should be equidistant on both sides of the road—this helps maintain the symmetry of the maneuver.

The S-turn maneuver begins when the aircraft is directly over the road. At this point the bank is initiated and is your steepest bank because the ground speed is the fastest here. As the aircraft continues to the crosswind point, the bank gradually shallows to become medium and you encounter the maximum crab angle. Turning to upwind, the bank gradually decreases to the most shallow as the air-

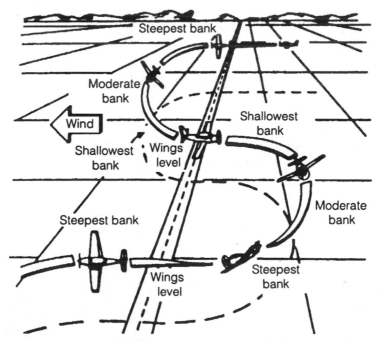

Fig. 4-2 *S-turns across a road.*

craft reaches the 180-degree point. It occurs just as the aircraft reaches the road, with the aircraft exactly perpendicular to the road.

The bank is reversed for the turn in the opposite direction. The aircraft is still upwind, so the shallow bank should be held until the aircraft turns far enough to cause you to gradually steepen the bank when you arrive at the crosswind point, with medium bank and maximum crab angle. From crosswind, continue to steepen the bank gradually until you arrive back over the road with the steepest bank for this half of the maneuver. You should arrive back over the road just as you complete the 180-degree turn, not before or after. The two half circles just completed form one complete S-turn.

Of course, you don't have to stop with just one S-turn. In fact, it's probably better practice to put together several in one direction and then reverse your course and go back down the road in the opposite direction. Be cautious and don't become so engrossed that you forget to watch for other aircraft. Seldom are you as alone as you think you are. In fact, given the scope and area of the sky, it is amazing how many midair collisions and near misses we have in the United States each year.

The S-turn may be made any size the pilot desires. The only thing governing the size of the half circles is your initial bank. The steeper the initial bank, the smaller the circle. The shallower the initial bank, the larger the circle. The symmetry of the S-turn is kept uniform by going twice as far down the road as you go out from the road. This symmetry ensures a perfect half circle.

The list of common errors in the execution of an S-turn would be headed by the pilot hurrying the turn from upwind. At this point, the ground speed is the slowest and there is a tendency not to hold the shallow bank long enough for the aircraft to fly away from the road as far as it was on the downwind side. The best way to overcome this is to pick reference points of equal distance from both sides of the road and then, without cutting corners, fly to them.

Crossing the road before or after the maneuver is completed is another common error. The aircraft should cross the road just as the 180-degree turn is completed and roll into a turn in the opposite direction. Since most of the time you will be flying with your head out of the cockpit for visual reference, a brief glance at the altimeter every so often confirms any altitude change you might not be aware of.

The S-turn is one of the better coordination exercises because you have to use visual as well as instrument reference, feel for the aircraft, and all your senses, with the possible exception of taste and smell. While you are doing all this turning from one direction to the other, always remember to watch for the other aircraft as well as monitor headings, altitude, airspeed, and bank. This constant monitoring helps improve the ability to think and act quickly and accurately. When taken seriously, the S-turn is a very exacting and demanding maneuver worth perfecting.

Rectangular patterns

The rectangular pattern is another ground track maneuver that involves not only varying the bank to correct for wind drift, but a great deal of crabbing into the wind to keep the aircraft on the desired path (Fig. 4-3). In the S-turn and the turn about a point, a little crabbing was done as the aircraft reached the crosswind point. But in those maneuvers, the aircraft was turning throughout the entire maneuver. In this maneuver, there is a lot of straight-and-level flight with crabbing the only means for correcting wind drift.

The rectangle pattern is flown parallel to, and equidistant from, the field or group of fields used to make up the rectangle. As in all ground track maneuvers, the entry is made downwind and the altitude should again be about 600 feet above obstructions.

Fly the aircraft to the point that is parallel to the downwind corner of the field. You are flying parallel to the field on a downwind heading. You will have to crab as necessary to maintain equal distance from the field. At the downwind corner of the field, make your first turn. Because the turn is from downwind to crosswind, the bank should begin steep and gradually shallow out to a medium bank as the aircraft reaches the crosswind point. The turn should be complete with all crab established at a point directly parallel to the corner of the field. This first turn is more than 90 degrees because of the crab angle required to maintain a straight ground track. Continue on, crabbing to hold a constant distance from the field on the downwind side, until you reach a point directly parallel to the next corner of the field.

Upon reaching the second corner, the aircraft is still crosswind. The initial bank is medium, gradually decreasing to shallow as you reach the point where you are into the wind and parallel to the corner of the

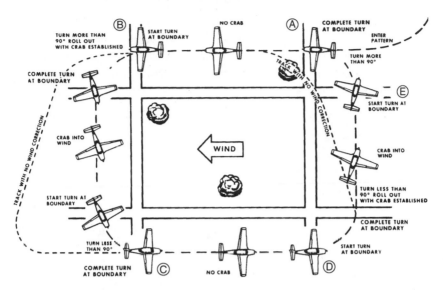

Fig. 4-3 *Rectangular pattern: A) Enter downwind at desired distance from rectangle; B) Downwind to crosswind turn, steep-to-medium bank, more than 90-degree turn, crab as necessary to maintain ground track; C) Crosswind to upwind turn, medium-to-shallow bank, less than 90 degree turn; D) Upwind to crosswind turn, shallow-to-steep bank, less than 90-degree turn, crab as necessary to maintain ground track; and E) Crosswind to downwind turn, medium-to-steep bank, more than 90-degree turn. (Note: All turns begin and end perpendicular to field corners.)*

field. This turn is less than 90 degrees because the turn began with the aircraft in a crabbed situation, pointing slightly into the wind, and it ended up pointing directly upwind (into the wind). All of these points assume the wind was on your nose. If not, you will have to crab, as necessary, to hold your constant distance from the field.

At the third corner, the aircraft is pointed into the wind, so the initial bank should be shallow and gradually increased to medium as the aircraft reaches the crosswind position. This turn is less than 90 degrees because the turn was started into the wind. The turn is completed crosswind, requiring a crab angle to hold your flight path parallel to the field.

Upon reaching a point parallel to the fourth corner, remembering you are still crosswind, start out with a medium bank and gradually increase it to steep as you turn downwind. This turn should be more than 90 degrees because you started the turn with a crab into the wind, and the turn ended up directly downwind. If you are directly downwind, no crab should be required to hold your pattern an equal distance from the field on this leg. This process completes one circuit of a rectangular pattern.

As you have probably already surmised, the rectangular pattern has a direct relationship to the normal traffic pattern. Since all good traffic patterns are rectangular, you must learn how this maneuver transfers to the traffic pattern. Many pilots fail to allow for wind effect and fly very erratic traffic patterns.

Advanced ground reference maneuvers

Commercial pilot test standards require candidates to demonstrate a couple of variations on the turn around a point. The first is simply two turns around a point in opposite directions, the second, a variation of method wherein altitude variations are used to control ground track, instead of discrete adjustments to bank. The details of each follow.

Eights around pylons

Figure 4-4 illustrates an eight around pylons. As you can see, two reference points, in this case trees, are used as the center of the circle for each half of the figure eight. This maneuver gives the pilot the chance to perform wind correction through turns to both the left and right while maintaining a constant altitude. The two points should be between one-half and one mile apart, and this distance can be adjusted depending on the speed of the plane you are flying and the speed of the wind. As with other ground reference maneuvers, you should perform eights around pylons away from homes, farms, livestock, etc., to avoid annoying those on the ground.

The line between the pylons you choose should be perpendicular to the direction of the wind. Fly the maneuver at pattern altitudes and enter between the two ground reference points on a downwind heading. In our example you will be making your initial turn to the right. Like turns about a point, you want to maintain a constant distance from the pylon you are flying around, but now you will need to transition smoothly from flying one circle, then another, each in opposite directions. Since we are entering on a downwind heading, the bank will initially be steep, then become more moderate as we

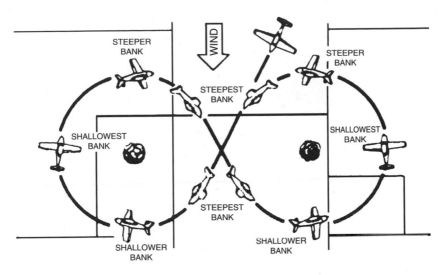

Fig. 4-4 *Eights around pylons.*

turn to a crosswind heading. It will shallow further at the upwind heading portion of the maneuver, then increase in bank as we turn to crosswind, then downwind once again. Crab, bank, and roll rate will apply in this maneuver just as they did in S turns and turns about a point, helping you maintain a constant distance from the point while you compensate for the effects of wind on your ground track.

Eights around pylons help you learn to judge your distance from an object on the ground, compensate for the effects of wind, and give you experience in dealing with the more complex task of judging wind and how it affects your ground track as you move between two points. It is not uncommon for pilots to focus so much on one aspect of the maneuver that they ignore other portions of it. Some may have a tendency to lose altitude, or to let the wind blow them from the desired course, but with a little practice the maneuver is not difficult to master.

Pylon eights

The objective of the pylon 8 (Fig. 4-5) is to develop the ability to fly your aircraft while you are dividing your attention between the airplane's flight path and the ground reference points. In other words, the FAA wants you to be competent enough to manage your aircraft while your attention is diverted outside for any extended period of time.

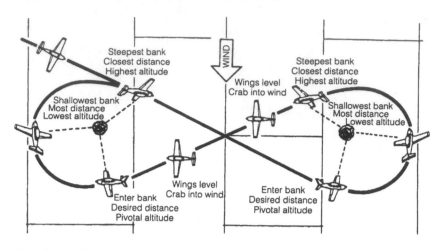

Fig. 4-5 *Pylon 8.*

The pylon 8 involves flying your aircraft in two circles about points on the ground, forming an 8 lying parallel with the ground, while you keep your aircraft's lateral axis (wingtip to wingtip) on the point. However, this is a ground reference maneuver as opposed to a ground track maneuver because you do not make any correction for your ground track. What I mean by not having any ground track correction is that you will make absolutely no correction for wind during the execution of the pylon 8. You drift where you drift and concentrate on holding your wingtip on the pylon.

However, during the pylon 8, if there is any wind blowing you will have a continuous change in your altitude. Your altitude will change as your ground speed changes. Hence, a new term, "pivotal altitude," has been created. Let me explain.

Pivotal altitude is actually a function of how fast you are traveling over the ground. In calm air, there is an altitude at which you could fly your aircraft about a point, keeping that point directly off the tip of your wing and your altitude would never change. This is *pivotal altitude*. But the wind is usually blowing, and this causes your ground speed to change, so you have to change your altitude to make up for the ground speed differential as you go from upwind to downwind. The faster your ground speed, the higher your pivotal altitude; the slower your ground speed, the lower your pivotal altitude.

To find your initial pivotal altitude, square your ground speed and divide by 15. For instance, if your aircraft has a ground speed of 100 knots, square it and you get 10,000 knots. Divide the 10,000 by 15 and your pivotal altitude is 666 feet. Now, add this to your terrain elevation and you have an indicated altitude to fly at to initiate the pylon 8.

Does it sound a bit difficult? Let's run through a set of pylon 8s to try to simplify it for you. To initiate your pylon 8s, first arrive at your pivotal altitude and pick two pylons in an open area. A line drawn between the pylons should be perpendicular to the wind, or crosswind, because you will want to enter the maneuver downwind at a 45-degree angle to, and in between, the pylons.

Okay, you have pivotal altitude attained, pylons picked, and you're heading downwind between the pylons at a 45-degree angle. As the pylon arrives at the wingtip reference, roll into your bank and place the pylon the same distance above or below your wingtip as the horizon would be in straight-and-level flight. Now, here is the key word that unlocks pylon 8s: anticipate. Anticipation of the ground speed changing as you turn from downwind to upwind allows you to be ready to change your pivotal altitude as need arises (Fig. 4-6).

As you turn from your downwind heading, the ground speed drops and your pylon appears to move forward as seen from its relationship to the wingtip reference. When this occurs, simply move the elevator control forward to maintain the pylon on its original position. When the pylon ceases moving, you are at a pivotal altitude for that

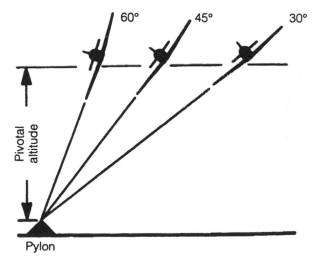

Fig. 4-6 *For a given pivotal altitude, distance from the pylon increases as bank decreases.*

particular ground speed and are ready to anticipate the next wind and ground speed change (Fig. 4-7).

Continue around the pylons, keeping in mind where the wind is in relation to your position, and anticipate moving the elevator in the same direction as the pylon moves. When you have turned about the first pylon to the point that the second pylon is at approximately a 45-degree angle to your aircraft, roll out of the turn, and fly straight and level until the second pylon is directly off the wingtip; then roll into the second half of the maneuver. Flying time between the pylons, straight and level, should be about three to five seconds so you have time to clear the area of any other traffic. Remember, as you practice your pylon 8s, you will be very busy with your wingtip reference, so make very sure you use the time to clear the area wisely.

The most common error pilots make in practicing this maneuver is to use their rudder to yaw the aircraft to hold the pylon on the wingtip reference point. While this works, it completely ruins the point of the maneuver and only results in an uncoordinated trip around the py-

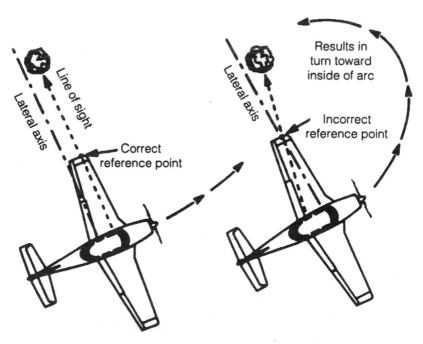

Fig. 4-7 *The correct reference point will have you looking exactly parallel with the lateral axis. This illustrates the same point as viewed from different seats.*

lons. Use rudder only to roll into and out of the pylon 8. This way, the bad habit of skidding the aircraft during the maneuver is never started.

I have heard student pilots talking to each other blasting the pylon 8 as a waste of time, money, and fuel. And, to a point, I suppose they are correct. You will never pylon 8 your way to Chicago or do a pylon 8 in the traffic pattern. But the point they miss is the value of the pylon 8 in learning to fly the airplane by feel. The ability to safely aviate while one's attention is drawn outside is invaluable. And the coordination learned from varying the bank and pitch while maintaining a circular flight path is something that wouldn't hurt any pilot.

Advanced flight maneuvers

The following maneuvers are required for the Commercial Pilot practical test. They are useful for developing advanced maneuvering and situational awareness skills.

Chandelle

The *chandelle* is an advanced training maneuver that requires a great deal of coordination and preplanning in order to be success-fully completed. In addition to its training status, the chandelle also has some very practical applications in everyday flight. For instance, because the chandelle is basically a 180-degree climbing turn, you might use it to turn up and away from another aircraft, up and away from adverse weather, or, possibly, to reverse your direction in a canyon you discover you cannot outclimb. It has more far-reaching possibilities than just another maneuver to learn to please the examiner at checkride time.

To initiate your practice of a chandelle, climb to at least 1500 feet above the ground because the maneuver requires the completion to be near stall speed, and minimum stall recovery altitude is 1500 feet AGL. Line up crosswind, because all turns should be into the wind in order to remain in the practice area. As far as the aircraft is concerned, it really doesn't matter which direction you turn. You won't gain any more altitude one way or the other. In addition, the aircraft should be at, or below, V_a (maneuvering speed) to initiate the maneuver. Depending on the aircraft, you might have to dive or slow down slightly in order to attain this recommended entry speed.

Pick a minimum of three reference points in order to complete the chandelle using outside visual references: one directly ahead, one at the 90-degree point, and one at the 180-degree point where the maneuver will be completed (Fig. 4-8).

When you are at least 1500 feet above the ground, at V_a, set up crosswind, and with your three reference points, you are ready to begin the maneuver. Roll into a moderate bank (about 30 degrees) and begin to smoothly increase your pitch. Maintain your 30-degree bank and steadily increase your pitch until you reach the highest point of pitch at the 90-degree point of the turn. Somewhere around the 45-degree point you will have to begin to add a bit of right rudder to correct the left-turning tendency caused by torque and P-factor. This application of torque correction during turns in both directions helps in making the turn rates constant, which is very important in the overall coordination process.

Upon reaching the 90-degree point, maintain a constant pitch attitude and begin to roll out of the 30-degree bank proportional to your rate of turn. Continue to correct for the effects of torque as you keep your pitch attitude constant through the second 90 degrees of the maneuver. Your rollout should be timed so that as you arrive at the 180-degree point, the wings will just be coming level and your airspeed will be just above a stall.

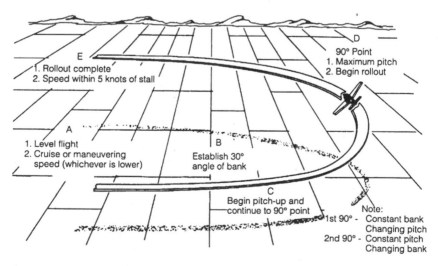

Fig. 4-8 *Chandelle.*

Note: Any pitch change after reaching the 90-degree point is evidence of improper pitch control and planning.

To maintain a constant pitch attitude while you are turning from the 90-degree to the 180-degree point, bring the elevator control back slightly to compensate for the loss of airspeed and the resultant control ineffectiveness. You see, as your airspeed slows, you will need more elevator input to maintain the desired pitch attitude. And a great deal of the success of this maneuver depends on maintaining constant pitch during the 90-through-180-degree portion.

The most common error made in executing the chandelle, other than the obvious pitch and bank errors, is a tendency for many pilots to hurry the maneuver. Do not hurry this maneuver. Chandelles done as slowly and as smoothly as possible are usually the best. This results in the greatest gain of altitude, and more importantly, helps you feel the vast difference in control effectiveness as your aircraft slows from maneuvering speed down to just above a stall. Also, watch for that torque and P-factor correction mentioned previously. Proper coordination goes a long way in helping your chandelles become meaningful and precise.

Lazy 8

The *lazy 8* is another advanced training maneuver requiring preplanning, coordination, and timing. One interesting aspect of this maneuver is that there are almost as many ways of performing it as there are pilots. However, two common denominators emerge: The lazy 8 is a superb training maneuver for everyone, and the lazy 8 can be unequalled in frustration for those who do not understand the various aspects of it. For those unfortunate pilots who draw an instructor who doesn't personally understand a lazy 8, life can be pretty miserable.

The lazy 8 is really a fairly simple maneuver if you are well schooled in your basics and the use of outside visual references. The lazy 8 is a compound training maneuver combining just about all of the aspects involved in flight and garners its name from the fact that the nose of the aircraft will scribe an 8 lying on its side as the maneuver unfolds. If you had a pencil attached to the spinner on the nose of your aircraft, the climbs, turns, and descents of this constantly changing ballet would draw an 8 lying on its side. In the performance of a lazy 8, the controls are always moving; bank is never constant, heading is never constant, and altitude is never constant. The lazy 8 is a graceful, ever-changing, symmetrical dance through the sky.

And one more important note: The lazy 8 is the only flight maneuver I am familiar with that cannot be done by rote. If the pilot performing the lazy 8 doesn't really understand them, it is abundantly evident from the beginning. I have had applicants for a commercial license try to sneak an ugly lazy 8 past me on more than one occasion. It is sort of like trying to hide a fire in the dark.

To begin your practice of the lazy 8, climb to at least 1500 feet and align your aircraft crosswind as you did for the chandelle. For the same reasons as the chandelle, all turns are made into the wind (Fig. 4-9).

Five outside visual reference points are needed to help you complete a lazy 8. These reference points have to be on the horizon directly in front of your aircraft and at each 45 degrees of turn. In other words, the five reference points begin with the first directly ahead and one at each of the 45-, 90-, 135-, and 180-degree points.

A well-executed lazy 8 begins and is completed at the same altitude and airspeed. For this reason, power selection is an important factor. You don't want too much power, which can cause you to gain more altitude than you can comfortably lose. On the other hand, too little

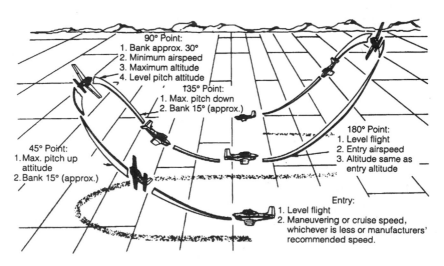

Fig. 4-9 *Lazy 8.*

power can cause too little altitude gain and destroys the symmetry of the lazy 8. So choose a power setting that allows you to begin at or slightly below maneuvering speed, and you should be very close.

Unlike the chandelle, where you began the roll and then initiated the pitch, the lazy 8 calls for a simultaneous initiation of both bank and pitch. And the bank and pitch are begun slowly. Remember the name—*lazy 8*. It's not an accelerated 8 or an abrupt 8.

Slowly and smoothly begin the pitch and bank simultaneously. The pitch and bank are continuously and slowly increased until your aircraft arrives at the 45-degree point of your turn as the highest pitch is reached and your bank is arriving at about 15 degrees. At this time, due to very slow airspeed and high angle of attack, continued application of bank can cause your aircraft to lose some of its vertical lift. The aircraft will then fly down through the horizon at about the 90-degree point of the turn as your bank reaches the 30-degree point.

As you pass through the 90-degree point, some backpressure is released and your bank is reversed so you arrive at the 135-degree point with your bank back to 15 degrees again and your nose as much below the horizon as it was above it at the 45-degree point. The purpose of the nose being equidistant above and below the horizon is to help ensure maximum symmetry during the maneuver.

From the 135-degree point, continue to slowly reduce the bank and adjust your pitch so that you arrive at the 180-degree point just as the wings come level and your airspeed and altitude return to their initial starting point. To complete the lazy 8, follow the same procedure in the opposite direction. Remember, it takes two 180-degree turns to properly complete a lazy 8.

Your timing and coordination are very important to the symmetry of the lazy 8. At no time during the maneuver should your controls be held constant. The pitch and bank are constantly changing during the climbing and descending turns, and corrections for torque and P-factor are needed during the climbing portions of the maneuver. This leads to a cross-controlled situation in the climbing turn to the left since the bank is continuously increasing as you turn in this direction. If right rudder is added to aid in overcoming torque (or no correction is made), then a situation is set up where you are either cross-controlled and coordinated or in a slipping turn.

There are three distinct errors common while learning to execute a lazy 8. The most common is to hurry the maneuver. As I mentioned previously, it is *not* called an accelerated 8; it should be done as slowly and as smoothly as possible for the best outcome.

The second most common error is for the longitudinal axis of the aircraft to pass through the horizon either too early or too late. It should fly through the horizon at exactly the 90-degree point. If the longitudinal axis passes through the horizon too early, you will usually complete the maneuver at an altitude much lower than the altitude at which you entered. This is because passing through the horizon early allows the aircraft more time to descend and it winds up using this time to descend more than it climbed, thus destroying the symmetry of the maneuver. Conversely, if the longitudinal axis passes through the horizon late, the maneuver will be completed at an altitude higher than the original. The reason for this is the exact opposite of passing through the horizon too early. The aircraft will not have as much time to descend, and this causes the maneuver to end up at an altitude higher than the one from which you started. You can cheat and force your aircraft to return to the original altitude, but the symmetry is destroyed and you make little gain in understanding the precision involved in mastering this maneuver.

The third most common error concerns power selection. The correct power setting is essential if the lazy 8 is to be performed with any degree of symmetry. For example, the power setting you choose on a day when you are the only person on board with half tanks of fuel and an outside air temperature of 35°F will be quite different than if you are loaded with full fuel on a very hot day. The reasoning behind this involves your power-to-weight ratio. If your aircraft is light and the day is cool, you require less power to lift the weight. So if you use too much power, you gain more altitude than you can lose without exceeding your entry airspeed, causing you to wind up higher at the end of the maneuver than you were at the entry. On the other hand, if you select too little power for a given day, you will not climb enough to make your lazy 8 symmetrical. So before you initiate your lazy 8, give some serious thought to what power setting should be right for the conditions you have on this particular day. Don't try to use the same power settings day in and day out; it won't work. Remember what I told you earlier. The lazy 8 is the only maneuver I know of that cannot be done by rote. The lazy 8 is either performed correctly or not at all.

Steep power turns

Figure 4-10 illustrates a steep power turn. The bank angle must be large enough that there is a tendency for "overbanking," or the bank actually steepens on its own and you must use opposite aileron to prevent the bank from becoming greater. For most general-aviation aircraft, the bank angle will be somewhere around 50 to 60 degrees. You do not want to exceed the recommended bank angle due to the load factors it may impose on the plane's structures. As the bank angle increases, the g-loads on the airplane increase. Once past 60 degrees of bank in level flight, the g-loads increase very rapidly, with the possibility of exceeding some plane's structural capability. At 60 degrees of bank, the load is 2 g's—twice normal weight. At 70 degrees it increases rapidly to about 3 g's. This is very near the limitation of most general-aviation aircraft in the normal category, so you must watch your bank angle when practicing steep power turns.

You are not using a reference point at the center of the turn and you do not need to be at pattern altitudes. In fact, until you become proficient at steep power turns you should maintain a higher altitude in case you stall the plane accidentally.

Enter the turn at or below maneuvering speed, V_a, to avoid overloading the airplane's structure. Roll smoothly into a 50- to 60-degree bank using coordinated rudder and aileron inputs. A noticeable

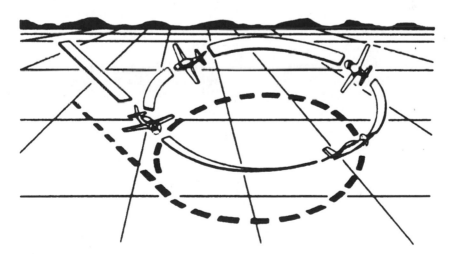

Fig. 4-10 *Steep power turn.*

amount of elevator backpressure will be needed to maintain a level altitude as the bank becomes steeper. This is where many pilots become uncomfortable with flying steep turns. As the bank becomes steeper and the elevator backpressure increases, so do the g-loads, as compared to normal turns. The elevator backpressure can become quite heavy, and many pilots hesitate pulling that hard. Many use rudder opposite the direction of the turn to hold the nose of the plane up and alleviate some elevator backpressure. The turn is then very uncoordinated, at a steep bank angle where the stall speed also increases, and in the perfect setup for a high-speed stall/spin situation. Use coordinated control inputs throughout the maneuver, and this potential problem can be reduced.

Once established in the turn, use slightly opposite aileron to compensate for the overbanking tendency steep turns cause. As you adjust the aileron, also adjust the rudder inputs. If the plane loses altitude and you are pulling hard on the elevator control, add more power. The maximum turning performance for a given aircraft will be reached when the radius of the turn is smallest and the rate of turn is highest. This will vary with the bank angle and airspeed of the plane. If power settings cannot be increased and the plane is descending, decrease the bank angle to hold altitude.

As the turn progresses, monitor the heading closely. Pilots have a tendency to fly past the desired rollout heading because the plane is turning so quickly. When rolling the plane back to level flight, remember to release elevator backpressure and, if necessary, reduce engine power. It is not uncommon for pilots to gain several hundred feet as they roll out of a steep power turn due to this mistake. The amount of lead necessary to roll out on heading will depend on the plane and the rate of turn, but be prepared to give a generous lead.

5

Takeoffs and landings

There are so many things that lead to good overall piloting ability. Taking off and landing happen to be two of the pieces that make up the overall picture. Granted, taking off and landing are important, but to do so with great skill and confidence, you must first master a host of other important and related tasks.

With today's tricycle-gear aircraft and miles of hard-surface runways, the normal takeoff is one of the easiest maneuvers to master. Almost all you have to do is point the aircraft straight down the runway, apply power, keep it straight until rotation speed is reached, and rotate. It's that simple. Or is it? If you are a robot, it is. If you are a pilot, there's a little more involved, such as knowing how to straighten it out if things suddenly go astray and remembering to keep one eye on the oil pressure and airspeed indicator while keeping the other eye on runway alignment of the aircraft, all at the same time. It goes on and on.

Ground operations

Before you can take off, you have to be able to get your aircraft safely started and out to the runway. While this may sound rather simplistic, it can be a source of problems to the careless or uninitiated.

The first and foremost considerations in ground operations are common sense and courtesy. For instance, when you are ready to fire up your engine, you always holler "Clear!" don't you? But do you wait a few seconds after you call out so some poor boob can actually move himself out of the propeller area? Applicants on a checkride are sometimes nervous and are especially susceptible to calling "Clear!" as they are turning the key to start the engine. I'll tell you this: Anyone in the way of the prop with one of these guys will be looking for his parts over a wide area. This is one example where thoughtlessness can turn deadly. Here's another:

An acquaintance of mine was in the process of starting his aircraft on his ranch, far away from the potential problems we mere mortals face at the local airport where we have to share space with the rest of humanity. As he prepared to start up his aircraft, it became clear to me that he wasn't going to clear the area before turning the starter switch. I asked him if he shouldn't clear the area even though we were in the boonies. He told me that, "Nah" he never needed to verbally clear at the ranch. There were only so many people around and he had them all accounted for. This particular day he started his aircraft and threw his favorite dog about 50 feet, straight into doggy heaven. We'll never know if old Scruffy would have responded to a loud call, but I'll bet he would have come around just to see what his master wanted. And you can believe any person over five years old will move in a hurry if warned of an impending startup. The point is that people and animals certainly deserve the chance to get out of the way, so slow down and give safety a helping hand.

If there's anyone standing behind you, they might also wish to move rather than be showered with a propeller blast, rock chips, dirty water, oil, or who knows what. A small amount of courtesy goes a long way towards a safe flight. And a safe flight literally begins when you strap into the aircraft.

Taxi procedures

As you begin to taxi, move away with only enough power to initiate forward movement, and then retard the throttle to a point that lets you taxi at about a fast walk. I said a fast walk. I have ridden with some pilots who taxi faster than a cheetah can run. And it gets pretty interesting when they come to a corner. Taxi only as fast as you are comfortable with, and never attempt to turn a corner too fast in an aircraft—they can, and will, tip over. Wouldn't you love to try to explain that at the next company party?

As you taxi, it is of vital importance that you position the controls to minimize the effects of wind as it relates to lift, directional control, and yaw tendency. (See Fig. 5-1.) The idea is to position the ailerons and elevator to generate the least amount of lift under the wings and tail during taxi, especially when the wind is from the rear. You want the wind to hold the wings and tail down rather than create any lift-

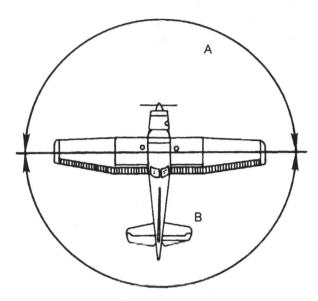

Fig. 5-1 *Selecting control positions during taxi. Imagine a line running through your aircraft from wingtip to wingtip, (the lateral axis): A) If the wind is coming from in front of the imaginary line, turn your control wheel towards the wind direction and hold neutral or up elevator; B) If the wind is coming from behind the imaginary line, turn your control wheel away from the wind direction and hold forward elevator.*

ing action that could overturn you during this critical phase. Remember that aircraft are not at home until they become airborne. In fact, most aircraft are rather awkward on the ground.

During the run-up prior to takeoff, you turn your aircraft directly into the wind. This will do two things for you:

1. It will provide maximum cooling for your engine;

2. It will maximize the aerodynamic flow of air over your control surfaces. In fact, if you have a bit of a breeze, when you test your controls for freedom of movement, the aircraft will react just as it would if airborne. Pull back on the yoke and the nose should rise slightly, etc.

Lastly, any run-up should be completed with the nosewheel pointed straight. If you stop during a turn, leaving the nosewheel crooked, the thrust from running up the engine can cause the bearings to get out of round. Then, when you take off or land, the nosewheel can shimmy very badly. If you have flown a plane in which the nosewheel vibrates violently at some point, you can usually attribute the cause to the plane having been run up with the nosewheel in a cocked position.

Normal takeoff

If you have learned your lessons well out in the practice area and can put to use the principles of slow flight, stalls, straight and level, and all the rest, the normal takeoff will hold few surprises for you. As I taxi onto the active runway, I always do a little mental jog I call my FFT check. It's a last check of the three most often overlooked items on your pretakeoff checklist. I call them the "killers" since a takeoff with one of these items set incorrectly can lead to dire consequences. FFT stands for fuel, flaps, and trim. As I think "fuel, flaps, and trim," I carefully check each one to be sure it is indeed in the takeoff setting. Some pilots have come to a sorry realization at a very inopportune time that one of these wasn't set correctly. One memorable event that cost many lives occurred to the crew of an MD-80 at Detroit, who attempted a takeoff with the flaps in the wrong position. This very preventable human mistake could have been avoided with a simple last-minute check that the fuel was on and that the flaps and the trim were set correctly.

When cleared, taxi onto the runway at the end. Don't waste 200 yards of that precious runway weaving back and forth to line up with the centerline. Go right to it. Line the longitudinal axis (nose-to-tail axis) with the centerline of the runway, and smoothly apply full power. You might need a little right rudder as the power is applied to overcome torque. Feel the controls begin to become effective as your speed increases. Glance quickly at the oil pressure and airspeed indicator, then get your eyes back outside where the action is. If you keep your eyes in the cockpit too long, there might be more action waiting for you out there than you want.

As the aircraft gains speed and the controls become more effective, use small control inputs to maintain your line straight down the runway. One of the most common errors in primary flight is the tendency to overcontrol, especially the rudder. However, be sure to use whatever it takes. Don't go the other way and be too timid with the controls either.

When you reach the rotation speed for your particular aircraft, smoothly apply a little backpressure and lift off (Fig. 5-2). The aircraft should be rotated so that you will be at an angle that produces a climb at about V_y (best rate-of-climb speed). If you find this a little difficult at first, don't feel alone. The proper amount of rotation will come with a little time and practice.

Okay, now you're airborne and climbing out at V_y. Trim the aircraft to help maintain airspeed while you glance around to verify that you are climbing straight out from the runway centerline. Any drift should be corrected for by the use of the crab technique. Remember that you learned to correct for wind when you learned ground track procedures. Now is the time to put that knowledge to practical use.

The FAA recommends climbing to at least 500 feet above the ground before turning out of the pattern. Another suggestion is to climb straight ahead until you reach the end of the runway. Unless otherwise directed by the tower, do whichever comes last.

The traffic pattern

The traffic pattern can be a busy environment, and maintaining an awareness of other aircraft and their relation to your plane can demand a great deal from the pilot. Airplanes fly at different airspeeds, and even different altitudes while in the pattern. Whenever you fly, you should be looking for traffic, but this becomes extremely important when you fly in the traffic pattern. The maneuvers covered

Fig. 5-2 *Everything checked, lined up, and ready to go.*

so far in the book become of fundamental importance due to the need for dividing your attention between looking outside the plane, monitoring your ground track, looking for other airplanes, and flying the approach. Knowing how to enter and exit the pattern can help you be more efficient as you fly near the airport.

Preparation for pattern entry

You should begin planning for your landing while you are still well away from the airport. In a controlled airport, you will need to listen to the approach and tower frequencies, normally at distances of 20 miles or greater, depending on the airspace around the airport. At uncontrolled fields you should begin listening to the Unicom frequency while you are at least 10 to 20 miles from the airport. However, in either case, you should develop a mental picture of the traffic, the runways in use, the flow of traffic pattern, and any other factors that can affect how you will fly in the pattern.

By planning ahead, you can increase the level of safety as you fly into the pattern. As you approach the airport, you should clearly state your intentions, altitude, and position from the airport. This is true whether you are flying into a controlled or uncontrolled field.

At controlled fields, the tower staff will direct you through the airspace, giving you headings and altitudes. You are not relieved of the need to look for other aircraft, though, so don't become complacent and assume you do not need to keep looking outside the plane. At uncontrolled fields you may want to overfly the airport before you actually enter the pattern to get an idea of the runway layout, look at the windsock to determine wind direction, or observe for other aircraft in the pattern. There are times when pilots do not use their radio, or the plane they are flying is not equipped with a radio, so even though you may not hear any other traffic over the radio, make sure you are looking for other aircraft as you fly over the airport or prepare to enter the pattern.

Whether you are being directed by the tower at a controlled field or are flying the pattern at an uncontrolled airport, make sure you check the wind sock and compare the information you are getting from it with what the tower or Unicom staff is telling you. No one will purposely give you wrong information when you ask for weather information, but you want to use the wind sock to confirm the wind direction as you fly through the pattern. On some airports the wind sock may be difficult to find, but do your best to locate it prior to landing. Figure 5-3 shows the typical wind direction indica-

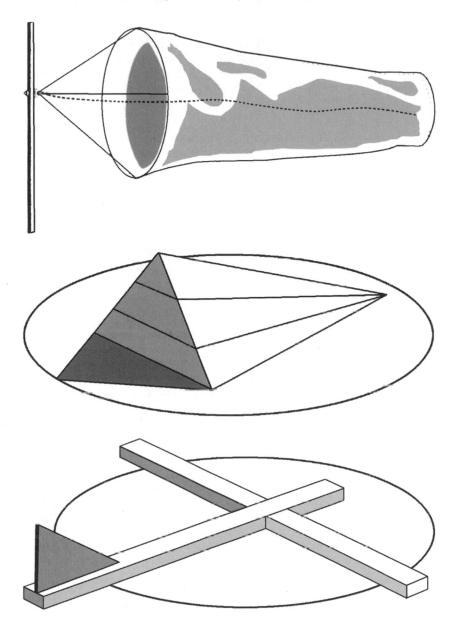

Fig. 5-3 *Some common wind direction indicators.*

tors that may be at airports. Both the wind tetrahedron and landing tee are designed to point into the wind, or in the same direction you should land the plane. At some airports they may also be lighted, making it easier to locate them at night. Any of these devices can give you an idea of the actual direction of the wind near the ground and runway. This can be important if you are making any type of a crosswind landing for judging the amount of crab or slip you are going to need during the approach and landing.

Most airports have traffic patterns that are 800 to 1500 feet AGL in altitude. You will need to consult an Airport Facility Directory or other reference to find out what the pattern altitude is at the airports you plan to use. Being at the correct altitude for the pattern will make it easier for pilots to find you and maintain proper separation.

Before discussing the legs of the pattern and the entry points, you should be aware that while normal patterns at uncontrolled airports are flown using left turns, known as a left-hand pattern, there are some airports that use right-hand patterns. Residential areas, ground obstructions, and towers in the vicinity of a left-hand pattern may cause the airport to set up a right-hand pattern. You will need to consult airport reference guides, available at most Fixed Based Operators (FBOs), to find out if an airport has right or left traffic patterns. Depending on the layout, some airports have right-hand patterns for one runway and left hand for another, so if you are flying into a field you are unfamiliar with, check with Flight Service or consult the airport reference guide before you take off.

Figure 5-4 shows the different legs of a pattern; you can see that they consist of the crosswind leg, the downwind leg, the base leg, and final approach. The figure also shows two runways and how the pattern is left hand in some cases, while right hand in others. Note the wind cone, which shows the direction that airplanes should land as a result of the wind direction.

The legs of the pattern are based on the direction of the wind. The crosswind leg is flown perpendicular to the active runway. The downwind leg is flown going with the direction of the wind, parallel to the runway. Realize that if the wind is not blowing directly down the runway, there will be a crosswind that will affect your ground track while flying the pattern. Separation from the runway while on downwind should in no case exceed the airplane's ability to glide to the runway in the event of an engine failure. On average, this distance is about one-half mile, although the pattern may grow substantially to accommodate more airplanes on a busy day.

When flying at a controlled airport, tower personnel will direct you regarding the headings and altitudes they want you to fly. They will normally use a standard pattern consisting of a downwind, base, and

THE TRAFFIC PATTERN

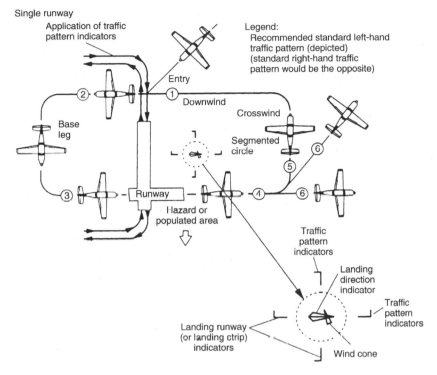

Fig. 5-4 *The traffic pattern.*

final, but traffic considerations at the airport may cause them to issue instructions that have you fly an abbreviated pattern, or in some cases, an extended pattern. It is not uncommon for the tower to issue the instruction for a pilot to fly an extended downwind and the tower will call base. This is normally done for spacing between aircraft, but is an example of how flying at a controlled airport can change the normal pattern.

Pattern entry

Figure 5-5 shows several pattern entry points. Position A in that illustration demonstrates a crosswind pattern entry. This is normally used when you are approaching the airport from a direction opposite the side of the runway the pattern is being flown, and provides an efficient method of merging with the flow of traffic. This pattern entry is accomplished by flying across the active runway at

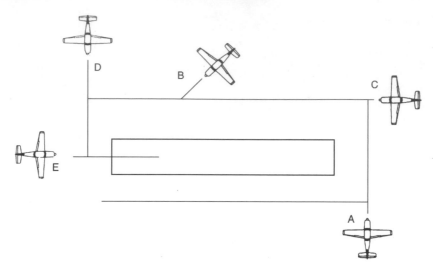

Fig. 5-5 *Pattern entry points.*

pattern altitude perpendicular to the runway. At the appropriate distance, you will then turn to the downwind leg of the pattern. The position of the crosswind leg should be such that aircraft taking off from the runway will not climb into you and you have the ability to maintain visual contact with departing traffic. This position is normally near the middle of the active runway but may vary depending on conditions. You should be careful when making a crosswind pattern entry for other reasons, however. If traffic volume in the pattern is at all busy, you may want to avoid a crosswind pattern entry due to the possibility of cutting off other traffic in the pattern as you make the turn from crosswind to downwind. Additionally, as you make the turn to downwind you will lose some visibility of traffic while you are turning away from them. These factors contribute to the fact that crosswind pattern entries should be used only when traffic volumes permit a safe entry.

Downwind entry

Downwind entry into the pattern is the most common entry point for uncontrolled fields. There are two types of downwind entry listed in Fig. 5-5, points B and C. Of the two entry types, B is the preferred method for reasons of safety and traffic flow and is the one most recommended by FAA publications. Position B is known as a 45-degree downwind entry and provides you with the best view of traffic in the pattern as you approach the downwind leg on a 45-degree angle. This method also provides for a relatively easy turn from the entry angle to the downwind leg and allows you to make any wind corrections without an excessively steep bank. Remember, as a rule of thumb you should not exceed banks of 30 degrees in the pattern, and the 45-degree downwind entry allows you to easily remain below this bank angle in most cases.

As you approach the pattern from position B, you should be scanning behind you on the downwind leg to assure that you are not cutting in front of other planes that may already be there. You will also be able to see aircraft on base and downwind legs from this point, giving you the ability to more easily keep track of aircraft in the pattern. Listening to the radio during your approach to the airport may indicate where to look for other planes as you fly this pattern approach.

An alternative to the downwind approach using a 45-degree angle is the straight in to downwind approach. Position C shows an airplane flying straight into the downwind leg. While this approach may make an efficient entry point, it can be more difficult to see other aircraft and maintain adequate separation.

Base entry

Position D in Fig. 5-5 illustrates a plane entering the pattern on a base leg. Once again, this entry point can allow a pilot entering from that direction an efficient method of entering the flow of traffic. Like other entry points, it can pose problems for the pilot in maintaining proper spacing with other planes in the pattern. Be very careful not to cut in front of other planes that are on downwind when you enter the pattern on the base leg; this not only poses a safety hazard but also does not promote a "good citizen" mentality when flying with other airplanes.

Final entry

Final entry to the pattern, also known as a straight-in approach, is depicted as position E in Fig. 5-5. Although the most efficient way to reach the runway, the straight-in approach is perhaps the worst method for traffic avoidance. Most pilots approaching the runway tend to focus on the asphalt, and not other airplanes, and pilots in the pattern already will also be looking at the runway as they maneuver through the base and final turns. Not only is it risky, but very rude, if your straight-in approach cuts somebody else off.

Pattern exits

As you probably have already guessed, there are a number of standard pattern exits that you should be aware of and use to improve aircraft separation and visibility. In this section we will discuss each of those options. Like pattern entry points, this is a guideline on possible ways to exit the pattern. Each situation may require that you adapt your path through the pattern to meet a given situation. Figure 5-4 depicts several suggested methods for departure from the traffic pattern. In every case, the airplane is maneuvered in the pattern until it is pointed approximately in the intended direction, and flown away. As ever, flight in the vicinity of other aircraft requires the pilot's rapt attention.

Normal landing

There's an old axiom in aviation that says, "A good landing is usually preceded by a good approach." This statement is still true today. A normal landing, the one you learn before going on to the other advanced types of landings, is like all other landings. It is merely the end result of a transitory period from the approach, through the flare, to ground contact. But before you can execute this transition to a landing, you have to arrive over the runway threshold at approximately the correct altitude and with proper alignment. Therefore, you first need to execute a good approach before you can attempt a good landing. And you can either land or you can arrive. Arrivals usually bring forth much laughter from your flying comrades and another gray hair for your instructor. Sometimes arrivals strain the landing gear, bend props, flatten tires, etc.

The normal landing begins long before the actual touchdown. It starts with your preplanning for your pattern, airspeed, traffic spacing, flap usage, and all the basic flight techniques you have previously learned. "A good landing is no accident," says the FAA.

On the downwind leg, you should go through the prelanding checklist, set your aircraft up at the proper distance from the runway, check traffic, and note that you are at traffic pattern altitude. You should be exactly on your altitude, not close to it. About halfway down the runway on the downwind leg, pull on the carb heat so it has time to work. Remember that the heat from the exhaust warms the carburetor when you pull on the carb heat, and the air is not exactly hot enough to melt steel. Even a blowtorch takes some time to melt ice.

When you are directly abeam of the approach end of the runway, reduce your power to an approach setting and trim your aircraft to maintain the recommended approach speed. After you have set up your glide and the airspeed is definitely in the white arc (flap-operating range), lower the first 10 degrees and retrim the aircraft as necessary to maintain approach speed.

Now comes one of the most important decisions you have to make to keep your traffic pattern uniform: When do you turn onto base leg (Fig. 5-6)? A good rule of thumb is to turn base as soon as you arrive at a point where the end of the runway is at a 45-degree angle behind the wingtip nearest the runway. This procedure prevents you from getting into the bad habit of turning over a certain tree, house, bend in the road, etc. Remember that particular reference point won't be available at another airport.

Fig. 5-6 *This pilot has waited too long to turn from base leg to final approach, a very common error. This is a prime area for an approach-to-landing stall.*

After you turn to base leg, another important judgment must be made: Are you too high, too low, or just right? This is called the key point in the approach. If you are just right, add another 10 degrees of flap, retrim as necessary, and continue on. And one other important thing— use your power, as you need it. That is why the throttle moves. Small power changes, introduced at the exact time they are needed, do much to smooth out the approach. Don't be timid with the power.

Let's break here for a moment and try to answer one of the most frequently asked questions concerning the traffic pattern. What do you do if you are too high or too low? The answer is about 50% common sense, 48% experience, and 2% instruction. Since there are so many possible combinations of errors, here are some general pointers.

If you are gliding in with constant power, constant airspeed, and flaps full down, and you are getting low, the first thing to do would be to add some power. Chances are this solution would take care of the problem. Fly the aircraft up to the point where you reintercept the glide path, reduce the power back to approach setting, and continue on to land. If a little power doesn't work, try a lot. Use it as you need it. I find it much preferable to add a lot of power and land on the runway than to maintain a beautiful constant glide with the airspeed and attitude constant and land in the mud. It's hard to taxi that way.

Most people don't have too much trouble deciding what to do when they are too low. I guess it's the ground coming up at them that shakes them into action. And if you find yourself too high, there are many things you can do to alleviate the situation. However, when some students are too high, it's an entirely different matter. They just sit there. I guess they figure it will come down sometime. It will, probably in an orchard or subdivision. So you see, there is as much reason to act if your aircraft is about to overshoot as there is if you are too low. If you don't put it on the runway, the results are usually the same—trouble.

If you are too high and have not yet put down all of your flaps, add some more flaps to increase your rate of descent. Or you can opt to remove some, or even all, of your power. Maybe you need to do both, add flaps and take off some power. Yet another option, used less and less these days, is a slip. A moderate forward slip will increase your sink rate a great deal and turn a possible overshoot into a workable approach. Of course, if you see you are still going to overshoot, nothing replaces the go-around, try-it-again method. In any case, don't just sit there and wait to see what happens; take action.

Back on base leg, we left ourselves in the key position with 20 degrees of flap and normal approach speed. The next step is to start the turn to final early enough so you can make a shallow turn from base leg to final approach, keeping the runway in sight throughout the turn. The most common error in this area is a pilot who starts the turn too early and makes the turn so shallow that he or she actually angles toward the runway instead of squaring the corner to complete the rectangular pattern. This results in actually arriving at the runway threshold at an angle to the runway instead of being aligned with it. All this does is further complicate an already difficult situation for the student pilot.

Cotter-pin approach

The opposite of this procedure occurs when the pilot starts the turn to final approach too late, tries to utilize the normal bank, and proceeds to complete what I call the cotter-pin approach. He overshoots the runway centerline and then has to steepen the bank beyond safe limits in order to turn back to intercept the runway centerline. Viewed from above, it looks like a cotter pin. And it can become very dangerous very quickly.

A seemingly harmless item like turning from base leg to final approach too late catches the unwary pilot in a situation of deteriorating control command, increasing bank and stall speed, and decreasing airspeed. If I ever wanted a recipe for a potential disaster, this would be it. And the innocent thinking of the unknowing pilot goes something like this:

"Oops, I turned too late—guess I'll just steepen my bank a little and get back on final to the runway. I remember my instructor told me as long as I didn't exceed 30 degrees of bank in the pattern, I'd be okay."

"Wait a minute, this still isn't getting it. I'm not turning fast enough to line up. I bet if I add just a tad of inside rudder (in the direction of the turn), my rate of turn will increase and I'll be okay. I certainly don't want anyone to see me go around. They say it's a sign of a poor pilot."

"Oh, oh. My addition of rudder has made my bank begin to increase beyond 30 degrees. They say that's not good. I'll just add a little bit of opposite aileron to counteract the rudder."

"Darn, still not getting turned quickly enough. I know. I remember from ground school that if I want to increase the rate of turn, all I have to do is add backpressure. Here goes. DAMN!" The airplane stalls, spins, and probably crashes. And it can happen about that fast.

The pilot in the above scenario has been tempted, trapped, and most likely killed by a combination of hangar talk, misunderstanding of flight dynamics, and poor association and correlation of aviation procedures. He has used some right ideas in the wrong places, ignored the bank-to-stall ratio, and allowed some interesting, but incorrect, precepts to become fact.

Look at the procedure used. He turned too late—added to his problems by adding inside rudder with opposite aileron (a slip), and then added the final piece to the recipe by pulling some more backpressure.

Think about this. The airplane is banked at 30 degrees—airspeed at about 70 knots. A cross-controlled slip is added that further reduces airspeed and increases stall speed. Then cross-controlled, at a slow airspeed, the pilot adds some backpressure. This is one of the best ways on earth to describe the entry to a snap roll, an aerobatic maneuver of immense fun and requiring a lot of skill. It is not designed to be performed by fledgling pilots on final approach!

The solution to this very real and present problem is to plan ahead and start your turn from base to final early enough so you can utilize the ground track procedures you learned earlier. You'll wind up with an ever-decreasing bank as you reach the final approach course.

On final, keep aligned with the runway centerline and add the remainder of your flaps as you need them. Maintain a constant airspeed and attitude. Power is reduced as you no longer need it, until, if all goes well, you reduce the power to idle in the landing flare.

If you have the approach pattern down pat, the landing should come rather easily. As I said before, it is just a transition from the normal approach to the landing attitude. This transition usually begins at about 25 to 30 feet above the runway with you slowly increasing the backpressure as you continue to sink. If you have too much airspeed or pull back too rapidly, you might actually climb a little. You don't want to do that, so increase the backpressure gradually. In a good landing, you almost think of it as trying to hold the aircraft off the ground through increased backpressure. You are transitioning the aircraft from a nose-low approach attitude through level flight to a slightly nose-high attitude at touchdown. If everything is right, you reach the point where you run out of backpressure just an instant before the touchdown occurs on the main wheels. Since you are nose high, the nose gear will still be up off the runway at touchdown. As your speed decreases following touchdown, allow the nose gear to lower to contact the runway and thereby provide you with more positive directional control (Fig. 5-7).

The landing is far from complete merely because you are on the ground. Many aircraft accidents occur during the rollout following touchdown. Keep your eyes outside of the cockpit, making sure you are maintaining runway centerline during the slowdown process. You can turn off the carb heat and bring the flaps up after you have cleared the active runway and are positive everything is under control. More than one beautiful approach and landing has been spoiled by the pilot fumbling with something inside the cockpit too soon after touchdown, only to be rudely awakened by the sound of runway lights shattering as the aircraft wandered off the runway. Remember that the control effectiveness steadily decreases as your speed decreases—just the opposite of takeoff. The slower you go, the more control movement it takes to get the job done. Use care but don't overcontrol in this phase either. Use whatever it takes—no more and no less.

Most of the errors common to landings, outside of a poor traffic pattern, have to do with your sight reference points. Nobody can show you exactly where to look, but you can gain some insight in where not to look:

Fig. 5-7 *A normal landing with full stall, full flaps, and full back stick.*

- Do not look down and to the side while landing. Speed blurs vision, and you will not be able to tell if you are five feet or five inches above the runway by looking down and to the side.

- Don't look too close in front of the aircraft during the landing flare. Pilots who do so have a tendency to flare high, stall, and drop in.

- Don't look too far ahead of your aircraft during the landing flare. Due again to depth perception, this causes many pilots to run the aircraft into the ground with little or no flare.

- Once you are in the flare, don't look inside the cockpit for any reason. You've got to see where you are going.

Now that you know where not to look, where do you look? During the short final and flare, look approximately as far ahead of your aircraft as you would if you were in a car traveling at the same rate of speed. If you don't drive, I guess it's trial and error.

One other very common problem associated with landing is learning to pull the yoke straight back. I have had students who would begin to flare perfectly, and then at about 10 feet up, turn and try to take it to the tie-down area. The problem lies in flare technique. In most aircraft, the flare requires the coordinated use of the hand, wrist, elbow, and shoulder. If you try to flare using only your hand, wrist, and elbow, the tendency is to twist the elbow up and out, causing the ailerons to be deflected to the right. It also causes a right-turn tendency that can spoil an otherwise decent approach and flare. Conversely, if you pull your elbow in toward your body as you flare, you will experience a left-turning tendency. So watch for these common errors, and if you have trouble in this area, sit in the aircraft and practice pulling the yoke straight back.

Crosswind landing

It seems that not too many people pay much attention to slips anymore. I guess the advent of flaps and spoilers has caused the need for the slip to be pushed to the rear of the list of things some pilots believe they ought to learn. However, they still have a very real place in the mind of the complete pilot. The FAA has only recently reintroduced the forward slip to a landing as a required maneuver for the private pilot checkride.

Some say, "If you have flaps, you don't need to slip, and anyway, it's unsafe to slip with your flaps down." To this I politely say, "Nonsense." Unless your particular aircraft is placarded against slipping with flaps, go ahead. And even in the aircraft that are placarded concerning slipping with flaps, most don't prohibit them. The placards usually say, "Avoid slips with flaps extended." Many aircraft flight manuals don't mention the fact one way or the other. If your aircraft is prohibited from slipping with flaps, don't. Otherwise, why not? It just might get you into a field you might have otherwise overshot. It can be especially true in an emergency situation.

Generally, there are two varieties of the slip: the forward slip and the side slip. Although the two are very much alike in the manner in which they are executed, the forward slip usually requires much larger doses of control input. The one facet in which both are alike is the fact that you must have your controls crossed. Crossed controls means you hold your ailerons in one direction and rudder in the other—something most pilots find rather uncomfortable until they have executed many slips.

If you have any wind, the slip should be done into the wind (see Fig. 5-8). If the wind is from your left, the left wing is the one put down. But, if all you do is put down the left wing, what will happen? You'll turn left. To keep the aircraft from turning, add opposite rudder. Now you are cross-controlled—slipping.

To put these slips to work for you, you need to know what each one is for and what the desired results should be. The forward slip is used mainly for altitude loss. The sideslip is used to align the aircraft with the runway and to touch down in a crosswind.

The forward slip

Let's set up a situation where both types of slips are used on one final approach and landing. Throw in a crosswind to complete the setup. You have turned from base leg to final approach and find yourself very high. You are landing on runway 18 and the wind is from 120 degrees at 10 knots. You already have full flaps and power at idle, yet you are still going to overshoot. Now what? If you are very high, you will have to go around; however, if you are only a little high, the forward slip just might be the ticket (Fig. 5-9).

Lower your wing into the wind as you apply opposite rudder. In this case, you need to lower your left wing and add right rudder. You

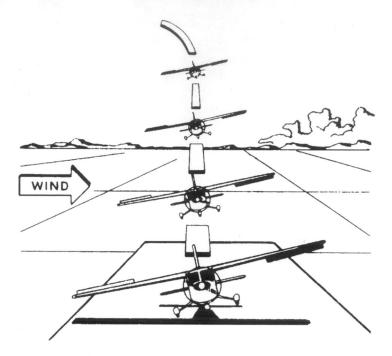

Fig. 5-8 *The sideslip approach to a crosswind landing.*

want the nose to swing past direct runway alignment, say 15 to 20 degrees to the right. You will still be tracking straight down the runway, but your nose will be pointing at a heading of about 200 degrees or so. This is the signature of a forward slip: Your heading changes while your ground track remains the same.

Now comes the important part: Don't let your airspeed build up much over approach speed. If you do, all the gain of the slip will be lost because once you get down, you will have to bleed off the airspeed and you will float much farther, negating the advantage you gained from slipping.

Continue to slip until you are down to the point where you reintercept your glide path. Then, come out of the slip and proceed with your approach. You get out of the slip by reversing your aileron and rudder, bringing your aircraft back to level flight with the ailerons, and using left rudder to swing the nose back to direct runway alignment. Most likely, since you have a crosswind, you should continue to bring your nose past direct runway alignment into a crab condition to take care of the crosswind ground track.

Fig. 5-9 *A forward slip as viewed from the cockpit. The wing is down into the wind, and the airplane is actually pointing at the terminal buildings while the ground track remains straight down the runway.*

Now, you have lost the unwanted altitude, are still tracking straight down the runway, and will be ready to complete the crosswind landing.

The sideslip

The crosswind landing is accomplished using the sideslip. In this slip, the longitudinal axis of your aircraft remains parallel to the flight path. In the sideslip, lower your wing into the wind and add opposite rudder as you did in the forward slip (Fig. 5-10). Add only enough rudder to maintain your track straight down the runway. You sort of lean into the wind like you used to do when you rode your bike in a strong crosswind. You still go straight but are banked to correct for the crosswind. It probably will take a little practice before you become proficient and comfortable with sideslips and remember to use the controls as you need them. You can't just put the wing down a certain number of degrees and leave it there because the wind will probably increase and decrease in intensity. So use whatever it takes to hold your runway alignment.

Fig. 5-10 *A sideslip as viewed from the cockpit. The wing is still down into the wind, but the aircraft is now aligned with its longitudinal axis parallel to the runway. This is the correct way to land in a crosswind from the right.*

The crosswind landing is accomplished exactly as any other landing, except you keep your sideslip throughout the approach, flare, and touchdown (Figs. 5-11 through 5-13). Remember, after touchdown, you're not through yet. Maintain your windward wing down and use your rudder as necessary to maintain runway heading. As you slow down, add more aileron into the wind because the controls become less and less effective as your speed slows. In fact, as you reach your slowest speed, you will want to have full aileron into the wind to correct for the crosswind, just as if you were taxiing. By the way, that is taxiing.

Soft-field takeoff and landing

Soft-field takeoff and landing techniques are used to get into or out of fields made soft by mud, snow, tall grass, or even just wet grass. And this technique is also very good for fields where the surface is rough. In fact, soft-field procedures are used almost anytime you are not using a smooth, hard-surfaced runway.

Fig. 5-11 *Crosswind landing approach. Wing down into the wind with enough rudder to hold the aircraft straight down the runway.*

Fig. 5-12 *Just prior to touchdown in a left crosswind. Wing is still down and rudder is used as needed to maintain alignment with the runway.*

Fig. 5-13 *Crosswind touchdown. Upwind wing down, rudder as required to maintain alignment; upwind wheel touches first.*

Soft-field takeoff

For a soft-field takeoff (Fig. 5-14), set your flaps as per the manufacturer's recommended flap setting. (Many low-wing aircraft manufacturers recommend using no flaps since the flaps on low-wing aircraft are very close to the ground and can easily be damaged by mud, rocks, etc.) Since the surface is soft, it is very important that once you start your aircraft moving, you keep it moving. Even taxiing should be done with full aft yoke to offset any noseover tendency.

The pilot's first concern during the soft-field takeoff is to transfer the weight from the wheels to the wings as rapidly as possible. The flaps help do this, but at the start of the takeoff run, you should have the yoke full back to help lift the nosewheel off of the ground and reduce the drag created by the soft field. Since this technique gives you a much higher angle of attack than other takeoffs you have been used to, once the nose begins to come up, you must be prepared to relax a bit of the backpressure so you don't smack the tail into the ground. That is very easy to do, and the loud bang the tail makes as it strikes the runway usually scares a pilot into shoving the stick forward, sometimes a bit too far. The nosewheel might then get swallowed up by the soft surface you are trying so hard to get out of.

Once you have the nose at the desired attitude and are rolling down the runway, keep the nose attitude constant until the aircraft is off of the ground. The liftoff should occur at a much slower-than-normal airspeed because of the high angle of attack. For this reason, once the aircraft becomes airborne, the angle of attack must slowly and smoothly be reduced to near-level flight attitude as the aircraft accelerates toward the normal climb speed. You should maintain the near-level attitude, in ground effect, until you reach your nominal

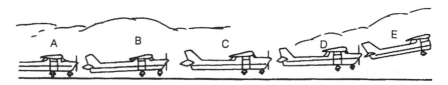

Fig. 5-14 *Soft-field takeoff: A) Smoothly apply full power and raise nosewheel clear of impeding substance; B) Lift off at slower than normal airspeed; C) Gently level off; D) Remain in ground effect until reaching V_x or V_y, depending on possible obstacles; and E) Upon reaching desired speed, gently pitch up and climb out normally.*

climb speed (either V_x or V_y, depending on whether or not there is an obstruction to be cleared).

After accelerating in ground effect to the desired speed, a normal climbout should be initiated and the flaps brought up once established in a climb. Don't be in a hurry to bring up the flaps. They are providing lift, and if brought up too soon, they could cause a momentary sink that could lead to trouble.

One other important point concerning soft-field takeoffs, or any takeoffs for that matter, is to be sure you lift off and maintain a straight track down the runway. If you have a slight crosswind, resist the tendency to crab into the wind as soon as you lift off. If the wind were to die, you could settle back to the runway in a crabbed configuration, and the result could be one of the shortest flights on record. Use the wing-low sideslip method mentioned earlier for crosswind takeoffs and landings. Keep your aircraft in this sideslip while heading straight down the runway and until you are positively airborne.

Soft-field landing

The approach to the soft-field landing (Fig. 5-15) should be made at normal approach speed. The touchdown should be as slow as possible in order to minimize the noseover tendency. Unless you are in a low-wing aircraft and the manufacturer recommends no flap landings in a soft field, this usually means an approach and landing with full flaps. The touchdown should be with flaps as recommended, full stall, and full back stick. A small amount of power can be used in the

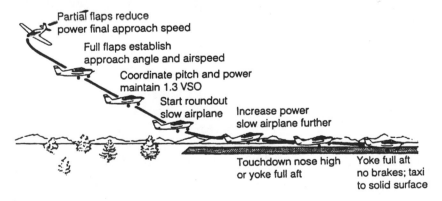

Fig. 5-15 *Soft-field landing.*

flare to help bring the nose up and provide the momentum necessary to prevent a noseover.

Keep the yoke in the full-back position during the rollout and be ready to use power to help you through any really soft spots like snow drifts or deep mud. When you are safely on firm ground and slow enough to taxi, you can take the carb heat off and bring up your flaps. Use caution as you taxi.

An important item to remember during the conduct of soft-field operations is that it is nearly impossible to judge when snow or mud get deeper. All one can see is the top of the surface and there is no way of knowing where the drifts or sinkholes begin. Soft-field operations should be handled with great care and forethought, not with haphazard preparation or spontaneous whims.

Short-field takeoff and landing

The short-field takeoff and landing is used for one of two purposes:

1. To get you into or out of fields that are actually physically short.
2. To take off or land in fields that have some type of obstruction that reduces their effective useful length.

In either case, the short-field technique should be used to take off or land safely.

Short-field takeoff

The very first consideration for anyone contemplating a short-field takeoff is to find the distance it will take for your aircraft to accelerate, lift off, and clear any obstacles. The first place for you to look would be in your aircraft's flight manual in the performance section. Here you will find a chart disclosing takeoff distance information allowing for such conditions as type of runway surface, temperature, and wind. The figures will show you the required distance both to lift off and to clear a 50-foot-high obstacle for a given set of conditions. And remember, the distance in the flight manual is determined under controlled circumstances. In other words, these distances are not an absolute guarantee of the distance it might take you to get airborne and over the obstacle.

Assuming you are taking off from a short field (Fig. 5-16) or a field with an obstruction, you should strive to learn to fly your aircraft by

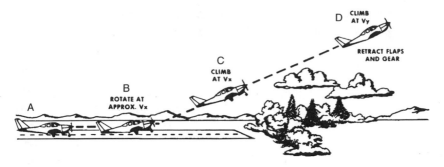

Fig. 5-16 *Short-field takeoff: A) Smoothly apply full power; B) Rotate at V_x airspeed; C) Maintain V_x airspeed until clear of obstruction; and D) Lower pitch to V_y and climb out normally.*

airspeed and attitude control rather than letting your instincts take over. If you are caught trying to clear an obstruction or take off before the aircraft is ready to fly, the results can be less than desirable. If you try to force your aircraft into the air before it is ready to fly, it might drop back onto the runway and actually lengthen the time needed to clear the obstruction.

When learning the short-field takeoff procedures, use every available inch of runway as you line up to practice short-field takeoffs. Any runway left behind you might as well be in another country. It will be of no use to you.

Utilize the manufacturer's recommended flap setting if the aircraft is so equipped. Line up, using the entire runway, and smoothly, but firmly, add full power. Don't hold the brakes. Get going! Studies have shown that, unless you are in a turbine-powered aircraft, holding the brakes while running the engine up to maximum RPM before brake release does nothing to enhance the short-field take-off distance.

Smoothly apply full power and let the aircraft accelerate until you arrive at the best angle-of-climb airspeed (V_x). Keep the aircraft directly parallel to the centerline and let it seek its own pitch attitude. That is, don't force the nose down while you are accelerating because it will likely give you a negative angle of attack and increase the distance it takes you to arrive at V_x. When you arrive at V_x, rotate and maintain that attitude until you have safely cleared any obstruction. After clearing the obstruction, the angle of attack should be reduced and the aircraft allowed to resume its normal climb speed (V_y). The flaps should be left alone until you have safely cleared any obstacle, accelerated to V_y and then they should be brought up very slowly.

Short-field landing

Short-field landings (Fig. 5-17) should be practiced assuming a 50-foot obstacle exists on the approach end of the runway. The approach to a short-field landing should be made with power as needed and at a speed no slower than 1.3 V_{so} (1.3 times the power-off stall speed with your gear and flaps down). A competent pilot should be able to execute the approach as if it were "descending slow flight." The key to final approach is one of obtaining a clear mental and visual picture of a straight line from your position on final, over the obstruction to the point of intended touchdown. Once you have set up this mental picture, attain the desired airspeed and control the descent using coordinated power, flaps, and proper pitch attitude. If it works out right, your power should be slowly reduced until it reaches idle in the landing flare.

While on final approach, if you see that you are going to be low, use additional power to reestablish your glide path. If you find yourself a bit high, merely reduce your power a little in order to attain a higher rate of descent until you arrive back at your proper approach angle and proceed normally. If you are way off either way, too low or too high, you would probably be wise to go around and try it again.

The touchdown should come at minimum controllable airspeed, with power at idle, and with little or no float. Your flaps should be

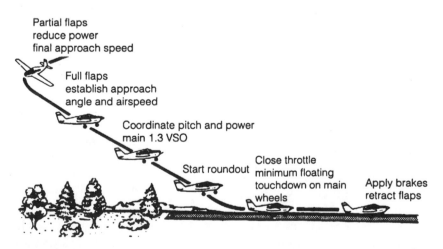

Fig. 5-17 *Short-field landing.*

retracted as soon as possible after touchdown to place all of the weight on the wheels and aid in braking.

Aside from improper pitch and power control, there are two very common and potentially dangerous errors in the execution of the short-field approach and landing. One is the tendency to lower the nose after clearing the obstruction, which results in an increase in airspeed and causes the aircraft to float, using more runway than you would have used if the attitude had been held constant all the way to the flare. As you might imagine, this could lead to an overshoot that could be as deadly as landing short.

The second error is reducing the power to idle as you cross over the obstruction. This procedure can be very dangerous because if you are approaching at a very low airspeed, the sudden loss of thrust can lead to an immediate stall. Keep your attitude constant down to the flare, and use your power as you need it. Do not reduce your power to idle until you are in the landing flare and are very close to the runway.

Strong or gusty winds can cause the short-field approach to become a little treacherous. It is wise to carry a little more power and airspeed in these conditions. A strong wind will slow the ground speed of your aircraft so a little more airspeed will bring about the desired results with an added amount of safety. Gusty winds are even more troublesome. Again, the power and airspeed should be a little higher than normal. On the approach, if the wind you are riding suddenly dies, the aircraft is likely to sink rapidly and leave you little or no time to recover. More than one pilot has been the victim of this bit of treachery. Remember, there is a huge difference between a short-field landing and landing short of the field.

If you have trouble with normal short-field approaches, return to approaches at 1.3 V_{so} and gradually work down to slower speeds. Don't try to force proficiency from yourself before you are ready to handle it. Short-field techniques require a high degree of reflex and feel for the aircraft. These are things that cannot be taught by the best instructor. They must be acquired by trial, practice, and time.

Go-around

One of the most important facts a truly competent pilot carries in mind is that not all approaches can be successfully completed. Perhaps you have misjudged the wind or are too high or too low to comfortably complete the approach. Or maybe someone has pulled out onto the runway in front of you. In any case, the prudent thing to do is to go around and try it again.

Some pilots feel a go-around is a sign of poor pilot technique or a lack of piloting skills or just plain embarrassing. Wrong. A go-around, done because it is needed, shows the pilot is thinking well and deserves an "A" for exercising good judgment.

Generally speaking, a go-around is merely the transition from approach configuration back to normal climb attitude. A little dose of common sense, along with some dual, should help you realize the ease and importance of this action. If you have misjudged your approach, the go-around can be done as you continue your track straight down the runway. If someone has pulled onto the runway in front of you, you should move off to the side in order to keep the other traffic in sight as you continue your go-around. Since you will probably be in the left seat, this would mean moving off to your right, just to the side of the runway, not into the next county.

Let's say you're on short final with full flaps and low power and you have to execute a go-around. Proceed with these steps:

- First, add full power as you begin the transition through level flight to a climb attitude.
- Second, take off the carb heat in order to develop full power.
- Third, bring the flaps up to the manufacturer's recommended go-around setting. This decreases your drag and allows you to accelerate better.

After completing all of these steps, make sure you are at least at V_x airspeed holding your own, or better still, climbing. When you are sure the airspeed, heading, and altitude are stable, retract the remainder of your flaps, allow the airspeed to accelerate to V_y, and continue a normal climbout.

In the event the situation requires a sidestep to avoid another aircraft, the go-around would be completed as shown above, except you would have to make a shallow turn. Unless the situation requires your immediate action, I believe it is better to get powered up and cleaned up before attempting the turn. Then, you will not get involved in a low-airspeed, low-power, flaps-down bank that can prove to be troublesome, even for a very experienced pilot.

There is one other problem that might necessitate a go-around—a bounced landing. If you haven't done any yet, you will. Most of the time, the bounce is slight and you can recover by adding a little power, lowering the nose to level flight attitude, and then flaring again—most of the time.

But if you really bounce it by dropping in from about 15 feet and spreading out your spring-steel gear to the point it groans, snaps back into position, and sends you about 30 feet back toward the sky, you just might want to consider a go-around. Let's face it; you're already up there, so you might as well go around. When this happens, stay calm and smoothly add full power. Lower your nose to level flight attitude and accelerate to climb airspeed as you slowly retract your flaps to the manufacturer's recommended go-around setting. You will probably experience a little sink as you bring the flaps up, but don't pull the nose up. Leave it at about level flight attitude so you can gain speed more quickly. Complete the go-around by heading straight down the runway, and climb out normally.

Ground effect

Figure 5-18 shows the changes in airflow that take place as a plane flies in close proximity to the runway. As the airflow moves past the trailing edge of the wing, a downward angle is imparted to it. When you are generating lift within close proximity to the ground, the flow of air at the trailing edge of the wing is affected and does not bend downward as sharply. The effect on the airflow becomes greatest when the plane is less than one-quarter of the wing span's distance above the surface. In fact, induced drag, or drag generated as a result of lift generation, is reduced by 47.6% when within one-tenth the distance of the wing span to the ground. This value is reduced to 23.5% at one-quarter span and drops to only about 1.4% at one

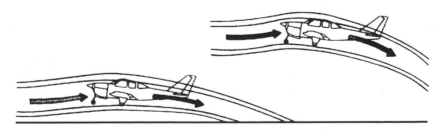

Fig. 5-18 *Ground effect.*

full span's distance. The vertical component of the airflow around the wing is affected by the runway, reducing the wing's downwash, upwash, and wingtip vortices. As a result there is a smaller rearward lift component and less induced drag generated by the wing.

Ground effect also generates an increase in the local air pressure at the static source, causing a lower-than-normal indicated airspeed and altitude. During takeoff this can make it seem as though the plane is able to fly at lower-than-normal airspeeds. As the plane leaves ground effect after takeoff, several things take place. Due to the increase in induced drag and the reduction of the coefficient of lift, the wing will require an increase in the angle of attack to generate the equivalent amount of lift. Additional thrust will also be needed as a result of the increase in induced drag. Finally, the change in air pressure around the static source will cause the airspeed indicator to register an increase in the indicated airspeed. There will also be a decrease in the aircraft's stability, along with a nose-up change in moment.

Night landings

At night you lose many of the visual references that are available during the day—ground reference points, trees, fences, and buildings that you might use to help judge your altitude above the ground. Instead, you must fly your approach using the movement of the runway lights to tell you whether you are above or below the glide path, on the centerline of the runway, and where you are touching down on the runway. However, with a little practice, night landings are just as easy as landing during the daytime.

Approach planning

At an airport that has much traffic in the pattern you should turn your landing light on as you get near the airport's vicinity to improve the chances of other traffic being able to see you. In lower traffic areas you may want to wait until you are on downwind before turning the landing light on to help conserve its life. As you come abeam the runway lights you are using for the touchdown point, bring your power back just as you would during a day approach, setting up the correct airspeed and descent values.

Once you have pulled the power back, trimmed the plane, and set the appropriate amount of flaps, you will continue on the downwind heading until you reach the key position we discussed previously. At that point begin your turn to base, letting other traffic know what you are doing over the radio.

As you fly the base leg, add flaps and maintain airspeeds just as you would during the day. You may find that night approaches make it more difficult to maintain your attitude visually from the horizon. Cross-check your external attitude reference with the instruments inside the plane. The artificial horizon, vertical speed indicator, and airspeed indicator become very important instruments during night approaches. Make sure you don't bury your head in the cockpit, however, and ignore looking out the windows of the plane. If the runway you are landing on has VASI (Visual Approach Slope Indicator), you can also use this as a reference for touching down in the correct spot on the runway. At this point we are on final approach, our airspeed is set, we've gone through the prelanding checklists, and it's time to think about when to begin the flare.

Height judgment/flare

During night landings you will have a limited field of view as you get close to the runway. Day landings allow you to see an almost unrestricted view of the runway and airport, but the runway illuminated by the landing light will be the only area you will have a clear view of. On very bright, moonlit nights you may gain some additional illumination, but this is often not the case. This means you must learn how to judge your height above the runway based on other references.

The first useful references are the runway lights themselves. Using peripheral vision, you may note the runway lights moving up on the side posts as you descend closer to the runway. As you descend low enough that the runway becomes illuminated by your landing lights, you have reached a point that you can use the runway itself as a visual reference.

Practicing night landings is the best way to become proficient at them and to develop a technique that works for you and the plane you fly. If you are at all rusty at night landings, or uncomfortable with them, find a qualified flight instructor to help you become safer and more proficient. With enough practice you will find that it is possible to accurately judge your height above the runway without any landing light at all.

6

Basic instrument flying

Flying by reference to instruments alone is very different from using them in conjunction with outside observations. While you should never go looking for instrument conditions, knowing basically how to use instruments could be very handy if the unexpected takes place. This chapter is by no means comprehensive, but offers only basic guidance for flight in instrument conditions.

Instruments

The basic flight instruments were discussed previously in regards to their simple functions. The following discussion illuminates the particulars of using these instruments for blind flight, especially as they relate to each other.

Airspeed indicator

In addition to airspeed, the indicator may be a useful cross-reference with the artificial horizon and vertical speed indicator to verify that they are also working correctly. If in a dive, the airspeed will increase. In a climb it will decrease as the nose rises. If you keep power settings constant, a drop in airspeed could mean the nose is rising. Cross-referencing this information with the artificial horizon, altimeter, and vertical speed indicator will help determine a faulty instrument in the system. If the other gauges give an indication that the plane is in level flight but the airspeed indicator is decreasing or increasing, it could be a sign of failure.

During a climb or descent, the attitude indicator can be used to establish pitch, but the airspeed indicator is used to refine the attitude until airspeeds match desired values, such as V_x or V_y in a climb. When pilots become disoriented under instrument conditions, it is not uncommon for the nose to drop and the airspeed to build rapidly. By monitoring the airspeed indicator, this problem can be avoided. The airspeed indicator rises to prominence in pilot attention while the airplane is climbing or on approach and landing phases of flight.

Attitude indicator

The attitude indicator (AI), sometimes called artificial horizon, is one of the most useful, and most abused, instruments in the panel. Both bank and pitch attitudes are available from the AI, making it the center of the pilot's attention—sort of replacing the view out the windshield. Unfortunately, most pilots have a tendency to focus almost exclusively on the AI to the point of excluding the other instruments. This tendency to "fixate" on a single instrument—even if it's a particularly valuable one, like the AI, should be avoided.

Altimeter

Next to the AI, the altimeter probably occupies the majority of a pilot's attention, especially while cruising straight and level. In theory, a pilot flying instruments at a safe altitude should handily avoid collision with anything on the ground. Couple that with air traffic controller demands, and pilots watch the altimeter almost religiously.

Turn and bank indicator

Figure 6-1 shows a turn and bank indicator, also known as a turn and slip indicator. You can see that it is made up of a vertical white bar, called a needle, and a black sphere in a glass tube, called the ball. If the ball is centered and the needle is located on the center mark, the plane is in a wings-level attitude. Better yet, if the wings are level, the plane is NOT turning. Although simple, this piece of information may be very useful in cross-checking the compass and directional dyro. But if the needle is to the left or right of the center marker, the plane is turning (i.e., changing heading). This can be a useful indicator of the plane's attitude and a good cross-check of the artificial horizon. If the needle is on the left or right "dog-house," (or the index mark corresponding to a bank on instruments which depict an airplane) and the ball is centered, the plane is making a turn in that direction at 3 degrees per second of heading change. When flying under instrument conditions, turns are normally made at the standard rate. The timing and layout for most instrument approaches are set up for standard rate turns.

Fig. 6-1 *Turn and bank indicator.*

Like the artificial horizon, the turn and bank indicator is driven by a gyroscope.

Directional gyro

The DG does not inherently know magnetic directions unless slaved to a magnetic compass. In simple installations, where the DG is un-slaved, pilots must set the initial heading on the DG to match the compass before flight, and regularly cross-check compass and DG indications. After the heading is set, the gyroscope in the DG attempts to maintain its alignment with that initial setting. When the plane turns, the card with the compass headings rotates within the DG, showing the airplane's current directional heading. The DG will maintain a steady indication of the plane's heading that is not subject to the bobbing and weaving of a magnetic compass.

Vertical speed indicator (VSI)

The VSI is especially useful to pilots for identifying vertical trends. Since the altimeter is graduated in large quantities, the smallest indication usually being 20 feet, holding altitude precisely is vastly simplified by a glance at the VSI, which may depict an up or down trend not yet visible on the altimeter.

Basic instrument scan

This is for emergency purposes only, and not intended to provide the reader with more than a rudimentary understanding of instrument flight. Proper planning and preflight checking can prevent the majority of accidental entry into IFR or minimal visibility flight conditions. But for those situations where you encounter weather that requires flight by reference to instruments, it is necessary that you understand their function and how to fly properly. Here are the basics of flying on instruments and the best methods for extricating yourself from the situation.

Upon encountering instrument weather conditions, continue to fly the airplane. Panicking and letting the airplane get into an uncontrolled attitude is one of the fastest ways to crash. Begin a regular scan of the instruments as soon as you encounter minimal visibilities and maintain the proper flight attitude. Then contact air traffic control (ATC). While this may seem like asking for trouble, ATC can help establish your position, direct you to an airport that may be in a more favorable weather situation, and keep you clear of other legal IFR traffic.

The primary reference instrument in IFR conditions will be the attitude indicator. Your scan will move to the directional gyro, the altimeter, and airspeed indicator, then back to the AI, essentially working a "T" pattern of instruments. The turn and bank indicator and the vertical speed indicator should also be checked periodically, but use of the other instruments will help you hold the desired attitude and heading. Avoid the tendency to focus on one instrument exclusively, normally the attitude indicator. Keep the scan smooth and constant. This will help you avoid blocking out important information regarding the airplane's flight status.

Once you have established contact with ATC and have the scan working and the plane in the desired flight attitude, relax as much as possible. Students learning to fly the plane on instruments often establish a "death grip" on the control yoke. This prevents them from feeling the airplane and making smooth inputs on the controls. When flying on instruments, make a conscious effort to use small, smooth control inputs when making changes in bank, pitch, or direction. Pilots often fall into the trap of using too much control input for a given change, then overcompensating once they realize they have gone too far. It is not at all uncommon for pilots to bank well past 30 degrees, then realize as the plane reaches 45 degrees of bank that they are losing altitude, the airspeed is increasing, and they have flown past the heading they wanted to establish. This gets them overcontrolling in an attempt to get back to where they want to be. Be patient. Make small adjustments to the attitude and wait for a response on the other instruments. If a bigger change is needed, make it in increments. Use the AI like the view out the windshield, and note that very small adjustments there go a long way toward speed, altitude, and heading changes. When desired flight parameters are reached, keep the attitude constant. If you keep the control inputs small and scan the gauges rhythmically, you will notice the results of the control inputs and be able to make any needed corrections quickly and efficiently.

Climbs by reference to instruments

ATC may have you initially climb after you contact them to reduce the chances of encountering hazards on the ground. If that is the case, use the same climb power settings and pitch attitudes that you would during a VFR climb. To begin the climb, add power to climb settings, then use the AI to establish a gradual nose-up pitch. For most airplanes, a 3-degree pitch-up angle from level on the artificial horizon would be a good place to start. Watch the airspeed as the nose pitches up, and if the airspeed drops below the correct value, slightly reduce the backpressure on the control yoke. Keep the inputs small to avoid overcontrolling the airplane. Once the airspeed is established at the correct climb value, hold a constant pitch attitude to maintain the airspeed. Monitor the altimeter and directional gyro to make sure you do not overshoot the desired altitude or stray from the heading ATC has assigned. Don't forget to keep the ball centered through proper rudder use during the climb. As you approach the assigned altitude, slowly lower the nose to a level attitude by using the artificial horizon. As the speed builds, reduce power to cruise settings and trim for level flight at that altitude. Keep scanning the gauges; don't focus on just the attitude indicator or the altimeter.

Straight-and-level flight and turns

To maintain straight-and-level flight, establish the correct power setting, then trim the plane to hold a constant altitude with neutral pressure on the elevator control. Proper use of trim can help reduce your workload and fatigue factor. Use the attitude indicator as the primary pitch and bank reference, but also scan the directional gyro for proper heading, and check the altimeter.

Use the turn and bank indicator to establish coordinated, standard-rate turns. Using standard-rate turns whenever you change headings will avoid steep banks and possible disorientation. It will also reduce the chances of turning past the desired heading. Remember to use a slight amount of backpressure on the control yoke to maintain constant altitude during turns. It is quite common for new instrument students to overbank the plane during turns and allow the nose to drop. Keep the control inputs small and smooth, and scan the gauges.

Descent by reference to instruments

Use the artificial horizon to initially pitch for the descent. If necessary, reduce the engine power setting to avoid gaining too much airspeed. Use the vertical speed indicator in conjunction with the airspeed indicator to establish a safe rate of descent. If you notice the airspeed is becoming too high, raise the nose of the plane and/or reduce power. Use approximately a 500-foot-per-minute descent rate and medium power settings to avoid excessive airspeeds.

Like the climb, a descent should be at a constant airspeed and rate of descent. Use a constant pitch attitude and power setting. Decide on the altitude to reach to before you start the descent, then begin raising the nose to a level attitude and increasing power as you approach the target altitude. If you are setting up for landing, it is a good idea to configure for landing before beginning the descent to reduce the workload on the way down. Continue to scan the airspeed indicator, artificial horizon, altimeter, and directional gyro during the descent. Again, it is common for pilots to focus on just one instrument during descents.

Instrument navigation

If you become immersed in instrument conditions, descending until you can see the ground and navigating via VFR procedures may not be an option. In severe weather cases, the clouds, rain, snow, or fog may completely obscure the surface and not allow you to descend. The same holds true for executing a 180-degree turn to get out of the weather, or climbing above it. In those instances ATC may direct you to navigate via instruments to an area of better visibility or to an airport. The controller will assign headings or "vectors" to better weather conditions. In some cases, the controller's directions will be in the simple form of "turn right, turn left, or stop turn." In any case, make small control inputs and do exactly as the controller directs.

7

Emergency procedures

Emergencies come in all shapes and sizes. They can range from fire to complete engine failure, loss of an aircraft component, radio failure, letting your maps blow out the window, or almost anything else you can dream up. It would be impossible to try to cover all the different possibilities here in this chapter. However, they all have one thing in common. They all need action. The right action. In most emergency situations, the pilot must be conditioned to respond quickly and correctly.

Some emergencies are much more serious than others. And what seems to be an emergency to one person might be only an annoyance to someone else. The defining element of each emergency is whether it seems to be an emergency to the pilot. So if you believe you have an emergency, then by golly it is! At least to you.

You are the key to resolving any problem. If you keep your head, keep flying the airplane, and work through the problem, you will improve your chances of a successful end to your flight.

It is much easier to stay out of emergencies than to get out of them once you get into a bad situation. The following information is only a guideline; it cannot be a "cookbook" recipe to everything that flying might throw at you.

Sometimes, as in an engine failure at high altitude, you have plenty of time to get organized and proceed accordingly. Other times you must make split-second decisions. Let's look at the problem that usually comes to mind first for most pilots—engine failure. This particular emergency can be very dangerous or no big deal, depending on where it happens and the pilot's readiness, training, and reaction to the engine failure. Whether the failure is total or only a partial power loss also makes quite a difference.

Generally speaking, there are three distinct actions you should complete, in order, if you suffer power loss:

- Set up a glide.
- Make a thorough cockpit check.
- If you cannot get a restart, turn into the wind (or crosswind) and land on the best available surface. Never land downwind if you can help it.

Glide

Let's go through the reasoning behind these three actions and determine why they should come one at a time, in this exact order. If you are anywhere other than on the ground (which is the best place to experience a power loss) what is the most important thing you have going for you? Altitude. Altitude gives you time. It gives you time to think, act, and call to ask for help or just to let someone know you are having problems. It gives you time to ready yourself and your aircraft for a possible emergency landing and time to pick the best possible spot to put down. There is even time to attempt a restart. Altitude buys you time, so in order to obtain the most time, instead of letting the aircraft descend rapidly in a cruise descent, which eats up huge chunks of altitude, set up a glide using the manufacturer's recommended best glide speed, save that altitude, and use it to your advantage. That's number one.

Cockpit check

Number two is a thorough cockpit check. Having set up a glide, and assuming you are at a reasonably high altitude, you should have time to investigate and maybe find and correct the cause of the power loss. Maybe you need carb heat to get rid of a small amount of carburetor ice, or maybe you need to switch to another fuel tank (one with some fuel in it).

You would be surprised at the number of very major accidents that have been caused by some very minor problems. Many of these problems could have been overcome with a proper cockpit check. When you perform your cockpit check, you should go to the most likely causes of power loss such as carb heat, fuel valve, mixture, mags, and the primer. However, don't go for them in a random pattern that might cause you to overlook something very important. Do it in a systematic fashion. If time permits, use your emergency checklist. Chances are that your mind will be in some degree of shock from the sudden loss of power, but with a written checklist you are less apt to forget something (Fig. 7-1).

Turn into the wind

Number three is to turn into the wind, or crosswind, and find the best place you can to put it down. This step should be taken if all else fails and you cannot get a restart. In this event, land as slowly as possible and with as much control as possible. The cardinal rule of aviation emergencies is, "If you know you are going to have to land, or crash, go in with as much control as possible." Then, you will have some control over your destiny.

Emergency landings

If you have the unpleasant experience of having to make an un-scheduled landing in an emergency, how do you look for the best

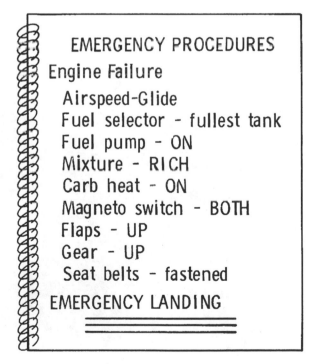

Fig. 7-1 *A sample emergency checklist.*

place to land? The first choice and the most obvious is an airport. Why put it in a field right beside a runway? (Don't laugh; it's been done.) The second choice is a road, vacant of traffic and void of power lines. It is your responsibility to take every option open to you to prevent any injury to innocent people on the ground. The third choice is a field, either hay or wheat, freshly cut or early in the season so it will be short, reducing the chances of a noseover after touchdown. All the rest follow in a rather random order, as none of them are very conducive to a successful emergency landing.

If you ever know that you are going to have to land in a field of any kind, you will be far better off to land with the rows rather than *across* them. This rule also applies to plowed ground. Anything that decreases your chances of remaining upright increases your chances of bodily injury. Other than innocent people on the ground, the most important thing for you to save is yourself. (Fig. 7-2).

Let's say we've checked all of the important steps (fuel, carb heat, mags, etc.) and the engine will not restart. Now we need to get serious about the emergency landing and step through that part of the emergency checklist. Following is an example of typical steps in an emergency-landing checklist:

- Airspeed: Set to best glide.
- Mixture: Idle cutoff.

Fig. 7-2 *The field may look good from the air, but sometimes things don't always go as planned.*

- Fuel selector: Off.
- Ignition switch: Off.
- Wing flaps: As required.
- Master switch: Off.
- Doors: Unlatched prior to touchdown.
- Touchdown: Soft-field attitude.
- Brakes: Apply as needed.

This checklist is not comprehensive and should not be used in place of the one your plane is equipped with. But it does show the steps that typically need to be taken as you make your approach to landing. The reason you shut down fuel and the master switch is to reduce the possibility for fire. Without fuel or an ignition source, it will be more difficult for a fire to accidentally start. Unlatching the door(s) prior to touchdown helps to keep them from becoming jammed in the closed position if the plane should flip over onto its back, or become bent if the plane strikes something after touchdown.

When landing in extremely rugged terrain, such as a forest or rock area, you will want to touch down at the absolute minimum airspeed to help reduce the chances of injury. In some cases it may be best to actually stall the airplane just above the trees or ground to let it drop just about vertically onto the landing area. This reduces the forward momentum of the plane and helps let the plane settle to the surface as gently as possible. This landing technique is one you probably will not get to practice but has saved the lives of pilots who have used it in real emergencies where the only landing area was on top of a forest of trees. The plane is normally damaged a great deal, but the occupants can often walk away uninjured. As you can see, emergency landing situations can be dealt with in a controlled manner that improves the chances you will land safely with little or no damage to yourself, your passengers, or the airplane. The key is to keep thinking and to not let the situation get away from you. With altitude on your side, you have time to set up the airplane and pick a favorable landing spot. Even landings in rough terrain can be dealt with if you prepare well.

Power failure

If you suffer a power failure during the takeoff roll, abort the takeoff. Reduce power (if you have any) to idle and stop. Don't try to struggle into the sky with an aircraft that is not developing full power. It's suicide.

If you suffer a power failure after liftoff, then there are several significant variables that must be considered. These items include the amount of runway remaining, altitude at the time of failure, and possible obstructions in the flight path. A general rule recommended by the FAA and taught by most flight instructors is that if you are not at least 500 feet above the surface, continue on straight, take whatever comes, and land as slowly as possible. Most pilots cannot safely perform the 180-degree plus turn and return to land safely below 500 feet. Many have tried; few have made it. Most stall during the turn or run out of airspeed, altitude, and ideas, all at the same time. The result? The probability is a much more severe crash than likely would have occurred had they continued straight ahead and landed the aircraft as slowly and with as much control as possible.

Zone of decision

If you are between 400 and 1000 feet above the ground, the area often called the *zone of decision,* you will still have to consider the factors of altitude, runway availability, obstructions, and safe landing areas. There are so many things you could do that making the correct decision is sometimes very difficult. Your first question is, where can you land with the best chance of doing so safely? And that choice is probably where you should go. Forget the what ifs and the maybes. Go to the spot you know offers the best chance for a safe landing.

In this zone of decision, you should have time to set up your glide, make a cockpit check, and providing you cannot obtain a restart, pick the most suitable place to put down. You have to practice your emergency techniques until they are automatic reflex actions. Emergency techniques should be almost mechanical. They have to be. Sometimes you just don't have time to get out the book.

When you are flying alone, taking off, landing, or whatever, look around and ask yourself, "What would I do if?" Try it. It's good fun and will make you aware of things you have never dreamed. Sometimes I find myself at home, sitting in my chair, thinking of a new situation, and asking myself, "What would I do if?" It's almost as good as getting some instruction.

If you experience a power failure at an altitude above 1000 feet, follow the three rules talked about previously. You should have time to utilize your emergency checklist and be certain you have checked every possible emergency procedure before making the decision to land. It is especially true if you happen to have the misfortune of being over some rather hostile terrain (Fig. 7-3). Above all, take your time and try not to panic. Panic has probably caused some routine emergencies to terminate with unnecessarily severe consequences.

Flight instructors know that a student who reacts in a poor or erratic manner when practicing simulated emergency landings will most likely have problems if the emergency ever becomes a reality. All of us have our heart rate quicken when confronted by a threat. Some people handle this stress much better than others. Actually, this quickening of the heart rate and the increased flow of adrenaline is our body's way of getting us geared up to meet the challenge. Some pilots react in a calm, efficient manner while others seem to use the extra blood and adrenaline in order to do *something,* whether right or wrong. They panic. All their emotions are so caught up in the instinct for survival that they forget their training, lose their thought process, and usually wind up acting incorrectly. The results are often fatal.

A pilot should be able to react within the scope of training and experience in a calm, efficient manner when faced with an emergency.

Fig. 7-3 *This Cessna 185 would have a very hard time finding a suitable landing area in this rugged terrain.*

This action is the mark of a professional. Once you give up being the pilot-in-command and become what amounts to a passenger, you are most certainly in deep, deep trouble.

There are many other types of aviation emergencies you might encounter. Most of you, however, will probably never come into contact with an emergency of any sort because flying is safe. And it's becoming safer every day. Pilots are being better trained and aircraft and their power plants are becoming more reliable. Although there are more people flying more aircraft now than ever before, the percentage of accidents versus hours flown is declining.

Accuracy landings

Although accuracy landings are fun to practice and should be on the private pilot checkride, they have a much more serious purpose. They are invaluable in an actual emergency situation.

The most common procedure for practicing accuracy landings is to pick a spot on the runway and then attempt to land on or within 200 feet beyond the spot. Any landing made short of this spot must be considered a failure. In an actual emergency, if you fail to make the field by even a few feet, the chances of your striking a fence or some other object are greatly increased.

To begin practice on accuracy landings, enter downwind at the normal pattern altitude and pick a spot on the runway that is fairly prominent. Select a spot that allows for some margin of error—not the very end of the runway, for example. When your aircraft reaches a point adjacent to the spot, reduce power to idle. The pattern should remain rectangular and as normal as possible. The use of flaps, slips, or slight S-turning is encouraged as long as you don't get too carried away and start using dangerous maneuvers to arrive at the spot.

It's best to turn onto final approach a little high, if possible, because you can always lose altitude by the methods just mentioned. Be careful that you don't stay so high that you have a problem with an overshoot. Remember, this is an emergency procedure. In real life you will probably get only one chance. On the other hand, if you are too low, there is no way on earth to get your aircraft onto the runway without power.

Always remember to aim a little short of the spot to allow for the float during the flare. When you turn onto final, watch the spot of intended landing. If the spot appears to move toward you, you are too high. If it appears to move away from you, you are too low. Of course, if you are way too low or high, you might not make the runway at all. It has to be within reason and it also depends on how far out you are from touchdown, airspeed, flap setting, etc. This method is as foolproof as any I have ever seen, and with a little practice you can land on a given spot with little difficulty.

Accuracy landings should be practiced with and without power, in full-flap and no-flap configurations. They should be practiced until you are confident that you can land on a given spot from any altitude and power setting. Then, continue to practice them often so you don't lose your touch. The knowledge that you can put your aircraft down on a given spot from any altitude at any time is one of great comfort and security.

Spiral

Spirals are useful to maintain orientation during prolonged descending turns. They help increase your ability to get your head out of the cockpit and still control the airspeed and bank. In the event of an actual emergency, the spiral is probably the best way to lose altitude and remain close to the point of intended landing. A spiral is also useful when coming down through a hole in the clouds. It prevents the possibility of illegal IFR flight.

Spirals are essentially a high-altitude emergency technique (Fig. 7-4). Whether spiraling about a point to an emergency landing or coming down through a broken cloud layer, the spiral usually indicates the need for some type of prompt action.

Spirals may be divided into two categories. The first is useful when coming down through a hole in the clouds or whenever a rapid descent is called for. Since this spiral does not include ground tracking, your airspeed and bank control are your main concerns.

To begin your practice for this particular spiral, attain a fairly high altitude, close the throttle, set up a normal glide, and perform a cockpit check. Although this might not be an emergency situation, it is very important to maintain these habits. After clearing the areas below, begin the spiral using about 50 degrees of bank. In this particular spiral, since ground track is not the major purpose, closely monitor the airspeed and bank control through the desired number of turns. As a

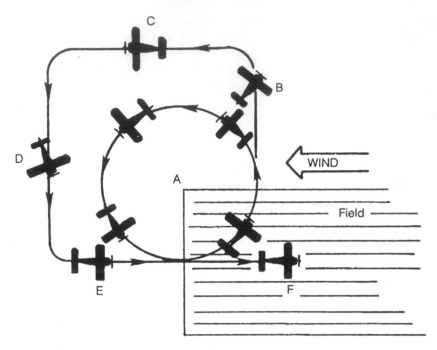

Fig. 7-4 *Emergency Spiral; A0 Spiral about the downwind corner of the downwind side of the field to stay close to your point of intended landing; B) Break out of the spiral several hundred feet above a normal pattern altitude; C) Turn downwind slightly higher than normal; D) Turn base slightly higher than normal; E) Turn to final slightly higher than normal. Use slips or flaps, as necessary, to land on intended spot; and F) Land short using correct technique (soft-field, rough field, etc.).*

good safety practice, you should not spiral down lower than 1500 feet above the ground. Continue your practice until you can perform at least three turns and keep your airspeed and bank constant.

The second type of spiral is more in the category of ground tracking. Incorporate the methods utilized in the previous spiral, but use a point on the ground about which you will spiral. It then becomes only a matter of doing a turn about a point in a gliding turn. Vary your bank as necessary to maintain a constant distance from the pivotal point. Upon reaching pattern altitude or slightly above, leave the spiral and enter a normal traffic pattern to the place of intended landing.

The use of the traffic pattern is wise because the more familiar the situation seems, the easier it will be for you to land on the intended spot. Most people have more problems landing out of a straight-in approach than from a normal traffic pattern. If you add a margin for human error due to the fact that an emergency situation might cause less than superior performance, you will see the need for establishing a normal traffic pattern. Set up the spiral about the downwind corner on the downwind side of the landing area you are trying to get into. It places you in the best possible position to execute a near normal traffic pattern.

Lost

We can end up lost for many reasons: Stronger winds than expected can blow us off course; radio navigation equipment can fail or give erroneous readings; we could have picked the wrong heading to fly through misreading the map as we planned our flight. Using an out-of-date sectional chart could have us dialed into a VOR using the wrong frequency; there are more reasons for getting lost than can be listed. Whatever the reason, though, we need to cross-check what we are doing often enough to assure that we do not have a single point of failure as we plan, then execute, our cross-country flights.

Use current charts and weather information as you plan your cross-country flights to help assure that your planning is accurate. Frequently verify the radio navigation equipment by cross-checking the information against what you see outside the window and on your sectional chart. By using several methods of figuring out your position, you can prevent having one failure cause a real emergency.

But what happens if you have done all of these things and you suddenly do end up lost? What can you do to help get yourself "found" again? First and foremost, keep flying the airplane. If you quit flying the airplane and let the heading or altitude get away from you, you have already lost the battle. If you are unsure of your position, you actually have several options. Many of us do not want to call and ask for help, egos being what they are, but that is a very easy way of getting out of the situation. There are very few areas in the country any more that are beyond radio communication. If you are equipped with a radio, you may ask for help to locate you, and the controllers will be more than happy to tell you where to go. If equipped with a VOR, you may cross-check your bearing from two or more VOR stations and triangulate your position. More and more aircraft are flying with GPS, which is an incredible piece of equipment. They often have a "nearest" function, which provides headings and distance to the nearest airport. If you have a properly functioning GPS on board, you should not get lost, but if you do it should be able to help you get to an airport with little difficulty.

The worst lost situations are when you are low on fuel, with poor weather conditions at night. Proper management of fuel and monitoring weather as your flight progresses can help you stay out of this particularly bad scenario. Who among us has not thought, "I can get there before things get too bad," trying to squeak into an airport before nightfall or weather makes the situation dangerous. "I want to sleep in my own bed tonight," has probably caused more cross-country related accidents than we realize. Don't set yourself up for low-fuel, poor-visibility, cross-country flights trying to stretch the range of the plane. Common sense can do much to keep your options open.

Weather-related emergencies

There are many steps you may take to improve the safety of the flight where weather is a factor. If you are in icing conditions, talk to air traffic and let them know what is going on, then ask for an altitude that will provide warmer temperatures. If you are in unexpected thunderstorms, let ATC know that. It may be impossible for you to hold altitude with updrafts and downdrafts that can be present around a thunderstorm cell. In these cases, fly the airplane at or below maneuvering speed, and maintain attitude. There is not much any plane can do to overcome the up and downdrafts, but if you keep the plane level and don't let the airspeed get away from you, it improves the chances that you will fly through the cell safely. It can be very rough around thunderstorms, and you want to make sure your seat belt is very tight around you. Even then you may find your head banging up against the cabin ceiling.

Being caught as a VFR pilot in IFR conditions can be very unnerving. As a VFR pilot you use the instruments, but you are very reliant on the ground for navigation and location fixes. Losing that reference forces you to fly strictly on instruments. If you find yourself in this situation, keep flying the airplane, maintain a level attitude, and contact ATC, then let them know what your situation is. They would much rather help you get to an airport safely than have you poking through clouds and not be communicating with them. Scan all of the instruments, using that information to fly the airplane at the correct heading and altitude assigned. Maintain a safe airspeed, altitude, and heading, and you've got the plane under control.

Weather is always a major factor in aviation. If pilots respect that and use the available resources to determine what the weather is expected to do, they will avoid most weather-related surprises. But meteorologists are not able to forecast weather with 100 percent accuracy. Conditions can change rapidly, and pilots should combine what they have been told with what they see. If you see towering cumulus clouds on a warm, summer afternoon, it may be a precursor to thunderstorms. If the temperature is dropping as you encounter precipitation, you may end up in icing conditions. The combinations are endless. Notice what is going on and recognize that weather can change rapidly and that you may need to adjust your flight plans accordingly. Getting to your destination a few hours later because you waited for weather, or diverted around it, is much better than the alternative of not making it at all.

Accident avoidance

A large portion of this discussion will involve what is known as the *poor judgment (PJ) chain.* No pilot intentionally sets himself or herself up for an accident or makes poor decisions with that intent in mind. But accidents do happen, and there are ways to minimize your chances of ending up in one. The PJ chain plays a role in these situations.

FAA publication FAA-P-8740-53, "Introduction to Pilot Judgment," discusses the poor judgment chain in great detail. Essentially the FAA document states that most accidents are the result of a series of events. The example used in this document is a pilot that is noninstrument rated, has a schedule restraint, and is running late. This pilot has limited adverse weather experience but decides to fly through an area of possible thunderstorms at dusk. Due to his lack of instrument experience, combined with the darkness, turbulence, and heavy clouds, he becomes disoriented and loses control of the airplane.

As you read that little example, you probably identified several poor decisions the pilot made that got him into trouble. Time was a big motivator for the pilot, and his decision to press on when the weather conditions were questionable. He was behind schedule and had the classic case of "get-home-itis" that afflicts every pilot from time to time. He also had minimal instrument experience and made a conscious decision to fly into a known thunderstorm area at dusk. The pilot had several opportunities to avoid the accident that resulted from the series of events and decisions that were made along the way.

First, the pilot could have decided to wait until conditions improved and the flight could have been made safely. This would have necessitated that the pilot overcome the pressure of meeting the schedule, but it is always better to arrive late than to never arrive. Second, the pilot could have altered the route of flight in order to fly around the area of thunderstorms. Like waiting, this would have resulted in arriving later than desired, but at least they would have arrived. Once the pilot encountered weather, he or she could have altered course to get out of that area, either turning back or in a direction away from the storm. Too often, though, pilots "lock on" to an idea and are hard pressed to consider other options as an alternative to the one they have decided on. Finally, once the pilot encountered inclement weather, he should have trusted the instruments and maintained control of the plane, as opposed to becoming disoriented and losing control of the aircraft ("Introduction to Pilot Judgment," p. 1).

Two major principles play a role in the poor judgment chain: Poor judgment increases the probability that another will follow, and judgments are based on information pilots have about themselves, the aircraft, and the environment. Pilots are less likely to make poor judgments if this information is accurate. Essentially, this means that one poor judgment increases the availability of false information, which might then negatively influence judgments that follow ("Introduction to Pilot Judgment," p. 3).

As a pilot continues further into a chain of bad judgments, alternatives for safe flight decrease. One bad decision might prevent other options that were available at that point from being open in the future. For example, if a pilot makes a poor judgment and flies into hazardous weather, the option to circumnavigate the weather is automatically lost. ("Introduction to Pilot Judgment," p. 3). By interrupting the poor judgment chain early in the decision-making process, the pilot has more options available for a safe flight. Through delaying making a good decision, the pilot may reach a point at which there are no good alternatives available. Get into the habit of making the best decision you can based on the information available, and do not be afraid to change your mind as additional information becomes known.

Three mental processes of safe flight

The same FAA document also covers three mental processes related to safe flight. These processes include automatic reaction, problem resolving, and repeated reviewing. Good pilots are actually performing many activities at the same time while they fly. Altitude, heading, and attitude are all constantly monitored and the airplane is adjusted to maintain the desired values. After a period of time, pilots no longer think consciously about what they need to do with their hands and feet to make the airplane do something; it just happens as they automatically make the controls move in the proper way to fly the plane the way they want. This is known as *automatic reaction*.

Problem solving is a three-step process that includes:

1. Uncover, define, and analyze the problem.
2. Consider the methods and outcomes of possible solutions.
3. Apply the selected solution to the best of your ability.

Through taking these steps, you will improve your ability to understand problems and resolve them. By correctly determining the actual cause of a problem, rather than misunderstanding it, you can aid in making better decisions to resolve it. The poor judgment chain can be avoided or broken through the use of good problem-solving skills.

The last mental process is repeated reviewing. This is the process of "continuously trying to find or anticipate situations that might require problem resolving or automatic reaction." Part of this skill includes using feedback related to poor decisions. As you fly, you need to constantly be aware of the factors that affect your flight, including yourself, the plane, weather, and anything else that could be a factor. Through remaining "situationally aware," you will be better informed and able to analyze the actual conditions you are flying in ("Introduction to Pilot Judgment," p. 5). An old aviation axiom sums up the foregoing: Superior pilots use superior judgment to avoid situations requiring superior skill.

Five hazardous attitudes

The last topics in the "Introduction to Pilot Judgment" we will cover are five attitudes that can be hazardous to a safe flight. These attitudes include:

1. Antiauthority: "Don't tell me!"

2. Impulsivity: "Do something—quickly."
3. Invulnerability: "It won't happen to me."
4. Macho: "I can do it."
5. Resignation: "What's the use?"

Each of these attitudes can get in the way of a pilot making a good decision. Overinflated ego, lack of confidence in their abilities, and the need to prove themselves are just a few personality traits that pilots can exhibit that can cloud their judgment. This poor judgment mindset is not because these pilots intend to make bad decisions, but because they are influenced by behavioral traits that negatively influence their ability to make a good decision. We all have these traits to one degree or another, but how much we let them influence us is a big factor in whether we can see through them to make a good decision. While you are monitoring the plane, weather, and other aspects of your flight, also monitor yourself. If you find that any of these traits are getting in the way of a good decision, it's time to take a step back and rethink the situation.

Weighing the risks

Take two pilots who decide to fly under a railroad bridge over a river. The first hops in the plane, heads out to the bridge, and flies underneath it. The second takes a boat out to the bridge, inspects it for wires that may run below, measures its height and width, and observes the terrain around the area. The plane is then measured, winds checked, and the pilot develops a method of practicing skills required to fly under the bridge. Finally, when all is safe, the pilot flies under the bridge. Both pilots executed the same maneuver, but the second pilot planned carefully, and waited until it was safe to fly. The first demonstrated relatively poor judgment in flying under the bridge without all the facts.

How many pilots don't have all the facts when they fly? Are they within weight and balance limits? How much fuel will they actually burn during the flight? What is the weather along the route of flight? The list of the factors that a pilot should understand is very large, but manageable. Whether the flight is a hop around the patch or a cross-country from New York to Los Angeles, you need to examine all the factors involved and weigh the risks.

Use your checklist. We all forget things, and using a checklist will help us avoid making embarrassing, if not serious, errors in operating the airplane. Every retractable-gear airplane has a checklist with some reference to the gear being down before landing, yet every year pilots manage to land with the gear up. We become comfortable with a plane that we fly on a regular basis, and it is not uncommon for anyone to become somewhat complacent.

You should also know the airplane you are flying. General Chuck Yeager once stated that knowing the systems of the test aircraft he flew gave him the ability to recover from situations other pilots may not have been able to. Learn about the fuel system, how the emergency gear extension works, and everything else you can read about. Poke around the airplane to verify how systems work. This knowledge could be very useful in an emergency situation.

Finally, know your limitations. Many accidents are caused because pilots push the airplane or themselves beyond what they are capable of. If you are not comfortable with your proficiency in a certain area, get instruction from a qualified flight instructor to gain experience and confidence. Do you feel good about your crosswind landing techniques, or does the plane tend to stray across the runway as you land? How about stalls? Most of us fly in a routine manner, to a small number of airports that we become accustomed to. Are you comfortable flying into a runway that has a real obstacle at the end of it or is truly a short runway? These are the things that cause accidents. Pilots get into situations that are not really dangerous but are beyond their level of experience.

Realize that when you are tired, sick, or under stress, your mind may not be clearly focused on flying, and you are more prone to making mistakes under these conditions. If you don't feel good about making a flight, for whatever reason, don't make it. Wait until the conditions become satisfactory before you fly.

8

Transitioning to multiengine airplanes

Multiengine aircraft can be deceiving. To the uninitiated, the multiengine aircraft appears twice as safe as its single-engine counterparts. Some say that with two engines a pilot has twice the insurance, twice the speed, and twice the redundancy that a single-engine pilot has. They also have twice the chance of having an engine fail.

And herein lies the tale. The multiengine aircraft is a thing of wonder, an aircraft that usually carries more weight, flies faster and farther, and offers the redundant safety of a spare engine. But all of these good points go out the window very quickly when an engine is lost on a multiengine aircraft. A multiengine aircraft operating on one engine can turn into a beast of a different color. And it can be a killer. In the hands of the unskilled or the careless, the multiengine aircraft operating on one engine is tantamount to playing Russian roulette with about five bullets in the chamber.

It's not that everyone who flies a multiengine aircraft is playing with fire. Quite the contrary. Treated with respect and not pushed beyond the laws of physics, a multiengine aircraft can be a very safe mode of transportation.

The increasing complexity of multiengine airplanes dictates the importance of a thorough checkout for pilots who change from one make or model airplane to another with which they are not familiar. The similarity of the operating controls in most airplanes leads many persons to believe that full pilot competency can be carried from one type of airplane to another, regardless of its weight, speed, performance characteristics, and limitations, or how many engines they possess (see Fig. 8-1). The importance of acquiring a thorough knowledge of an unfamiliar airplane and the inefficiency

Fig. 8-1 *Moving up to a twin-engine aircraft like this Cessna Corsair only requires motivation, time, and money.*

of trial-and-error methods of learning to fly that airplane have been well established. So the pilots desiring to add a multiengine rating to their certificates will need some good dual instruction prior to applying for the multiengine checkride.

In order to really learn the operating characteristics of the new aircraft, do not limit familiarization flights to the mere practice of normal takeoffs and landings. It is extremely important to learn the limitations, and become thoroughly familiar with the stall performance, minimum controllability characteristics, maximum performance techniques, and all pertinent emergency procedures, as well as all normal operating procedures.

The pilot transitioning to multiengine should study and understand the airplane's flight and operations manual. A thorough understanding of the fuel system, electrical and/or hydraulic system, empty and maximum allowable weights, loading schedule, normal and emergency landing gear and flap operations, and preflight inspection procedures, is essential.

The transition from training type single-engine airplanes to larger and faster multiengine airplanes may be the pilot's first experience in airplanes equipped with a constant-speed propeller, a retractable landing gear, and wing flaps. And all airplanes having a constant-speed propeller require that the pilot have a thorough understanding of the need for proper combinations of manifold pressure (MP) and propeller revolutions per minute (RPM), which are prescribed in the airplane manufacturer's manuals.

The instructor should include in the checkout of any aircraft at least a demonstration of takeoffs and landings and in-flight maneuvers with the airplane fully loaded. Most four-place and larger airplanes handle quite differently when loaded to near-maximum gross weight, as compared with operations when lightly loaded. Weight and balance should also be figured for various loading conditions.

It is very important that the transitioning pilot readily accept the flight instructor's evaluation of performance during the checkout process. It is inadvisable to consider oneself qualified to accept responsibility for the airplane before the checkout is completed; half a checkout may prove more dangerous than none at all.

Normal multiengine procedures

The increased complexity of multiengine airplanes requires a very systematic approach to preflight inspection prior to entering the cockpit, and a dedicated use of checklists for all operations.

Preflight visual inspections of the exterior of the airplane should be conducted in accordance with the manufacturer's operating manual. The procedures set up in the flight manuals usually provide for a comprehensive inspection, item by item, as with all aircraft, irrespective of their complexity. The transitioning pilot should have a thorough briefing and understanding of each item in the preflight inspection, as it may be quite different from the pilot's former aircraft.

Checklists

As with any aircraft, the multiengine aircraft comes replete with checklists. And they can be relatively complex and lengthy. The multiengine checklist is divided into common areas of operation, as are all airplanes, but will have a few areas, such as continued flight after an engine failure, that single-engine airplanes do not have.

Multiengine aircraft also characteristically have more controls, switches, dials, and instruments than single-engine airplanes. Some of these items carry a serious penalty if not properly positioned before flight. This teaches the would-be multiengine pilots to carefully consider the results of their actions, or inactions. For instance, the multiengine pilot has two of everything, including fuel selector knobs, mixture controls, etc., which give the pilot twice the opportunity to miss an important item from the checklist.

If you have the luxury of having a front-seat passenger who can read without missing a line, you might consider having them read you the checklist. But then you are placing the direct responsibility for the safe conduct of your flight in the hands of a layman. Although it might be them who misses an important item, everyone on board might pay the penalty.

The pilot of the multiengine aircraft should develop the habit of actually touching the control or device and repeating the instrument reading or proscribed control position in question, under the careful observation of the pilot calling out the items on the checklist. Even when no copilot is present, the pilot should form the habit of touching, pointing to, or operating each item as it is read from the checklist.

In the event of an in-flight emergency, the pilot should be sufficiently familiar with emergency procedures to take immediate action instinctively, when it is required, to prevent more serious situations. However, as soon as circumstances permit, the emergency checklist should be reviewed to ensure that all required items have been checked.

Taxi procedures

Although ground operation of multiengine airplanes may differ in some respects from the operation of single-engine airplanes, the taxiing procedures also vary somewhat between those airplanes with a nosewheel and those with a tailwheel-type landing gear. With either of these landing-gear arrangements, the difference in taxiing multiengine airplanes that is most obvious to a transitioning pilot is the capability of using asymmetrical power between individual engines to assist in directional control (see Fig. 8-2).

Tailwheel-type multiengine airplanes are often equipped with tailwheel locks that can be used to advantage for taxiing in a straight line, especially in a crosswind. The tendency to weathervane can also be neutralized to a great extent in these airplanes by using more power on the upwind engine, with the tailwheel lock engaged and the brakes used as necessary.

Fig 8-2 *Asymmetrical, or unequal power is often used on multiengine aircraft to aid in taxi procedures.*

Braking

On nosewheel-type multiengine airplanes, the brakes and throttles are used mainly to control the momentum, and steering is done principally with the steerable nosewheel. The steerable nosewheel is usually actuated by the rudder pedals, or in some larger airplanes by a separate hand-operated steering mechanism.

Brakes may be used, as with any airplane, to slow down, stop, or turn tighter than normal while taxiing. When initiating a turn, though, they should be used cautiously to prevent overcontrolling of the turn. No airplane should be pivoted on one wheel when making sharp turns, as this can damage the landing gear, tires, or the airport pavement. All turns should be made with the inside wheel rolling, even if only slightly.

Brakes should be used as lightly as practicable while taxiing to prevent undue wear and heating of the brake discs. When brakes are used repeatedly or constantly, they tend to heat to the point that they may either lock or fail completely. Also, tires may be weakened or blown out by extremely hot brakes. And any abrupt use of brakes in multiengine as well as single-engine airplanes is evidence of poor pilot technique. The FAA frowns on this during checkrides.

Due to the greater weight of multiengine airplanes, effective braking is particularly essential. Therefore, as the airplane begins to move forward when taxiing is started, the brakes should be tested immediately by depressing each brake pedal. If you find the brakes are weak, it may be wise to have them checked.

Looking outside the cockpit while taxiing becomes even more important in multiengine airplanes. Since these airplanes are usually somewhat heavier, larger, and more powerful than single-engine airplanes, they often require more time and distance to accelerate or stop, and provide a different perspective for the pilot. While it usually is not necessary to make S-turns to observe the taxiing path, additional vigilance is necessary to avoid obstacles, other aircraft, or bystanders.

Trim tabs

The trim tabs in a multiengine airplane serve the same purpose as in a single-engine airplane, but their function is usually more important to safe and efficient flight. In addition to the elevator trim tab, multiengine aircraft always have a very effective rudder trim tab to help alleviate the asymmetrical thrust produced when being flown with only one engine running. This is because of the greater control forces required to handle the asymmetrical thrust with one engine inoperative.

In some multiengine airplanes it taxes the pilot's strength to overpower an improperly set trim tab on takeoff, go-around, or with one engine inoperative. Many fatal accidents have occurred when pilots took off or attempted a go-around with the airplane trimmed for something other than normal flight. Therefore, prompt retrimming of the elevator and rudder trim tabs in the event of an emergency go-around from a landing approach, especially from a single-engine approach, is essential to the success of the flight.

Multiengine airplanes, like all airplanes, should be retrimmed in flight for each change of attitude, airspeed, power setting, or single-engine configuration. Without changes in trim, the airspeed will suffer and the flight profile will often be somewhat like a roller coaster.

Performance characteristics

Normal takeoff

There is virtually no difference between a takeoff in a multiengine airplane and one in a single-engine airplane. The controls of each class of airplane are operated the same; the multiple throttles of the multiengine airplane normally are treated as one compact power control and can usually be operated simultaneously with one hand.

It is very important that the pilot have a plan of action to cope with engine failure during takeoff. It is equally important that just prior to takeoff the pilot mentally review takeoff procedures, especially procedures pertaining to losing an engine. This mental review should consist of the engine-out minimum control speed (V_{mc}), the best all-engine rate of climb speed (V_y), the best single-engine rate of climb speed (V_{yse}), and what procedures will be followed if an engine fails prior to reaching minimum control speed. This first speed (V_{mc}) is the minimum airspeed at which safe directional control can be maintained with one engine inoperative and one engine operating at full power.

The multiengine pilot's primary concern on all takeoffs is the attainment of at least the engine-out minimum control speed prior to liftoff. Until this speed is achieved, directional control of the airplane in flight will be impossible after the failure of an engine, unless power is reduced immediately on the operating engine. If an engine fails before the engine-out minimum control speed is attained, the pilot has no choice but to close both throttles, abandon the takeoff, and direct complete attention to bringing the airplane to a safe stop on the ground.

The multiengine pilot's second concern on takeoff is the attainment in the least amount of time of the single-engine best rate-of-climb speed (V_{yse}). This is the airspeed that will provide the greatest rate of climb when operating with one engine out and feathered (if possible), or the slowest rate of descent. In the event of an engine failure, the single-engine best rate-of-climb speed must be held until a safe maneuvering altitude is reached, or until a landing approach is initiated. The V_{yse} speed is frequently shown on the airspeed indicator as a blue line across the indicator.

The engine-out minimum control speed (V_{mc}) and the single-engine best rate-of-climb speed (V_{yse}) are published in the airplane's FAA-approved flight manual. These speeds must be addressed mentally by the pilot before every takeoff (see Fig. 8-3).

Crosswind takeoffs

Crosswind takeoffs are performed in multiengine airplanes in basically the same manner as those in single-engine airplanes. At the beginning of the takeoff roll, less power can be used on the

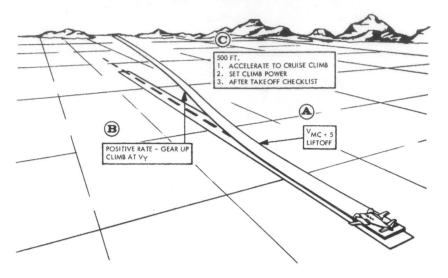

Fig. 8-3 *A multiengine takeoff profile requires reaching $V_{mc} + 5$ knots prior to liftoff.*

downwind engine to overcome the tendency of the airplane to weathervane into the wind, and then full power applied to both engines as the airplane accelerates to a speed where better rudder control is attained.

Stalls

As with a single-engine airplane, the pilot should be familiar with the stall and minimum controllability characteristics of the multiengine airplane being flown. The larger and heavier airplanes have slower responses in stalls and recoveries and in maneuvering at critically slow speeds due to their greater weight. The practice of stalls in multiengine airplanes, therefore, should be performed at altitudes sufficiently high to allow recoveries to be completed at least 3000 feet above the ground.

It usually is inadvisable to execute full stalls in multiengine airplanes because of their relatively high wing loading. Therefore, practice should be limited to approaches to stalls (imminent), with recoveries initiated at the first physical indication of the stall. As a general rule, however, full stalls in multiengine airplanes are not necessarily violent or hazardous.

The pilot should become familiar with imminent stalls entered with various flap settings, power settings, and landing gear positions. It should be noted that the extension of the landing gear will cause little difference in the stalling speed, but it will cause a more rapid loss of speed in a stall approach.

Power-on stalls should be entered with both engines set at approximately 65 percent power. Takeoff power may be used provided the entry speed is not greater than the normal lift-off speed. Stalls in airplanes with relatively low power loading using maximum climb power usually result in an excessive nose-high attitude and make the recovery more difficult. Additionally, some t-tail type multiengine aircraft tend to blank out their tails if the nose attitude is too high.

Because of possible loss of control, stalls with one engine inoperative or at idle power and the other developing effective power are not to be performed during multiengine flight tests nor should they be practiced by applicants for multiengine ratings. This is a dangerous practice and should be avoided at all times.

The same techniques used in recognition and avoidance of stalls of single-engine airplanes apply to stalls in multiengine airplanes. The pilot must be familiar with the characteristics that announce an approaching or imminent stall, the signals that the aircraft sends, and the proper technique for a positive recovery.

As with all aircraft, the increase in pitch attitude for stall entries should be gradual to prevent the airplane from climbing at an abnormally high nose-up attitude at the time the stall occurs. It is recommended that the rate of pitch change results in a one-knot-per-second decrease in airspeed. In all stall recoveries, the controls should be used very smoothly, avoiding abrupt pitch changes.

Slow flight

Smooth control manipulation is important in all flight at minimum or critically slow airspeeds. As with all piloting operations, a smooth technique permits the development of a more sensitive feel of the controls with a keener sense of stall anticipation. Flight at minimum or critically slow airspeeds gives the pilot an understanding of the relationship between the decreasing control effectiveness and airspeed. The slower the aircraft travels through the air, the less effective will be the controls.

Generally, the technique of flight at minimum airspeeds is the same in a multiengine airplane as it is in a single-engine airplane. Because of the additional equipment in the multiengine airplane, the transitioning pilot has more to do and observe, and the usually slower control reaction requires better anticipation. Care must be taken to observe engine temperature indications for possible overheating, and to make necessary power adjustments smoothly on both engines at the same time.

Approaches and landings

Multiengine aircraft characteristically have steeper gliding angles than single-engine aircraft because of their relatively high wing loading and greater drag with wing flaps and landing gear when extended. For this reason, power is used throughout the approach to shallow the approach angle and prevent a high sink rate.

The accepted technique for making a stabilized landing approach is to reduce the power to a predetermined setting during the arrival descent so the appropriate landing gear extension speed (V_{lo}) will be attained in level flight as the downwind leg of the approach pattern is entered. With this power setting, the extension of the landing gear (when the airplane is on the downwind leg opposite the intended point of touchdown) will further reduce the airspeed to the desired traffic pattern airspeed. When within the maximum speed for flap extension (V_{fe}), the flaps may be partially lowered, if desired, to aid in reducing the airspeed to traffic pattern speed.

The prelanding checklist should be completed by the time the airplane is on base leg so that the pilot may direct full attention to the approach and landing. In a powered approach, the airplane should descend at a stabilized rate, allowing the pilot to plan and control the approach path to the point of touchdown. Further extension of the flaps and slight adjustment of power and pitch should be accomplished as necessary to establish and maintain a stabilized approach path. The stabilized, powered approach should allow for the altitude to be controlled with power adjustments and the airspeed to be controlled with small pitch inputs.

The airspeed of the final approach should provide the engine-out best rate-of-climb speed (V_{yse}) until the landing is assured, because that is the minimum speed at which a single-engine go-around can be made if necessary. In no case should the approach speed be less than the critical engine-out minimum control airspeed (V_{mc}). If an engine should fail suddenly and it is necessary to make a go-around from a final approach at less than that speed, a catastrophic loss of control would occur. As a rule of thumb, after the wing flaps are extended the final approach speed should be gradually reduced to V_{mc}+5 knots or the airspeed recommended by the manufacturer.

The roundout or flare should be started at sufficient altitude to allow a smooth transition from the approach to the landing attitude. The touchdown should be smooth, with the airplane touching down on the main wheels and the airplane in a tail-low attitude as the power is reduced to idle. Directional control on the rollout should be accomplished primarily with the rudder and the steerable nosewheel, with discrete use of the brakes applied only as necessary for crosswinds or other factors.

Crosswind landings

The crosswind landing technique in multiengine airplanes is little different from that required in single-engine airplanes. The only significant difference lies in the fact that because of the greater weight, more positive drift correction must be maintained before the touchdown.

It should be remembered that the FAA requires that most airplanes have satisfactory control capabilities when landing in a direct crosswind of at least 20 percent of the landing configuration stall speed ($0.2\ V_{so}$). Thus, an airplane with a power-off stalling speed of 60 knots has been designed to handle at least a direct crosswind of 12 knots ($.2 \times 60$) on landings. Though skillful pilots may successfully land in much stronger crosswinds, poor pilot technique may cause serious damage in less wind.

The two basic methods of making crosswind landings, the slipping approach (wing-low), and the crabbing approach may be combined. The essential factor in all crosswind-landing procedures is touching down without drift, with the heading of the airplane parallel to its direction of motion. This will result in minimum side loads on the landing gear.

Go-around procedure

The complexity of modern multiengine airplanes makes a knowledge of and proficiency in emergency go-around procedures particularly essential for safe piloting. The emergency go-around during a landing approach is inherently critical because it is usually initiated at a very low altitude and airspeed with the airplane configured for landing.

Unless absolutely necessary, the decision to go around should not be delayed to the point where the airplane is ready to touch down. The more altitude and time available to apply power, establish a climb, retrim, and set up a go-around configuration, the easier and safer the maneuver becomes. When the pilot has decided to go around, immediate action should be taken without hesitation, while maintaining positive control and accurately following the manufacturer's recommended procedures.

Go-around procedures can vary with different airplanes, depending on their weight, flight characteristics, flap and retractable gear systems, and flight performance. And there are several go-around procedures that can apply to most multiengine airplanes. Although there may be slight differences, most twins follow the same basic procedures. When the decision to go around is reached, takeoff power should be applied immediately and the descent stopped by adjusting the pitch attitude to avoid further loss of altitude.

The flaps should be retracted only in accordance with the procedure prescribed in the airplane's operating manual. Usually this will require the flaps to be raised to near takeoff position.

After a positive rate of climb is established, the landing gear should be retracted, best single-engine rate-of-climb airspeed obtained and maintained, and the airplane trimmed for this climb. The procedure for a normal takeoff climb should then be followed.

The basic requirements of a successful go-around, then, are to power up, pitch up to arrest the descent, then attain and maintain the best rate-of-climb airspeed, clean the aircraft up, and proceed.

At critically slow airspeeds, retracting the flaps prematurely can cause an unanticipated loss of altitude. Rapid or premature retraction of the flaps should be avoided on go-arounds, especially when close to the ground, because of the careful attention and exercise of precise pilot technique necessary to prevent a sudden loss of altitude. It generally will be found that retracting the flaps only halfway or to the specified approach setting decreases the drag a relatively greater amount than it decreases the lift.

The FAA approved *Airplane Flight Manual* or *Pilot's Operating Handbook* should be consulted regarding landing gear and flap retraction procedures because in some installations simultaneous re-traction of the gear and flaps may increase the flap retraction time, and full flaps create more drag than the extended landing gear.

Multiengine engine-out procedures

From the would-be multiengine pilot's point of view, the primary difference between a light-twin and a single-engine airplane is the potential problem involving an engine failure. The information that follows is designed to highlight that one potentially deadly issue of what happens when one engine fails.

V speeds

Before operating techniques in light twin-engine airplanes can be fully discussed, the subject of "V" speeds and their relevance to mul-tiengine flight must be addressed. "V" speeds such as V_{mc}, V_{xse}, and V_{yse} are the main performance speeds the light-twin pilot needs to know in addition to the other performance speeds common to both twin-engine and single-engine airplanes. The airspeed indicator in twin-engine airplanes is usually marked (in addition to other nor-mally marked speeds) with a red radial line at the minimum con-trollable airspeed with the critical engine inoperative (V_{mc}), and a blue radial line at (V_{yse}), the best single-engine rate-of-climb airspeed with one engine inoperative (see Fig. 8-4).

V_x is the speed for best angle of climb. At this speed the airplane will gain the greatest amount of altitude for a given distance of forward travel. This speed is used for obstacle clearance with all engines op-erating. However, this speed is very different when one engine is in-operative, and is referred to as V_{xse}, or best angle of climb airspeed (single-engine).

V_y is the speed for the best rate of climb. This speed will provide the maximum altitude gain for a given period of time with all engines operating. However, this speed, too, will be different when one en-gine is inoperative and is referred to as V_{yse}, or best rate of climb (single-engine).

V_{mc}

V_{mc} is the minimum control speed with the critical engine inopera-tive. The term V_{mc} can be defined as the minimum airspeed at which the airplane is controllable when the critical engine is made sud-denly inoperative, and the remaining engine is producing takeoff

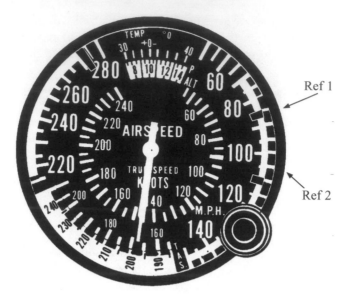

Fig. 8-4 *A typical twin-engine airspeed indicator contains two extra colored radials; the one at V$_{mc}$ is red and the one at V$_{yse}$ is blue.*

power. The Federal Aviation Regulations under which the airplane was certificated, stipulate that at V$_{mc}$ the aircraft must be able to:

- Stop the turn, which results when the critical engine is suddenly made inoperative within 20 degrees of the original heading, using maximum rudder deflection and a maximum of 5-degrees bank into the operative engine.

- After recovery, maintain straight flight with not more than a 5-degree bank (wing lowered toward the operating engine).

This does not mean that the airplane must be able to climb or even hold altitude. It only means that a heading can be maintained. The principle of V$_{mc}$ is not at all mysterious. It is simply that at any airspeed less than V$_{mc}$, air flowing along the rudder is such that application of rudder forces cannot overcome the asymmetrical yawing forces caused by takeoff power on one engine and a powerless windmilling propeller on the other.

Many pilots erroneously believe that because a light-twin has two engines, it will continue to perform at least half as well with only one of those engines operating. Wrong! There is nothing in FAR, Part 23, governing the certification of light-twins, which requires an airplane to maintain altitude while in the takeoff configuration and with one engine inoperative. In fact, many of the current light-twins are not required to do this with one engine inoperative in any configuration, even at sea level.

Single-engine performance

When one engine fails on a light-twin, performance is not only halved, it is actually reduced by 80 percent or more. The performance loss is greater than 50 percent because an airplane's climb performance is a function of the thrust horsepower that is in excess of that required for level flight. When power is increased in both engines in level flight and the airspeed is held constant, the airplane will start climbing. The rate of climb will depend on the power added (which is power in excess of that required for straight-and-level flight). When one engine fails, however, it not only loses power, but the drag increases considerably because of asymmetric thrust and the operating engine must then carry the full burden alone. To do this, it must produce 75 percent or more of its rated power. This leaves very little excess power for climb performance, especially in a hot, humid weather condition or at high altitude.

As an example, an airplane that has an all-engine rate of climb of 1860 FPM and a single-engine rate of climb of 190 FPM would lose almost 90 percent of its climb performance when one engine fails. Quite a difference!

Nonetheless, the light-twin does offer obvious safety advantages over the single-engine airplane (especially in the en-route phase) but only if the pilot fully understands the real options offered by that second engine in the takeoff and approach phase of flight.

Engine-out emergencies

In general, the operating and flight characteristics of modern light-twins with one engine inoperative are excellent. These airplanes can be controlled and maneuvered safely as long as sufficient airspeed is maintained. Note the key word is airspeed. Airspeed is life to a multiengine pilot operating on only one engine. However, to utilize the safety and performance characteristics effectively, the pilot must have a sound understanding of single-engine performance and limitations resulting from unequal thrust.

A pilot upgrading to a multiengine airplane should practice and become thoroughly familiar with the control and performance problems that result from the failure of one engine during any flight condition. This is really the major issue of multiengine flight. Proficiency in all the control operations and precautions must be demonstrated on multiengine rating flight tests.

The feathering of a propeller should be demonstrated and practiced in all airplanes equipped with propellers that can be feathered and unfeathered safely in flight. If the airplane used is not equipped with feathering propellers, or is equipped with propellers that cannot be feathered and unfeathered safely in flight, one engine should be secured (shut down) in accordance with the procedures in the FAA approved *Airplane Flight Manual* or the *Pilot's Operating Handbook*.

Propeller feathering

When an engine fails in flight, the movement of the airplane through the air tends to keep the propeller rotating, much like a windmill. Since the failed engine is no longer delivering power to the propeller to produce thrust, but instead, is absorbing energy to overcome friction and compression of the engine, the drag of the windmilling propeller is significant and causes the airplane to yaw toward the failed engine. Most multiengine airplanes are equipped with "full-feathering propellers" to minimize that yawing tendency.

The blades of a feathering propeller may be positioned by the pilot to such a high angle that they are streamlined in the direction of flight. In this feathered position, the blades act as powerful brakes to assist engine friction and compression in stopping the windmilling rotation of the propeller. This is of particular advantage since a stopped propeller creates the least possible drag on the airplane and reduces the yawing tendency. As a result, multiengine airplanes are easier to control in flight when the propeller of an inoperative engine is feathered.

Practice feathering of propellers should be performed only under such conditions and at such altitudes and locations that a safe landing on an established airport could be accomplished readily in the event of difficulty in unfeathering the propeller. Most instructors never feather a propeller below 3000 feet above the ground or away from an airport.

Engine-out practice

The following procedures are useful to teach proper technique for the multiengine pilot to cope with an inoperative engine. At a safe altitude (minimum 3000 feet above terrain) and within gliding distance of a suitable airport, an engine may be shut down with the mixture control or fuel selector. At lower altitudes, however, engine shutdown should be simulated by reducing power by means of the throttle to the zero thrust setting. The following procedures should then be practiced.

- Set the mixture and propeller controls full forward. Then both throttles should be increased to full throttle for maximum power to maintain at least V_{mc}.
- Retract wing flaps and landing gear to minimize drag.
- Determine which engine is failed, and then verify it by closing the throttle to the dead engine.
- Bank about 5 degrees into the operative engine.
- Determine the cause of failure, or feather the inoperative engine.
- Turn toward the nearest airport.
- Secure (shut down) the inoperative engine in accordance with the manufacturer's checklist and check for engine fire.
- Monitor the engine instruments on the operating engine, and adjust power, cowl flaps, and airspeed as necessary.
- Maintain altitude and an airspeed of at least V_{yse} if possible.

The pilot must be proficient in the control of heading, airspeed, and altitude, in the prompt identification of a power failure, and in accuracy of shutdown and restart procedures as prescribed in the FAA-approved *Airplane Flight Manual*.

There is no better way to develop skill in single-engine emergencies than by continued practice. The techniques and procedures of single engine operation do not necessarily remain at a consistently high level. Like any other skill, lack of review and practice tends to erode fundamental procedures, and skills may be lost. And some engine-out emergencies can be so critical that there may be no safety margin for lack of skill or knowledge. Unfortunately, many light-twin pilots never practice single-engine operation after receiving their multiengine rating.

The pilot should practice and demonstrate the effects of engine-out performance at various configurations of gear, flaps, and both; the use of carburetor heat; and the failure to feather the propeller on an inoperative engine. Each configuration should be maintained at best engine-out rate-of-climb speed long enough to determine its effect on the climb (or sink) achieved.

The critical engine

"P-factor" is present in multiengine airplanes just as it is in single-engine airplanes. P-factor is caused by the unequal thrust of rotating propeller blades when the airplane is in a nose-high, power-on configuration. It is the result of the descending blade having a greater angle of attack than the ascending blade when the relative wind striking the blades is not aligned with the thrust line (as in a nose-high attitude).

In many U.S. designed light-twins, both engines rotate to the right (clockwise) when viewed from the cockpit, and both engines develop an equal amount of thrust. At high angles of attack and high-power conditions, the downward moving propeller blade of both engines develop more thrust than the upward moving blade. This asymmetric propeller thrust or "P-factor," results in a center of thrust positioned to the right side on both engines (see Fig. 8-5). The turning (or yawing) force of the right engine is greater than the left engine because the center of thrust is much farther away from the center line of the fuselage, giving it a longer arm.

Thus, when the right engine is running and the left engine is inoperative, the turning (or yawing) force is greater than in the opposite situation of an operating left engine and an inoperative right engine. In other words, directional control may be difficult when the left engine (the critical engine) is suddenly made inoperative.

It should be noted that many newer twin-engine airplanes are equipped with engines turning in opposite directions; that is, the left engine and propeller turn clockwise and the right engine and propeller turn counterclockwise. With this arrangement, the thrust line of either engine is the same distance from the center line of the fuselage, so there will be no difference in yaw effect between loss of left or right engine. This effectively eliminates this type of airplane having a critical engine, and it makes them much safer when an engine is lost.

V_{mc} demonstrations

Multiengine instruction must include a demonstration of the airplane's engine-out minimum control speed (V_{mc}). The engine-out

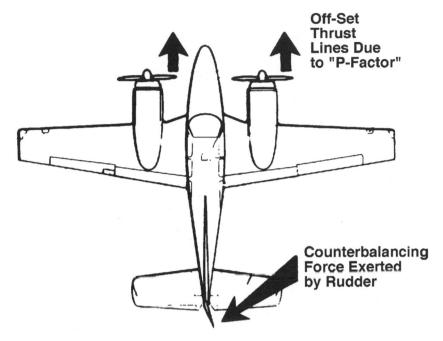

Fig. 8-5 *On twins whose engines both turn in the same direction, the forces of p-factor produce a more pronounced left-turning tendency since the force is double that of a single-engine aircraft.*

minimum control speed given in the Airplane Flight Manual is determined during original airplane certification under conditions specified by the FAA. These conditions normally are not duplicated during pilot training because they consist of the most adverse situations for airplane type certification purposes.

Basically, when one engine fails the pilot must overcome the asymmetrical thrust created by the operating engine by setting up a counteracting moment with the rudder. When the rudder is fully deflected, its yawing power will depend on the velocity of airflow across the rudder, which in turn is dependent on airspeed. As the airplane decelerates it will reach a speed below which the rudder command will no longer balance the unequal thrust and directional control will be lost. This is V_{mc}.

V_{mc} is shown on the airspeed indicator as a red radial line. The indicated V_{mc} speed is as high as it will ever get. It can be lower, but never higher. V_{mc} is determined with the aircraft in a configuration which will create the worst possible conditions relative to the sudden loss of one engine. These conditions are:

- Takeoff power on both engines.
- Most rearward center of gravity.
- Flaps in the takeoff setting.
- Landing gear retracted.
- Propeller windmilling.

The V_{mc} demonstrations should be performed at an altitude of at least 3000 feet above the surface, and preferably near an airport. One demonstration should be made while holding the wings level and the ball centered, and another demonstration should be made while banking the airplane at least 5 degrees toward the operating engine to establish "zero sideslip." These maneuvers will demonstrate the engine-out minimum control speed for the existing conditions and will emphasize the necessity of banking into the operative engine. No attempt should be made to duplicate V_{mc} as determined for airplane certification.

After the propellers are set to high RPM, the landing gear is retracted, and the flaps are in the takeoff position, the airplane should be placed in a climb attitude and airspeed at V_{yse}. With both engines developing takeoff power, power on the critical engine (usually the left) should then be reduced to idle. After this is accomplished, the airspeed should be reduced about one knot per second with the elevators until directional control no longer can be maintained.

At this point, recovery should be initiated by simultaneously reducing power on the operating engine and reducing the angle of attack by lowering the nose. Allow the airspeed to build back just above V_{mc} and then increase the power on the good engine as the aircraft is banked about five degrees toward the good engine. Slowly raise the nose to level flight attitude, maintain V_{yse}, and set the rudder trim tab to aid in directional control.

Should indications of a stall occur prior to reaching this point, recovery should be initiated immediately by reducing the angle of attack. In this case, a minimum engine-out control speed demonstration is not possible under existing conditions.

If it is found that the minimum engine-out control speed is reached before indications of a stall are encountered, the pilot should demonstrate the ability to control the airplane and initiate a safe climb in the event of a power failure at the published engine-out minimum control speed.

Lateral lift

During engine-out flight the large rudder deflection required to counteract the asymmetrical thrust also results in a "lateral lift" force on the vertical fin. This lateral "lift" represents an unbalanced side force on the airplane that must be countered either by allowing the airplane to accelerate sideways until the lateral drag caused by the sideslip equals the rudder "lift" force or by banking into the operative engine and using a component of the airplane weight to counteract the rudder-induced side force.

In the first case, the wings will be level, the ball in the turn-and-slip indicator will be centered, and the airplane will be in a moderate sideslip toward the inoperative engine. In the second case, the wings will be banked 3–5 degrees into the good engine, the ball will be deflected one diameter toward the operative engine, and the airplane will be at zero sideslip.

The sideslipping method has several major disadvantages:

- The relative wind blowing on the inoperative engine side of the vertical fin tends to increase the asymmetric moment caused by the failure of one engine.
- The resulting sideslip severely degrades stall characteristics.
- The greater rudder deflection required to balance the extra moment and the sideslip drag cause a significant reduction in climb and/or acceleration capability.

Flight tests have shown that holding the ball of the turn-and-slip indicator in the center while maintaining heading with wings level drastically increases V_{mc} as much as 20 knots in some airplanes. (Remember, the value of V_{mc} given in the FAA-approved flight manual for the airplane is based on a maximum 5-degrees bank into the operative engine.) Banking into the operative engine reduces V_{mc}, whereas decreasing the bank angle away from the operative engine increases V_{mc} at the rate of approximately 3 knots per degree of bank angle.

Flight tests have also shown that the high drag caused by the wings-level, ball-centered configuration can reduce single-engine climb performance by as much as 250 FPM, which may be more than is available at sea level in a nonturbocharged light twin.

Banking at least 5 degrees into the good engine ensures that the airplane will be controllable at any speed above the certificated V_{mc}, that the airplane will be in a minimum drag configuration for best climb performance, and that the stall characteristics will not be degraded. Engine-out flight with the ball centered is never correct.

For an airplane with nonsupercharged engines, V_{mc} decreases as altitude is increased. Consequently, directional control can be maintained at a lower airspeed than at sea level. The reason for this is that since power decreases with altitude, the thrust moment of the operating engine becomes less, thereby lessening the need for the rudder's yawing force. Since V_{mc} is a function of power (which decreases with altitude), it is possible for the airplane to reach a stall speed prior to loss of directional control.

Stall prior to V_{mc}

It must be understood that there is a certain density altitude above which the stalling speed is higher than the engine-out minimum control speed. When this density altitude exists close to the ground because of high elevations or temperatures, an effective flight demonstration of V_{mc} is impossible and should not be attempted. When a flight demonstration is impossible, the instructor should emphasize orally the significance of the engine-out minimum control speed, including the results of attempting flight below this speed with one engine inoperative, the recognition of the imminent loss of control, and the recovery techniques involved.

Relationship of V_{mc} to CG

V_{mc} is greater when the center of gravity is at the rearmost allowable position. Since the airplane rotates around its center of gravity, the moments are measured using that point as a reference. A rearward CG would not affect the thrust moment, but would shorten the arm to the center of the rudder's horizontal "lift," which would mean that a higher force (airspeed) would be required to counteract the engine-out yaw.

Generally, the center of gravity range of most light twins is short enough so that the effect on the V_{mc} is relatively small, but it is a factor that should be considered. Many pilots would only consider the rear CG of their light-twin as a factor for pitch stability, not realizing that it could greatly affect the controllability with one engine out.

There are many light-twin pilots who think that the only control problem experienced in flight below V_{mc} is a yaw toward the inoperative engine. Unfortunately, this is not the whole story.

With full power applied to the operative engine, as the airspeed drops below V_{mc}, the airplane tends to roll as well as yaw into the inoperative engine. This tendency becomes greater as the airspeed is further reduced. Since this tendency must be counteracted by aileron control, the yaw condition is aggravated by adverse yaw (the "down" aileron creates more drag than the "up" aileron). If a stall should occur in this condition, a violent roll into the dead engine may be experienced. Such an event occurring close to the ground would be disastrous. This can be avoided by keeping the airspeed above V_{mc} at all times during single-engine operation. If the airspeed should fall below V_{mc} for whatever reason, power must be reduced on the operative engine, and the nose must be lowered to regain airspeed above V_{mc} (see Fig. 8-6). At that point, power may be reapplied to the operating engine and the airplane can be carefully controlled.

Zone of decision

The most critical time for an engine-out condition in a twin-engine airplane is during the few seconds immediately following

Fig. 8-6 *The rudder trim saves a lot of legwork when flying with one engine out.*

liftoff while the airplane is accelerating to a safe engine-failure speed.

This "zone of decision" is bounded by the point at which V_y is reached and the point where the obstruction altitude is reached. An engine failure in this area demands an immediate decision to abort the flight, land and stop, or to continue the flight. Any indecision in this zone will probably end tragically. Beyond this decision area, the pilot has but one choice, to continue with the takeoff, maneuver carefully to whatever altitude can be attained, and work carefully through the process of returning to the airport. Here is where patience is a true virtue. Don't attempt to force performance from the aircraft or get in a hurry to get back. If the aircraft is flying and has no obstacles in front of it, altitude can usually be coaxed from the aircraft.

Although most twin-engine airplanes are controllable at a speed close to the engine-out minimum control speed, the performance is so far below optimum that continued flight following takeoff may be marginal or impossible. A more suitable recommended speed, termed by some aircraft manufacturers as minimum safe single-engine speed (V_{sse}), is that speed at which altitude can be maintained while the landing gear is being retracted and the propeller is being feathered.

Upon engine failure after reaching the safe single-engine speed on takeoff, the twin-engine pilot (having lost one-half of the normal power) usually has a significant advantage over the pilot of a single-engine airplane. This is because, if the aircraft has single-engine climb capability at the existing gross weight and density altitude, there may be the choice of stopping or continuing the takeoff. This compares with the only choice facing a single-engine airplane pilot who suddenly has lost half of the normal takeoff power. They have to stop!

Factors in a go no-go decision

If one engine fails prior to reaching V_{mc}, the decision has already been made to close both throttles and bring the airplane to a stop on the runway. If engine failure occurs after becoming airborne, the pilot must decide immediately whether to land or to continue the flight.

If the decision is made to continue the flight, the airplane must be able to at least hold its altitude with one engine inoperative. This requires acceleration to V_{yse} if no obstacles are involved, or to V_{xse} if obstacles are a factor.

To make a correct decision in an emergency of this type, the pilot should have considered the runway length, field elevation, density altitude, obstruction height, headwind component, and the airplane's gross weight prior to takeoff (Fig. 8-7). This would not be an enviable position to be in without a prior plan.

Accelerate/stop distance

The "accelerate-stop distance" is the total distance required to accelerate the twin-engine airplane to lift-off speed and, assuming failure of an engine at the instant that speed is attained, bring the airplane to a stop on the remaining runway.

The "accelerate-go distance" is the total distance required to accelerate the airplane to a specified speed and, assuming failure of an engine at the instant that speed is attained, continue takeoff on the remaining engine to a height of 50 feet.

For example, assume that with a temperature of 80 degrees F, a calm wind at a pressure altitude of 2000 feet, a gross weight of 4800 pounds, and all engines operating, the airplane being flown requires 3525 feet to accelerate to 105 MPH and then be brought to a stop. Let's also assume that the airplane under the same conditions

Fig. 8-7 *A thorough preflight will do more to prevent a multiengine flight from becoming a single-engine flight than almost anything else.*

requires a distance of 3830 feet to take off and climb over a 50-foot obstacle when one engine fails at 105 MPH.

With such a slight margin of safety (305 feet), it would be better to discontinue the takeoff and stop if the runway is of adequate length, since any slight mismanagement of the engine-out procedure would more than outweigh the small advantage offered by continuing the takeoff. At higher field elevations the advantage becomes less and less until at very high density altitudes a successful continuation of the takeoff is extremely improbable.

Other factors in takeoff planning

Proficient pilots of light-twins plan the takeoff in sufficient detail to be able to take immediate action if one engine fails during the take-off process. They are thoroughly familiar with the airplane's performance capabilities and limitations, including accelerate-stop distance, as well as the distance available for takeoff, and will include such factors in their plan of action. For example, if it has been determined that the airplane cannot maintain altitude with one engine operative (considering the gross weight and density altitude), the seasoned pilot will be well aware that should an engine fail right after liftoff, an immediate landing will have to be made in the most suitable area available. The sane pilot will not attempt to maintain altitude at the expense of a safe airspeed. This amounts to suicide.

Consideration will also be given to surrounding terrain, obstructions, and nearby landing areas so that a definite direction of flight can be established immediately if an engine fails at a critical point during the climb after takeoff. It is imperative, then, that the takeoff and climb path be planned so that all obstacles between the point of takeoff and the available areas of landing can be cleared if one engine suddenly becomes inoperative.

Airspeed control

The twin-engine airplane must be flown at precise airspeeds if maximum takeoff performance and safety are to be obtained. For example, the airplane must lift off at its specific lift-off airspeed, accelerate to V_y airspeed, and climb with maximum permissible power on both engines to a safe single-engine maneuvering altitude. Prior to that, if an engine fails, a different airspeed must be attained immediately, usually V_{yse}. This airspeed must be held precisely because only at this airspeed will the pilot be able to obtain maximum performance from the airplane. To understand the factors involved in proper takeoff planning, a further explanation of this critical speed follows, beginning with the liftoff.

I have said several times the light-twin can be controlled satisfactorily while firmly on the ground when one engine fails prior to reaching V_{mc} during the takeoff roll. This is possible by closing both throttles, by proper use of rudder, brakes, and nosewheel steering. If the airplane should be airborne at less than V_{mc}, however, and suddenly loses all power on one engine, it cannot be controlled satisfactorily. Thus, on normal takeoffs, liftoff should never take place until the airspeed reaches and exceeds V_{mc}. The FAA recommends a minimum speed of V_{mc} plus 5 knots before liftoff.

An efficient climb procedure is one in which the airplane leaves the ground at V_{mc} +5 knots, accelerates quickly to V_{yse} (best rate-of-climb speed, single-engine), and then accelerates to V_y. The climb at V_y should be made with both engines set to maximum takeoff power until reaching a safe single-engine maneuvering altitude, a minimum of 500 feet above field elevation. At this point, power may be reduced to the allowable maximum continuous power setting (METO-maximum except takeoff) or less, and any desired en-route climb speed then may be established.

Airspeed versus altitude

In the event of engine failure, a pilot who uses excessive speed on takeoff will discover suddenly that all the energy produced by the engines has been converted into speed. Although airspeed is important, excessive airspeed is not nearly so important as is altitude, especially below 500 feet, or so.

Improperly trained pilots often believe that the excess speed can always be converted to altitude, but this theory is not correct. Available power is always wasted in accelerating the airplane to an unnecessary speed. Also, experience has shown that an unexpected engine failure so surprises the unseasoned pilot that proper reactions are often extremely slow in coming. By the time the initial shock wears off and the pilot is ready to take control of the situation, the excess speed has dissipated and the airplane is still barely off the ground. From this low altitude, the pilot still has to climb, with an engine inoperative, to whatever height is needed to clear all obstacles and get back to the approach end of the runway. The prospect for a success-ful conclusion from this mismanagement of airspeed is dim.

The laws of physics dictate that the airplane will expend less energy to fly in level flight than it will to climb. Therefore, if the total energy of both engines is initially converted to enough height above the ground to permit clearance of all obstacles (safe maneuvering alti-tude), the problem is much simpler in the event an engine fails. If some extra altitude is available, it can always be traded for airspeed or gliding distance when needed.

There is a fine line between too much or too little airspeed or alti-tude. Usually both are the pilot's friends, and one usually comple-ments the other. But, in the time just after liftoff, until the aircraft reaches about 500 feet, or so, altitude is more important than air-speed. However, the pilot cannot place so much emphasis on alti-tude that airspeed strays far from the desired amount. There must be a balance between enough altitude and enough airspeed that is brought about by paying strict attention to all revealing information.

Consequently, during takeoff and early climb, the pilot should al-ways be ready for any eventuality by keeping one hand on the con-trol wheel and the other hand on the throttle. The airplane must remain on the ground until V_{mc}+5 knots is reached so that a smooth transition to the proper climb speed can be made. THE AIRPLANE SHOULD NEVER LEAVE THE GROUND BEFORE V_{mc} IS REACHED.

If an engine fails before leaving the ground, it is advisable to dis-continue the takeoff and stop. If an engine fails after liftoff, the pilot will have to decide immediately whether to continue flight, or to close both throttles and land. However, waiting until the engine fail-ure occurs is not the time for the pilot to plan the correct action. The action must be planned before the airplane is taxied onto the run-way. The plan of action must consider the density altitude, length of the runway, weight of the airplane, and the airplane's accelerate-stop distance, and accelerate-go distance under these conditions. Only on the basis of these factors can the pilot decide intelligently what course to follow if an engine should fail.

To reach a safe single-engine maneuvering altitude as safely and quickly as possible, the climb with all engines operating must be made at the proper airspeed, usually V_y. That speed will provide for:

- Good control for the airplane in case an engine fails.

- Quick and easy transition to the single-engine best rate-of-climb speed if one engine fails.
- A fast rate of climb to attain an altitude that permits adequate time for analyzing the situation and making decisions.

To make a quick and easy transition to the single-engine best rate-of-climb speed, in case an engine fails, the pilot should climb at some speed greater than V_{yse}, probably V_y. If an engine fails at less than V_{yse}, it would be necessary for the pilot to lower the nose to increase the speed to V_{yse} in order to obtain the best climb performance. If the climb airspeed is considerably less than this speed, it might be necessary to lose valuable altitude to increase the speed to V_{yse}. Another factor to consider is the loss of airspeed that may occur because of erratic pilot technique after a sudden, unexpected power loss. Consequently, the normal initial two-engine climb speed should not be less than V_y.

Normal takeoff procedures

After run-up and pretakeoff checks have been completed, the airplane should be taxied into takeoff position and aligned with the runway. If the crew consists of two pilots, the pilot in command should brief the other pilot on takeoff procedures prior to receiving clearance for takeoff. This briefing consists of at least the following:

- Minimum controllable airspeed (V_{mc}).
- Rotation speed (V_r).
- Lift-off speed (V_{lof}).
- Single-engine best rate-of-climb speed (V_{yse}).
- All-engine best rate-of-climb speed (V_y).
- What procedures will be followed if an engine failure occurs prior to V_{mc}?

Both throttles then should be advanced simultaneously to takeoff power, and directional control maintained by the use of the steerable nosewheel and the rudder. Brakes should be used for directional control only during the initial portion of the takeoff roll when the rudder and steerable nosewheel are ineffective.

As the takeoff progresses, flight controls are used as necessary to compensate for wind conditions. Liftoff should be made at no less than V_{mc} +5 knots. After liftoff, the airplane should be allowed to accelerate to the single-engine best rate-of-climb speed, V_{yse}, and then accelerated to the all-engine best rate-of-climb speed, V_y, and then to climb maintaining this speed with takeoff power until a safe maneuvering altitude is attained.

The landing gear may be raised as soon as practicable, but not before reaching the point from which a safe landing can no longer be made on the remaining runway. The flaps should be retracted as directed in the airplane's operating manual. Upon reaching a safe maneuvering altitude, the airplane should be allowed to accelerate to cruise climb speed before power is reduced for normal climb power, cowl flaps are adjusted, and trims tabs are reset.

Short-field or obstacle takeoff

When it is necessary to take off over an obstacle or from a critically short field, the procedures will be altered slightly from a normal takeoff. For example, the initial climb speed that will provide the best angle of climb for obstacle clearance is V_x rather than V_y. Additionally, every inch of runway should be utilized because runway behind you is never useful.

Generally, use the aircraft manufacturer's recommended airspeed, flap, and power settings. However, if the published best angle-of-climb speed (V_x) is less than V_{mc} +5, then the prudent pilot uses no less than V_{mc} +5 knots for liftoff and initial climb.

During the takeoff roll as the airspeed reaches the best angle-of-climb speed, or V_{mc} +5, whichever is higher, the airplane should be firmly rotated to establish an angle of attack that will cause the airplane to lift off and climb at that specified speed. At an altitude of approximately 50 feet, or after clearing the obstacle, the pitch attitude can be lowered gradually to allow the airspeed to increase to the all-engine best rate-of-climb speed. Upon reaching safe maneuvering altitude, the airplane should be allowed to accelerate to normal, or en-route climb speed and the power reduced to the normal climb power settings, cowl flaps adjusted, and aircraft retrimmed.

Engine failure during takeoff roll

If an engine should fail during the takeoff roll before becoming airborne, the pilot should close both throttles immediately and bring the airplane to a stop. The same procedure is recommended if, after becoming airborne, an engine should fail prior to having reached the single-engine best rate-of-climb speed (V_{yse}). An immediate landing is inevitably safer because the altitude loss required to increase the speed to V_{yse} usually will preclude obstacle avoidance.

The pilot must have determined before takeoff what altitude, airspeed, and airplane configuration must exist to permit the flight to continue in event of an engine failure (see Fig. 8-8). The pilot also should be ready to accept the fact that if engine failure occurs before these required factors are established, both throttles must be closed and the situation treated the same as engine failure on a single-engine airplane. That is, the pilot will have to land straight ahead and as slowly as possible to enhance the chances of surviving the forced landing. If it has been predetermined that the engine-out rate of climb under existing circumstances will be at least 50 feet per minute at 1000 feet above the airport, and that at least the engine-out best angle-of-climb speed has been attained, the pilot may decide to continue the takeoff.

If the airspeed is below the engine-out best angle-of-climb speed (V_{xse}) and the landing gear has not been retracted, the takeoff should be abandoned immediately.

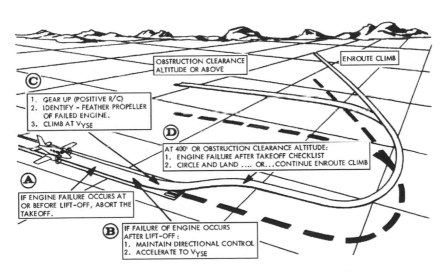

Fig. 8-8 *If an engine fails prior to attaining V_{mc}, abort the takeoff (A). An engine failure after reaching V_{mc} requires preplanning, proficiency, and patience.*

Engine failure during climbout

If the engine-out best angle-of-climb speed (V_{xse}) has been obtained and the landing gear is in the retract cycle, the pilot should climb at the engine-out best angle-of-climb speed (V_{xse}) to clear any obstructions, and thereafter stabilize the airspeed at the engine-out best rate-of-climb speed (V_{yse}) while retracting the landing gear and flaps and maintaining aircraft control. But at all costs, fly the airplane. Don't trade the job as pilot for one as passenger during an emergency; it is usually fatal.

If the decision is made to continue flight, the single-engine best rate-of-climb speed should be attained and maintained. Even if altitude cannot be maintained, it is best to continue to hold that speed because it would result in the slowest rate of descent and provide the most time for executing the emergency landing. After the decision is made to continue flight and a positive rate of climb is attained, the landing gear should be retracted as soon as practical.

If the airplane is barely maintaining altitude and airspeed, it is wise to avoid any attempt to turn the aircraft. When a turn is made under these conditions, both lift and airspeed decrease and an altitude loss is inevitable. Consequently, continue straight ahead whenever possible until reaching a safe maneuvering altitude. Pilots have died for lack of patience in this most gut-wrenching of times.

When an engine fails after becoming airborne, the pilot should hold heading with rudder and simultaneously roll into a bank of at least 5 degrees toward the operating engine. In this attitude the airplane will tend to turn toward the operating engine, but at the same time, the asymmetrical power resulting from the engine failure will tend to turn the airplane toward the "dead" engine. The result is a partial balance of those tendencies and provides for an increase in airplane performance as well as easier directional control.

The best way to identify the inoperative engine is to note the direction of yaw and the rudder pressure required to maintain heading. To counteract the asymmetrical thrust, extra rudder pressure will have to be exerted on the operating engine side. To aid in identifying the failed engine, some pilots use the expression "idle foot-idle engine," meaning the engine paired to the pilot's foot that is not pressing the rudder is the dead engine. Never rely on tachometer or manifold pressure readings to determine which engine has failed. After power has been lost on an engine, the manifold pressure will indicate the approximate atmospheric pressure, which often looks remarkably the same as climb power. Don't be fooled by this.

Experience has shown that the biggest problem is not in identifying the inoperative engine, but rather in the pilot's actions after the inoperative engine has been identified. In other words, a pilot may identify the "dead" engine and then attempt to shut down the good one. This leads to a rather quiet airplane and a very rapid pulse.

To avoid this mistake, the pilot should verify that the dead engine has been identified by s-l-o-w-l-y retarding the throttle of the suspected engine before shutting it down and feathering the propeller. Must I explain why slowly is the proper speed for this action? Think about it.

When demonstrating or practicing procedures for engine failure on takeoff, the feathering of the propeller and securing of the engine should be simulated rather than actually performed, so that the engine may be available for immediate use if needed.

In all cases, the airplane manufacturer's recommended procedure for single-engine operation should be followed. The general procedure listed below is not intended to replace or conflict with any procedure established by the manufacturer of any airplane. It can be used effectively for general training purposes and to emphasize the importance of V_{yse}. It should be noted that this procedure is concerned with an engine failure on a takeoff where obstacle clearances are not critical. If the decision is made to continue flight after an engine failure during the takeoff climb, the pilot should maintain directional control at all times, maintain V_{yse}, and:

- Check that all mixture controls, prop controls, and throttles (in that order) are at maximum permissible power settings.
- Check that the flaps and landing gear have been retracted.
- Decide which engine is inoperative (dead).
- Raise the wing on the suspected "dead" engine at least 5 degrees.
- Verify the "dead" engine by retarding the throttle of the suspected engine. (If there is no change in rudder forces, then that is the inoperative engine.)
- Feather the prop on the "dead" engine (verified by the retarded throttle).
- Declare an emergency if operating from a tower-controlled airport. Advise the tower of your intentions.

Engine failure en route

Normally, when an engine failure occurs while en route at cruising flight altitude, the situation is not as critical as when an engine fails on takeoff. Having plenty of altitude and airspeed, the pilot can take time to determine the cause of the failure and remedy the condition, if possible. If the condition cannot be corrected, the single-engine procedure recommended by the manufacturer should be accomplished and a landing made as soon as practical (see Fig. 8-9).

A primary error during any engine failure is the pilot's tendency to perform the engine-out identification and shutdown too quickly, resulting in improper identification or incorrect shutdown procedures. The sudden realization that this is not a simulated failure, often associated with an actual failed engine, often results in confused and hasty reactions.

When an engine fails during cruising flight, the pilot's first problem is to maintain a rational thought pattern, then to manage remaining altitude and airspeed to be able to continue flight to the point of intended landing. How far they can fly is dependent on the density altitude, gross weight, and obstructions. When the airplane is above its single-engine service ceiling, altitude will be lost, at least down to the single-engine service ceiling. Remember, the single-engine ser-

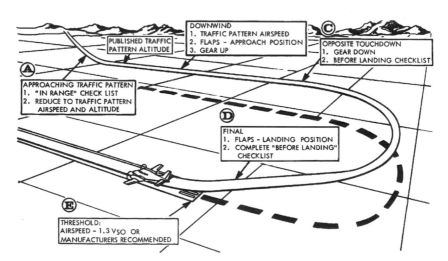

Fig. 8-9 *About the only difference between a single-engine approach to landing and a multiengine approach to landing is a slightly higher approach, and waiting to lower flaps until the airport is assured.*

vice ceiling is the maximum density altitude at which the single-engine best rate-of-climb speed will produce 50 FPM rate of climb. This ceiling is determined by the manufacturer on the basis of the airplane's maximum gross weight, flaps, and landing gear retracted, critical engine inoperative, and with the propeller feathered.

Although engine failure while en route in normal cruise conditions may not be critical, it will darn sure get the pilot's attention. It is a good practice to add maximum permissible power to the operating engine before securing or shutting down the failed engine. If it is determined later that maximum permissible power on the operating engine is not needed to maintain altitude, then reduce the power and save the engine.

However, with reduced power on the good engine, if the airspeed decreases too much, this could present a serious performance problem, especially if the airspeed should drop below V_{yse}. If this occurs, add whatever power is needed to the operating engine. In short, do not hesitate to use the power if you need it.

Altitude should be maintained if it is within the capability of the airplane. In an airplane not capable of maintaining altitude with an engine inoperative under existing circumstances, the airspeed should be maintained within 65 knots of the engine-out best rate-of-climb speed (V_{yse}) so as to conserve altitude as long as possible to reach a suitable landing area.

After the landing gear and flaps are retracted and the failed engine is shut down and everything is under control, it is a good idea to communicate with the nearest ground facility to let them know the flight is being conducted with one engine inoperative. ATC facilities are able to give valuable assistance if needed, particularly when the flight is conducted under radar observation, which is nearly everywhere these days.

Good judgment would dictate, of course, that a landing be made at the nearest suitable airport as soon as practical rather than continuing the flight to a distant destination.

During cold-weather, engine-out practice using zero thrust power settings, the engine may cool to temperatures considerably below the normal operating range. This factor requires caution when advancing the power at the termination of single-engine practice. If the power is advanced rapidly, the engine may not respond and an actual engine failure may be encountered. This can be helped by closing the cowl flaps during periods of prolonged zero thrust on the idled engine.

This lack of heat on the idled engine is particularly important to remember when practicing engine-out approaches and landings. A good procedure is to slowly advance the throttle to approximately one-half power, then allow it to respond and stabilize before advancing to higher power settings. This procedure not only guarantees a steady supply of power, but results in much less wear on the engines of the aircraft.

Restarts after feathering require the same amount of care, primarily to avoid engine damage. Following the restart, the engine power should be maintained at the idle setting or slightly above until the engine is sufficiently warm and is receiving adequate lubrication.

Engine failure checklists

Although each aircraft must be operated in accordance with the POH, the following checklists are presented to familiarize the would-be multiengine pilot with the actions that are typically required when an engine fails.

Engine failure during cruise

1. Mixtures—As required for flight altitude.
2. Propellers—Full forward.
3. Throttles—Full forward.
4. Landing gear—Retracted.
5. Wing flaps—Retracted.
6. Inoperative engine—Determine.
7. Establish at least 5-degree bank—Toward operative engine.
8. Inoperative engine—Secure.
   a. Throttle—Close.
   b. Mixture—Idle cut-off.
   c. Propeller—Feather.
   d. Fuel selector—Off.
   e. Auxiliary fuel pump—Off.
   f. Magneto switches—Off.
   g. Alternator switch—Off.
   h. Cowl flap—Close.

9. Operative engine—Adjust.
   a. Power—As required.
   b. Mixture—As required for flight altitude.
   c. Fuel selector—As required.
   d. Auxiliary fuel pump—On.
   e. Cowl flap—As required.
10. Trim tabs—Adjust bank toward operative engine.
11. Electrical load—Decrease to minimum required.
12. As soon as practical—Land.

Airstart (after shutdown)

Airplanes without propeller unfeathering system:

1. Magneto switches—On.
2. Fuel selector—Main tank (feel for detent).
3. Throttle—Forward approximately one inch.
4. Mixture—As required for flight altitude.
5. Propeller—Forward of detent.
6. Starter button—Press.
7. Primer switch—Activate.
8. Starter and primer switch—Release when engine fires.
9. Mixture—As required.
10. Power—Increase after cylinder head temperature reaches 200 degrees F.
11. Cowl flap—As required.
12. Alternator—On.

Airplanes with propeller unfeathering system:

1. Magneto switches—On.
2. Fuel selector—Main tank (feel for detent).
3. Throttle—Forward approximately one inch.
4. Mixture—As required for flight altitude.
5. Propeller—Full forward.
6. Propeller—Retard to detent when propeller reaches 1000 RPM.
7. Mixture—As required.
8. Power—Increase after cylinder head temperature reaches 200 degrees F.

9. Cowl flap—As required.

10. Alternator—On

Engine-out approach and landing

Essentially, an engine-out approach and landing is the same as a normal approach and landing. Long, flat approaches with high-power output on the operating engine and/or excessive threshold speed that results in floating and unnecessary runway use should be avoided. Due to variations in the performance limitations of many light twins, no specific flight path or procedure can be proposed that would be adequate in all engine-out approaches. In most light-twins, however, a single-engine approach can be accomplished with the flight path and procedures almost identical to a normal approach and landing.

During multiengine training, the pilot should perform approaches and landings with the power of one engine set to simulate the drag of a feathered propeller (zero thrust), or if feathering propellers are not installed, the throttle of the simulated failed engine set to idle. With the "dead" engine feathered or set to "zero thrust," normal drag is considerably reduced, resulting in a longer landing roll. Allowances should be made accordingly for the final approach and landing.

The final approach speed should not be less than V_{yse} until the landing is assured; thereafter, it should be at the speed commensurate with the flap position until beginning the roundout for landing. Under normal conditions the approach should be made with full flaps; however, neither full flaps nor the landing gear should be extended until the landing is assured. With full flaps the approach speed should be 1.3 V_{so} or as recommended by the manufacturer.

The pilot should be careful not to lower the flaps too soon on a single-engine approach. Once they have been extended it may not be possible to retract them in time to initiate a go-around. Most light-twins are not capable of making a single-engine go-around with full flaps. In fact, a single-engine go-around in light-twins under the best of circumstances is not something one wants to have happen. Do not place yourself in the position of having to go around. Plan your approach and get it right the first time! (Fig. 8-10.)

Fig. 8-10 *At this point, it's perfectly all right to exhale. You've earned it!*

Part Two

Navigate

Flying provides a bird's eye view of the world unequaled by any other mode of transportation. Surprisingly, it can still be difficult to find your way—especially when the weather will not cooperate, or the land below is not familiar...

9

Chart history

We define *map* as a graphical representation of the physical features, whether natural, artificial, or both, of the Earth's surface, by means of signs and symbols at an established scale on a specified projection with a means of orientation. A *chart* is a special-purpose map designed for navigation. And specifically, an *aeronautical chart* is designed to meet the requirements of aerial navigation.

Pilot's directions

Commercial aviation in the United States was launched on August 12, 1918, with the initiation of regular airmail service between Washington and New York. Pilots were forced to use railroad and road maps, or pages from atlases. As late as 1921, with transcontinental airmail operations day and night, no aeronautical charts existed. Pilots noted times and courses between prominent landmarks. If they were lucky, they flew two trips behind veteran pilots: if not two trips, just one. Notes from various pilots were assembled and published by the Post Office Department. These *Pilot's Directions* contained distances, landmarks, compass courses, and emergency landing fields, with services and communications facilities at principal points along the route in a narrative form:

> *Hazelhurst Field, Long Island.—Follow the tracks of the Long Island Railroad past Belmont Park racetrack, keeping Jamaica on the left. Cross New York over the lower end of Central Park.*

> *Newark, N.J.—Heller field is located in Newark and may be identified as follows: The field is $1^1/4$ miles west of the Passaic River and lies in the V formed by the Greenwood Lake Division and Orange branch of the New York, Lake Erie and*

Western Railroad. The Morris Canal bounds the western edge of the field. The roof of the large steel hanger is painted an orange color.

These narrative checkpoints covered the routes at 10- to 25-mile intervals.

With the passage of the Air Commerce Act in 1926, the Department of Commerce became responsible for the production of aeronautical charts for the nation's airways. The first chart published in 1927 was a strip map that covered the air route from Kansas City to Moline, Illinois. These early charts depicted prominent topographical features for visual flying and contained the locations of the newly installed airway-lighted beacon system for night operations. The strip map concept was extended throughout the late 1920s to other lighted airways between major airports.

The original airway beacon was a 24-inch rotating searchlight containing a parabolic mirror. It was powered by a 110-volt, 1000-watt lamp that produced approximately 1,000,000 candlepower. Beacons were established at intervals of 10 to 15 miles along the airway. Rotating at six revolutions per minute, they produced a clear flash every 10 seconds. Beacons were supplemented by green or red coded flashes; green for beacons at landing fields (sites were numbered from west to east, or south to north, depending upon the direction of the airway); red or green course lights pointed in the direction of the airway.

Lights flashed a Morse-code letter that identified the site. For simplicity, letters that contained fewer elements, dots and dashes, than numbers, were used. A sequence of letters evolved: W-U-V-H-R-K-D-B-G-M. These letters formed the popular mnemonic, "When Undertaking Very Hard Routes, Keep Direction By Good Methods." This sequence was repeated every 100 miles. Figure 9-1 shows a strip map for the Los Angeles-to-Las Vegas route in March 1931 with the airway beacons.

The strip map had a scale of 1:500,000, the same as today's sectional charts. Topography was shown as contours and color tints. Cultural features included railroads, highways, cities, and towns, and prominent electric transmission lines. Airports were shown by type: military or civilian. The lighted airways and new low-frequency radio ranges were depicted.

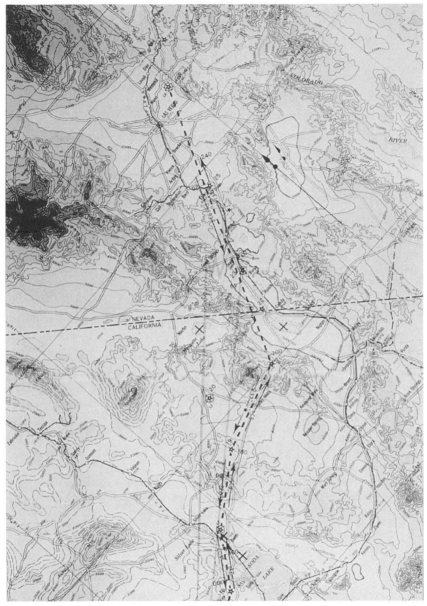

Fig. 9-1. *This Los Angeles to Las Vegas strip map, dated March 1931, shows one of the lighted airways of the period.*

Fig. 9-1. *Continued.*

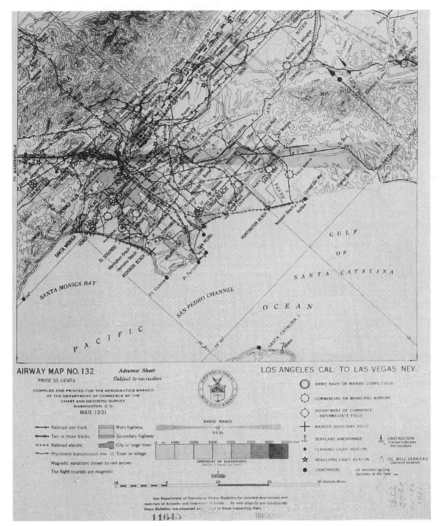

Fig. 9-1. *Continued.*

Supplemental aeronautical information was published in the *Domestic Air News* until 1929, when it was replaced by *Air Commerce Bulletins*. These publications contained official aviation information assembled and distributed by the Department of Commerce. These were the forerunners of today's *Airport/Facility Directory* and *Notice to Airmen* publications. In 1932, the free bulletin series described airports, intermediate landing fields, and meteorological conditions in the various states, along with the low-frequency A/N radio ranges for air navigation. By 1940, bulletins contained notices to airmen, air navigation radio aids, danger areas in air navigation, and a directory of airports. They were prepunched notebook size and included airway radio facility charts.

Sectional charts

With more flying conducted away from established airways, it became apparent that a system of charts was needed to provide complete coverage. Recommendations of the Committee on Aerial Navigation Maps in 1929 prompted the Coast and Geodetic Survey to develop a series of 92 sectional aeronautical charts for the United States. Sectionals were perhaps the best topographic maps; however, strip maps continued to be published until 1932 when the initial series of 31 sectionals was completed.

Figure 9-2 shows the June 1932 Los Angeles Sectional Chart. Sectionals contained the same general information as strip maps. Lighted airways as well as the low-frequency radio ranges were shown. This chart refers the user to see Airway Bulletin No. 2, "Descriptions of Airports and Landing Fields in the United States," for detailed information on airports and landing fields. The Air Corps used strip maps until 1935 when the maps were officially discontinued in favor of the new sectional aeronautical charts. Aeronautical charts for Alaska were nearly completed by the beginning of World War II.

New charts were required as aircraft became more reliable and instruments for blind flying with ground navigational systems were developed. A pioneer of instrument charts was E. B. Jeppesen, founder of today's Jeppesen Sanderson Company.

Jeppesen joined Boeing Air Transport, the predecessor of United Airlines, as an airmail pilot in 1930. He worked his way up to the route between Cheyenne, Wyoming, and Salt Lake City, one of the more dangerous because of terrain and changeable weather conditions. Appalled by the lack of navigational information for pilots, Jeppesen began making notes about every bit of navigational information, compiling data on airports, slopes, obstacles, and drainage patterns. (Drainage patterns pertain to the overall appearance of features associated with water, such as shorelines, rivers, lakes, and marshes, or any similar feature.) He developed data for the routes between airports and in 1934 published his first *Airway Manual*. The manual included routes via the new radio navigation aids plus the individual airport flight patterns.

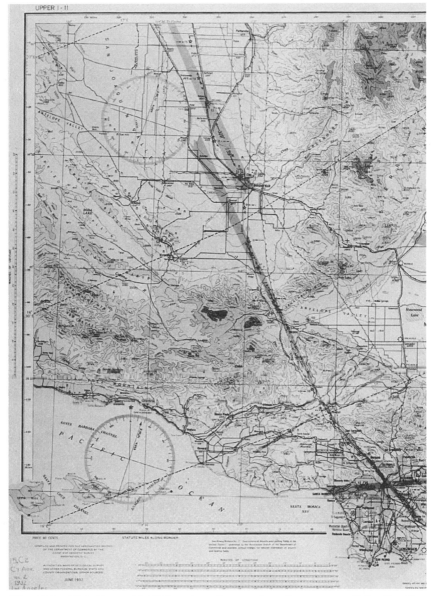

Fig. 9-2. *This June 1932 Los Angeles sectional was one of the first in the second generation of aeronautical charts.*

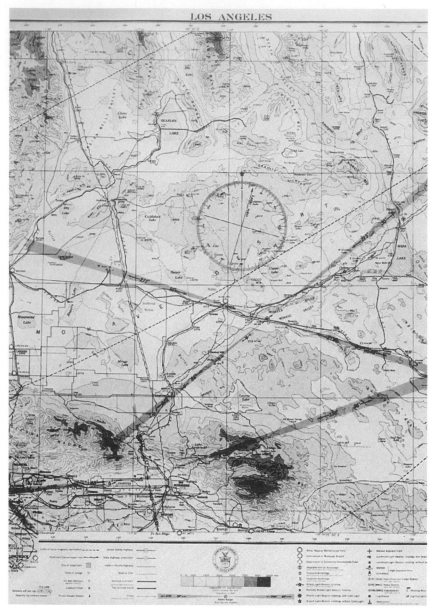

Fig. 9-2. *Continued.*

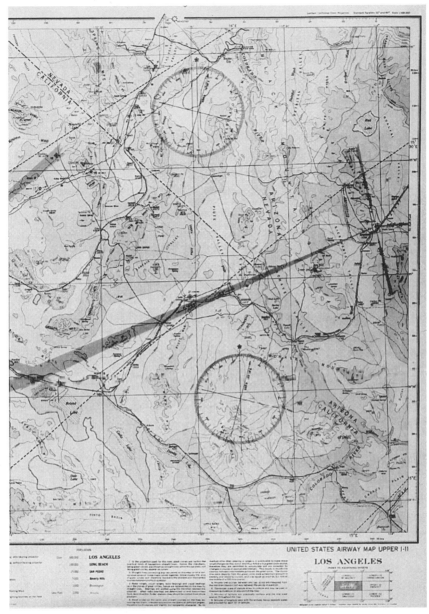

Fig. 9-2. *Continued.*

The Air Corps also recognized the need for specialized air charts with the advent of instrument flying and radio navigation. In 1937, the Army issued its first radio facility charts, which were used through the end of World War II.

Wartime

World War II compressed a quarter-century of peacetime aeronautical chart development into a few years. The need for all types of charts was urgent and insatiable. The term "aeronautical chart" became firmly established during this time. Previously, charts were referred to as air maps, aeronautical maps, flight maps, or aeronautical charts. Most charts were variations of those in existence prior to 1939, which saved precious time during the war. Charts rolled off the presses by the millions. At its peak, the production of charts in the St. Louis plant reportedly reached 10 tons a day.

New radio navigation aids were developed for en-route position finding during the war. The Coast and Geodetic Survey started developing a series of radio direction-finding charts in 1939 to cover the United States. The *long range navigation* (loran) system was established on the East Coast in 1941. New charts were developed to accommodate this new navigation system and a new series of *world aeronautical chart* (WAC) scale maps for the Western Hemisphere were developed and subsequently completed for the rest of the world in 1943. Additional series for world planning and world long-range charts were initiated. In 1942, the first of a new series of instrument approach and landing charts was distributed by the geodetic survey.

The aeronautical chart service used the Lambert conformal projection; the Navy's Hydrographic Office employed the Mercator projection. For planning and operating within a global system, new projections were introduced and old systems adapted. The Lambert conformal was preferred for its accuracy in air navigation for most parts of the world; however, for navigation in polar regions, charts using the *transverse Mercator* or *polar stereographic* projections were selected.

Postwar

By the end of the war, responsibility for revising and distributing world aeronautical charts covering the continental United States was turned over from the military to the Coast and Geodetic Survey. International conferences began shortly after the war and established criteria and requirements to serve the needs of aircraft engaged in international flights.

From the middle to late 1940s, the U.S. Air Force published *Instrument Let Down* publications. These procedures were bound volumes consisting of four charts to each page produced by the Coast and Geodetic Survey. A new series of radio facility charts was introduced by the survey in 1947, superseded two years later by a series of 59 standard radio-facility charts covering the entire United States.

Jeppesen introduced the Standard Instrument Approach Procedures in 1947. Prior to this time, instrument approach procedures were designed by individual operators, for their own use, then approved by the Civil Aeronautics Authority (CAA), which was the predecessor to the Federal Aviation Administration (FAA). Jeppesen and the CAA developed a program where the CAA would provide standard approach procedures and authorize operators to use those procedures. The first instrument landing system (ILS) approach chart was developed in 1948, followed a year later with the first very-high-frequency omnidirectional radio range (VOR) approach chart. Supplemental flight information documents were introduced in the early 1950s to keep pace with an increasing amount of navigational information.

Dunlap and Associates was given a contract by the U.S. Office of Naval Research in 1951 to study aeronautical charts and other graphical aids to navigation. Dunlap reported "two trends in the development of aeronautical charts. First, charts have become more complex because there has been a tendency to add new information to already existing charts. Second, there has been developed a wide variety of aeronautical charts so that the pilot must go to many sources to gather the information he or she needs. Both trends have been due to the increasing complexity of flight." Also during this period the *Airman's Guide, Directory of Airports, and Flight Information Manual* replaced *Airway Bulletins*.

Simplified navigation

The Coast and Geodetic Survey introduced a new family of aeronautical charts in the early 1950s to simplify high-speed jet and transport navigation as well as lightplane visual flying. The series included planning, radio facility, approach and landing, as well as visual charts. These included *jet navigation charts* (JNC) for visual flying and the issuance of new experimental approach and landing charts for instrument operations. The new format permitted two procedures to be printed on one side of the sheet; it was hoped that the more than 1100 instrument approach charts could be reduced to approximately 400. The smaller-size sheets were also easier to handle in the aircraft. Specialized charts were also developed and tested: *operational navigation charts* (ONC), *global navigation charts* (GNC), and various *long-range navigation charts* (loran, CONSOL, and CONSOLAN).

The radio-facility charts proved unsatisfactory for jet aircraft flying at speeds faster than 500 knots. Consequently, in 1953, the Aeronautical Chart and Information Center introduced a new series of radio facility experimental charts that covered an area 750 nautical miles by 250 nautical miles and folded to $4^{1}/_{2} \times 9$ inches for convenient handling.

The Aeronautical Chart and Information Center introduced the *Flight Information Publication-Planning* (FLIP) in 1958 to further eliminate nonessential material. This publication contained charts and textual data necessary for flight planning. FLIP, and other publications of this type, provide supplemental information that cannot be printed on the chart due to space constraints. The *Airman's Information Manual* (AIM) replaced the *Airman's Guide* in 1964. Since then, the AIM has gone through various evolutionary stages, at times consisting of four documents, to its present form of *Basic Flight Information and ATC Procedures, Airport/Facility Directory,* and *Notices to Airmen.* In 1995, the *Airmen's Information Manual* was renamed the *Aeronautical Information Manual,* merely a sign of the times.

Standard instrument departure (SID) procedure charts were introduced in 1961. Standard terminal arrival route (STAR) procedure charts came in 1967. A year later, VORTAC (VOR with distance measuring TACAN equipment) area navigation (RNAV) en-route charts were introduced; VORTAC RNAV approach charts were developed in 1971.

Area navigation prompted the National Oceanic and Atmospheric Administration (NOAA) to produce a series of high-altitude RNAV charts crisscrossing the United States with RNAV jet routes. These proved to be of limited value, however, and were subsequently discontinued. Jeppesen still offers RNAV en-route charts that, in effect, allow the pilot to design specific routes.

Sectional and world aeronautical charts evolved through several stages in the 1960s and 1970s. Mostly because of economic considerations, information was eventually printed on both sides, which reduced the total number of charts but unfortunately eliminated what many pilots considered useful information printed on the reverse side. Much of this general information was transferred to various sections of the AIM. Profile descent charts were published in 1976 and loran RNAV approach charts were introduced in 1990.

The 1990s

In 1992, the National Oceanic and Atmospheric Administration proposed the elimination of world aeronautical charts. The government wanted pilots to switch to Defense Mapping Agency Operational Navigation Charts (ONCs). Various aviation organizations, citing safety and pilot demand, lobbied for the continuation of this series. Fortunately, the government listened, and this series is still available.

Implementation of complex airspace configurations around major airports, Class B Airspace fostered development of the *terminal area chart* (TAC) for improved presentation of Class B dimensions and better resolution of ground references. (Many pilots remember that Class B airspace originated as the terminal control area's upside-down wedding cake.)

[Pilots had previously paid a nominal fee for charts; however, the government in the face of increasing deficits decided pilots should pick up more of the tab, and prices for charts and publications skyrocketed. At one point it was proposed that users pay the development as well as printing costs, which would have put the price of charts almost out of sight for many pilots. Fortunately, aviation organizations, notably the Aircraft Owners and Pilots Association (AOPA), pressured the government into a compromise. Many American's still think the prices are outrageous, but considering the information available, and the cost of charts in other countries, pilots in the United States are still getting a bargain.]

On April 4, 1991, a new "Grand Canyon Visual Flight Rules (VFR) Aeronautical Chart" became available. The chart depicts communications and minimum altitudes for flight over the canyon in accordance with Special Federal Aviation Regulations. Jeppesen introduced VFR routing charts for the Los Angeles area in the early 1990s, and similar charts for San Diego and San Francisco were introduced in 1994.

The military's global positioning system (GPS) of satellite navigation became available for civilian use, and testing of GPS approaches began in 1993. The first stand-alone GPS approaches were published in 1994.

With the rapid adaptation of GPS, the U.S. Coast Guard, which runs the loran radio navigation transmitting stations, in 1994 proposed to turn off the navigational aid after 1996. Again, however, aviation groups protested. At the present, it appears loran will be operational into the twenty-first century, but not with the number of nonprecision approaches promised.

Preparation of visual and instrument charts for the United States is one responsibility of the National Ocean Service (NOS), which is part of the National Oceanic and Atmospheric Administration (NOAA) within the Department of Commerce. Charts for most of the world are available from the Defense Mapping Agency's Combat Support Center. Commercially prepared charts and other navigational publications are also available. Many individual states of the United States, and many international countries also produce aeronautical charts and publications.

10

Projections and limitations

Only a globe can accurately portray locations, directions, and distances of the Earth's surface. Although not actually a sphere, but a spheroid, the Earth only approximates a true sphere due to the force of rotation that expands the Earth at the equator and flattens it at the poles. This elliptical nature of the Earth is a concern to cartographers; for most practical purposes of navigation, the Earth can be considered a sphere. On the Earth, meridians are straight and meet at the poles; parallels are straight and parallel (Fig. 10-1). Meridian spacing is widest at the equator and zero at the poles; parallels are equally spaced. Scale is true for every location. Because a globe is not possible for practical aeronautical charts, mathematicians and cartographers have devised numerous systems, known as projections, to describe features on the Earth in the form of a plane or flat surface. A *map projection* is a system used to portray the sphere of the Earth, or a part thereof, on a plane or flat surface.

Locations on the Earth are described by a system of latitude and longitude coordinates (Fig. 10-1). By convention, latitude is named first, then longitude. The reference point of latitude is the equator, with latitude measured in degrees north and south of the equator. Longitude, as discussed in Chap. 1, is measured east and west of the prime meridian (Greenwich meridian).

Any point on the Earth can be described using the system of latitude and longitude in degrees, minutes, and seconds. With the sophistication of today's navigational systems, aeronautical charts and publications express latitude and longitude in degrees, minutes, and hundredths of minutes. A degree is an arc that is $1/360$ of a circle; therefore, a point with latitude 47°N longitude 122°W would be the intersection of the parallel 47° north of the equator and the meridian 122° west of the Greenwich meridian. Degrees can be further subdivided into minutes ('), which represent $1/60$ of a degree, and seconds

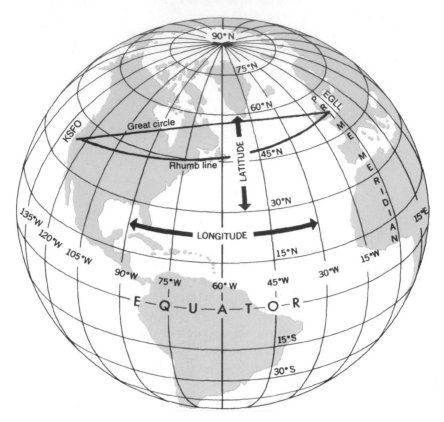

Fig. 10-1. *Only on a globe are areas, distances, directions, and shapes true.*

("), which represent ¹/₆₀ of a minute. For example, the Seattle-Tacoma International Airport is located N47°26.28′ W122°18.67′ (47 degrees 26.28 minutes north and 122 degrees 18.67 minutes west).

Each degree of latitude equals 60 nautical miles (nm). Because meridians meet at the poles, a degree of longitude decreases in length with distance north or south of the equator; therefore, only for the special case of the equator does a degree of longitude equal 60 nautical miles.

Projections

The goal of the map projection is to accurately portray true areas, shapes, distances, and directions. This includes the condition that lines of latitude are parallel and meridians of longitude pass through the Earth's poles and intersect all parallels at right angles.

Areas. Any area on the Earth's surface should be represented by the same area at the scale of the map. These projections are termed *equal-area* or *equivalent*.

Distances. The distance between two points on the Earth should be correctly represented on the map. These projections are termed *equidistant*.

Directions. The direction, or azimuth, from one point to other points on the Earth should be correct on the map. These projections are termed *azimuthal* or *zenithal*.

Shapes. The shape of any feature should be correctly represented. The scale around any point must be uniform. These projections are termed *conformal*.

Because it is only possible to obtain all these properties on a globe, the cartographer must select the projection that preserves the most desired properties based on the chart's use. Figure 10-2 illustrates the four projections most often used on aeronautical charts: Mercator, transverse Mercator, Lambert conic conformal, and polar stereographic. Table 10-1 compares the different projections used on aeronautical charts.

Mercator projection

The Mercator projection transfers the surface of the Earth onto a cylinder tangent at the Earth's equator. In the Mercator projection, meridians and parallels appear as lines crossing at right angles. Meridians are parallel on the Mercator projection, unlike meridians on the Earth that meet at the poles. This results in increasingly exaggerated areas toward the poles. Scale, which is the relationship between distance on a chart and actual distance, changes with latitude.

The advantage of the Mercator is that a straight line on this projection crosses all meridians at the same angle. This allows the navigator to set a constant course from one point to another. A course crossing all meridians at a constant angle is known as a *rhumb line*. Figure 10-1 shows a rhumb line from San Francisco (KSFO) to London (EGLL).

Table 2-1.
Chart projections

Projection	Lines of longitude (Meridians)	Lines of latitude (Parallels)	Graticule spacing	Scale	Uses
Globe and Earth	Straight and meet at poles	Straight and parallel	Meridian spacing maximum at equator, zero at poles; parallels equally spaced.	True	Impractical for navigation
Mercator	Straight and parallel	Straight and parallel	Meridian spacing equal; parallel spacing increases away from equator.	True only along equator; distortion increases away from equator.	Dead reckoning, celestial
Transverse Mercator	Curved and concave	Concave arcs	Meridian increase away from tangent meridian; parallels equally spaced.	True along line of tangency only.	Polar navigation
Lambert conic conformal	Straight converging at poles	Concave arcs	Parallels equally spaced.	True along standard parallels.	Pilotage, dead reckoning
Polar stereographic	Straight radiating from the pole	Concentric circles unequally spaced	Conformal	Increases away from pole.	Polar navigation

A rhumb line is not normally the shortest distance between two places on the surface of the Earth. The *great-circle* distance is always the shortest distance between points on the Earth. A great circle is an arc projected from the center of the Earth through any two points on the surface (Fig. 10-1). A great circle, unlike the rhumb line, crosses meridians at different angles, except in two special cases: where the two points lie along the equator or the same meridian (in both cases, the rhumb and great circle coincide).

For practical purposes at low latitudes, rhumb and great-circle distances are nearly identical. As latitude and distance increase, differences become increasingly significant, as illustrated in Fig. 10-1; however, the projection is conformal in that angles and shapes within any small area are essentially true.

Transverse Mercator projection

The transverse Mercator projection rotates the cylinder so that it becomes tangent to a meridian (Fig. 10-2). These projections are used in high north and south latitudes (high-numbered latitudes toward the north and south) and where the north-south direction is greater than the east-west direction. All properties of the regular Mercator are preserved, except the straight rhumb line. Parallels are no longer straight, becoming curved lines; meridians become complex curves. The projection is conformal. The line of true scale is no longer the equator, but the central meridian of the projection, where the cylinder is tangent.

Distances are true only along the central meridian selected by the cartographer, or else along two lines parallel to it, but all distances, directions, shapes, and areas are reasonably accurate within 15° of the central meridian. Distortion of distances, directions, and size of areas increases rapidly outside the 15° band.

Lambert Conformal Conic projection

A projection widely used for aeronautical charts is the *Lambert conformal conic,* with two standard parallels (Fig. 10-2). As the name implies, a cone is placed over the Earth; the cone intersects the Earth's surface at two parallels of latitude. Scale is exact everywhere along the standard parallels, but scale decreases between the parallels and scale increases beyond the parallels. Shape and

Mercator

Central meridian (selected by mapmaker)

Great distortion in high latitudes

Examples of rhumb lines (direction true between any two points)

Equator touches cylinder if cylinder is tangent

Reasonably true shapes and distances within 15° of equator

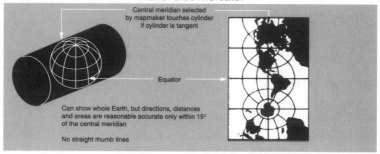

Transverse Mercator

Central meridian selected by mapmaker touches cylinder if cylinder is tangent

Equator

Can show whole Earth, but directions, distances and areas are reasonable accurate only within 15° of the central meridian

No straight rhumb lines

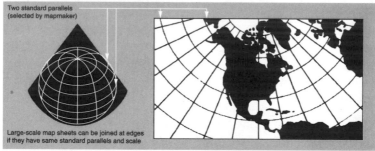

Lambert conic conformal

Two standard parallels (selected by mapmaker)

Large-scale map sheets can be joined at edges if they have same standard parallels and scale

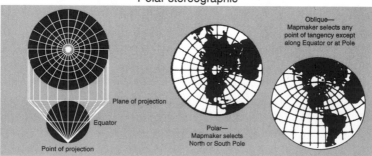

Polar stereographic

Oblique—Mapmaker selects any point of tangency except along Equator or at Pole

Plane of projection

Equator

Point of projection

Polar—Mapmaker selects North or South Pole

Fig. 10-2. *The cartographer selects the projection that preserves the most desired properties based on the charts used.*

area distortion is minimal at the standard parallels, but the distortion increases away from the standard parallels.

All meridians are straight lines that meet at a point beyond the map; parallels are concentric circles. Meridians and parallels intersect at right angles. The chart is considered conformal because scale is almost uniform around any point; scale error on any chart is so small that distances can be considered constant anywhere on the chart. A straight line from one point to another very closely approximates a great circle.

Polar stereographic projection

The standard Lambert is too inaccurate for navigation above a latitude of approximately 75° to 80°. The *polar stereographic projection* is sometimes used for polar regions. A plane tangent to the Earth at the pole provides the projection (Fig. 10-2). Meridians are straight lines radiating from the pole, and parallels are concentric circles. A rhumb line is curved and a great-circle route is approximated by a straight line. Directions are true only from the center point of the projection. Scale increases away from the center point. The projection is conformal, with area and shape distortion increasing away from the pole.

Horizontal datum

Cartographers need a defined reference point upon which to base the position of locations on a chart. This is known as the horizontal datum, or *horizontal constant datum* or *horizontal geodetic datum*. The horizontal datum used as a reference for position is defined by the latitude and longitude of this initial point. Prior to 1992, the horizontal datum for the United States was located at Meades Ranch, Kansas, referred to as the North American Datum 1927 (NAD 27).

With the introduction of geodetic satellites for mapping the Earth's surface and satellite navigation systems for innumerable applications, there were recommendations to revise NAD 27. Beginning October 15, 1992, the horizontal geodetic referencing system applied to all charts and chart products was changed from the North American Datum of 1927 to the *North American Datum of 1983* (NAD 83). This resulted in differences of only approximately 1000 feet between NAD 27 and NAD 83 positions. The greatest coordinate shifts occurred in

Alaska and Hawaii where latitude was moved by as much as 1200 feet and longitude by up to 950 feet. In the conterminous United States, the maximum change was approximately 165 feet in latitude and 345 feet in longitude.

Limitations

In addition to the limitations of chart projections, cartographers and chart users are faced with the problems of scale, simplification, and classification. Finally, chart users, especially pilots, are faced with the crucial issue of current information. With visual charts only updated annually or semiannually and instrument charts updated every 56 days, the pilot must understand the system used to provide the latest information that is effective between routine chart revisions.

Scale

Charts provide a reduced representation of the Earth's surface. Recall that scale defines the relationship between a distance on a chart and the corresponding distance on the Earth. Scale is generally expressed as a ratio. The numerator, customarily 1, represents chart distance, and the denominator, a large number, represents horizontal ground distance. For example, 1:500,000 (sometimes written 1/500,000) states that any single unit, whether inch, foot, yard, statute mile, nautical mile, or kilometer on the chart, represents 500,000 units on the ground. That is, 1 inch on the chart equals 500,000 inches on the Earth.

Chart makers provide scales for conversion of chart distance to statute or nautical miles or kilometers. Manufacturers of aeronautical plotters provide scales for standard aeronautical charts; however, the pilot must be familiar with the plotter used and chart scale for accurate calculations.

While flying in the annual Hayward-Bakersfield-Las Vegas air race, I use sectional charts (1:500,000), except in the Las Vegas area where a terminal area chart (1:250,000) is available. Sure enough, in my haste, I measured a leg using the wrong scale. This is disastrous in a race where the finishing order is based upon time in seconds and fuel in tenths of a gallon.

The smaller the scale of a chart, the less detail it can portray. For example, a chart with a scale of 1:1,000,000 cannot provide the detail of a chart with a scale of 1:250,000. Charts with a smaller scale increase the size of the area covered (assuming a constant size), but reduce the detail that can be shown.

Simplification and classification

Because scale reduces the size of the Earth, information must be generalized. Making the best use of available space is a major problem in chart development. The detail of the real world cannot be shown on the chart. The crowding of lines and symbols beyond a specific limit renders the chart unreadable, yet the amount of information that might be useful or desirable is almost unlimited. The smaller the chart scale, the more critical and difficult the problem; therefore, the cartographer is forced to simplify and classify information.

Simplification is the omission of detail that would clutter the map and prevent the pilot from obtaining needed information. The necessity for detail is subjective and not all will agree on what should, or should not, be included. The inclusion of too much detail runs the risk of confusing the reader by obscuring more important information. For example, the chart producer might have to decide whether to include a prominent landmark, the limit of controlled airspace, or a symbol indicating a parachute jump area. The problem of simplification has led directly to the use of aeronautical publications, such as the *Airport/Facility Directory*.

The *Airport/Facility Directory* is divided into seven booklets that cover the United States, including Puerto Rico and the Virgin Islands. Alaska is covered by the *Alaska Supplement,* and areas of the Pacific are covered by the *Pacific Chart Supplement.* These directories are a pilot's manual containing data on airports, seaplane bases, heliports, navigational aids, communications, special notices, and operational procedures. They provide information that cannot be readily depicted on charts: airport hours of operation, types of fuel available, runway widths, lighting information, and other data. The directories are also a means of updating charts between chart issuances. Directories, charts, and other related publications may be obtained directly from NOAA. Free catalogs are available.

Classification is necessary in order to reduce the amount of information into a usable form. The cartographer must classify towns, rivers, and highways of different appearance on the ground into a common symbol for the chart. The pilot must then be able to interpret this information.

To maximize the amount of information on a chart, the cartographer uses symbols. Symbol shape, size, color, and pattern are used to convey specific information. The pilot must be able to interpret these symbols. Lack of chart symbol knowledge can lead to misinterpretation, confusion, and wandering into airspace where a pilot has no business.

Currency

With the ever-changing environment, charts are almost outdated as soon as they are printed and become available. A pilot's first task when using any chart or publication is to ensure its currency.

Visual charts are revised and reissued semiannually or annually. Changes to visual charts are supplemented by the "Aeronautical Chart Bulletin," in the *Airport/Facility Directory,* revised every 56 days (eight weeks).

NOS IFR charts contained in the *Terminal Procedures Publication* (TPP) are published every 56 days. A 28-day midcycle change notice volume contains revised procedures that occur during the 56-day publication cycle. These changes are in the form of new charts. The subsequent publication of the TPP incorporates change notice volume revisions and any new changes since change notice issuance.

Visual and instrument charts are further supplemented by the FAAOs *Notice to Airmen* system (NOTAMs). The NOTAM publication is published every 14 days and supplements the "Aeronautical Chart Bulletin" of the *Airport/Facility Directory,* and TPP and change notice volumes.

Aeronautical information not received in time for publication is distributed through the FAA's telecommunications systems. The information includes unanticipated or temporary changes or hazards when their duration is for a short period or until published. A NOTAM is classified into one of three groups:

- NOTAM (D)
- NOTAM (L)
- FDC NOTAM

NOTAM (D)s consist of information that requires wide distribution and pertains to en-route navigational aids, civil public-use landing areas listed in the *Airport/Facility Directory,* and aeronautical data that relates to IFR operations.

NOTAM (L)s include information that requires local dissemination, but does not qualify as a NOTAM (D), such as bird activity, moored balloons, airport beacons, taxiway lights, and the like.

FDC NOTAMs consist of information that is regulatory in nature pertaining to charts, procedures, and airspace. This includes such items as temporary flight restrictions and revisions to visual and instrument charts.

NOTAMs from each category are routinely provided as part of a standard flight service station (FSS) weather briefing. Nonautomated flight service stations only provide FDC NOTAMs for locations within 400 miles of the facility, the pilot must request FDC NOTAMs in areas beyond 400 miles. NOTAMs, except (L), are also available through direct user access terminals (DUAT) and other commercial vendors of weather information. When FDC NOTAMs are associated with a specific facility identifier, they are included as part of the DUAT briefing; however, most en-route chart changes are not associated with a specific facility identifier. DUAT users are faced on every briefing with a disclaimer:

> *FDC NOTAMs that are not associated with an affected facility identifier will now be presented unless you specifically choose to decline to receive such information.*

It's almost like looking for the proverbial "needle in the haystack," but to be safe, these NOTAMs must be reviewed. Once published, unlike the FSS where the pilot has the option to request the data, DUAT users must remember that:

> *Published FDC NOTAM Data are not available and must be obtained from other publications/charts/etc.*

Pilots using NOS charts must be aware of these limitations and plan their flight briefings accordingly. This might mean a call to the FSS specifically for any pertinent FDC NOTAMs. Loran and GPS NOTAMs are available on DUATs, but only available from the FSS on request.

Once a new chart becomes effective, the NOTAMs are canceled, including those carried in the *Notice to Airmen* publication and "Aeronautical Chart Bulletin" of the *Airport/Facility Directory.* Pilots are presumed to be using current charts.

A major advantage of commercial suppliers of chart services is a more timely revision schedule than is available through government products. Immediate and short-term changes to the National Airspace System (NAS) must still be obtained, but this information is normally provided as part of an FSS standard briefing or DUAT briefing.

It is not within the scope of this book to provide a detailed explanation of the NOTAM system. For the reader who would like additional information on this subject, including decoding, translating, and interpreting NOTAMs, refer to McGraw-Hill's Practical Flying Series book, *Pilot's Guide to Weather Reports, Forecasts, and Flight Planning.*

A British friend with whom I flew in England in the middle 1960s was astonished by the frequent revisions of U.S. aeronautical charts. The British civil charts that I had to fly with were infrequently updated and only contained about a fifth of the information of U.S. charts. For example, all NAVAIDs were shown by a single symbol without any indication of the type or frequency. And during the 1960s, the English charts cost about $5 in American currency.

NOS has made every effort to ensure that each piece of information shown on NOAA's charts and publications is accurate. Information is verified to the maximum extent possible. According to NOS, "You, the pilot, are perhaps our most valuable source of information. You are encouraged to notify NOAA, National Ocean Service, of any discrepancies you observe while using our charts and related publications. Postage-paid chart correction cards are available at FAA Flight Service Stations for this purpose (or you may write directly to NOAA, at the address below). Should delineation of data be required, mark and clearly explain the discrepancy on a current chart (a replacement copy will be returned to you promptly)."

National Ocean Service
NOAA, N/CG31
6010 Executive Blvd.
Rockville, MD 20852
1-800-626-3677

NOS emphasizes: "Use of obsolete charts or publications for navigation may be dangerous. Aeronautical information changes rapidly, and it is vitally important that pilots check the effective dates on each aeronautical chart and publication to be used. Obsolete charts and publications should be discarded and replaced by current editions." One pilot called an FSS and requested a briefing from Bishop to Santa Cruz, California. The briefer explained that the airport was closed. The pilot responded, "Oh, I must be using an old chart." Indeed, the airport had been closed for over two years. There are no valid reasons for using obsolete charts.

The use of current charts and publications, and obtaining a complete preflight briefing, cannot be overemphasized. It's like using the restroom before a flight; we know we should, but sometimes it's a little inconvenient. Failure, however, often leads to a very uncomfortable flight. Only by understanding the system of charts and publications can fliers ensure they meet their obligation as pilots in command.

11

Visual chart terminology and symbols

Terms and symbols discussed in this chapter not only apply to standard visual charts (world aeronautical charts, sectional charts, and terminal area charts), but most of the terms and symbols carry over to planning charts, instrument charts, and charts available through the Defense Mapping Agency. While reading this chapter, the reader should keep in mind chart limitations and any problems facing the cartographer, specifically scale, simplification, and classification as discussed in Chap. 10.

Topography and obstructions

The elevation and configuration of the Earth's surface are of prime importance to visual navigation. Cartographers devote a great deal of attention to portraying relief and obstructions in a clear, concise manner.

Terrain

Three methods are used on aeronautical charts to display relief: *contour lines, shaded relief,* and *color tints*. Contour lines, as the name implies, connect points of equal elevation above mean sea level (MSL) on the Earth's surface. Contours graphically depict terrain and are the principal means used to show the shape and elevation of the surface. Contours are depicted by continuous lines—except where elevations are approximate, then with broken lines—labeled in feet MSL. On sectional charts, basic contours are spaced at 500-foot intervals, although intermediate contours might be shown at 250-foot intervals in moderately level or gently rolling areas. Occasionally, auxiliary contours portray smaller relief features at 50-, 100-, 125-, or 150-foot intervals.

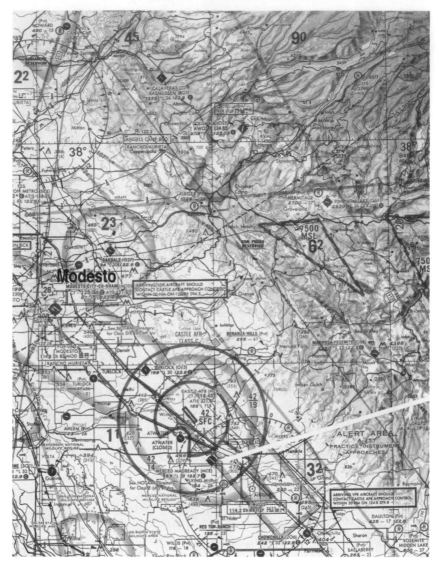

Fig. 11-1. *Chart relief is represented by contour lines, shaded relief, and color tints.*

Figure 11-1 shows how contours, shaded relief, and color tints depict terrain. Contours show the direction of the slope, gradient, and elevation. For example, in Fig. 11-1, valley floors have little or no gradient, while the mountains have steep gradients. The contours are labeled with their elevation. Shaded relief depicts how terrain might appear from the air. The cartographer shades the areas that would appear in

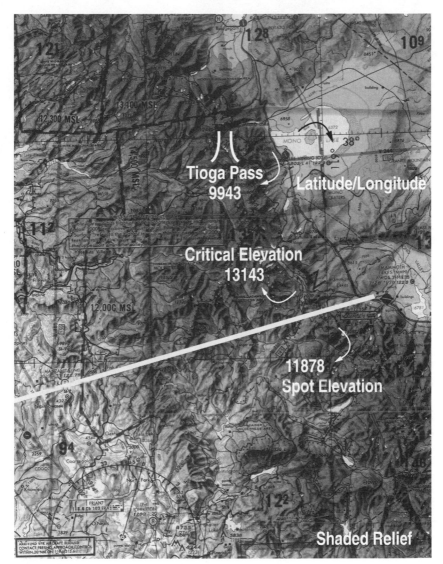

Fig. 11-1. *Continued.*

shadow if illuminated from the northwest. Shaded relief enhances and supplements contours by drawing attention to canyons and mountain ridges (Fig. 11-1). Color tints depict bands of elevation. These colors range from light green for the lowest elevations to dark brown for higher elevations. Color tints in Fig. 11-1 range from light green in the valley to dark brown over the mountain ranges, supplementing the contours and enhancing recognition of rapidly rising terrain.

In addition to contours, shading, and tints, significant elevations are depicted as *spot* elevations, *critical* elevations, and *maximum* elevation figures. Spot elevations represent a point on the chart where elevation is noted. They usually indicate the highest point on a ridge or mountain range. A solid dot depicts the exact location when known. An "x" denotes approximate elevations; where elevation is known, but location approximate, only the elevation appears, without the dot or "x" symbol. Critical elevation is the highest elevation in any group of related and more-or-less similar relief formations. Critical elevations are depicted by larger elevation numerals and dots than are used for spot elevations. Figure 11-1 illustrates the difference between spot and critical elevations.

Maximum elevation figures (MEF) represent the highest elevation, including terrain and other vertical obstacles—natural and constructed—bounded by the ticked lines of the latitude/longitude grid on the chart. Depicted to the nearest 100-foot value, the last two digits of the number are omitted. The center of the grid in the upper right portion of Fig. 11-1 shows 128. The MEF for this grid is 12,800 feet MSL. This figure is determined from the highest elevation or obstacle, corrected upward for any possible vertical error (including the addition of 200 feet for any natural or man-made obstacle not portrayed), then rounded upward to the next higher hundred-foot level; therefore, almost all MEFs will be higher than any elevation or obstacle portrayed within the grid on the chart. Pilots should note that these figures cannot take into account altimeter errors and should be considered as any other terrain elevation figure for flight-planning purposes.

Latitude and longitude are labeled in degrees. Lines of latitude and longitude are subdivided by lines representing 10 minutes, and half lines representing 1 minute of arc. Because longitude represents the same distance anywhere on the Earth—unlike latitude, which decreases toward the poles—one minute of longitude anywhere on the Earth equals one nautical mile (nm); therefore, lines of longitude can be used for quick estimates of distance.

After large earthquakes in southern California in 1971 and 1994, perhaps the Los Angeles sectional chart should have contained the comment: "CAUTION—Terrain elevations subject to change without notice."

Other topographical relief features considered suitable for navigation are contained in Fig. 11-2. They include lava flows, sand and gravel

RELIEF
UNSURVEYED AREAS   Label appropriately as required   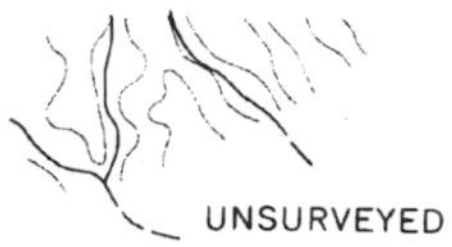
DISTORTED SURFACE AREAS   
LAVA FLOWS   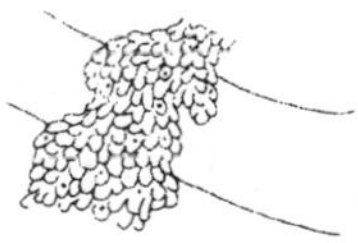
SAND OR GRAVEL AREAS   
SAND RIDGES   **To Scale**   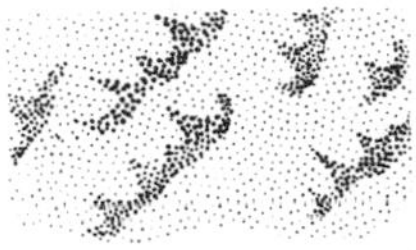
SAND DUNES   **To Scale**   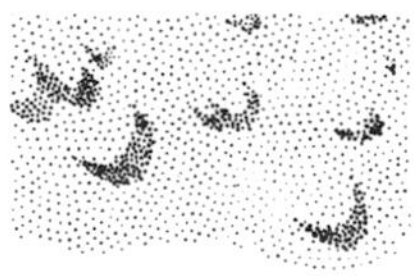

Fig. 11-2. *Supplemental relief features aid in visual flying.*

RELIEF
SHADED RELIEF 
ROCK STRATA OUTCROP 
QUARRIES TO SCALE 
STRIPMINES, MINE DUMPS AND TAILINGS   To Scale
CRATERS 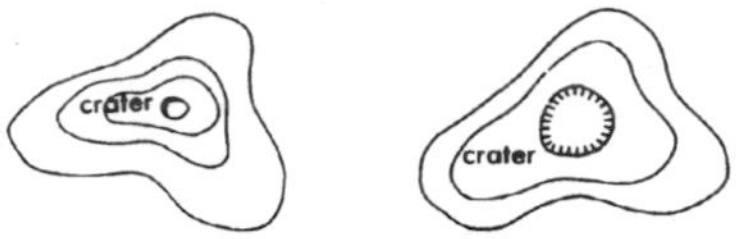
ESCARPMENTS, BLUFFS, CLIFFS, DEPRESSIONS, ETC. 
LEVEES AND ESKERS 

Fig. 11-2. *Continued.*

areas, rock strata and quarries, mines, craters, and other relief information usable for visual checkpoints.

Lava flows, sand ridges, and sand dunes are quite pronounced when seen from the air, and they make excellent checkpoints, especially if they are isolated by other terrain features. Unfortunately, most of these features only appear in the western United States. Strip mines and large quarries also make excellent checkpoints because of their visibility. Large craters, where they appear, also make excellent checkpoints.

Hydrography

Hydrography pertains to water and drainage features. Hydrographic features on aeronautical charts are represented in blue: Streams, rivers, or aqueducts are depicted by single blue lines; lakes and reservoirs are depicted by a blue tint. Small dots or "hatching" indicate where streams and lakes fan out (or are not perennial) or where reservoirs are under construction.

Figure 11-3 shows how shorelines, lakes, streams, reservoirs, and aqueducts are depicted. Shorelines usually make excellent checkpoints, except where they are relatively straight without features. Pilots need to pay attention to shoreline orientation. For example, most people assume that California's coastline is north-south; however, in certain areas, such as around Santa Barbara, the coast is actually east-west. This has led to much confusion for student pilots and others unfamiliar with the area. Lakes usually make good checkpoints, especially when their shape is unique or they are dammed, as discussed in Chap. 12.

Caution needs to be exercised with all lakes, perennial and nonperennial. A perennial lake contains water year round; a nonperennial lake is intermittently dry, usually during the dry season. There can be confusion during periods of drought when perennial lakes will be dry. Other discrepancies result from human decisions to drain, expand, or abandon reservoirs. If at all possible, streams should only be used to support other checkpoints. That is, there should be other landmarks that establish position that are supported by the position of the stream. Perennial and nonperennial streams should be treated with the same cautions as perennial and nonperennial lakes. Reservoirs are similar to lakes and can be treated in the same way; however, reservoirs are usually perennial.

HYDROGRAPHY	
STREAMS (Continued) **Fanned Out** Alluvial fan	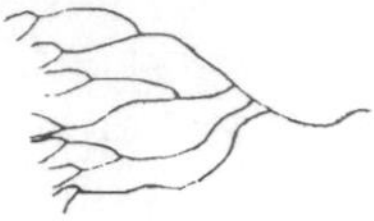
Braided	
Disappearing	
Seasonally Fluctuating with undefined limits	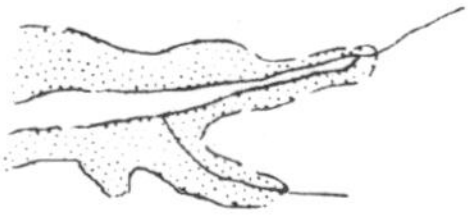
with maximum bank limits, prominent and constant	
Sand Deposits In **and Along Riverbeds**	
WET SAND AREAS Within and adjacent to desert areas	

Fig. 11-3. *Shorelines, lakes, streams, reservoirs, and aqueducts have landmark value, with specific limitations.*

HYDROGRAPHY
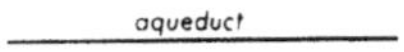 **AQUEDUCTS**   ————— aqueduct —————
Abandoned or Under Construction   _ abandoned aqueduct _ _ 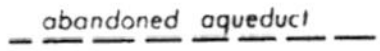
 **Underground**   _ _ _ underground aqueduct _ _
Suspended or Elevated   
Tunnels   ——→ - - - - - ←—— 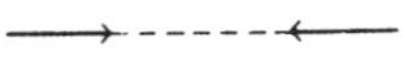
Kanats   Underground aqueduct   with air vents   underground aqueduct   o— —o— —o— —o— —o 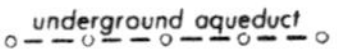

Fig. 11-3. *Continued.*

HYDROGRAPHY	
SHORELINES **Definite**	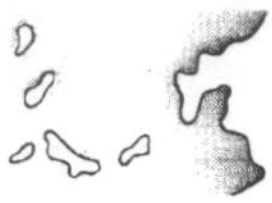
Fluctuating	
Unsurveyed Indefinite	
Man-made	
LAKES Label as required **Perennial** When too numerous to show individual lakes, show representative pat- tern and descriptive note.	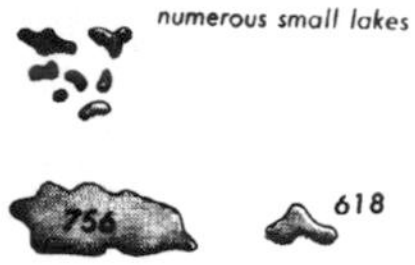
Nonperennial (dry, intermittent, etc.) Illustration includes small perennial lake	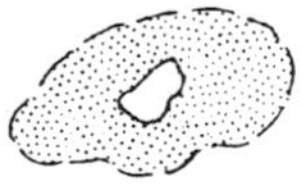

Fig. 11-3. *Continued.*

HYDROGRAPHY	
RESERVOIRS **Natural Shorelines**	
Man-made Shorelines Label when necessary for clarity	
Too small to show to scale	
Under Construction	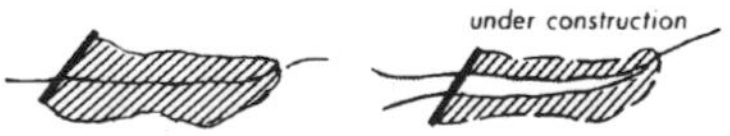
STREAMS **Perennial**	
Nonperennial	

Fig. 11-3. *Continued.*

HYDROGRAPHY	
SALT EVAPORATORS AND SALT PANS MAN EXPLOITED	 salt pans
SWAMPS, MARSHES AND BOGS	
HUMMOCKS AND RIDGES	
MANGROVE AND NIPA	 mangrove
PEAT BOGS	 peat bog
TUNDRA	tundra
CRANBERRY BOGS	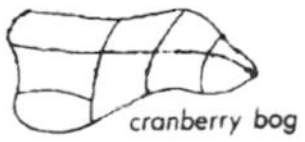 cranberry bog

Fig. 11-4. *Some hydrographical features warn of danger, such as swamps, others provide supplemental landmark value.*

HYDROGRAPHY	
FLUMES, PENSTOCKS AND SIMILAR FEATURES	 ————— flume —————
Elevated	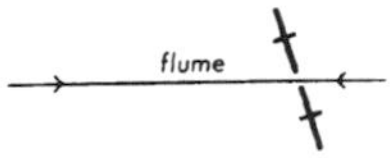 flume
Underground	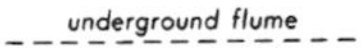 underground flume
FALLS	
Double-Line	falls
Single-Line	 falls
RAPIDS	
Double-Line	rapids
Single-Line	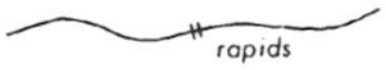 rapids

Fig. 11-4. *Continued.*

Figure 11-4 describes other hydrographical features contained on aeronautical charts. Among them are symbols for swamps, marshes, and bogs. A swamp is nothing more than a lake with trees growing out of it—an emergency landing there could be catastrophic. Pilots

HYDROGRAPHY	
RICE PADDIES Extensive areas indicated by label only.	
LAND SUBJECT TO INUNDATION	
SPRINGS, WELLS AND WATERHOLES	
GLACIERS	
Glacial Moraines	
ICE CLIFFS	
SNOWFIELDS, ICE FIELDS AND ICE CAPS	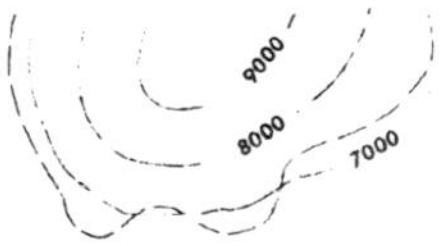

Fig. 11-4. *Continued.*

HYDROGRAPHY	
CANALS	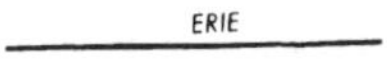 ————————— ERIE —————————
To Scale	 ▬▬▬▬▬▬▬▬
Abandoned or Under Construction	 _ _ _ abandoned _ _ _
To Scale	══ ══ abandoned ══ ══
SMALL CANALS AND DRAINAGE/IRRIGA-TION DITCHES **Perennial**	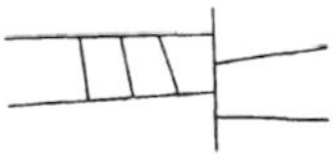
Nonperennial	
Abandoned or Ancient	 abandoned
Numerous Representative pattern and/or descriptive note.	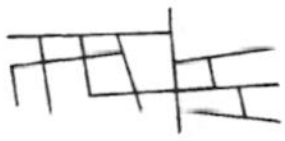
Numerous	*numerous canals and ditches*

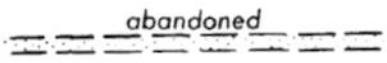

Fig. 11-4. *Continued.*

flying over unfamiliar terrain would be well advised to seek the advice of local pilots or the FSS that is responsible for that region.

Tundra describes a rolling, treeless, often marshy plain, usually associated with arctic regions. Hummocks and ridges describe a wooded tract of land that rises above an adjacent marsh or swamp. Mangroves are any of a number of evergreen shrubs and trees growing in marshy and coastal tropical areas; a nipa is a palm tree indigenous to these areas. Bogs are areas of moist, soggy ground, usually over deposits of peat. Flumes, penstocks, and similar features depict water channels used to carry water as a source of power, such as a waterwheel. Pilots flying in northwestern Montana and especially Alaska can expect to see glaciers and glacial moraines (debris carried by the glacier), ice cliffs, snow and ice fields, and ice caps. Other than canals, the other features in Fig. 11-4 might be difficult to verify and should normally only be used to support other checkpoint features.

Figure 11-5 depicts the remaining hydrographical features contained on aeronautical charts. Ice peaks, polar ice, and pack ice are features restricted to polar and arctic regions. Boulders, wrecks, reefs, and underwater features are displayed because they have certain landmark value. Some of these features might be small and difficult to identify.

One other topographical feature should be mentioned. Mountain passes are depicted by black curved lines outlining the pass. The name of the pass and its elevation are shown. Tioga Pass in Fig. 11-1, upper right, is shown with an elevation of 9943 feet MSL.

Culture

Constructed land features include roads and highways, railroads, buildings, canals, dams, boundary lines, and the like. Many landmarks that can be easily recognized from the air, such as stadiums, racetracks, pumping stations, and refineries, are identified by brief descriptions adjacent to a small black square or circle marking exact location. Depictions might be exaggerated for improved legibility.

Figure 11-6 contains a description of railroads, roads, bridges, and tunnels shown on aeronautical charts. Differences between sectionals and WACs are noted. Single-track railroads have one crosshatch; double and multiple railroads have a double crosshatch. Railroads often make excellent checkpoints. A word of caution. Numerous railroads emanate like spokes from many large cities. Pilots navigating exclusively by the "iron compass" have become hopelessly confused when they inadvertently took the wrong track—pardon the pun.

Never navigate solely by one landmark. Major highways (category 1) also make excellent checkpoints, but they do suffer from the same problems as the railroad. Secondary roads (category 2, and especially secondary category 2) are often difficult to positively identify, especially when flying over sparse areas of desert or plains. Bridges, viaducts, and causeways are often very good checkpoints.

HYDROGRAPHY	
ICE PEAKS	
FORESHORE FLATS Tidal flats exposed at low tide.	
ROCKS – ISOLATED **Bare or Awash**	
WRECKS **Exposed**	
REEFS – ROCKY OR **CORAL**	
MISCELLANEOUS **UNDERWATER FEA-** **TURES NOT OTHER-** **WISE SYMBOLIZED**	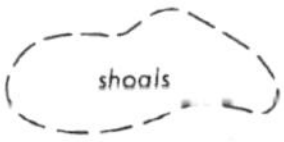 shoals
FISH PONDS AND **HATCHERIES**	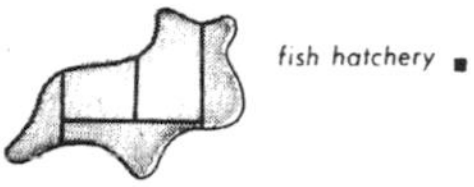 fish hatchery
Intentionally blank	Intentionally blank

Fig. 11-5. *Most ice features are only applicable to arctic regions.*

HYDROGRAPHY

ICE

Permanent Polar Ice

Pack Ice

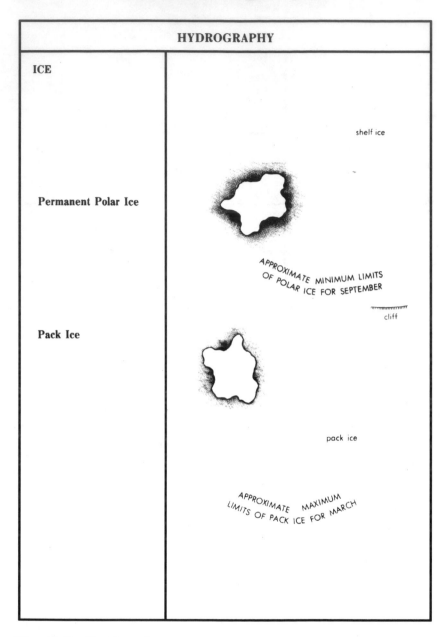

shelf ice

APPROXIMATE MINIMUM LIMITS
OF POLAR ICE FOR SEPTEMBER

cliff

pack ice

APPROXIMATE   MAXIMUM
LIMITS OF PACK ICE FOR MARCH

Fig. 11-5. *Continued.*

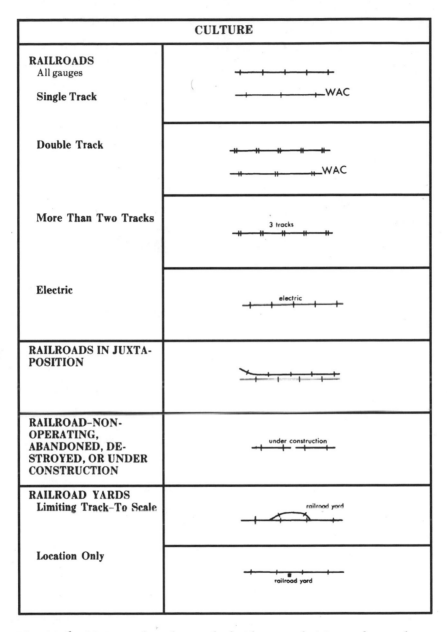

Fig. 11-6. *Major railroads, roads, bridges, and cities make good visual landmarks.*

CULTURE	
ROAD NAMES	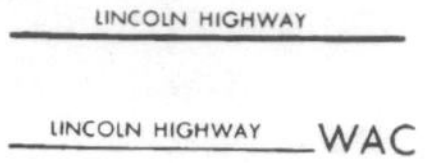 LINCOLN HIGHWAY LINCOLN HIGHWAY ____ WAC
ROADS – UNDER CONSTRUCTION	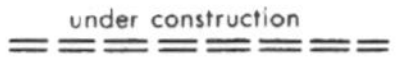 under construction == == == == == == ==
BRIDGES AND VIADUCTS **Railroad**	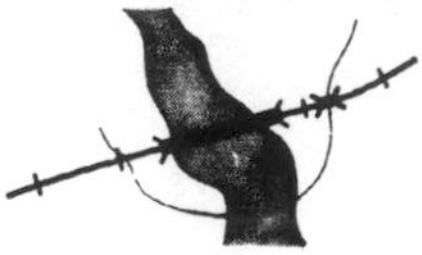
Road	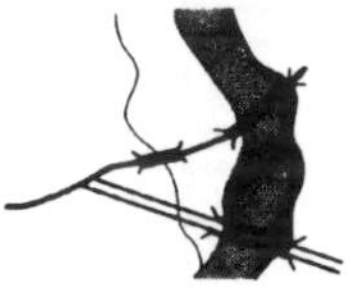
OVERPASSES AND UNDERPASSES	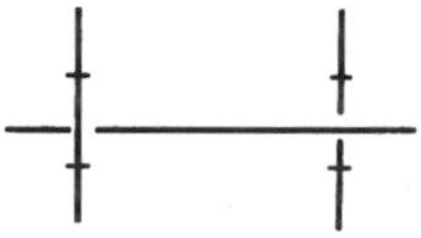
CAUSEWAYS	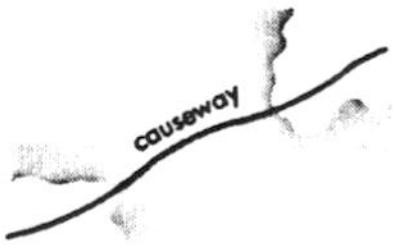

Fig. 11-6. *Continued.*

CULTURE	
RAILROAD STATIONS	
RAILROAD SIDINGS AND SHORT SPURS	

ROADS
  Dual Lane
  Category 1
	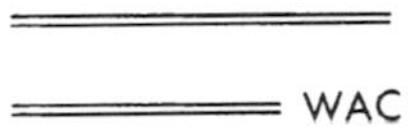 WAC

Primary **Category 2**	WAC
Secondary **Category 2**	
TRAILS Category 3 Provides symbolization for dismantled railroad when combined with label "dismantled railroad."	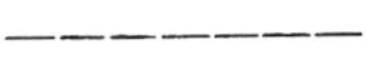

ROAD MARKERS
  U.S. route no.
 	
  Interstate route no.
 	
  Air Marked Identification Label
	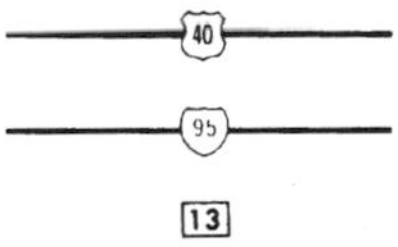

Fig. 11-6. *Continued.*

CULTURE	
TUNNELS – ROAD AND RAILROAD	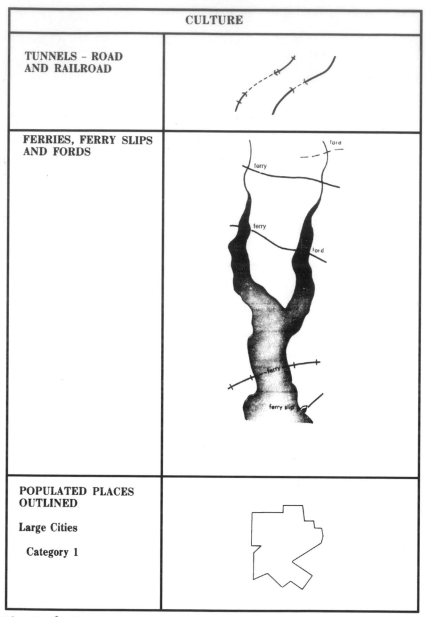
FERRIES, FERRY SLIPS AND FORDS	
POPULATED PLACES OUTLINED **Large Cities** **Category 1**	

Fig. 11-6. *Continued.*

Figure 11-7 shows populated areas (large cities from Fig. 11-6), boundaries, water features, and miscellaneous cultural features. Large and medium cities are shown by their outlines as they appear on the ground. This helps significantly with identification. Towns and villages are only represented by a small circle. Especially where

CULTURE	
POPULATED PLACES OUTLINED **Cities and Large Towns** Category 2	 □ WAC not shown
POPULATED PLACES **Towns and Villages** Category 3	○
PROMINENT FENCES	—x—x—x—x—
BOUNDARIES International	— —— — — —— — —
State and Provincial	— — — — — —
Convention or Mandate Line	US–RUSSIA CONVENTION LINE OF 1867 — — — — — — —
Date Line	INTERNATIONAL (Monday) — — — — — DATE LINE (Sunday)
Time Zone	PST +8 (+7 DT) = UTC MST +7 (+6 DT) = UTC WAC not shown

Fig. 11-7. *Towns and villages, secondary roads, and small cultural features should only be used as supplemental landmarks.*

CULTURE	
WEIRS AND JETTIES	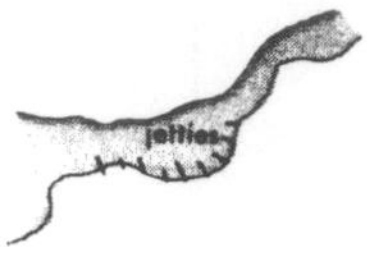
SEAWALLS	
BREAKWATERS	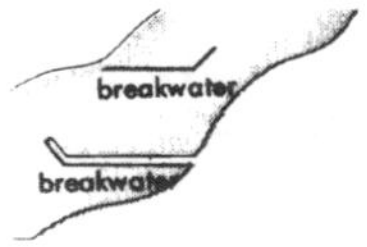
PIERS, WHARFS, QUAYS, ETC.	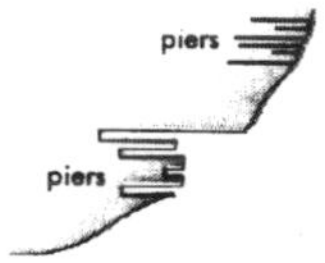
MISCELLANEOUS CULTURE FEATURES	■ stadium
FORTS CEMETERIES	■ fort ■ cemetery
OUTDOOR THEATER	

Fig. 11-7. *Continued.*

several towns or villages are in the same general area, this symbol-ogy makes them hard to positively identify. Political boundaries are shown using standard map symbols. Cultural coastal features are de-picted because of their landmark value. Small mines and quarries are shown by a small crossed-picks symbol.

CULTURE	
MINES AND QUARRIES Shaft Mines and Quarries	⚒
POWER TRANS- MISSION, TELEPHONE & TELEGRAPH LINES	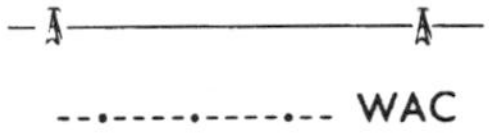 —Å————————Å— --·--·----·---- WAC
PIPELINES	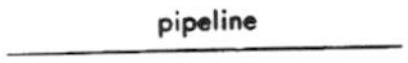 pipeline
Underground	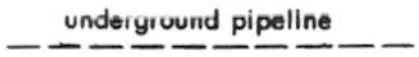 underground pipeline
DAMS	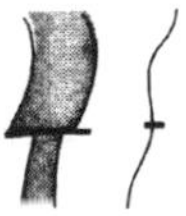
DAM CARRYING ROAD	
PASSABLE LOCKS	 locks
SMALL LOCKS	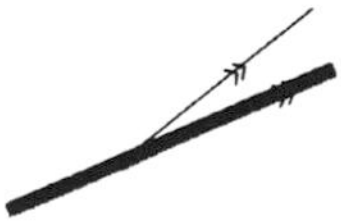

Fig. 11-7. *Continued.*

CULTURE	
WELLS **Other Than Water**	o oil well
RACE TRACKS	⊂⊃
LOOKOUT TOWERS Air marked identification	Ⓐ P-17 (Site number) 618 (Elevation base of tower)
LANDMARK AREAS	dark area
TANKS	● water ● gas
COAST GUARD **STATION**	✦ CG
AERIAL CABLEWAYS, **CONVEYORS, ETC.**	aerial cableway aerial cableway ■–––––■ ■–––––––––■ WAC
Intentionally blank	Intentionally blank

Fig. 11-7. *Continued.*

Pilots should pay particular attention to the symbol for aerial cable-ways, conveyers, and the like, which are formally called catenaries (Fig. 11-7). The catenaries depicted on aeronautical charts are cables, power lines, cable cars, or similar structures suspended between peaks, a peak and valley below, or across a canyon or pass. A cable-way is normally 200 feet or higher above terrain, which poses a very serious hazard to low-flying aircraft; the cable might be marked with orange balls or lights.

Cultural features are not revised as often as aeronautical information; therefore, especially in areas of rapid metropolitan development, cultural features as seen from the air might differ from those depicted on the chart.

Power transmission lines (high-tension lines) are depicted for their landmark and safety value. Often, transmission lines can be used to verify the identification of other landmarks. Although not normally qualifying as an obstruction, their depiction alerts pilots flying at low altitudes to this sometimes almost invisible hazard. Transmission lines are shown on a chart as small black towers connected by a single line (Fig. 11-1).

Obstructions

Charted obstructions normally consist of features extending higher than 200 feet AGL. Objects 200 feet or lower are charted only if considered hazardous, for instance close to an airport where they might affect takeoffs and landings. Federal Aviation Regulations require that airplane pilots, even when flying over sparsely populated areas, cannot "operate closer than 500 feet to any…structure," except for takeoff or landing; therefore, objects 200 feet high or lower, except in the case of an emergency, should have no operational impact.

Sectional charts contain a caution note: "This chart is primarily designed for VFR navigational purposes and does not purport to indicate the presence of all telephone, telegraph, and power transmission lines, terrain, or obstacles which may be encountered below reasonable and safe altitudes." The fact that objects of 200 feet can exist without the requirement to be charted should be a sobering thought when considering scud running. (*Scud* are shreds of small detached clouds moving rapidly below a solid deck of higher clouds. Pilots who attempt to negotiate these conditions ("scud running") often fly at extremely low altitudes and in low visibility. These operations are usually very hazardous to the health of the pilot, passengers, and those on the ground.) Also keep in mind that many obstructions have guy wires that extend some distance outward from the structure. How about helicopter pilots? Yes, helicopters can operate at less than these minimums "if the operation is conducted without hazard to persons or property on the surface." Helicopter, balloon, and ultralight pilots need to understand these charting limitations and plan their flights accordingly.

Obstacles lower than 999 feet AGL are shown by the standard obstruction symbol as illustrated in Fig. 11-8. Obstacles 1000 feet or higher above ground level are shown by the elongated obstruction symbol (Fig. 11-8). Certain cultural features that can be clearly seen from the air and used as checkpoints might be represented with pictorial symbols (black) with elevation data in blue, such as the prominent Golden Gate Bridge.

Height of the obstacle above ground level and mean sea level elevation of the top of the obstacle are shown when known or when they can be reliably determined. The height above ground level is shown in parentheses below the mean sea level elevation of the obstacle, which is a logical arrangement:

2468

(1200)

The height of the obstacle is 1200 feet AGL (1200). The top of the obstacle is 2468 feet MSL (2468). The height above ground level might be omitted in extremely congested areas to avoid confusion. Within high-density groups of obstacles, only the highest obstacle in the area

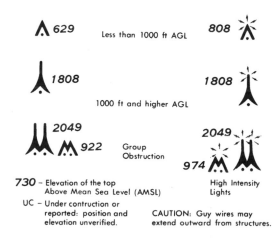

Fig. 11-8. *Normally, only obstructions higher than 200 feet are charted.*

will be shown using the group obstacle symbol (Fig. 11-8). Obstacles under construction are indicated by the letters "UC" immediately adjacent to the symbol. When available, the eventual above-ground-level height of the obstruction will appear in parentheses. Obstacles with strobe lighting systems are shown as indicated in Fig. 11-8.

Aeronautical

Aeronautical information on visual charts consists of airports, radio aids to navigation (NAVAIDs), airspace, and navigational information. Airports are identified for:

- Size (runway lengths).
- Use (civil, military, private).
- Services (availability of fuel and repairs, which are typically detailed in a directory rather than on a chart).

The types and frequencies of NAVAIDs are depicted. Controlled and special-use airspace within the scope of the chart is shown (below 18,000 feet). Navigational information such as magnetic variation, airway intersections, and lighting aids are depicted.

Airports

Visual charts depict civil, military, and some private, landplane, helicopter, and seaplane airports (Fig. 11-9). Hard-surfaced runways of 1500 to 8000 feet are enclosed within a circle depicting runway orientation. All recognizable runways, including some that might be closed, are shown for visual identification. Hard-surfaced runways greater than 8000 feet do not conveniently fit in a circle; the circle is omitted, but runway orientation is preserved. Airports with other than hard-surfaced runways, such as dirt, sod, gravel, and the like, are depicted as open circles. Airports served by an FAA control tower (CT) or nonfederal control tower (NFCT) are shown in blue; all other airports are in magenta, which is a purplish red color. Tick marks around the basic airport symbol indicate the availability of fuel and that the airport is tended during normal working hours. Pilots should keep in mind that types of fuel and specific hours attended are contained in the *Airport/Facility Directory,* with changes or nonavailability of services mentioned in NOTAMs.

Restricted, private, and abandoned airports are shown for emergency or landmark purposes only. Pilots wishing to use restricted or private

AIRPORTS		
LANDPLANE-MILITARY Refueling and repair facilities for normal traffic. All recognizable runways, including some which may be closed, are shown for visual identification.	PAPAGO AAF *1270 ˙L 30* NAS MOFFETT CT - **118.3** *40 L 92*	 WAC
SEAPLANE-MILITARY Refueling and repair facilities for normal traffic.	NAS ALAMEDA *00 ˙L 100*	 WAC
LANDPLANE-CIVIL Refueling and repair facilities for normal traffic.	SCOTT VALLEY (CA06) *2728 ˙L 37 122.8* FSS SISKIYOU CO (SIY) *2648 L 75 123.0* SAN FRANCISCO (SFO) INTL CT - **120.5** ATIS 115.8 113.7 *12 L 106 123.0*	 WAC
SEAPLANE-CIVIL Refueling and repair facilities for normal traffic.	ESSEX SKYPARK (28B) *00 L 150*	 WAC
LANDPLANE CIVIL AND MILITARY Refueling and repair facilities for normal traffic.	SIOUX CITY (SIO) *1097 L 90 123.0* SANTA MONICA (SMO) CT — **120.1** * ⓒ ATIS 119.15 *175 L 82 122.95*	 WAC

Fig. 11-9. *Airport symbols provide information on size, use, and services.*

AIRPORTS	
SEAPLANE CIVIL AND MILITARY Refueling and repair facilities for normal traffic.	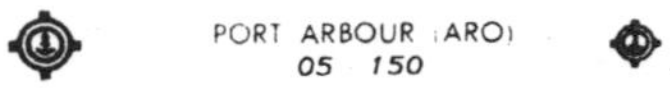 ⊕ PORT ARBOUR (ARO) ⊕ 05 150 <div align="right">WAC</div>

LANDPLANE- EMERGENCY No facilities or complete information is not available.
 	
Add appropriate notes as required: "closed, approximate position, existence unconfirmed."
	ELMA (WA22)
	20 L 21
	O PUBLIC USE – limited attendance or no service available O
	
	OMH (Pvt)
	200 - 17
	Ⓡ RESTRICTED OR PRIVATE – use only in emergency, or by Ⓡ
	    specific authorization
	
	AIRPORT
	- - - UNVERIFIED – a landing area available for public use but
	Ⓤ warranting more than ordinary precaution due to: Ⓤ
	    (1) lack of current information on field conditions, and/or
	    (2) available information indicates peculiar operating limi-
	    tations
	
	⊗ ABANDONED – depicted for landmark value or to prevent ⊗
	    confusion with an adjacent useable landing area. (Normally
	    at least 3000' paved) **WAC**

SEAPLANE- EMERGENCY No facilities or complete information is not available.	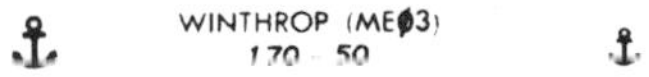 ⚓ WINTHROP (ME∅3) ⚓ 1.70 50 <div align="right">WAC</div>
HELIPORT (Selected)	 Ⓗ PENTAGON Ⓗ (ARMY) 995 <div align="right">WAC</div>
ULTRALIGHT FLIGHT PARK (Selected)	Ⓕ WAC not shown

Fig. 11-9. *Continued.*

landing facilities must obtain permission from that airport authority. A check of your insurance policy might also be in order. Some insurance policies restrict landings to public airports, except in emergencies. Airports are labeled unverified when available for public use, but warranting more than ordinary precautions due to lack of current information on field conditions, or available information indicates peculiar operating limitations. Selected ultralight flight parks appear only on sectional charts as an "F" within the airport circle.

Figure 11-10 decodes standard airport information. The circled letter "R" preceding the airport name indicates the availability of airport surveillance radar, and the airport location identifier follows the airport name: R Oakland (OAK). With airspace reclassification, Class D airspace replaced the control zone and airport traffic area. This all but eliminated the need for special airport traffic areas defined by FAR 93. Special airport traffic areas are indicated on the chart by the airport name placed within a box, for example, Anchorage. Two still exist in Alaska at Anchorage and Ketchikan. Specific requirements are contained in the *Alaska Supplement, Regulatory Notices.*

FSS indicates a flight service station on the field, and RFSS indicates a remote flight service station. These facilities might provide a local airport advisory at selected airports. The decision whether AFSSs will provide local airport advisories is still in question. Only the primary tower local control frequency appears. A star following the local control frequency indicates a part-time tower. Supplemental and additional frequencies, such as approach, secondary local control and ground frequencies, and tower hours of operation are contained on the end panels or margin of the chart and in the *Airport/Facility Directory.* Automatic terminal information service (ATIS), where available, is always shown. Supplemental frequencies, such as an aeronautical advisory station (UNICOM) or VFR advisory service, might be listed. The letter "C" within a circle indicates the common traffic advisory frequency (CTAF): not shown on WACs. This frequency is usually the tower frequency at airports with part-time towers. Airport elevation is always in feet above mean sea level and never abbreviated, as in Fig. 11-10, 285 feet MSL.

Runway length is the length of the longest active runway, including displaced threshold and excluding overruns. Runway length is shown to the nearest 100 feet, using 70 as the division point; a runway 8070 feet long is charted as "81," and a runway 8069 feet long is charted as "80." In the example, the longest runway is 7200 feet.

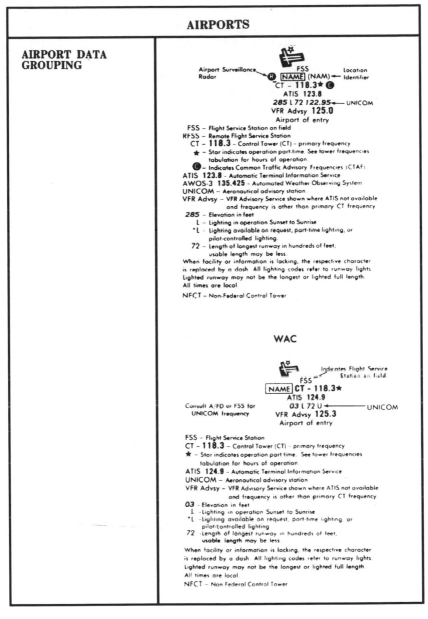

AIRPORTS

AIRPORT DATA GROUPING

Airport Surveillance Radar — FSS — Location Identifier
® [NAME] (NAM)
CT - 118.3★ ©
ATIS 123.8
285 L 72 122.95 — UNICOM
VFR Advsy 125.0
Airport of entry

FSS – Flight Service Station on field
RFSS – Remote Flight Service Station
CT - 118.3 – Control Tower (CT) - primary frequency
★ – Star indicates operation part-time. See tower frequencies tabulation for hours of operation.
© – Indicates Common Traffic Advisory Frequencies (CTAF)
ATIS 123.8 – Automatic Terminal Information Service
AWOS-3 135.425 - Automated Weather Observing System
UNICOM – Aeronautical advisory station.
VFR Advsy – VFR Advisory Service shown where ATIS not available and frequency is other than primary CT frequency
285 – Elevation in feet
L – Lighting in operation Sunset to Sunrise
*L – Lighting available on request, part-time lighting, or pilot-controlled lighting.
72 – Length of longest runway in hundreds of feet; usable length may be less.
When facility or information is lacking, the respective character is replaced by a dash. All lighting codes refer to runway lights. Lighted runway may not be the longest or lighted full length. All times are local.
NFCT – Non-Federal Control Tower

WAC

FSS — Indicates Flight Service Station on field.
[NAME] CT - 118.3★
ATIS 124.9
Consult A/FD or FSS for UNICOM frequency
03 L 72 U — UNICOM
VFR Advsy 125.3
Airport of entry

FSS – Flight Service Station
CT - 118.3 – Control Tower (CT) - primary frequency
★ – Star indicates operation part-time. See tower frequencies tabulation for hours of operation.
ATIS 124.9 – Automatic Terminal Information Service
UNICOM – Aeronautical advisory station
VFR Advsy – VFR Advisory Service shown where ATIS not available and frequency is other than primary CT frequency
03 – Elevation in feet
L – Lighting in operation Sunset to Sunrise
*L – Lighting available on request, part-time lighting, or pilot-controlled lighting
72 – Length of longest runway in hundreds of feet; usable length may be less.
When facility or information is lacking, the respective character is replaced by a dash. All lighting codes refer to runway lights. Lighted runway may not be the longest or lighted full length. All times are local.
NFCT – Non Federal Control Tower

Fig. 11-10. *Airport information normally provides enough data for the pilot to operate into or out of the field under the provisions of VFR.*

Airport lighting, indicated by the letter "L," operates sunset to sunrise, unless preceded by an asterisk, which indicates limitations exist. All lighting codes refer to runway lights. The lighted runway might not be the longest or lighted full length. Pilots must refer to the *Airport/Facility Directory* for specific limitations, such as pilot-controlled lighting. Other remarks are added as required, such as airport of entry. When information is not available, the respective character is replaced by a dash.

Radio aids to navigation

Figure 11-11 shows standard symbols for the very high frequency omnidirectional radio range (VOR), a VOR collocated with distance measuring equipment (DME), and a VOR collocated with a tactical air navigation (TACAN) facility. A TACAN provides azimuth information similar to a VOR, but on an ultra high frequency (UHF) band used by military aircraft, and distance information from the DME. When the NAVAID is located on an airport (Fig. 11-11), the type of facility (in this case VOR only) appears above the NAVAID box; otherwise, the

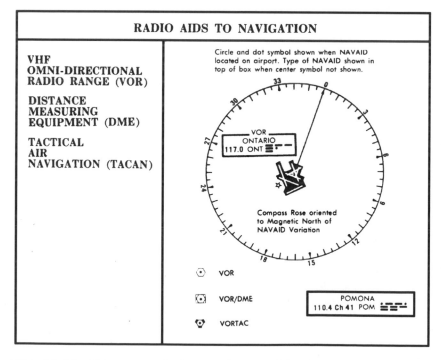

Fig. 11-11. *NAVAID symbols provide type of facility and frequency.*

appropriate symbol indicates the type of facility: VOR, VOR/DME, or collocated VOR and TACAN (VORTAC).

Note in Fig. 11-11 that the NAVAID box provides frequency and identification information for the Ontario VOR. The VOR frequency is 117.0 MHz, identification ONT, followed by a representation of its aural Morse code signal. The lower right corner of Fig. 11-11 shows a VORTAC NAVAID box. This is the Pomona VORTAC. The VOR frequency is 110.4 MHz, and the DME and TACAN channel (Ch) is 41. (Because VOR frequencies and DME channels are paired, when a pilot chooses the VOR frequency, the paired DME channel is automatically selected.) The identification of the NAVAID is POM, followed by the representation of the aural Morse code signal.

Locations were abbreviated with two letters in the early days of aviation, for instance NK was Newark. As the number of NAVAIDs and airports increased, three letter identifiers came into use. All VOR, VOR/DME, VORTAC, and many low-frequency radio beacons have three-letter identifiers.

Figure 11-12 contains other standard NAVAID and flight service station communication symbols. Low- and medium-frequency NAVAIDs are shown in magenta; VOR, VOR/DME, and VORTACs are in blue. Low- and medium-frequency radio ranges have been decommissioned. Nondirectional beacons, marine beacons, and broadcast-station symbols are shown.

Heavy-line boxes indicate standard *simplex* FSS communication frequencies 121.5 and 122.2 MHz; simplex means one-way radio communications in which the pilot transmits and receives on the same channel. Other FSS frequencies are printed above the box— for example, 123.6 (for local airport advisories) and FSS discrete frequencies. Routine communications should be accomplished on the station's discrete frequency. These frequencies are spread apart at individual facilities and locations to avoid frequency congestion with aircraft calling adjacent stations.

If a frequency is followed by the letter R (122.1R), the FSS has only receive capability on that frequency; therefore, the pilot transmits on 122.1 (or another designated frequency). The pilot must tune another frequency, usually the associated VOR, to receive voice communications from the FSS. This *duplex* communication requires the pilot to ensure that the volume is turned up on the VOR receiver. For example, in the upper right box of Fig. 11-12, the Prescott FSS has a

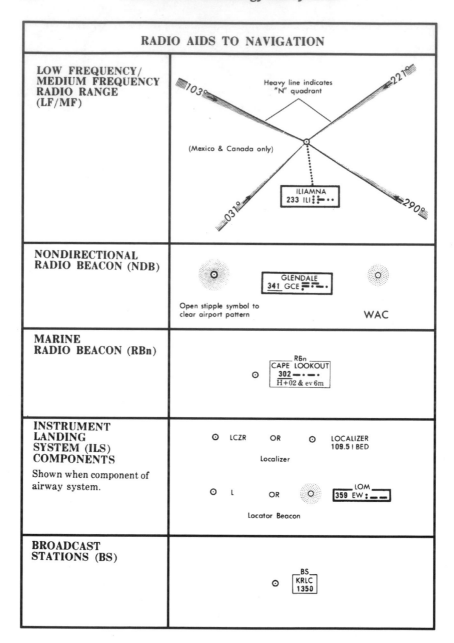

Fig. 11-12. *As well as NAVAID information, visual charts contain FSS communication frequencies.*

RADIO AIDS TO NAVIGATION

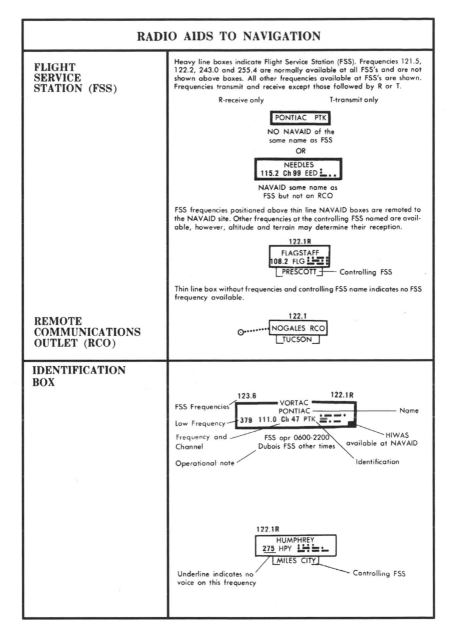

Fig. 11-12. *Continued.*

receiver located at the Flagstaff VOR on 122.1R, noted above the NAVAID box. A pilot wishing to communicate through the VOR would tune the transmitter to 122.1 MHz, and select Flagstaff, 108.2 MHz, on the VOR receiver. An FSS can transmit on many frequencies (VORs and remote outlets, for instance). With FSS consolidation, it is important for the pilot to *advise the FSS which frequency is being monitored in the airplane and the airplane's general location.* For example, "Reno Radio, Cessna four three three four echo, listening one two two point six, Ely, over."

Note that only selected frequencies are depicted on these charts. Because en-route flight advisory service (flight watch) has a common frequency of 122.0 MHz, the frequency is not shown. Pilots calling flight watch should always include their approximate location on initial contact. Approach control and air route traffic control center frequencies are also omitted. Other frequencies are on the chart's end panels and margins, in the *Airport/Facility Directory,* or are available from an FSS.

A small square in the lower right corner of the NAVAID box indicates hazardous in-flight weather advisory service (HIWAS) is available on the VOR frequency. The circled letter "T" means that a transcribed weather broadcast (TWEB) is transmitted over the VOR. An automated weather observation system and the frequency (AWOS-3 135.425) advertises the availability of this service. An underlined frequency indicates no voice communications available on that particular frequency.

Airspace

Controlled airspace has become rather complicated and, ironically, it affects the VFR pilot to a much greater degree than the IFR pilot. The reason why airspace has become so complicated can be explained by its evolution. In the early days of aviation, all flying was visual; there was no such thing as controlled airspace or air traffic control. It wasn't until the middle 1930s that blind flying using instruments became practical. Instrument flying came into its own after World War II with the development of navigational aids and communications.

The purpose of controlled airspace is to provide a safe environment for instrument operations. Controlled airspace established weather minimums. As one might expect, controlled airspace originally developed around airports where air traffic was congested. The next logical extension included the then-new electronic airway system.

As radios become more common, certain airspace required the pilot to establish radio communications with the controlling authority. As jet travel increased, all airspace in the contiguous United States above 14,500 feet became controlled, and flights above 24,000 feet required an IFR clearance. In the 1960s and 1970s, more and more airspace became controlled. Airspace that required a clearance for all aircraft was lowered to 18,000 feet, along with the airspace around major terminals. Specific communications and aircraft equipment requirements were established around smaller terminals.

The VFR pilot of the twenty-first century must contend with various weather minimums in an alphabet soup of controlled airspace, establish communications in certain areas, and make sure the aircraft has the required electronic equipment.

For our purposes, airspace on visual charts can be divided into two basic categories: controlled airspace and special-use airspace. Controlled airspace designated on visual charts is Class G, E, D, C, or B airspace. Their primary purpose is to protect IFR aircraft when weather conditions do not allow see-and-avoid separation. Class A airspace is not depicted. Special-use airspace is designated as prohibited, restricted, warning, alert, and military operations areas and military training routes. Figures 11-13 and 11-14 depict controlled airspace shown on visual charts.

With the preceding as a background briefing of sorts, let's see if we can make some sense out of the chaos. Areas with no air traffic control services are designated Class G airspace. Airspace not designated as Class E, D, C, or B on visual charts is Class G airspace. Refer to Fig. 11-1. The Mammoth Lakes Airport is in Class G airspace.

If we were to fly west from Mammoth toward Merced (the white line in Fig. 11-1), we would encounter a blue vignette encompassing airway Victor 230. (Note the number "115" in the box just below the airway designation. This is total mileage in nautical miles between NAVAIDs on the airway.) The blue vignette designates controlled Class E airspace at 1200 feet AGL. The dark edge of the vignette indicates the limit, and the vanishing edge is the direction of controlled airspace. Continuing west, we exit Class E airspace northwest of the airway.

When the base of Class E airspace is above 1200 AGL, the lower limit is printed on the chart. Continuing westbound, we come across another blue vignette (along the 119°30' longitude line). The base of this Class E airspace is 12,000 feet MSL.

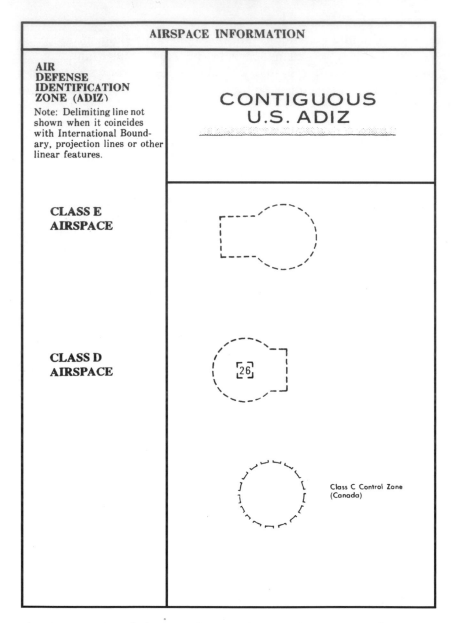

Fig. 11-13. *Visual charts must provide enough airspace information to allow the VFR pilot to operate safely within an ever increasingly complex system.*

AIRSPACE INFORMATION	
LOW ALTITUDE AIRWAYS VOR LF/MF Low altitude Federal Airways are indicated by center line. Only the controlled airspace effective below 18,000 feet MSL is shown.	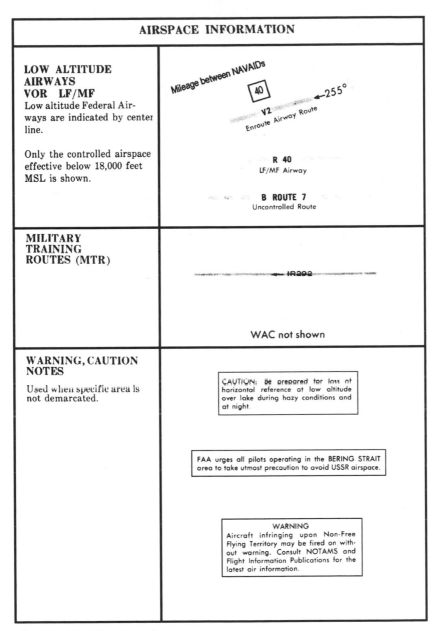
MILITARY TRAINING ROUTES (MTR)	IR292 WAC not shown
WARNING, CAUTION NOTES Used when specific area is not demarcated.	CAUTION: Be prepared for loss of horizontal reference at low altitude over lake during hazy conditions and at night. FAA urges all pilots operating in the BERING STRAIT area to take utmost precaution to avoid USSR airspace. WARNING Aircraft infringing upon Non-Free Flying Territory may be fired on without warning. Consult NOTAMS and Flight Information Publications for the latest air information.

Fig. 11-13. *Continued.*

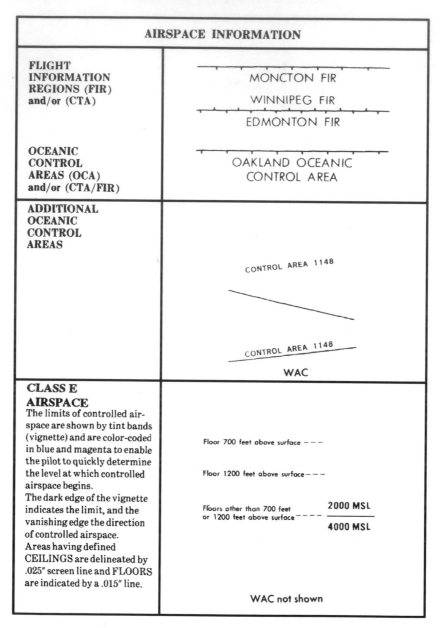

AIRSPACE INFORMATION

FLIGHT INFORMATION REGIONS (FIR) and/or (CTA)	MONCTON FIR WINNIPEG FIR EDMONTON FIR
OCEANIC CONTROL AREAS (OCA) and/or (CTA/FIR)	OAKLAND OCEANIC CONTROL AREA
ADDITIONAL OCEANIC CONTROL AREAS	CONTROL AREA 1148 CONTROL AREA 1148 **WAC**
CLASS E AIRSPACE The limits of controlled airspace are shown by tint bands (vignette) and are color-coded in blue and magenta to enable the pilot to quickly determine the level at which controlled airspace begins. The dark edge of the vignette indicates the limit, and the vanishing edge the direction of controlled airspace. Areas having defined CEILINGS are delineated by .025″ screen line and FLOORS are indicated by a .015″ line.	Floor 700 feet above surface − − − Floor 1200 feet above surface − − − Floors other than 700 feet or 1200 feet above surface − − − **2000 MSL** / **4000 MSL** WAC not shown

Fig. 11-13. *Continued.*

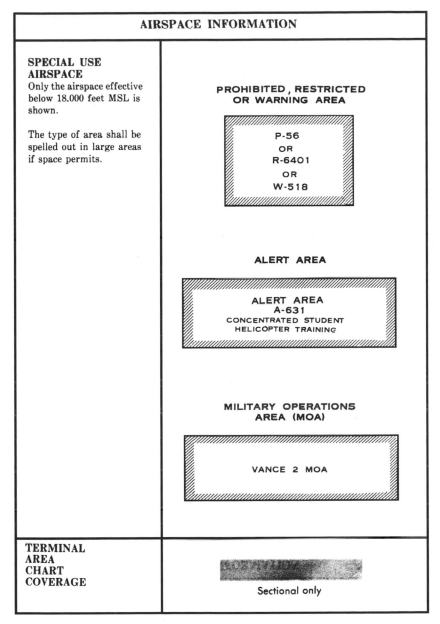

AIRSPACE INFORMATION

SPECIAL USE AIRSPACE
Only the airspace effective below 18,000 feet MSL is shown.

The type of area shall be spelled out in large areas if space permits.

PROHIBITED, RESTRICTED OR WARNING AREA

P-56
OR
R-6401
OR
W-518

ALERT AREA

ALERT AREA
A-631
CONCENTRATED STUDENT
HELICOPTER TRAINING

MILITARY OPERATIONS AREA (MOA)

VANCE 2 MOA

TERMINAL AREA CHART COVERAGE

Sectional only

Fig. 11-13. *Continued.*

The staggered thick-blue line (just east of Victor 165) indicates a change in the floor of Class E airspace. With no altitude specified, Class E airspace begins at 1,200 AGL.

From this point, we decide to fly direct to the El Nido VOR, near Merced. Class E airspace continues at 1200 AGL until we're about

AIRSPACE INFORMATION	
PARACHUTE JUMPING AREA	☺ WAC not shown
HANG GLIDING ACTIVITY	WAC not shown
GLIDER OPERATING AREA	 WAC not shown
ULTRALIGHT ACTIVITY	 WAC not shown
CLASS B AIRSPACE Appropriate notes as required may be shown.	LAS VEGAS CLASS B (Outer limit only shown on WAC)
	20 NM – – Distance from facility (TAC) 70 – – – – Ceiling of Class B in hundreds of feet MSL 50 – – – – Floor of Class B in hundreds of feet MSL 124.3 – – –ATC Sector Frequency WAC not shown CONTACT LAS VEGAS APPROACH CONTROL ON 121.1 OR 257.8 (TAC only)
MODE C AREA (See FAR 91.215/AIM) Appropriate notes as required may be shown.	MODE C 30 NM Distance from facility All mileages are nautical (NM)
TERMINAL RADAR SERVICE AREA (TRSA) Appropriate notes as required may be shown.	BILLINGS TRSA WAC not shown
	80 – – – – – Ceiling of TRSA in hundreds of feet MSL 40 – – – – – Floor of TRSA in hundreds of feet MSL SEE TWR FREQ TAB WAC not shown

Fig. 11-14. *As well as airspace information, visual charts include potential hazards, such as parachute jumping areas, glider, ultralight, and hang glider activity, and locations of high-speed military operations.*

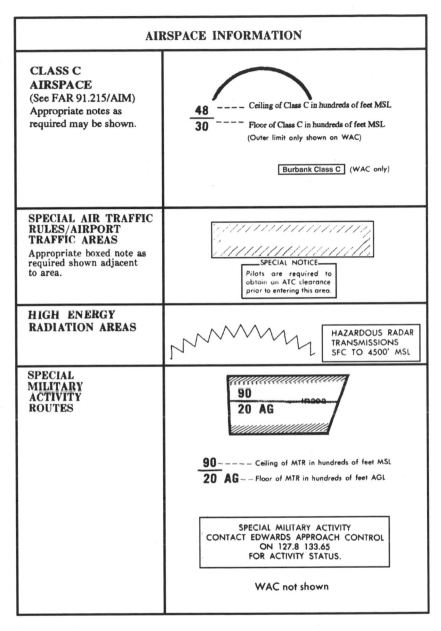

AIRSPACE INFORMATION

CLASS C AIRSPACE
(See FAR 91.215/AIM)
Appropriate notes as required may be shown.

SPECIAL AIR TRAFFIC RULES/AIRPORT TRAFFIC AREAS
Appropriate boxed note as required shown adjacent to area.

HIGH ENERGY RADIATION AREAS

SPECIAL MILITARY ACTIVITY ROUTES

Fig. 11-14. *Continued.*

two miles from the VOR, where we see a magenta vignette. The magenta vignette indicates that Class E airspace begins at 700 AGL.

If we were to proceed to the Merced Airport, we would encounter a magenta dashed line around the airport. The magenta dashed line means Class E airspace starts at the surface (surface-based Class E airspace). A magenta dashed line also designates surface-based Class E airspace associated with other types of airspace. For example, note the magenta dashed lines north of Castle AFB and south of Modesto.

At what altitude does Class E airspace begin over the Mammoth Airport? Unless designated at a lower altitude, Class E airspace begins at 14,500 MSL over the United States, except for airspace that is less than 1500 feet AGL (above mountains that are 13,000 or more feet MSL). (A sage pilot once observed that the first word in the Federal Aviation Regulations was "except.") Class E airspace extends upward to but not including 18,000 feet.

The purpose of weather minimums is to allow enough ceiling, visibility, and cloud clearance for VFR *and* IFR aircraft to "see and avoid." VFR weather minimums—that's what they are, minimums—evolved in much the same way as controlled airspace. VFR weather minimums, especially below 10,000 feet, are much the same as they were in the beginning days of the Piper Cub and DC-3. With high-performance airplanes of all sizes and capabilities flying in the same airspace, a weather "minimum" does not necessarily equate to "safe"!

As well as establishing weather minimums, Class D, C, B, and A airspace impose one or all of the following requirements:

- Communications
- AIRCRAFT EQUIPMENT
- ATC clearance
- Minimum pilot qualifications

Figure 11-15 shows a vertical cross-section of the airspace with VFR minimums and certain equipment requirements.

All Class D airspace is surface-based. The upper limit is normally 2500 feet AGL. As shown in Fig. 11-1, Modesto Class D airspace extends up to and includes 2600 feet MSL. This is indicated by the blue number "26" in the dashed blue box.

Class C airspace extends generally from the surface to 4000 feet AGL around airports with control towers and is served by a radar approach

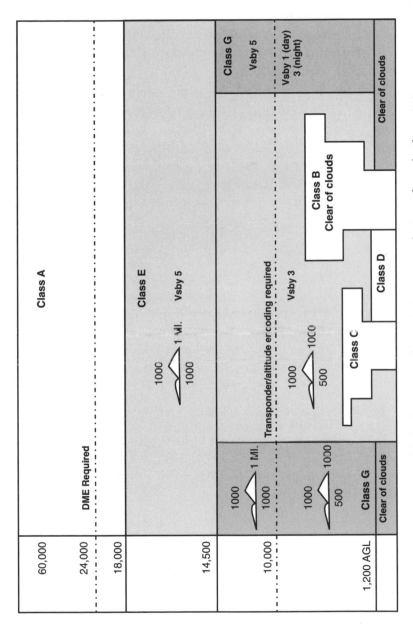

Fig. 11-15. *Airspace established VFR weather minimums and aircraft and pilot minimum equipment and qualifications.*

control. The boundaries of Class C airspace are individually tailored, based upon terrain and operational requirements. Class C airspace is charted using solid magenta lines. Various "shelves" exist beyond the surface-based airspace. Bases and tops of the "shelved airspace" are indicated in magenta (SFC/42 indicates "surface to 4200 feet MSL"; 14/42 indicates "1400 feet MSL to 4200 feet MSL"). Examine the Castle AFB Class C airspace in Fig. 11-1.

Class B airspace surrounds the busiest airports. Class B generally consists of the airspace from the surface to 10,000 feet MSL with various shelves, sometimes referred to as an "upside-down wedding cake." Class B airspace is charted using solid blue lines. Boundaries are defined by VOR radials, DME arcs, and prominent landmarks. Like Class C airspace, the bases and tops are charted. Like Class C airspace, bases and tops of Class B airspace are indicated, in this case in blue. See Fig. 12-14 in Chap. 12.

Occasionally, Class D airspace extends to the base of overlying Class C or Class B airspace. This is one reason for nonstandard Class D airspace tops; in any case, the height is specified on the chart. When Class C airspace terminates at the base of Class B airspace, it is indicated by the magenta "T" (T/15 base of Class C 1500-feet upper limit base of overlying Class B airspace). Notice in Fig. 12-14 the Oakland Class C airspace over the Hayward Airport. (The letter "T" is a leftover from the old airspace classification: the base of the terminal control area.)

Class A airspace consists of that area from 18,000 to 60,000 feet MSL. Class E airspace is that area above 60,000 feet MSL. Class F is not an airspace designation in the United States; however, Class F is an International Civil Aviation Organization (ICAO) airspace classification. Where applicable, IFR and VFR flight are permitted. Air traffic advisory service and flight information service are provided on request. An ATC clearance is not required.

Fixed-wing special VFR is normally available in surface-based controlled airspace. In certain high-density surface-based airspace, special VFR is prohibited. This is indicated in the airport data block by "NO SVFR." See the data block for San Francisco Class B airspace in Fig. 12-14.

Terminal radar service areas (TRSAs) designate airspace where traffic advisories, vectoring, sequencing, and separation of VFR aircraft are provided. TRSAs are designated Stage I, Stage II, or Stage III, which specify radar services that are available. The type of TRSA (Stage I, II, or III) can be found in the *Airport/Facility Directory*.

Visual charts depict special-use airspace (SUA) below 18,000 feet MSL. SUA consists of airspace where activities must be confined because they pose a hazard to aircraft operations. Prohibited, restricted, warning, alert, and military operations areas and military training routes are shown in Fig. 11-13, with special military activity routes in Fig. 11-14. Aircraft operations are prohibited within prohibited areas. These areas are established for security or other reasons associated with the national welfare. Aircraft operations are prohibited within restricted areas when the area is active. Restricted areas are established for unusual, often invisible, hazardous activities, such as artillery firing, aerial gunnery practice, or guided missile firing. Warning areas are established for the same hazards as restricted areas, but over international waters. Alert areas inform nonparticipating pilots of areas that might contain a high volume of training or unusual activity. Pilots should exercise extra caution within these areas.

Military operation areas (MOAs) alert pilots to military training activities. In addition to a possible high concentration of aircraft, military pilots might conduct aerobatic flight and operate at speeds in excess of 250 knots below 10,000 feet. High-speed low-level military operations are conducted along military training routes (MTRs). An MTR is designated IR when IFR operations are conducted within that route; VFR operations are designated VR. IR and VR routes operated at or below 1500 feet AGL will be identified by four-digit numbers (IR 1007, VR 1009). Operations that are conducted above 1500 feet AGL are identified by three-digit numbers (IR 205, VR 257). Special military activity routes alert pilots to areas where cruise missile tests are conducted.

Alert areas, like MOAs, advertise a high concentration of military activity. Figure 11-1 shows alert area A-251 near Castle AFB, which warns pilots of military practice instrument approaches.

Figure 11-14 contains symbols that alert pilots to parachute jumping, glider operations, and ultralight activity. An additional symbol has been added for hang-gliding activity. The symbol resembles a hang glider in flight. Where these symbols appear, pilots cannot expect to be alerted to the activity through NOTAMs. Details on the activity are normally found in the *Airport/Facility Directory*.

Navigational information

Navigational information consists of isogonic lines and values, local magnetic disturbance notes, aeronautical lights, airway intersection depictions, and VFR checkpoints. *Isogonic lines* (lines of equal magnetic declination for a given time) provide the pilot with the difference between true north and magnetic north in degrees. These lines and values are updated every five years. Local magnetic notes alert pilots to areas where the magnetic compass might be unreliable, often due to large deposits of iron ore (Fig. 11-16).

Aeronautical lights at one time were a primary means of navigation—the lighted airways of the 1930s. Light symbols are depicted on visual charts for their navigational value; however, because electronic aids have taken over the majority of navigational needs, many of the old and large airport beacons have been replaced with smaller units. This can lead to confusion. The Bakersfield, California, Meadows Airport has one of the small, less-intense beacons, while the nearby Shafter Airport has the large unit. Pilots, including Army helicopter pilots, regularly key on the Shafter beacon, and even land at the wrong airport. The larger lights can often be seen twice as far as the smaller units. Airport rotating or oscillating beacons are indicated by a star adjacent to the airport symbol and operate sunset to sunrise. Rotating lights (with flashing code identification) and course lights are from the lighted airway days, and almost all have been decommissioned. Symbols are depicted in Fig. 11-16.

Named intersections that can be used as reporting points are depicted on some visual charts (Fig. 11-17). The intersections consist of a five-letter name and might be difficult to pronounce. Use of intersections by VFR pilots is limited because the cross-fix radial headings are not shown. Intersections made up of VOR radials are shown in blue; those made up with low-frequency radio beacons are shown in magenta.

Visual ground signs and VFR checkpoints that are easily recognizable from the air are depicted (Fig. 11-16). Many visual signs are left over from the early days of aviation, have faded, and are of little value; others are large and prominent.

Special conservation areas a20re shown on sectional charts as depicted in Fig. 11-17. Landing, except in an emergency, is prohibited on lands or waters administered by the National Park Service, U.S. Fish and Wildlife Service, or U.S. Forest Service without authorization. All aircraft are requested to maintain a minimum altitude of 2000 feet above the surface. The Yosemite National Park, shown in Fig. 11-1, is one of these areas.

NAVIGATIONAL AND PROCEDURAL INFORMATION

ISOGONIC LINE & VALUE Isogonic lines and values shall be based on the five year epoch chart.	
LOCAL MAGNETIC NOTES **Unreliability Notes**	
COMPASS ROSETTE Shown only in areas void of VOR roses.	

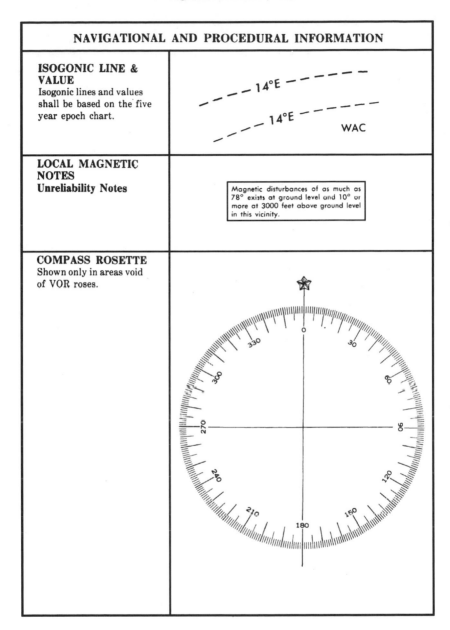

Isogonic line & value: `— — 14°E — — — — —` `— — 14°E — — — — —` WAC

Local magnetic notes: Magnetic disturbances of as much as 78° exists at ground level and 10° or more at 3000 feet above ground level in this vicinity.

Fig. 11-16. *Navigational and procedural information includes lines of magnetic variation and aeronautical lights.*

NAVIGATIONAL AND PROCEDURAL INFORMATION	
AERONAUTICAL LIGHTS **Rotating or Oscillating**	 Located at aerodrome In isolated location
Rotating Light with Flashing Code Identification Light	
Rotating Light with Course Lights and Site Number	
Flashing Light	Rotating Beacon

Fig. 11-16. *Continued.*

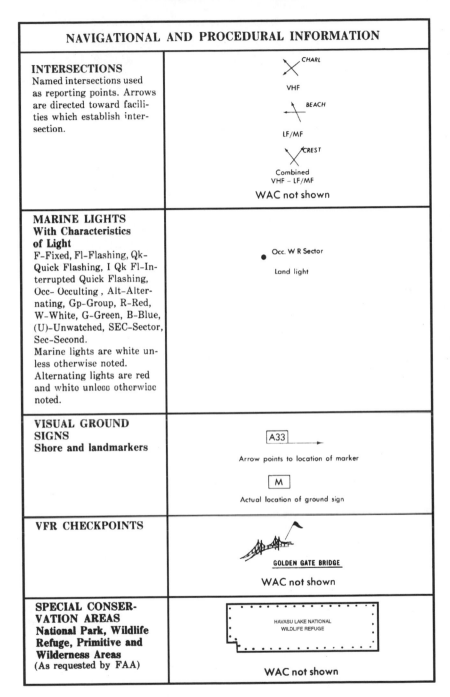

Fig. 11-17. *Visual charts contain airway intersections, visual ground signs, and VFR checkpoints to aid in visual navigation.*

12

Standard visual charts

The National Ocean Service (NOS) publishes several aeronautical charts specifically designed to assist the pilot with visual or VFR navigation: planning charts, sectional charts, terminal area charts (TACs), and world aeronautical charts (WACs). Interestingly, these charts are often more complex than those used for IFR navigation. Chapter 11 discusses the dozen or so types of airspace, plus a description of terminology and symbols common to visual charts; therefore, in this chapter only those terms and symbols unique to individual charts are presented.

Planning charts

Planning charts, as the name implies, are designed for the initial portion of flight preparation, as opposed to operational charts (sectionals, TACs, WACs) used for preflight planning and navigation. These charts are most useful for planning long trips, usually with multiple legs. Planning chart coverage is illustrated in Fig. 12-1.

IFR/VFR low-altitude planning chart

The IFR/VFR low-altitude planning chart combines many of the features previously found on the discontinued IFR/VFR wall planning chart and the flight-case planning chart. The 1:3,400,000 (1″ equals 47 nm) scale chart is designed for preflight planning below 18,000 feet.

This five-color, two-sided chart can be obtained either flat (35 × 36 inches) or folded (5 × 9 inches). The eastern half is on one side, and the western half is on the other side. For wall display, the purchase of two flat charts is required to show the conterminous states. This

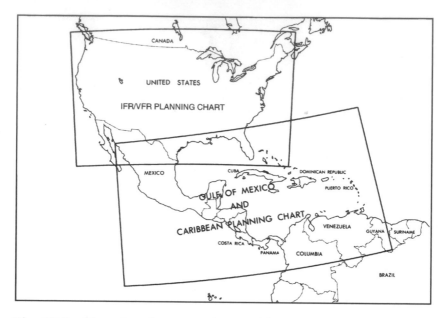

Fig. 12-1. *Planning charts are designed for the initial portion of flight preparation.*

chart is often found in planning rooms of flight schools and fixed-based operators (FBO).

Figure 12-2 contains airspace and navigational symbols unique to this chart. Features include:

- Low-altitude airways
- Navigational aids
- Airports (3000 feet paved or with an instrument approach)
- Tabulated airport airspace classification
- Maximum elevation figures
- Time zones
- ARTCC boundaries
- Sectional chart outlines
- IFR en-route chart outlines
- Special-use airspace
- State boundaries
- Large bodies of water
- Selected cities and towns

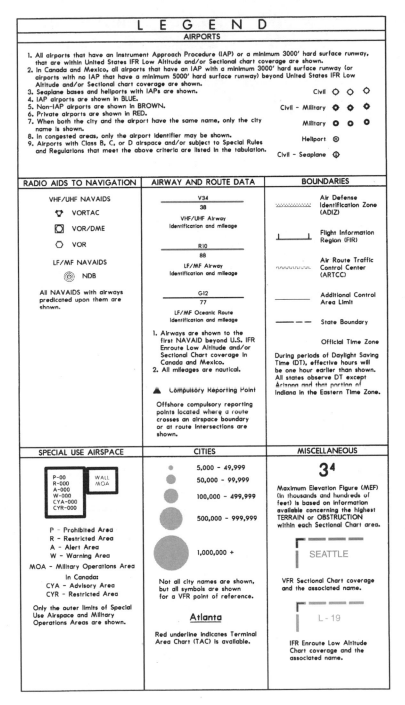

Fig. 12-2. *Many of these symbols are unique to the IFR/VFR planning chart.*

Gulf of Mexico and Caribbean planning chart

The Gulf of Mexico and Caribbean planning chart has a smaller scale than the IFR/VFR low-altitude planning chart. This chart is designed for preflight planning through and around the Gulf of Mexico and Caribbean. Intended to be used in conjunction with world aeronautical charts, it is printed on the back of the Puerto Rico-Virgin Islands VFR terminal area chart. The area of coverage is shown in Fig. 12-1. The chart is 60 × 20 inches, which can be folded to the standard 5 × 10, and has a scale of 1:6,192,178 (1 inch equals 85 nm). This chart is revised annually. Features include:

- Airports of entry
- Special use airspace below 18,000 feet
- Significant bodies of water
- International boundaries
- Large islands and island groups
- Capital cities and cities where an airport is located
- Selected other major cities
- Air mileage between airports of entry
- Index of world aeronautical charts
- Directory of airports, including facilities, servicing, and fuel
- Department of Defense requirements for civilian use of military airports
- Checklist for ditching
- Runway visual range (RVR) conversion table from feet to meters
- Emergency procedures

Charted VFR flyway planning charts

Charted VFR flyway planning charts are designed to assist pilots planning flights through or around high-density areas such as Class B and Class C airspace. These two-color charts are printed on the back of selected terminal area charts (TACs) with coverage corresponding to the TACs. The following TACs contain VFR flyway planning charts:

- Atlanta
- Baltimore-Washington

- Chicago
- Dallas-Fort Worth
- Denver
- Detroit
- Houston
- Los Angeles
- Miami
- Phoenix
- Salt Lake City
- San Diego
- Seattle

Charted VFR flyway planning charts, as the name implies, are not to be used for navigation, or as substitutes for the TAC or sectional charts. Features include:

- Airports
- NAVAIDs
- Special-use airspace
- Class B airspace
- Class C airspace
- Class D airspace
- VFR flyways (suggested headings and altitudes)
- Procedural notes
- Military training routes
- Selected obstacles
- VFR checkpoints
- Hydrographic features
- Cultural features
- Terrain relief designated as VFR checkpoints
- Crucial spot elevations

Using planning charts

A long cross-country flight is nothing more than a series of individual legs. Take, for example, a flight from Van Nuys, California, to Jamestown, New York, and return. The first consideration is the aircraft; the second consideration is the pilot and passengers. It doesn't do much good to plan 4-hour legs with 2-hour bladders. In a Cessna 150, 250-mile legs are comfortable; in a Bonanza, depending upon fuel load, 350- to 400-mile legs would be comfortable. In a Hughes 269 helicopter, with its speed and limited fuel, only 100-mile legs are practical. It all depends on the aircraft, fuel, and pilot/passenger comfort.

With average legs in mind, we proceed to the planning chart to factor in aircraft performance, terrain, airports and their services, and controlled and special-use airspace. I don't like flying over high, rough terrain, or unpopulated areas of deserts or swamps; avoidance is my choice, if at all possible. En-route destinations are selected for the services available. We can plan the flight to major airports, if we have the proper electronic equipment in the aircraft, or to uncontrolled fields. We have to avoid prohibited and restricted areas, and we might wish to avoid MOAs.

From the planning chart, we determine which sectional charts are needed. From these charts, we can either plan to avoid Class B airspace or obtain required terminal area charts to negotiate or avoid their heavy concentrations of traffic.

Sectional charts

Sectional charts are designed for visual navigation of slow- to medium-speed aircraft. These multicolored charts provide the most accurate means of pilotage—navigating the aircraft by means of ground reference—because of their scale. The chart is 20 × 60 inches, which can be folded to the standard 5 × 10 inches and has a scale of 1:500,000 (1 inch equals 7 nm). They are revised semiannually, except for some Alaskan charts that are revised annually. Sectional charts are named for a major city within the area of coverage. Figure 12-3 contains sectional chart coverage for the contiguous states, and Fig. 12-4 shows coverage for Alaska. Features include:

- Visual aids to navigation
- Radio aids to navigation
- Airports
- Controlled airspace
- Restricted areas
- Obstructions

- Topography
- Shaded relief
- Latitude and longitude lines
- Airways and fixes
- Other low-level related data

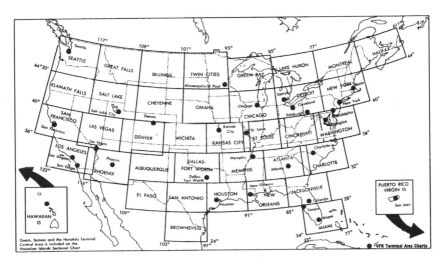

Fig. 12-3. *Sectional charts are named for a major city within their area of coverage.*

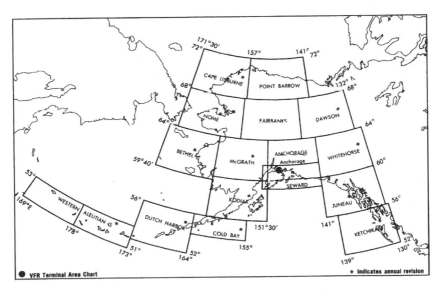

Fig. 12-4. *Sectional charts are revised semiannually, except for some Alaskan charts that are revised annually.*

The Hawaii sectional is the only chart of this series that is not oriented to true north. This is necessary to portray all the islands of the group on a standard size sheet, which also contains the Mariana and Samoa Island groups.

Before using any chart or aeronautical publication, a pilot's first task is to determine currency. That means reviewing the cover pages for effective and obsolescent dates. Figure 12-5 is the cover page and data panel from the Las Vegas Sectional Aeronautical Chart. The chart is a Lambert conformal conic projection with standard parallels 33°20′ and 38°40′, based on the North American Datum of 1983 (World Geodetic system 1984). Topographic data has been corrected to December 1994.

This is the 53d edition of the chart, which became effective March 30, 1995, including airspace amendments that would include changes in the lateral limits of controlled and special-use airspace, and other aeronautical data received by February 2, 1995, such as radio navigation aids and frequencies and radio communication frequencies and locations. Note that it cautions pilots to consult appropriate NOTAMs and flight information publications for supplemental data and current information. The statement is straightforward: "This chart will become OBSOLETE FOR USE IN NAVIGATION upon publication of the next edition scheduled for SEPTEMBER 14, 1995."

The margins of sectional charts contain data panels with supplemental control tower information, selected radar approach control frequencies, and SUA information (Fig. 12-5). The China Lake NWC (Naval Weapons Center) control tower operates between 6:30 a.m. and 10:30 p.m. Monday through Friday on frequencies 120.15 for civilian aircraft and 340.2 for military aircraft. Ground control (GND CON), automatic terminal information service (ATIS), and the availability of radar approaches are also noted. This information is followed by selected Class B, Class C, TRSA, and radar approach control facilities.

Data regarding special-use airspace includes altitudes, effective times, and controlling agency. The status of some restricted areas is routinely released to VFR and IFR operations when not in use. Pilots can obtain this information from the local flight service station or controlling agency, usually approach control or an air route traffic control center. Some restricted areas are reported via the NOTAM (D) category and will be part of a standard FSS or DUAT briefing.

LAS VEGAS
SECTIONAL AERONAUTICAL CHART
SCALE 1:500,000

Lambert Conformal Conic Projection Standard Parallels 33°20' and 38°40'
Horizontal Datum: North American Datum of 1983 (World Geodetic System 1984)
Topographic data corrected to December 1994

53 RD EDITION March 30, 1995
Includes airspace amendments effective March 30, 1995
and all other aeronautical data received by February 2, 1995
Consult appropriate NOTAMs and Flight Information
Publications for supplemental data and current information.
This chart will become, OBSOLETE FOR USE IN NAVIGATION upon publication of
the next edition scheduled for SEPTEMBER 14, 1995

PUBLISHED IN ACCORDANCE WITH INTERAGENCY AIR CARTOGRAPHIC COMMITTEE
SPECIFICATIONS AND AGREEMENTS, APPROVED BY:
DEPARTMENT OF DEFENSE • FEDERAL AVIATION ADMINISTRATION • DEPARTMENT OF COMMERCE

The horizontal reference datum of this chart is North
American Datum of 1983 (NAD 83), which for charting purposes is
considered equivalent to World Geodetic System 1984 (WGS 84)

CONTOUR INTERVAL 500 feet
Intermediate contour 250 feet
Auxiliary contours 100 foot intervals

HIGHEST TERRAIN elevation is 13063 feet
located at 38°59'N – 114°19'W

Critical elevation - - - - - - - - - - - - - - - • 4254

Approximate elevation - - - - - - - - - - - x 3200

Doubtful locations are indicated by omission
of the point locator (dot or "x")

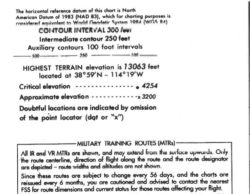

MILITARY TRAINING ROUTES (MTRs)

All IR and VR MTRs are shown, and may extend from the surface upwards. Only
the route centerline, direction of flight along the route and the route designator
are depicted – route widths and altitudes are not shown.

Since these routes are subject to change every 56 days, and the charts are
reissued every 6 months, you are cautioned and advised to contact the nearest
FSS for route dimensions and current status for those routes affecting your flight.

Routes with a change in the alignment of the charted route centerline will be
indicated in the Aeronautical Chart Bulletin of the Airport/Facility Directory.

Military Pilots refer to Area Planning AP/1B Military Training Route North and
South America for current routes.

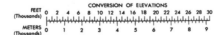

CONVERSION OF ELEVATIONS

Published at Washington, D.C.
U.S. Department of Commerce
National Oceanic and Atmospheric Administration
National Ocean Service

Fig. 12-5. *Chart cover pages display effective dates and data panels with supplemental information.*

CONTROL TOWER FREQUENCIES ON LAS VEGAS SECTIONAL CHART

Airports which have control towers are indicated on this chart by the letters CT followed by the primary VHF local control frequency. Selected transmitting frequencies for each control tower are tabulated in the adjoining spaces, the low or medium transmitting frequency is listed first followed by a VHF local control frequency, and the primary VHF and UHF military frequencies, when these frequencies are available. An asterisk (*) follows the part-time tower frequency remoted to the collocated full-time FSS for use as Local Airport Advisory (LAA) during hours tower is closed. Hours shown are local time. Ground control frequencies listed are the primary ground control frequencies.

Automatic Terminal Information Service (ATIS) frequencies, shown on the face of the chart are primary arrival VHF/UHF frequencies. All ATIS frequencies are listed below. ATIS operational hours may differ from control tower operational hours.

ASR and/or PAR indicates Radar Instrument Approach available.

"MON-FRI" indicates Monday thru Friday.

CONTROL TOWER	OPERATES	TWR FREQ	GND CON	ATIS	ASR/PAR
CHINA LAKE NAWS/ ARMITAGE FIELD	0630-2230 MON-FRI 1600-2230 SUN CLSD ALTN FRI	120.15 340.2	360.2	265.2	
GRAND CANYON NATIONAL PARK	1 APR-30 SEP 0600-2000 1 OCT-31 MAR 0700-1900	119.0	121.9	124.3	
INDIAN SPRINGS AF AUX	0600-1700 MON-FRI EXC HOL	118.3 358.3	118.3 275.8		
McCARRAN INTL	CONTINUOUS	119.9 257.8	121.1 (E OF RY 01R/19L) 121.9 (W OF RY 01R/19L) 319.95	ARR 132.4 DEP 125.6	
NELLIS AFB	CONTINUOUS	132.55 324.3	121.8 275.8	270.1	ASR/PAR
NORTH LAS VEGAS	0600-2000	125.7	121.7	118.05	

CLASS B, CLASS C, TRSA, AND SELECTED RADAR APPROACH FREQUENCIES

LAS VEGAS CLASS B	119.4 379.15 (360°-080°) 133.95 353.7 (280°-360°) 125.9 380.05 (W. OF INTERSTATE 15; S. OF 280°) 118.4 380.05 (E. OF INTERSTATE 15; S. OF 080°)
NELLIS AFB RADAR	124.95 279.7

SPECIAL USE AIRSPACE ON LAS VEGAS SECTIONAL CHART

Unless otherwise noted altitudes are   The word "TO" an altitude means "To and including."
MSL and in feet; time is local.   "MON-FRI" indicates "Monday thru Friday"
Contact nearest FSS for information.   FL—Flight Level
†Other time by NOTAM contact FSS   NO A/G – No air to ground communications

U.S. P—PROHIBITED, R—RESTRICTED, A—ALERT, W—WARNING, MOA—MILITARY OPERATIONS AREA

NUMBER	LOCATION	ALTITUDE	TIME OF USE	CONTROLLING AGENCY**
R-2502 N	FORT IRWIN, CA	UNLIMITED	CONTINUOUS	HI-DESERT TRACON, EDWARDS
R-2505	CHINA LAKE, CA	UNLIMITED	CONTINUOUS	HI-DESERT TRACON, EDWARDS
R-2506	CHINA LAKE SOUTH, CA	TO 6000	SR-SS MON-FRI	HI-DESERT TRACON, EDWARDS
R-2524	TRONA, CA	UNLIMITED	CONTINUOUS	HI-DESERT TRACON, EDWARDS
R-4806 E	LAS VEGAS, NV	100 AGL TO UNLIMITED	0500-2000 MON-SAT†	ZLA CNTR
R-4806 W	LAS VEGAS, NV	UNLIMITED	CONTINUOUS	ZLA CNTR
R-4807 A	TONOPAH, NV	UNLIMITED	CONTINUOUS	ZLA CNTR
R-4807 B	TONOPAH, NV	UNLIMITED	CONTINUOUS	ZLA CNTR

WARNING: R-4806 W AND R-4807 A AND R4807 B CONTAIN MANY UNEXPLODED BOMBS AND ROCKETS, AND OTHER ORDNANCE THAT MAY EXPLODE IF DISTURBED. NO AIRFIELD OR ACTIVE RUNWAY EXISTS WITHIN THESE AREAS; HOWEVER, SIMULATED AIRFIELDS UTILIZED AS BOMBING TARGETS DO EXIST. AIRCRAFT LANDINGS IN R-4806 W AND R-4807 A AND R4807 B ARE AT THE PILOTS OWN RISK.

R-4808 N	LAS VEGAS, NV	UNLIMITED	CONTINUOUS	NO A/G
R-4808 S	LAS VEGAS, NV	UNLIMITED	CONTINUOUS	ZLA CNTR
R-4809	TONOPAH, NV	UNLIMITED	CONTINUOUS	NO A/G
R-4816 N	DIXIE VALLEY, NV	1500 AGL TO BUT NOT INCL FL 180	0715-2330	ZOA CNTR
R-4816 S	DIXIE VALLEY, NV	500 AGL TO BUT NOT INCL FL 180	0715-2330	ZOA CNTR
R-6402 A	DUGWAY PROVING GROUND, DUGWAY, UT	TO FL 580	CONTINUOUS	ZLC CNTR
R-6405	WENDOVER, UT	100 AGL TO FL 580	CONTINUOUS	ZLC CNTR
R-6407	HILL AFB, UT	TO FL 580	CONTINUOUS	ZLC CNTR
A-481	NELLIS AFB, NV	7000 TO 17,000	SR-2200†	NO A/G

**ZLA-Los Angeles, ZLC-Salt Lake City, ZOA-Oakland

MOA NAME	ALTITUDE OF USE*	TIME OF USE†	CONTROLLING AGENCY**
AUSTIN 1	200 AGL	0800-2100 MON-FRI	ZLC CNTR
AUSTIN 2	200 AGL	0800-2100 MON-FRI	ZLC CNTR
DESERT	100 AGL	SR-SS MON-SAT	ZLA CNTR
GABBS CENTRAL	100 AGL	0715-2330	ZOA CNTR
GABBS N, S	100 AGL	0715-2330	ZOA CNTR
GANDY	100 AGL	0500-2000 MON-SAT	ZLC CNTR
ISABELLA	200 AGL	0600-2200 MON-FRI	HI-DESERT TRACON, EDWARDS
OWENS	200 AGL	0600-2200 MON-FRI	HI-DESERT TRACON, EDWARDS
PANAMINT	200 AGL	0600-2200 MON-FRI	HI-DESERT TRACON, EDWARDS
REVEILLE	100 AGL	INTERMITTENT SR-SS MON-SAT	ZLC CNTR
SALINE	200 AGL	0600-2200 MON-FRI	HI-DESERT TRACON, EDWARDS
SEVIER A	100 AGL TO 14,500	0500-2000 MON-SAT	ZLC CNTR
SEVIER B	100 AGL TO 9500	0500-2000 MON-SAT	ZLC CNTR
SEVIER C	14,500	BY NOTAM 6 HRS IN ADVANCE	ZLC CNTR
SEVIER D	9500	BY NOTAM 6 HRS IN ADVANCE	ZLC CNTR
SHOSHONE	200 AGL	0600-2200 MON-FRI	ZLA CNTR
SILVER	200 AGL TO 7000	INTERMITTENT BY NOTAM	ZLA CNTR
SUNNY	12,000	BY NOTAM 24 HRS IN ADVANCE	ZAB CNTR

*Altitudes indicate floor of MOA. All MOA's extend to but do not include FL 180 unless otherwise indicated in tabulation or on chart.
†Other time by NOTAM contact FSS.
**ZAB-Albuquerque, ZLA-Los Angeles, ZLC-Salt Lake City, ZOA-Oakland

Fig. 12-5. *Continued.*

Military operation areas (MOA) are confusing; they are not restricted areas. Their purpose is to alert pilots to the existence of military operations. Pilots should exercise additional vigilance while transiting active MOAs. Flight service station specialists often receive requests for published MOA activity. Pilots should first refer to the chart for this information. If the MOA is activated by NOTAM, automated FSSs will be able to provide its status. Nonautomated FSSs will only be able to provide MOA status within approximately 100 miles of the FSS; therefore, a pilot will have to check with FSSs en route for MOA activity beyond this distance. Like restricted areas, MOA status can also be obtained from the respective controlling agency.

With preliminary planning completed, it's time to open the sectionals or WACs and prepare a navigational log with routes, true courses, distances, and magnetic variation, along with communication and navigation frequencies. This can be done on any commercially available form, such as the preflight planner navigation log illustrated in Fig. 12-6.

Planning a long trip several days in advance, even several weeks, is not a difficult chore. What about the weather? The general weather patterns of the United States are well documented. For example, we know about the winter storms of the Midwest and East, the convective weather of the Midwest in the spring and summer, the heat of the southwestern deserts in summer, and coastal low clouds of the Pacific states in late spring, summer, and fall. If needed, a call to the area's FSS will often provide the general weather conditions for a certain area and time of year. Please, don't expect specifics.

How can we use this general weather for flight planning? With a VFR-only Cessna 150 flying out of the Los Angeles Basin, I always plan to depart on the first leg in the afternoon, after the fog clears. This also permits flight over the desert in the late afternoon or early evening when the turbulence has diminished. Departures are then planned early the next morning to avoid desert convective activity and turbulence. Another solution is to move the airplane inland out of the affected coastal areas. Time of year is also important. If you plan VFR-only winter operations, be prepared for delays. May, June, September, and October seem to have the best flying weather.

What about winds aloft? If we don't try to stretch our trip legs, 10 to 20 knots of wind either way shouldn't present a problem. For example, I planned a leg from Phoenix to Albuquerque; however,

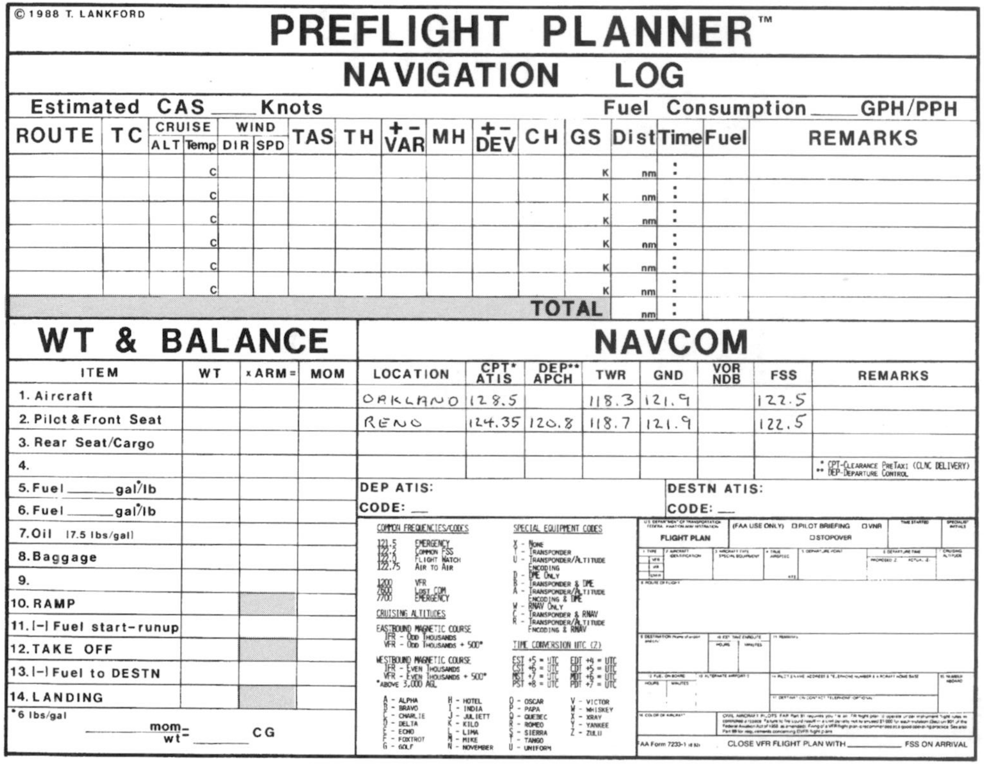

©1988 T. LANKFORD

PREFLIGHT PLANNER™

NAVIGATION LOG

Estimated CAS ____ Knots Fuel Consumption ____ GPH/PPH

ROUTE	TC	CRUISE ALT	CRUISE Temp	WIND DIR	WIND SPD	TAS	TH	+− VAR	MH	+− DEV	CH	GS	Dist	Time	Fuel	REMARKS
			C									K	nm	:		
			C									K	nm	:		
			C									K	nm	:		
			C									K	nm	:		
			C									K	nm	:		
			C									K	nm	:		
												TOTAL	nm	:		

WT & BALANCE

ITEM	WT	× ARM =	MOM
1. Aircraft			
2. Pilot & Front Seat			
3. Rear Seat/Cargo			
4.			
5. Fuel ____ gal*/lb			
6. Fuel ____ gal*/lb			
7. Oil (7.5 lbs/gal)			
8. Baggage			
9.			
10. RAMP			
11. [−] Fuel start-runup			
12. TAKE OFF			
13. [−] Fuel to DESTN			
14. LANDING			

*6 lbs/gal

mom/wt = ____ CG

NAVCOM

LOCATION	CPT* ATIS	DEP** APCH	TWR	GND	VOR NDB	FSS	REMARKS
OAKLAND	128.5		118.3	121.9		122.5	
RENO	124.35	120.8	118.7	121.9		122.5	
							* CPT-CLEARANCE PRETAXI (CLNC DELIVERY) ** DEP-DEPARTURE CONTROL

DEP ATIS:
CODE: __

DESTN ATIS:
CODE: __

COMMON FREQUENCIES/CODES
121.5 EMERGENCY
123.6 COMMON FSS
122.0 FLIGHT WATCH
122.75 AIR TO AIR
1200 VFR
7600 LOST COM
7700 EMERGENCY

CRUISING ALTITUDES
EASTBOUND MAGNETIC COURSE
IFR - ODD THOUSANDS
VFR - ODD THOUSANDS + 500*
WESTBOUND MAGNETIC COURSE
IFR - EVEN THOUSANDS
VFR - EVEN THOUSANDS + 500*
*ABOVE 3,000 AGL

A - ALPHA, B - BRAVO, C - CHARLIE, D - DELTA, E - ECHO, F - FOXTROT, G - GOLF, H - HOTEL, I - INDIA, J - JULIETT, K - KILO, L - LIMA, M - MIKE, N - NOVEMBER, O - OSCAR, P - PAPA, Q - QUEBEC, R - ROMEO, S - SIERRA, T - TANGO, U - UNIFORM, V - VICTOR, W - WHISKEY, X - XRAY, Y - YANKEE, Z - ZULU

SPECIAL EQUIPMENT CODES
X - NONE
T - TRANSPONDER
U - TRANSPONDER/ALTITUDE ENCODING
D - DME ONLY
B - TRANSPONDER & DME
A - TRANSPONDER/ALTITUDE ENCODING & DME
W - RNAV ONLY
C - TRANSPONDER & RNAV
R - TRANSPONDER/ALTITUDE ENCODING & RNAV

TIME CONVERSION UTC (Z)
EST +5 = UTC, EDT +4 = UTC
CST +6 = UTC, CDT +5 = UTC
MST +7 = UTC, MDT +6 = UTC
PST +8 = UTC, PDT +7 = UTC

FLIGHT PLAN (FAA USE ONLY) □ PILOT BRIEFING □ VNR □ STOPOVER

FAA Form 7233-1

CLOSE VFR FLIGHT PLAN WITH ____ FSS ON ARRIVAL

Fig. 12-6. *A navigation log with navcom section is ideal for noting frequencies to be used en route.*

because of the distance I could not tolerate any headwind component; therefore, I planned an alternate, using Gallup, New Mexico. If I were not on time at a specific point, about halfway, I would divert to the planned alternate. Fortunately, the winds were favorable, and I proceeded to the planned destination.

On another flight from Prescott to Albuquerque, things just weren't meant to be. Crossing Winslow, Arizona, the Cessna's ground speed never reached three digits. I changed the flight plan and proceeded to Gallup. Hoping that a stronger-than-forecast headwind will abate is folly.

Despite the best plans of mice and men, things go wrong. On a leg from Kalamazoo, Michigan, to Detroit, flight service advised of a thunderstorm over Jackson, Michigan. Further checking indicated that weather toward Toledo, Ohio, (southward) was clear. A slight diversion and pilotage navigation led safely to the new destination.

Study the charts in advance to determine best routes, comfortable legs, adequate services at destinations, and possible alternates. With everything planned, if a problem occurs, a pilot is in a much better position to evaluate the situation and develop a sound alternative.

Terminal area charts

Terminal area charts (TACs) replaced the local chart series beginning in the early 1970s. These multicolored charts depict Class B airspace, and the larger scale provides much more detail than is available with sectional charts. They are designed for pilots operating from airports within or near Class B airspace or transiting the vicinity. Charts are 20 × 25 inches, which can be folded to the standard 5 × 10 inches, and have a scale of 1:250,000 (1 inch equals 3 nm). TAC charts are revised semiannually. Charts are named for the Class B airspace they depict. Locations are shown in Fig. 12-3 and Fig. 12-4. Note that the Honolulu TAC is on the Hawaii Sectional Chart. Features include:

- Visual aids to navigation
- Radio aids to navigation
- Airports
- Controlled airspace
- Restricted areas
- Obstructions

- Topography
- Shaded relief
- Latitude and longitude lines
- Airways and fixes
- Other low-level-related data

Improved scale allows for a great deal of topographical detail. Along with depicted NAVAIDs, TACs should allow a pilot to safely navigate in the vicinity of, and remain clear of, Class B airspace. Even with this detail, in marginal weather conditions, new pilots might not have the experience to navigate in these areas. Remembering that FAR minimums are just that, minimums, not necessarily equating to safeness, each pilot must set standards based upon experience and training. This might mean avoiding terminal airspace altogether, only flying in clear weather, or obtaining additional training from a qualified instructor. Every new pilot planning to fly into Class B airspace or other congested airspace should make at least one trip with an instructor or an experienced pilot.

Helicopter route charts

Helicopter route charts are designed primarily to depict helicopter routes in and around major metropolitan areas. Since the helicopter charts have a longer life span than other NOS products, all new editions will be printed on a synthetic paper. Helicopter route charts effective in the spring of 1995, or later, will use the new paper—at an increase in cost, of course. These charts are available for the following locations:

- Baltimore-Washington (District of Columbia)
- Boston
- Chicago
- Houston
- Los Angeles
- New York

Scale is the same as a TAC's and dimensions are similar, except for the New York chart, which includes a larger scale inset of Lower Manhattan and the Hudson and East Rivers. The Boston chart has a downtown inset. The Chicago chart has an O'Hare and vicinity inset. Charts contain specific route descriptions as illustrated in Fig. 12-7,

HELICOPTER ROUTE CHART
BOSTON
SCALE 1:125,000

Lambert Conformal Conic Projection Standard Parallels 42°05' and 42°45'
Horizontal Datum: North American Datum of 1983 (World Geodetic System 1984)
Topographic data corrected to September 1994

3 RD EDITION December 8, 1994
Includes airspace amendments effective *December 8, 1994*
and all other aeronautical data received by October 13, 1994
Consult appropriate NOTAMs and Flight Information
Publications for supplemental data and current information.

This chart will become *OBSOLETE FOR USE IN NAVIGATION* upon
publication of the next edition. See Dates of Latest Editions.

PUBLISHED IN ACCORDANCE WITH INTERAGENCY AIR CARTOGRAPHIC COMMITTEE
SPECIFICATIONS AND AGREEMENTS, APPROVED BY:
DEPARTMENT OF DEFENSE • FEDERAL AVIATION ADMINISTRATION • DEPARTMENT OF COMMERCE

CONTROL TOWER FREQUENCIES ON BOSTON HELICOPTER ROUTE CHART

Airports which have control towers are indicated on this chart by the letters "CT" followed by the primary VHF local control frequency. Selected transmitting frequencies for each control tower are tabulated in the adjoining spaces, the low or medium transmitting frequency is listed first followed by a VHF local control frequency, and the primary VHF and UHF military frequencies, when these frequencies are available. An asterisk (*) follows the part-time tower frequency remoted to the collocated full-time FSS for use as Local Airport Advisory (LAA) during hours tower is closed. Hours shown are local time. Ground control frequencies listed are the primary ground control frequencies.

Automatic Terminal Information Service (ATIS) frequencies, shown on the face of the chart are primary arrival VHF/UHF frequencies. All ATIS frequencies are listed below. ATIS operational hours may differ from control tower operational hours.

ASR and/or PAR indicates Radar Instrument Approach available.

"MON-FRI" indicates Monday thru Friday.

CONTROL TOWER	OPERATES	TWR FREQ	GND CON	ATIS	ASR/PAR
BEVERLY	0700-2200	125.2	121.6	118.7	
BOIRE NF	0700-2100 APR 1-OCT 31	134.9	121.8	125.1	
	0700-1800 NOV 1-MAR 31				
HANSCOM	0700-2300	118.5 236.6	121.7	124.6	
LAWRENCE	0700-2200	120.0	124.3	126.75	
LOGAN INTL	CONTINUOUS	119.1 (RWYS 4R/22L & 9/27)	121.9	135.0	
		128.8 257.8			
NORWOOD MEM	0700-2200	126.0	121.8	119.95	
NAS SOUTH WEYMOUTH-SHEA	0700-2300	126.2 360.2	352.4		ASR/PAR

CLASS B, CLASS C, TRSA, AND SELECTED RADAR APPROACH CONTROL FREQUENCIES

FACILITY	FREQUENCIES	SERVICE AVAILABILITY
BOSTON CLASS B	124.4 279.6 (270°-090°)	CONTINUOUS
	124.1 343.6 (091°-269°)	
MOORE AAF RADAR	124.4 279.6	CONTINUOUS

SPECIAL USE AIRSPACE ON BOSTON TERMINAL AREA CHART

Unless otherwise noted altitudes are
MSL and in feet; time is local.
Contact nearest FSS for information.
†Other time by NOTAM contact FSS

The word "TO" an altitude means "To and including."
"MON-FRI" indicates "Monday thru Friday"
FL – Flight Level
NO A/G – No air to ground communications

U.S. P—PROHIBITED, R—RESTRICTED, A—ALERT, W—WARNING, MOA—MILITARY OPERATIONS AREA

NUMBER	LOCATION	ALTITUDE	TIME OF USE	CONTROLLING AGENCY**
R-4102A	FORT DEVENS, MA	TO BUT NOT INCL 2000	0800-2200 SAT †24 HRS IN ADVANCE	BOSTON APPROACH CONTROL
R-4102B	FORT DEVENS, MA	2000 TO 3995	0800-2200 SAT †24 HRS IN ADVANCE	BOSTON APPROACH CONTROL
W-103	CASCO BAY, ME	TO 2000	INTERMITTENT	ZBW CNTR

**ZBW-Boston

The horizontal reference datum of this chart is North American Datum of 1983 (NAD 83), which for charting purposes is considered equivalent to World Geodetic System 1984 (WGS 84).

ROUTE DESCRIPTIONS

NOTE: Helicopters planning flights to BOSTON and/or within 10.5 NM CONTACT BOSTON AIR TRAFFIC CONTROL TOWER ON FREQ. 121.75.

BAY ROUTE

Southern end of Nantasket Beach in Hull, via the coastline to the Long Island Bridge then to the Channel then to the Logan Helipad. NOTE: It is recommended that the Bay Route be used by multi-engine and float equipped helicopters due to the low altitudes occasionally imposed.

FENWAY ROUTE

NOTE: Entry Point is within the Norwood Class D Airspace. CONTACT NORWOOD TOWER ON FREQ. 126.0.

At the intersection of I 95 and I 93 and the Conrail Tracks in Norwood. Follow the Conrail Tracks to the Fens (passing over the Fens east of Fenway Park and west of the Prudential Building). Joining the Turnpike Route.

Fig. 12-7. *Helicopter route charts are designed primarily to depict helicopter routes in and around major metropolitan areas.*

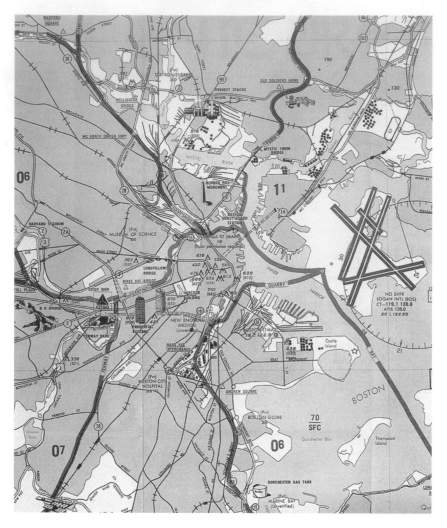

Fig. 12-7. *Continued.*

which includes symbols that are unique to this chart. In addition to specific helicopter routes, features include:

- Pictorial symbols of prominent landmarks
- Public, private, and hospital heliports
- NAVAID and communications frequencies
- Selected obstructions
- Roads
- Spot elevations

- Commercial broadcast stations
- Class B, C, and D airspace boundaries

Large metropolitan areas without published helicopter route charts often have local procedures that accomplish the same purpose. Letters of agreement between air traffic control facilities in these areas designate helicopter checkpoints, routes, and route names. Pilots planning operations in these areas should contact local pilots or ATC facilities for details on these procedures. This might require the pilot or operator to become a signatory to the letter of agreement, stating that he or she understands and will comply with its provisions.

World aeronautical chart/Gulf Coast VFR aeronautical chart

World aeronautical charts (WACs) are designed for visual navigation by moderate-speed aircraft and aircraft operating at higher altitudes, up to 17,500 feet MSL. Because of their smaller scale, these charts cannot show the detail of sectionals and TACs. For example, the limits of Class D and Class E airspace are not shown. WACs are normally not recommended for students or new pilots flying at slow speeds and low altitudes. A WAC would not be satisfactory while operating in the vicinity of Class B or Class C airspace. The charts are 20 × 60 inches, which can be folded to the standard size of 5 × 10 inches and have a scale of 1:1,000,000 (1 inch equals 14 nm). They are revised annually, except for a few Alaskan and Central American charts that are revised every two years. WACs are identified by a letter-number group. Areas of coverage are contained in Fig. 12-8 for the contiguous United States, Mexico, and the Caribbean, and Fig. 12-9 for Alaska. Features include:

- Visual aids to navigation
- Radio aids to navigation
- Airports
- Restricted areas
- Obstructions
- Topography
- Shaded relief
- Latitude and longitude lines
- Airways
- Other VFR-related data

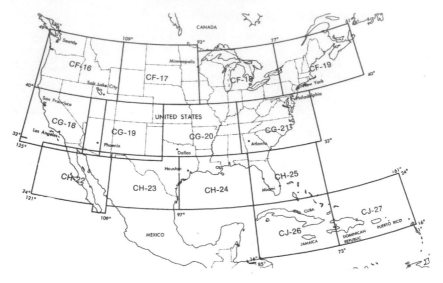

Fig. 12-8. *World aeronautical charts are designed for visual navigation of moderate-speed aircraft.*

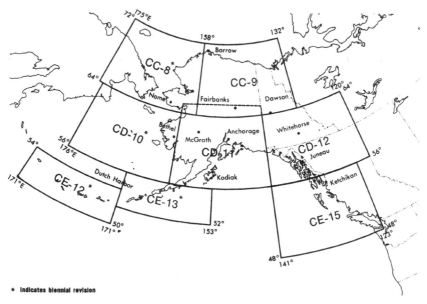

* Indicates biennial revision

Fig. 12-9. *World aeronautical charts are revised annually, except for a few in Alaska and Central America that are revised biennially.*

The U.S. Gulf Coast VFR aeronautical chart is designed primarily for helicopter operations in the Gulf of Mexico, usually serving the offshore oil and gas interests. The chart shows the same onshore features as the WAC, covering the Gulf Coast and extending south to 26°30′ N. The chart is 27 × 55 inches, which can be folded to the standard 5 × 10 inches, has a scale of 1:1,000,000, and is revised annually. Features include, in addition to those shown on WACs:

- Offshore mineral leasing areas and blocks
- Oil drilling and production platforms
- High-density helicopter activity areas

Some pilots prefer WACs for cross-country flying. Refolding sectionals is cumbersome, even at the speeds of the Cessna 150, and WACs reduce cabin clutter. One WAC covers about the same area as four sectionals, shaving approximately one-fourth off the chart bill. These are the major advantages of the WAC; the biggest disadvantage is scale. The following incident occurred in 1991.

It was the second day of the trip back from the Oshkosh fly-in. I had departed Gillette, Wyoming, for Pocatello, Idaho, in a Bonanza. Using a WAC chart, I planned to use VOR and pilotage navigation through Jackson Hole, Wyoming, and Idaho Falls, Idaho. The weather was not a factor, but visibility was restricted due to smoke from numerous forest fires. In this part of the country, even at 12,500 feet, I was still below the peaks. After passing what I identified as the Grand Tetons, I turned southwest toward Pocatello.

Well, you guessed it, I couldn't receive any VOR or establish communications with any facility. Continuing to fly down what I thought was the Snake River Valley, things didn't seem quite right. I followed an old aviation axiom: Follow a river or a road and it will normally bring you to a town, and hopefully an airport.

Even with two and a half hours of fuel, I decided it was time to resolve the issue of position. Because I was unable to establish communications on standard frequencies, I selected 121.5 MHz. I was not in distress, but there was a sense of urgency. As outlined in the *Aeronautical Information Manual,* I broadcast "PAN PAN PAN" followed by the aircraft identification. Almost immediately a military air evacuation flight responded. Based upon my assumed position, the Snake River Valley, the air evacuation flight provided a frequency for the Salt Lake Air Route Traffic Control Center.

After several tries, I was unable to establish communication. I had come across a small town with a good-sized airport. Unfortunately, there was no name on the airport. Because the transponder was being interrogated, I knew someone had us on radar. I selected 7700 and again broadcast "PAN PAN PAN." The air evacuation flight again responded. I asked the pilot to have the center look for a 7700 squawk. In a few moments, the pilot responded with another center frequency.

Calling center, the controller immediately responds, "Your position is six miles east of Big Piney." That left me with one minor question: Where is Big Piney? After a few moments shuffling the chart, I was on my way, although not by the route originally planned. As Maxwell Smart would say, "Missed it by that much!" Well, it isn't much on a WAC chart.

I usually back up WACs with sectionals. Some IFR pilots routinely carry WACs in case of electrical failure or other emergencies, when having a visual chart would be helpful.

Using Visual Charts

Consider consulting a flight planning chart for preliminary preparation of a long cross-country. These charts help determine which WACs, sectionals, and TACs are required. Specific charts will depend on the route, possible alternatives, and the mission.

Ensure chart currency; if a chart is to be revised next month, and the proposed flight is six weeks off, wait as long as possible for that new chart before finalizing your flight planning. Charts can be obtained from many sources; most pilot supply stores and FBOs carry charts, they are available through subscriptions from a number of sources, and they can be obtained from the government. Smaller FBOs normally only carry charts for their immediate vicinity.

Pilot supply stores might carry charts for the entire United States and most of North America. Pilots should become familiar with outlets in their area and determine which charts are readily available. Obtain all the charts that might be required. It's very embarrassing to end up needing a chart and not having it. Pilots usually realize this in the air or at an airport store that has a limited chart selection. It's always better to have too many charts than too few.

Bridging the gaps

Planning flights across chart boundaries can become a problem because sectionals and WACs are printed on both sides. Plotting these routes is accomplished in the following manner (Fig. 12-10). There are approximately two minutes of latitude overlap between the north and south sides of each chart, with more overlap in certain cases. The pilot must determine this overlap and either visually note the position or draw two match lines that are common to both sides. To draw the match line on the north side, connect the latitude tick marks of the most southern minute of latitude and on the south side connect the marks of the most northern minute of latitude. These match lines must have the same latitude on north and south sides.

On the side of the chart having the terminal (departure or destination) nearest the match line, place a sheet of paper so that one edge corresponds to the match line and the other edge intersects the terminal airport. Mark the edge of the paper at terminal point 1, and label it "Mark A." Then make another mark on the chart extending from the match line to the edge of the chart, and label it "Mark B" as shown in Fig. 12-10.

Turn the chart over and transfer "Mark B" to the other side of the chart, be sure to extend the mark to the match line. Align the sheet of paper to the match line with the corner of the sheet at the transferred "Mark B." With a plotter, or other straightedge, align terminal point 2 with "Mark A," and draw a line from the match line, now called "Point C," to terminal point 2 as shown in Fig. 12-10.

Turn the chart over again and transfer "Point C" to the other side of the chart. This can be done by measuring the distance from "Mark B" to "Point C." With a plotter, draw a line from "Point C" to terminal point 1 as shown in Fig. 12-10.

A direct course now consists of the line segment from terminal point 1 to "Point C" on one side of the chart, and from "Point C" to terminal point 2 on the other side of the chart. Be careful with this procedure. Errors occur from not properly considering the overlap area, incorrectly transferring "Mark B" or "Point C," or including the overlap when measuring total distance. This procedure is a little complex, but can be mastered with a little practice.

Oakland to Reno

Now we're ready to apply the principles and knowledge from Chap. 11 (terminology and symbols) and the preceding portion of this chapter. Let's plan a pilotage flight from the Oakland International Airport to the Reno Tahoe International Airport. Review the WAC in Fig. 12-11 for initial planning. For relatively short flights,

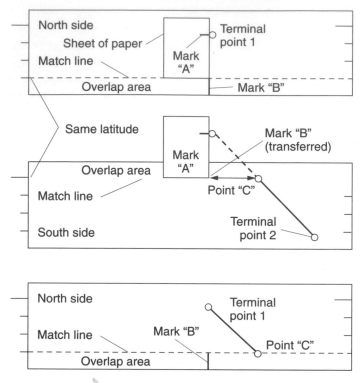

Fig. 12-10. *Plotting a course from the front to back side of charts can be cumbersome.*

WACs might serve as excellent planning charts because of their scale and detail. For our flight, there is an added advantage in that we can view the entire route on the WAC; on the sectional, departure and destination airports are on different sides of the chart.

The Oakland airport is located within Class C airspace, under Class B airspace. Terrain is mostly flat through the Sacramento Valley. We will come close to the McClellan Class C airspace in the Sacramento area. Should we choose to verify our route with NAVAIDs, we see that the Sacramento, Hangtown, Squaw Valley, and Mustang VORs are along our course. If you haven't noticed or are new to flying, you might not be aware that in the early days, NAVAIDs located in the vicinity of airports usually carried the same name as the associated airport. For example, the VOR just northeast of Reno was called Reno and the Squaw Valley VOR was called Lake Tahoe. To prevent any confusion or misunderstanding in our computerized ATC system, a program evolved to change the names of NAVAIDs that are

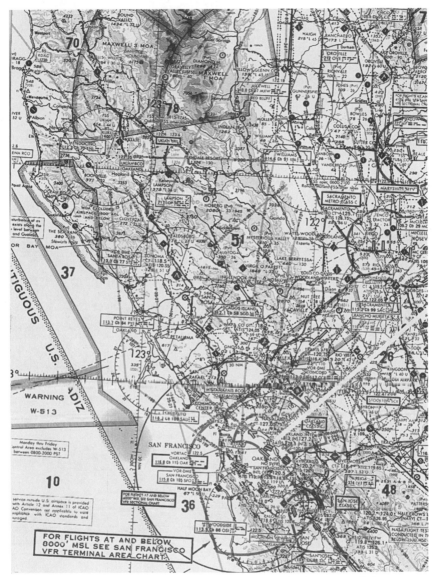

Fig. 12-11. *WACs are often helpful with initial planning of short trips.*

not collocated with the airport of the same name. Policy dictates that names come from the immediate area. A diversion to ward off cross-country boredom might be reviewing airport VOR names and determining the reference. If the reference is not obvious, perhaps a hidden meaning can be explained by a pilot in the area: for example, "S"qua"W" Valley "R"esort (SWR).

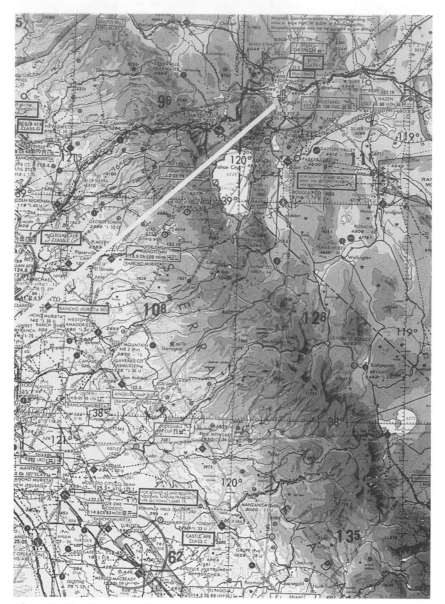

Fig. 12-11. *Continued.*

For the portion of the route over the Sierra Nevada Mountains, maximum elevation figures (MEF) indicate terrain from 9600 to 11,100 feet MSL. This might pose a problem for low-performance aircraft, especially considering weather and density altitude. Finally, we see that the destination is also located within Class C airspace. We might want to consider the note just north of Reno that states, "Magnetic disturbance exists in the area extending 50 miles or more N.W., W, & S.W. of Reno, Nevada. Magnetic compass might not be accurate at low altitude." Apparently gold and silver *are* in those hills.

From this preliminary view, we would also check for any SUA, such as restricted areas and MOAs, or any other high-density areas. It is worth noting that MTRs do not appear on WACs. Notes on the WAC reveal that a San Francisco TAC is necessary for flight below 8000 feet MSL, which is the top of the Class B airspace. The WAC refers us to the San Francisco Sectional for flight below 4100 feet in the Sacramento area and below 8400 feet in the Reno area, which are the upper limits of the Class Bravo airspace. The WAC would be perfectly acceptable for a flight in good weather from Livermore (east of the San Francisco Class B airspace) to, for example, Carson City (south of Reno).

We could fly east of Livermore and follow the highway to Stockton. The Stockton Airport has a control tower, so we'll want to stay above its Class D airspace or obtain the required clearance. Proceeding northeast of Stockton toward the large reservoirs, we could pick up the highway and proceed south of the Lake Tahoe Airport, which also has a tower, then to Douglas County, and finally into Carson City. This would be unwise in poor weather because it is too easy to misidentify landmarks from the WAC, pick the wrong canyon, and end up boxed in; it has happened.

Let's move on to the sectional. A direct route has been established on the sectional chart, as illustrated in Fig. 12-12 (south portion) and Fig. 12-13 (north portion). A direct flight will take us out over the coastal mountains where the MEF is 4200 feet. We will have to negotiate the Oakland Class C airspace and San Francisco Class B airspace. We review the sectional, end panels, and margin. In good weather, the sectional provides enough information to negotiate Class C and Class B airspace. From the sectional, we see that once out of the Class C airspace, we will have to remain below 4500 MSL and 6000 MSL, respectively, to clear the Class B airspace shelves along our direct route.

Once clear of the Class B airspace, the course will take us over the Sacramento River Delta. Just southwest of Franklin are a number of towers that, from the symbols, extend above 1000 feet. From the chart, these towers extend to 2001 feet MSL.

Terrain starts to rise over the Sierra Foothills. Critical elevations range generally between 8000 and 11,000 feet. Donner Pass, just west of Truckee, California, has an elevation of 7088 feet. The

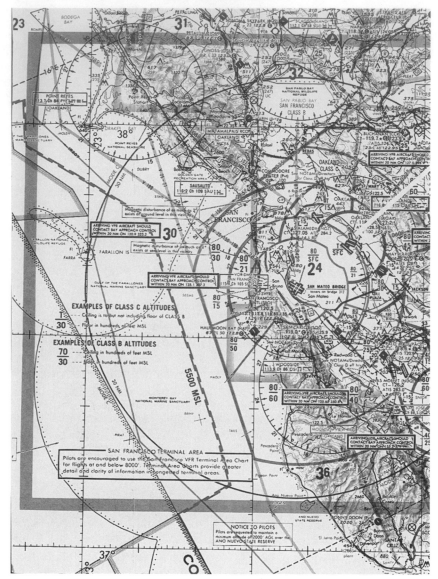

Fig. 12-12. *The sectional is satisfactory for operating outside of Class B airspace, but operations close to or underneath Class B airspace require the appropriate terminal area chart.*

Spooner Summit Pass, on the east central side of Lake Tahoe, has an elevation of 7146 feet. East of the passes, it's generally all downhill through the valleys to Reno, which has a field elevation of 4412 feet.

There appears to be a long runway on the west side of Lake Tahoe, next to the Homewood Seaplane Base. A "gotcha" flight instructor-

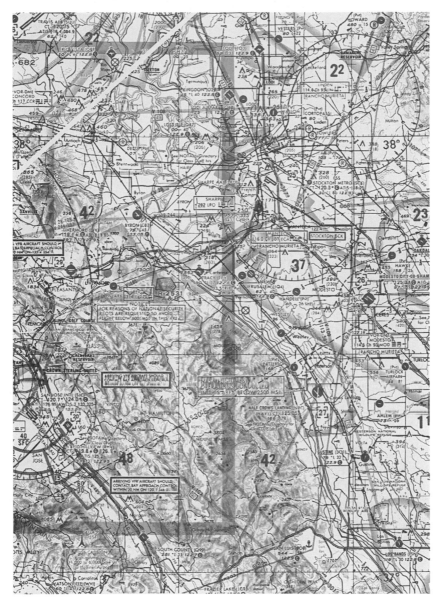

Fig. 12-12. *Continued.*

to-student question is "What does this symbol represent?" The an-swer is at the end of this chapter.

Reno's Class C airspace extends from the surface to 8400 feet MSL. From the chart or end panel, we determine the approach frequency is 120.8. The chart also provides ATIS (124.35) and tower (118.7) frequencies; ground frequency for this airport is 121.9. The only

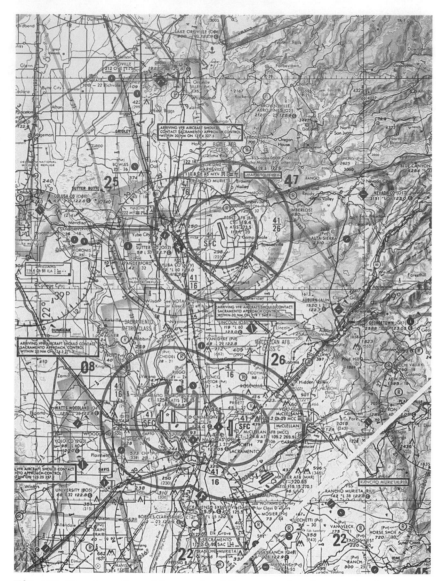

Fig. 12-13. *A sectional is required for operating into or in the vicinity of Class C airspace.*

other frequency we might need is the Reno FSS, which the chart indicates is 122.5.

From our review of the sectional, we determine there are no military training routes along our proposed course. Any MTR numbers would have been noted: four-digit numbers for routes flown at or below 1500 feet, and three-digit numbers for routes flown above

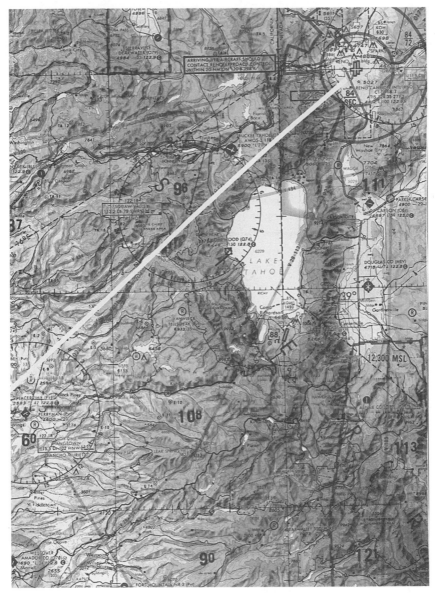

Fig. 12-13. *Continued.*

1500 feet AGL. When flight altitudes are above 1500 feet AGL, you can disregard any routes with four digits. Call an FSS for operational details along any conflicting routes.

Now for the flight from Livermore, California, to Carson City, Nevada, using the sectional: Weather is marginal, ceilings are between 1000 and 3000 feet, and visibility is 3–5 miles. We'll definitely need the sectional to establish the limits of controlled airspace. We will observe all FARs as they apply to controlled airspace, distances from clouds, and minimum safe altitudes. Departing east of Livermore, we could cruise at 1500 feet MSL and even lower after crossing into the valley and through the Stockton area. Livermore has Class D airspace. Class E airspace based at 700 feet exists to the east, and along the highway to Stockton, Class E airspace starts at 1200 feet AGL; therefore, at our altitude, communication is required to transit Stockton's Class D airspace.

We could then proceed northeast of Stockton, California, toward the Comanche and Hogan Reservoirs. Then follow the highway through Carson Pass, which has an elevation of 8650 feet. The base of Class E airspace through this area is 1200 feet AGL, as indicated by a blue vignette. Where controlled airspace begins at other than this altitude, the chart is labeled in feet above mean sea level. For example, south of Minden, Nevada, the floor of Class E airspace is 12,300 MSL. From the Carson Pass we could then follow the highway into Carson City.

(This is not a recommendation to fly at low altitudes through mountainous areas. FAA: "The decision as to whether a flight can be conducted safely rests solely with the pilot." This depends upon a pilot's training and experience.)

Could we fly from Livermore to Stockton with ceilings of 500 to 1000 feet and visibilities ranging from 1 to fewer than 3 miles? Technically, yes. This flight would require a special VFR clearance out of the Livermore Class D airspace and into the Stockton Class D airspace. En route we would be required to remain clear of clouds below 700 feet AGL—that's the floor of Class E airspace east of Livermore—and then below 1200 feet to the boundary of the Stockton Class D airspace. Could this flight be conducted at night under the same conditions? No. At night, even in Class G airspace, 3 miles visibility and standard distance from clouds is required.

When flying with reduced ceilings and low visibilities, it's often a good idea to slow down, which I have done before. My destination was Crescent City, California. The VOR was out of service, which precluded an instrument approach. The coastal stratus tops were about 1500 feet, hugging the hills. Bases were reported at 600 feet with a visibility of 2 miles. I found a small hole in the stratus and obtained a special VFR clearance. To allow more time to see and avoid obstacles or anything else, I slowed the Mooney to 100 knots. There is no rational reason to be flying in marginal weather at 160 knots.

When flying in marginal weather, be very careful not to impose on someone's airspace, especially if a preplanned route is altered. I participated in the 1991 Hayward-Bakersfield-Las Vegas Air Race. The event began with overcast ceilings between 1500 and 2500 feet AGL. The first checkpoint was the Pine Mountain Lake Airport in the Sierra foothills at an elevation of 2900 feet. Most of us ended up bypassing this checkpoint because of the weather. Using pilotage, our crew flew to the next plotted checkpoint south of Pine Mountain to resume the preplanned course. Many pilots, after abandoning Pine Mountain, headed straight for the next checkpoint, apparently neglecting to consider the Castle AFB Class C airspace or communicate with the appropriate airspace controllers.

Back to the flight planning: San Francisco's terminal area chart will be used to navigate the first part of the flight (Fig. 12-14). The Oakland airport data block reveals ATIS on 128.5 MHz and the north field tower on 118.3 MHz—notice the capital "N" in parentheses. The letters "FSS" at the top of the data indicate a flight service station by the same name on the field. Oakland International is, in effect, two airports. The south field serves air carriers, and the north field serves general aviation.

The end panel also notes that the ground control frequency is 121.9. The Oakland VORTAC box indicates that the Oakland FSS has a discrete frequency of 122.5. Appropriate Class C airspace and Class B airspace frequencies are contained on the chart. It's usually not necessary to obtain departure control frequencies because ATC will normally assign the appropriate frequency upon departure. The navcom portion of the preflight planner and navigation log in Fig. 12-6 has been filled out for the trip.

Take advantage of the details available on the TAC where they're available, especially in marginal weather. This chart would be useful for the Livermore to Carson City trip. From the chart, east of Livermore, the transmission lines that cross the course can be seen more clearly. The transmission lines can be used with the railroad to update progress of the flight. The TAC shows that one of the railroad tracks goes through a tunnel on its second crossing of the highway, which is not seen on the sectional chart. Contours indicate that the approximate elevation of the pass between the Livermore Valley and the San Joaquin Valley is above 1000 feet, but lower than 1500 feet.

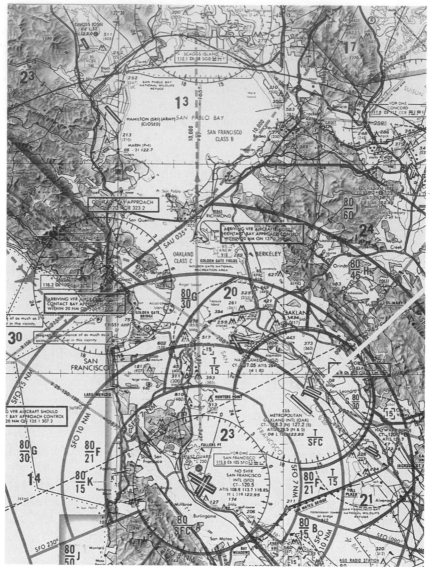

Fig. 12-14. *TACs provide the detail necessary for operating in congested terminal airspace.*

Flight time

I obtained a DUAT weather briefing in the morning prior to departure and filed a VFR flight plan through the computer service. Ground control issued a clearance with heading information and altitude restrictions. The tower provided an ATC frequency for the Class C airspace, and I contacted Bay Departure Control.

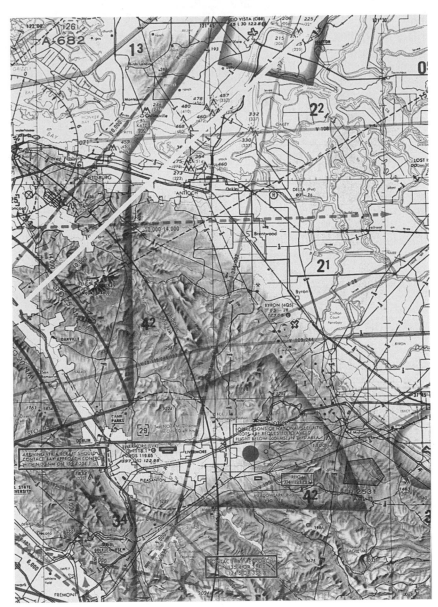

Fig. 12-14. *Continued.*

Departure vectored me south of a direct route at 2500 feet to clear other traffic. Just east of the Class C airspace, I requested and received clearance to climb VFR through the Class B airspace to the initial cruising altitude of 5500 feet. In the vicinity of Danville, California, clear of the Class B and Class C airspace, I terminated radar service and switched to Oakland Radio on 122.5 to open the flight plan. I was south of Mt. Diablo, correcting the heading toward the planned flight course, and I contacted Oakland Flight Watch on 122.0 for a weather update over the Sierra Nevada mountains.

Checkpoint considerations

A primary checkpoint consists of a topographical feature, or set of features, that cannot be mistaken for any other place in the same general area. Often three or more secondary features can be combined to form a primary checkpoint. Secondary checkpoints are small towns, streams, a single road or railroad, mountain range, or any other feature that could be mistaken for similar features in the same general area.

A primary checkpoint could be a single feature, such as Arizona's Meteor Crater. It is large and unique and cannot be mistaken for any other feature in the same area. Several smaller features can be combined to form a primary checkpoint. Unless the features are relatively unique, such as an airport and adjacent town, three features should be used: town, highway, or railroad. One feature, or even two, can be mistaken; therefore, reliance upon one or two features should be avoided if at all possible.

Pilots should consider the availability of suitable checkpoints during flight planning. This is the time to select an alternate route if you're not comfortable with what's available. If you do get lost, or even think you're lost, call for assistance before a relatively simple flight assist becomes an accident.

Back on course

Returning to our flight again: The TAC shows that Mt. Diablo has a critical-elevation symbol of 3849 feet. There is an obstruction on the peak 285 feet AGL, 3865 feet MSL. Note the MEF of 4200 feet because another known feature in this quadrangle apparently requires this higher figure. Mt. Diablo is like a beacon to the bay area, often above the haze and fog. Figure 12-15 shows Mt. Diablo as seen from the east with the haze and bay in the background. (Mt. Diablo at one time sported an airway beacon because of its prominence.)

When well clear of bay-area airspace and traffic congestion, switch from the TAC to the sectional chart for navigation. Still a little south

Fig. 12-15. *Single prominent features can be used as primary checkpoints.*

of course, Fig. 12-16 shows the town of Antioch, California, and the Sacramento River Delta as the river flows toward the bay. These areas make good primary checkpoints because of the relation of water, towns, and roads. A word of caution: Be very careful flying in these areas at low altitudes, especially in low visibilities because the numerous obstruction symbols on the chart represent many power lines that are stretched across the water.

Figure 12-17 shows the town of Rio Vista, California. This town, waterway, roads, and bridge make an excellent checkpoint. Remember, if at all possible, stay away from sparse checkpoints, such as a small town and a road. Almost every small town has a road running through it, and especially in flat country, many of the roads are parallel. Try to select checkpoints with three or more features, such as Fig. 12-17.

An antenna farm is approximately 8 miles northeast of Rio Vista. The chart indicates a maximum antenna elevation of approximately 2000 feet MSL; strobe lights help, but are still hard to see, especially in haze or fog. Realize that large towers usually have guy wires that extend diagonally to the ground for support, which can easily snag an aircraft

Fig. 12-16. *Several secondary features can be combined to make a primary checkpoint.*

Fig. 12-17. *The relationship between the town, waterway, roads, and bridge, makes Rio Vista an excellent checkpoint.*

flown by an unsuspecting pilot. Flying at low altitudes in low-performance aircraft, complying with all FARs, of course, these features are not necessarily hazardous with good visibility at a minimum 3 miles.

Along the northern portion of the sectional in Fig. 12-13, the course runs just south of Sunset Sky Ranch. A four-lane highway and a railroad track crossing with a town to the north is another excellent checkpoint. Finding a place where a road, railroad, or river crosses another feature can "fix" the aircraft position. Figure 12-18 shows the view from the airplane while cruising at 5500 feet MSL above the McClellan AFB Class C airspace shelf above the former Mather AFB.

An FSS communication box just south of course reads RANCHO MURIETA RIU. Routine communications, such as position reporting, flight plan updating, or other FSS services, are possible on the standard FSS frequency 122.2 MHz. Also notice the Hangtown VOR-DME data box; Rancho radio has receive-only at that site on 122.1. We could contact Rancho radio by transmitting on 122.1 and listening on the Hangtown VOR frequency 115.5. Always advise the FSS which frequency is being monitored.

Figure 12-19 shows a view of the Sacramento Valley behind the airplane, and the coastal range of mountains is in the distance. Often

Fig. 12-18. *Locating the point where a road and railroad cross can fix the aircraft's position.*

Fig. 12-19. *When flying over flat farmlands, few verifiable checkpoints are available.*

over flat farmland few landmarks are available; select verifiable checkpoints when planning and keep careful track of your position while flying through such an area.

Approaching the Sierra foothills, we begin a climb to 9500 feet. Off to our left is Folsom Lake, shown in Fig. 12-20. These lakes often make excellent checkpoints; the shape of the lake and the dam are easily verifiable with the chart. Compare the presentation of these lakes on the WAC in Fig. 12-11; with the sectional in Fig. 12-13.

A little farther along and to the right is another lake (Fig. 12-21); again, lake and dam shapes are verifiable. The chart shows many small lakes along the Sierra foothills. Care must be exercised selecting them as checkpoints because they are numerous and their shapes are similar.

Approaching the crest of the Sierra Nevada mountains, the terrain rises rapidly and the aircraft is buffeted by some turbulence. The lowest terrain is south of a direct course, over the center of Lake Tahoe. As we approach the lake, we are greeted by the scene in Fig. 12-22. We can tell that we will clear the crest of the mountains because the terrain beyond appears to be descending in relation to the crest of the mountains. (We were over the lake and decided to continue through Spooner Summit Pass, then over Carson City, and into Reno.)

Fig. 12-20. *Lakes and dams make easily verifiable checkpoints.*

Fig. 12-21. *Be careful about using lakes as checkpoints when many similar lakes are in the same general area.*

Fig. 12-22. *The airplane is above the crest of the mountains because the terrain behind appears to be descending in relation to the crest.*

East of Lake Tahoe, we again contact flight watch and file a pilot report regarding conditions over the mountains. Frequent and objective pilot reports cannot be overemphasized, even if the weather and the ride are clear and smooth. Reporting en route is a timely practice that makes current information available to briefers and forecasters, and ultimately to other pilots.

The VFR flight plan was closed over Carson City via Reno radio prior to switching to another frequency while approaching the destination. The ATIS report is noted prior to contacting approach control for entry into the Reno Class C airspace. Recall that frequencies are already listed; therefore, we concentrate on flying the airplane, looking for traffic, and navigating, rather than fumbling with the chart trying to find a frequency. We pass Steamboat and follow the highway to the "Biggest Little City in the West" (Fig. 12-23).

Coastal cruise

The value of a TAC cannot be overemphasized. Let's take a flight from Livermore, California, to Half Moon Bay, California, on the coast southwest of San Francisco (Fig. 12-14). A direct flight will take

Fig. 12-23. *When the destination is in sight, consider closing the flight plan before switching to approach control or the tower.*

us through the Oakland Class C and the San Francisco Class B airspace. Departing Livermore we will probably wish to stay below 4000 feet because of aircraft at 4000 feet flying inbound to Oakland and Hayward as indicated by the IFR arrival route symbol (blue aircraft on dashed blue line).

The Class C airspace in the vicinity of Hayward extends from 1500 feet MSL to the base of the Class B airspace. We can either skirt south of the Oakland Class C airspace or contact approach control on 135.4 for Class C airspace services through the area. Don't forget about the Hayward Class D airspace. The San Francisco Class B airspace sector C extends from 2500 feet to the top of the Class B airspace at 8000 feet: sector B from 1500 feet. Crossing the bay, we will need to obtain clearance through either the San Carlos or Palo Alto Class D airspace. This is often easier than trying to obtain clearance through the Class B airspace. Tower controllers might assign aircraft specific routes to avoid the airport traffic patterns.

Beyond the Class D airspace, the base of the Class B airspace in sector D is 4000 feet. Now we can climb safely over the hills and proceed into Half Moon Bay, where the base of the Class B airspace is 5000 feet. We need to be very careful of minimum safe altitudes and required cloud clearance and visibility, and keep a sharp eye out for other traffic. Pilots of high-performance aircraft might wish to slow down in congested airspace in the vicinity of Class C or Class B airspace.

Previously in this chapter, the reader was asked about the symbol adjacent to the Homewood Seaplane Base on the west side of Lake Tahoe. Look northwest-to-southeast from just above the Squaw Valley NAVAID box toward just south of the Lake Tahoe Airport, then look south-southwest of the Alpine County Airport. You should be able to see large letters S-I-E-R-R-A (the A might not be visible). The so-called "runway" is the "I" in SIERRA. You might want to have some fun and ask your pilot friends, or better yet an instructor, to decode this symbol.

13

Supplemental visual charts

Pilots have access to a number of supplemental visual charts through the Defense Mapping Agency (DMA), NOS, and other sources, notably Canadian charts published by the Canada Map Office, Department of Energy, Mines, and Resources. DMA maintains a public sales program through the NOS and publishes a catalog that contains descriptions, availability, prices, and ordering procedures for DMA-produced aeronautical products, primarily covering foreign regions.

Canada produces charts similar to U.S. counterparts: WACs, VFR navigation charts (sectionals), and VFR terminal area charts. A Canadian chart catalog and products are available by mail or through various authorized dealers located at airports and cities throughout North America.

DMA visual charts

The Defense Mapping Agency produces a series of visual navigational charts, mainly in support of military missions. Some maps might be adapted for civil flight planning and navigation: global navigation and planning charts, loran charts, operational navigation charts, tactical pilotage charts, and joint operations graphical charts.

Global navigational and planning charts

Global navigational and planning charts (GNCs) are designed for flight planning, operations over long distances and en-route navigation in long-range high-altitude high-speed aircraft. GNC scale is 1:5,000,000 (1 inch equals 69 nm). Sheet size is approximately 42 × 58 inches. Polar regions are transverse Mercator projections, and other regions are the Lambert conformal conic projections.

The global navigation chart series serves as the basis for production of global loran navigation charts (GLCC) and spacecraft tracking charts (NST). Features include:

- Principal cities
- Towns
- Drainage
- Primary roads
- Primary railroads
- Prominent culture
- Shaded relief
- Spot elevations
- NAVAIDs
- Airports
- Restricted areas

Figure 13-1 contains an excerpt from a global navigational chart. Shaded relief contains tints indicating relatively flat and steep areas, along with spot and critical elevations; however, contours and gradient tints are not included. Cities of strategic or economic importance, major towns, primary road and railroad networks, and other significant cultural features are displayed. Hydrography includes open-water vignette, coastlines, major lakes, and major rivers.

In Fig. 13-1, a cautionary note warns: "Before using this chart, consult the current DMA Aeronautical Chart Updating Manual (CHUM)/CHUM Supplement, and the latest Flight Information Publications (FLIPs) and Notices to Airmen (NOTAMs) for vital updating information." The CHUM and FLIP, in effect the military version of the *Airport/Facility Directory,* are discussed in a later chapter. Figure 13-1 also contains an index to the GNC series for the Northern Hemisphere.

These charts can be used for wall display because one sheet covers the United States, Canada, and part of Alaska. They are suitable as planning charts due to relief and major cultural features, for example, plotting a flight from San Francisco to Denver. The chart would show that a direct route would be over rough, sparsely populated terrain. The lower terrain would be through Reno, Battle Mountain (Wyoming), Elko (Nevada), Salt Lake City, then through southern Wyoming to Cheyenne, and south to Denver. This was the original airmail route. An alternate, although much longer route, would be through Las Vegas, northern Arizona and New Mexico to Las Ani-

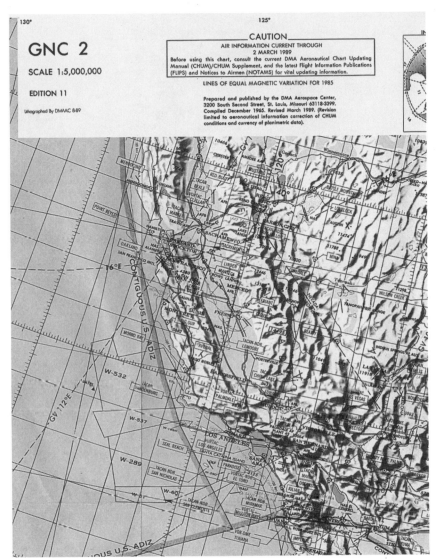

Fig. 13-1. *Global navigational charts are designed primarily for flight planning, operations over long distances, and en route navigation in long-range high-speed high-altitude aircraft.*

mas, then north to Denver. This route would generally be over lower terrain, perhaps a preferable route in the winter months.

Loran charts

Loran charts provide a plotting area where ground wave and sky wave correction values have been printed for loran navigation.

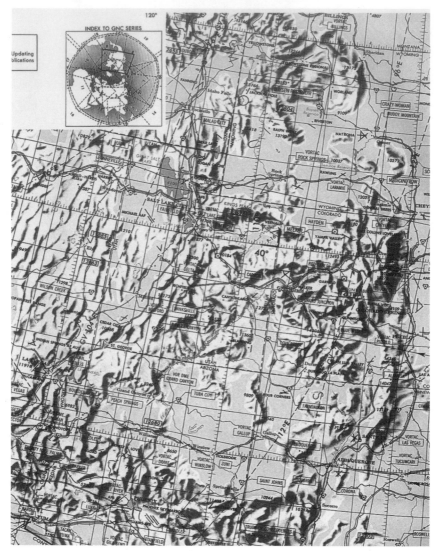

Fig. 13-1. *Continued.*

Loran lines on these charts furnish a constant time difference between signals from a master and slave loran station; however, with most of today's units, which incorporate microprocessors, loran units provide the pilot with direct position readout, along with course, speed, and distance to specified locations. Several loran charts are available from DMA: global Loran-C navigation charts, Loran-C coastal navigation charts, and Loran-C navigation charts.

Global Loran-C coastal navigation charts (GLCCs) are selected GNCs modified with Loran-C and Consol/Consolan overprints. Consol/Consolan is a long-range radio aid to navigation, the emissions of which, by means of their radio frequency modulation characteristics, enable bearings to be determined. They satisfy high-speed long-range navigation requirements over large expanses of water. Chart scale is 1:5,000,000 (1 inch equals 69 nm); sheets are approximately 42 × 58 inches. Polar charts are transverse Mercators, and lower latitudes are Lambert conformal conic projections. The charts include spot elevations, land tint, major cities, coastlines, and major lakes and rivers.

Loran-C coastal navigation charts (LCNC) are a series of three charts used for loran navigation for entry into the United States when a high degree of accuracy is required to comply with air defense identification and reporting procedures. They are also suitable for celestial navigation. Chart scale is 1:2,000,000 (1 inch equals 27 nm); sheet size is 35 × 57 inches. The chart provides spot elevations only, with land masses portrayed by a light gray tint. Principal cities and towns and international boundaries are shown, along with drainage areas and lake elevations.

Loran-C navigation charts (LCCs) are a series of four charts used for precise long-range polar loran navigation in support of weather reconnaissance, air search and rescue, and other operations in the Arctic area. The chart uses a transverse Mercator projection with a scale of 1:3,000,000 at the 90°E and 90°W meridian. The sheet is approximately 42 × 58 inches. Charts show spot elevations, major cities, railroads and roads, coastlines, and major lakes and rivers.

Jet navigation charts

Jet navigation charts (JNC) and universal jet navigation charts (JNU) are suitable for long-range high-altitude high-speed navigation. Chart scale is 1:2,000,000. Features include:

- Cities
- Major roads
- Railroads
- Drainage
- Contours
- Spot elevations

- Gradient tints
- Restricted areas
- NAVAIDs
- Broadcast stations
- Airports
- Runway patterns

Runway patterns are exaggerated so they can be more readily identified as visual landmarks.

JNCs are available for the world; three charts cover the United States. The charts that cover the United States can be combined into a reasonably sized wall map. These charts are ideal for planning purposes because they have better terrain information, which is important to pilots of low-performance aircraft.

Jet navigation charts for the Arctic (JNCA) serve the same purpose as JNCs, using a transverse Mercator projection for Arctic regions, with two additional charts covering the United States and Central America for training purposes. JNCAs have a scale of 1:3,000,000.

Operational navigation charts

Operational navigation charts (ONC) support high-speed radar navigation requirements at medium altitudes. Other uses include visual, celestial, and radio navigation. These charts have a scale of 1:1,000,000, the same as WACs. ONCs are available for all of the land masses of the world. Sheet size is approximately 42 × 58 inches, covering 8° of latitude.

Figure 13-2 is an example of an ONC that covers the same general area as the WAC excerpt in Fig. 12-11. Notice that ONCs do not provide communications or airways information, nor airspace information, except restricted, military operations, and alert areas. Air information is only current through the date stated on the chart. Pilots are advised to consult NOTAMs and FLIPs for the latest air information and the CHUM for other chart revision information.

Operational navigation charts are identified in the same manner as WACs. Starting at the North Pole with the letter "A," each successive row of charts uses the next letter of the alphabet through "X," which covers the South Pole. Each row of charts is labeled with a number

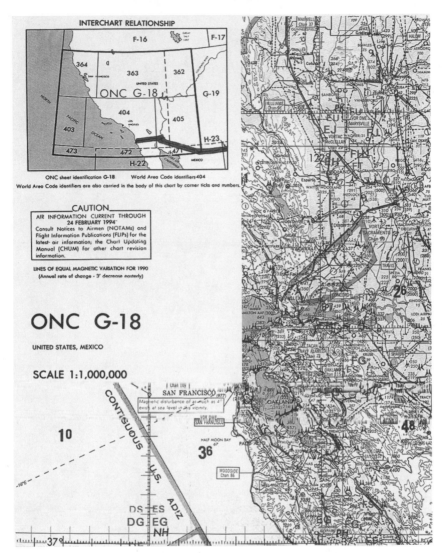

Fig. 13-2. *Operational navigational charts support high-speed navigational requirements at medium altitudes.*

that generally begins at the Prime Meridian, with subsequent numbers to the east. Because only land masses are charted in this series, chart G-18 will not necessarily be located under chart F-18. In Fig. 13-2, the chart joining G-18 to the north is F-16.

Because of the lack of aeronautical information, ONCs are not suitable for flight in the United States; however, these charts might be

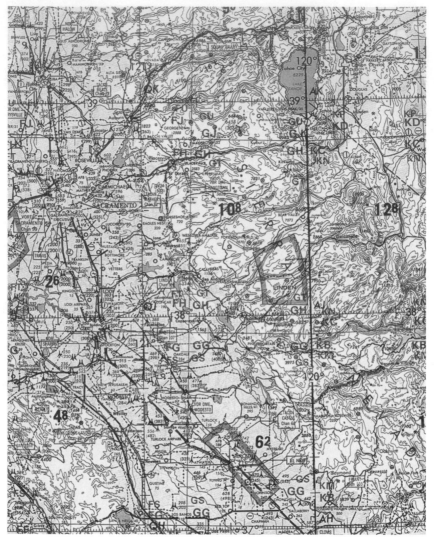

Fig. 13-2. *Continued.*

useful for pilots planning to fly in other countries where WACs are not available.

Tactical pilotage charts

Tactical pilotage charts (TPC) support high-speed, low-altitude radar, and visual navigation of high-performance tactical and reconnaissance aircraft at very low through medium altitudes. Tactical pilotage charts cover one-fourth the area of operational navigational charts. They are identified by the respective ONC letter and number and an additional letter representing the TPC (TPC G-18A). TPCs are not available for all areas of ONC coverage. TPCs have a scale of 1:500,000, the same as sectional charts.

Figure 13-3 contains an example of TPC covering the same general area as the sectional in Fig. 12-12. Like ONCs, tactical pilotage charts do not provide communications, airways and fixes, or controlled airspace. Airports, shaded relief, and topography are similar to the sectional.

Tactical pilotage charts are not suitable for navigation in the United States because of the lack of aeronautical information; however, these charts might be useful for pilots planning to fly in other countries. Some pilots like to obtain these charts when planning trips outside the United States just to get the lay of the land.

Joint operations graphics

Joint operations graphics-air (JOG-A) charts are suitable for preflight and operational functions. Scale is 1:250,000, the same as TACs. Figure 13-4 contains an example of a JOG-A, which covers the same general area as the TAC in Fig. 12-14. Communications, airways and fixes, and controlled airspace are not indicated.

JOGs are not available for sale outside of the United States. These charts could serve helicopter pilots or other operators where low-level navigation is required and NOS TACs are not available; however, they would have to be used in conjunction with the associated sectional for proper communications and to establish the limits of controlled airspace.

Military training route charts

The FAA issued a waiver to the Department of Defense (DOD) in 1967 to conduct various training activities below 10,000 feet MSL at speeds in excess of 250 knots. These activities included low-altitude navigation, tactical bombing, aircraft intercepts, air-to-air combat, ground troop support, and other operations in the interest of national defense. The number and complexity of these routes were to be limited to that considered absolutely necessary. Route widths vary from 2–10 nm. En-

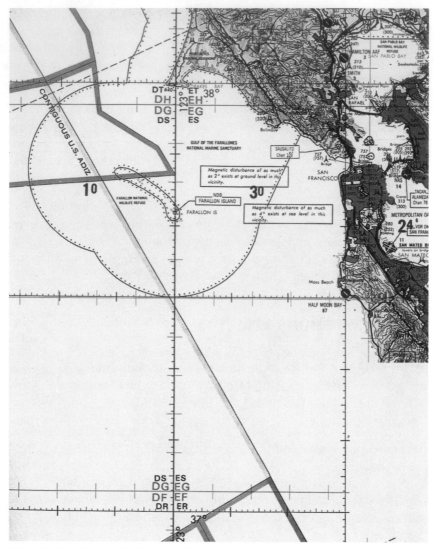

Fig. 13-3. *Tactical pilotage charts are designed for high-speed, low-altitude, radar, and visual navigation.*

route altitudes will be the minimum necessary for operational requirements, but in no case at altitudes less than those specified in FARs for minimum safe altitudes. The en-route altitudes range from 500 feet, or lower, to higher than 10,000 feet. Active times vary and are specified for each route, ranging from daylight hours Monday through Friday to continuous. Routes are designed to be clear of terminal airspace. Additionally, to the extent possible, routes remain clear of populated areas, controlled airspace, and uncontrolled airports.

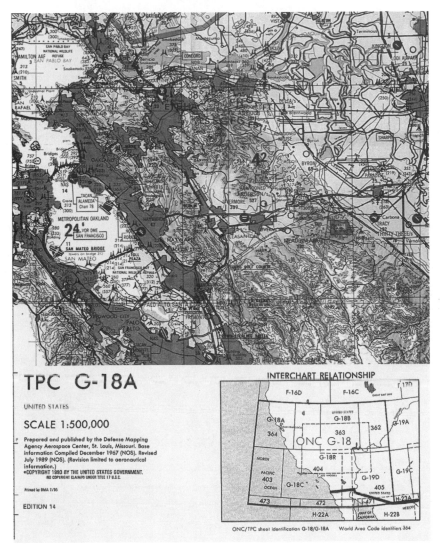

Fig. 13-3. *Continued.*

Military training routes fall into two categories:

- IFR military training routes (IR)
- VFR military training routes (VR)

VRs are only established when an IR route cannot accommodate the mission. IR routes might be flown in all weather conditions. VFR routes are only flown when forecast and encountered weather conditions equal or exceed 5 miles visibility and a 3000-foot ceiling.

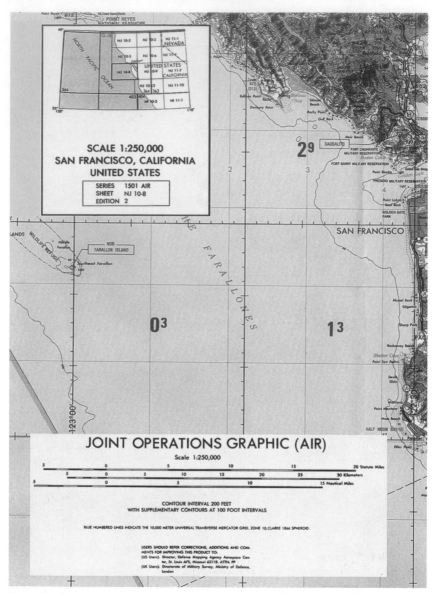

Fig. 13-4. *Joint operations graphics are suitable for preflight and operational functions.*

DMA publishes military training route (MTR) charts. The charts provide a visual depiction of routes, along with a specific route number. Three charts are published for the United States: western, central, and eastern. Charts are published every 56 days and are available by single copy or annual subscription. The Department of Defense

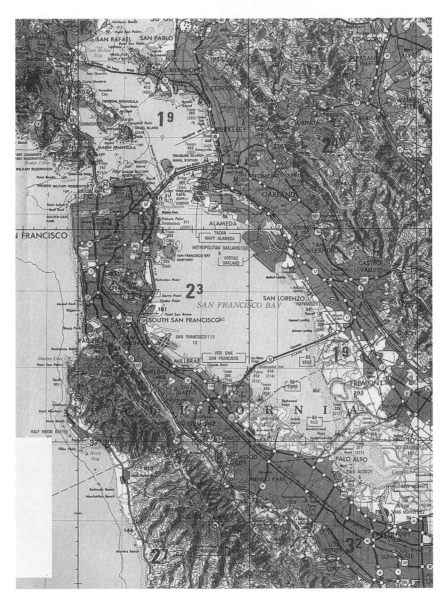

Fig. 13-4. *Continued.*

(DOD) provides these publications to flight service stations for use in preflight pilot briefings.

Pilots should review this information and acquaint themselves with routes located along planned flight paths and in the vicinity of airports from which they operate. Flight instructors, and especially flight schools, should be familiar with the routes that traverse their normal areas of operation. (Obtain the chart and post the information in the flight planning area.) Figure 13-5 contains an excerpt from an area planning military training route chart. Features include:

- Major airports
- NAVAIDs
- Flight service stations
- Military training routes
- Route altitudes
- Route hours of operation
- Special-use airspace
- Nuclear power plants
- Radioactive waste sites
- VFR helicopter refueling tracks
- Index to tactical pilotage charts

In addition to depicting IR and VR routes, the chart includes slow-speed low-altitude training routes (SRs). These routes are used for military air operations at or below 1500 feet at speeds of 250 knots or slower. Information about MTR activity is available from an FSS. Upon request, the specialist at an automated FSS will provide information on military training routes along the pilot's route. Please provide the specialist with the specific route numbers. Nonautomated FSSs will normally only have MTR route information within 100 miles of that FSS's location. Because the area of MTR activity provided by a nonautomated FSS is limited, pilots should routinely request this information en route. Additional information about MTRs is published in a DOD FLIP for North and South America, which is explained later.

Miscellaneous DMA charts

Aerospace planning charts (ASC) consist of six charts, at each scale, with various projections that cover the world. Chart scales are 1:9,000,000 and 1:18,000,000, with a sheet size of approximately 58 × 42 or 21 × 29 inches, respectively. These charts are designed for wall mounting and are useful for general planning, briefings, and studies. The charts do not contain contours, gradient tints, or aeronautical information. Also shown are cities of strategic or economic

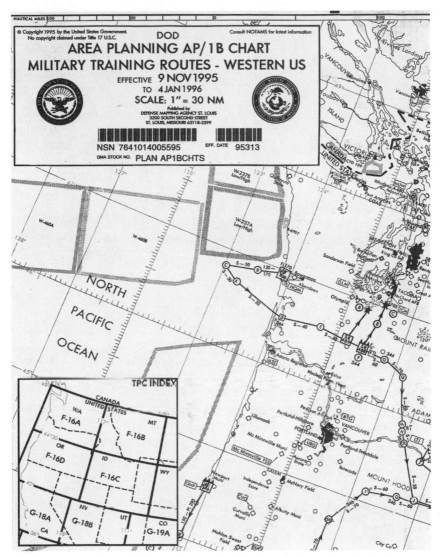

Fig. 13-5. *Military training route charts provide a detailed, visual depiction of low-level high-speed routes.*

importance, major towns, transportation networks, international boundaries, prominent landmarks, plus major lakes and rivers.

Oceanic planning charts (OPC) are designed for transoceanic flights by pilots who do not have a navigator. The charts can be used during preflight and in-flight planning and rapid in-flight orientation. Chart scales vary from 1:10,000,000 to 1:20,000,000. Sheet size is ap-

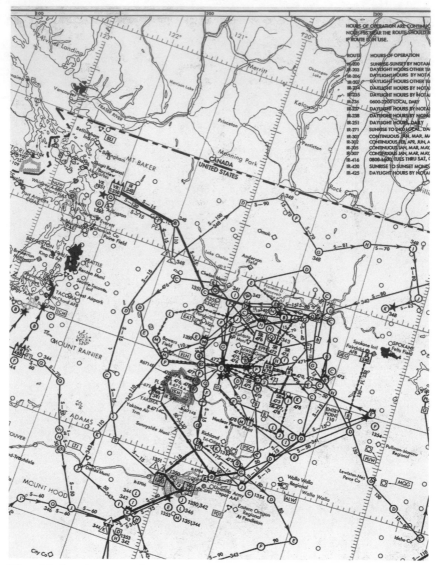

Fig. 13-5. *Continued.*

proximately 17 × 11 inches. Charts are available for the North Pacific and North and South Atlantic. The charts show international boundaries and continental outlines, selected radio aids, and no-wind equal-distance lines between selected diversion airports are shown.

Standard index charts (SIC) are graphics with index overprints for the major aeronautical chart series; sheet size is 28 × 48 inches, covering the world, with a scale of 1:35,000,000. SICs are available for the following chart series:

- Global navigational charts
- Jet navigation charts
- Operational navigational charts
- Tactical pilotage charts

Other charts

Special-purpose and supplementary charts are also available: airport obstruction charts, the Grand Canyon VFR aeronautical chart, Jeppesen VFR charts, Air Chart Systems charts, and state aeronautical charts.

Airport obstruction charts

NOS publishes an airport obstruction chart (OC), with a scale of 1:12,000, that graphically depicts FAR Part 77, "Objects Affecting Navigable Airspace." OCs provide data for computing maximum takeoff and landing weights of civil aircraft, for establishing instrument approach and landing procedures, and for engineering studies relative to obstruction clearing and improvements in airport facilities. Features include:

- Airport obstruction information
- FAR Part 77 surfaces
- Runway plans and profiles
- Taxiways and ramp areas
- Air navigation facilities
- Selected planimetry

(Planimetry shows man-made and natural features, such as woods and water, but does not include relief.)

Grand Canyon chart

NOS, in coordination with the FAA, has developed a Grand Canyon National Park chart. First published on April 4, 1991, this chart is designed to promote aviation safety and assist VFR navigation in this popular flight area. The Grand Canyon VFR aeronautical chart has a scale of 1:250,000, same as TACs, and will be revised as needed, probably once a year. The chart covers the procedures and restrictions required by Special Federal Aviation Regulation (SFAR) 50-2. One side of the chart is for noncommercial operations and the other side is for commercial air tour operations. Features include:

- SFAR operations below 14,500 MSL
- Flight-free zones, where aircraft operations are prohibited
- Corridors between flight-free zones
- VFR checkpoints
- Communications frequencies
- Minimum altitudes
- Navigational data

Jeppesen's VFR charts

Jeppesen has introduced a special series of VFR charts to assist pilots flying in the Los Angeles Basin, San Diego, and San Francisco Bay areas. These charts feature arrival and departure charts for airports with overflight and Class B airspace transition charts. Charts contain a detailed legend, airspace boundaries, and airport diagrams. Each chart is $8^1/2 \times 13^1/2$ inches and patterned after Jeppesen's area charts. Features include:

- Recommended flight tracks
- Recommended courses and radials
- VFR checkpoints
- Appropriate communications frequencies
- General terrain contours and elevations

Air charts systems

Howie Keefe developed Air Charts Systems to minimize the excessive work and cost of maintaining current charts. The VFR group includes *VFR Enroute, VFR Terminal Atlas,* and *Loran/GPS Navigator Atlas.* A feature of the service is 56-day updates. The system allows the VFR pilot to have en-route altitude, distances, and reporting points only available on IFR charts. Addresses and telephone numbers for this service are contained in Chap. 10.

State aeronautical charts

Many states publish aeronautical charts that cover the area within their boundaries, usually based upon the WAC scale of 1:1,000,000.

The reverse side of the chart often contains specific airport information, airport diagrams, and other useful aeronautical or tourist information. Charts might be available from local distributors of aeronautical charts or more often from that state's transportation department or affiliated aeronautics agency.

Canadian VFR charts

Canada produces and distributes pilotage charts. Similar to those used in the United States, the Canadian charts consist of world aeronautical charts, VFR navigational charts, and VFR terminal area charts. In addition, aeronautical planning, North Atlantic plotting, polar plotting, Canada-Northwestern Europe plotting, and Canada plotting charts are available.

Canadian WACs use the Lambert conformal conic projection and have a scale of 1:1,000,000. These charts serve the requirement of visual navigation for medium-speed medium-range operations. Coverage, type, and number are noted in Fig. 13-6.

Canadian VFR navigation charts (VNCs) are equivalent to U.S. sectionals. VNCs use the Lambert conformal conic projection with a scale of 1:500,000. They serve the requirements of visual navigation for low-speed short- and medium-range operations

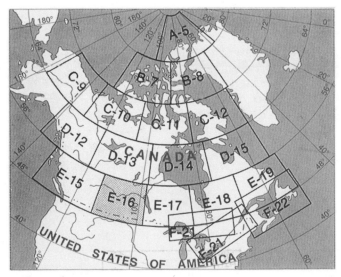

Fig. 13-6. *Canadian WACs serve the requirements of visual navigation for medium-speed, medium-range operations.*

and are suitable for basic pilotage and navigational training (Fig. 13-7). A special VNC has been developed for the Alaska Highway, and covers this route from Fort Nelson, Canada, to Northway, Alaska.

Canada also produces four VFR terminal area charts (VTA) for Montreal, Toronto, Winnipeg, and Vancouver using the transverse Mercator projection with a scale of 1:250,000, which is equivalent to TACs in the United States.

Topography and obstructions

Canadian charts display relief as contour lines, shaded relief, and color tints. Green color indicates flat or relatively level terrain, regardless of altitude above sea level. Significant elevations are depicted as spot elevations, critical elevations, and maximum elevation figures. An MEF on Canadian charts indicates the highest terrain elevation plus 328 feet, or the highest known obstruction elevation, whichever is higher.

Hydrography and culture symbols are similar to those of U.S. charts. Obstructions are also indicated in the same manner, with obstructions 1000 feet AGL or higher shown with a larger symbol. Obstruction elevation in feet above sea level (ASL) appears above the height in feet above ground level, which is enclosed with parentheses.

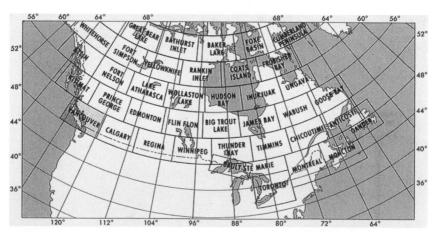

Fig. 13-7. *Canadian VFR navigation charts are designed primarily for low-speed, short, and medium-range operations.*

Navigational aids

Approved land airports in Canada having runways 1500 feet or longer are charted. Airport symbols might be offset for clarity of presentation. Airports with hard-surface runways are depicted using the runway layout. When the use of a particular radio frequency is mandatory, the airport name is followed by the letter "M." The appropriate frequency in the airport data is preceded by the letter "A" (airport traffic frequency). Other airport data and the availability of services are indicated in the same manner as U.S. charts. Airports where customs service is available are indicated by a broken-line box around the airport name.

Radio aids to navigation have the same general appearance as those used on U.S. charts. Heavy-line boxes indicate services similar to an FSS with standard frequencies of 126.7 and 121.5 MHz. Other frequencies are shown above the box. At remote facilities, the name of the controlling FSS appears in brackets below the NAVAID box. Control tower frequencies are not shown for all airports, Canadian WACs, and VNCs, nor are they available in tabulated form as on U.S. charts. These frequencies must be obtained from other sources, such as VTAs or the *Canada Flight Supplement,* which are discussed in Chap. 10. Tower frequencies, along with detailed flight procedures, are contained on Canadian VTAs.

Navigational information

Canada uses the standard International Civil Aviation Organization (ICAO) airspace classification system to describe airspace: A, B, C, D, E, F, and G. In Canada, the base of *Class A* airspace varies from 18,000 feet MSL in the Southern Control Area, to FL230 in the Northern Control Area, and to FL280 in the Arctic Control Area. All controlled high-level airspace terminates at FL600.

Class B airspace is controlled airspace within which only IFR and controlled VFR (CVFR) flights are permitted. It includes all controlled low-level airspace above 12,500 feet above sea level or the minimum en-route IFR altitude, whichever is higher. ATC procedures pertinent to IFR flights are applied to CVFR aircraft. Class B airspace terminates at the base of Class A airspace.

Class C airspace is controlled airspace within which IFR and VFR flight are permitted, but VFR flight requires a clearance from ATC to enter.

Class D airspace is controlled airspace within which both IFR and VFR flight are permitted, but VFR flights do not require a clearance from ATC to enter.

Class E airspace is airspace within which IFR and VFR flights are not subject to control, similar to uncontrolled airspace in the United States.

In Canada, *Class F* airspace is of defined dimensions within which activities must be confined because of their nature, or within which limitations are imposed upon aircraft operations that are not a part of those activities, or both. Special-use airspace in the United States is equivalent.

Canadian special-use airspace is designated alert (CYA), danger (CYD), and restricted (CYR). Alert-area activity is divided into one of the following:

- **A** Aerobatic
- **F** Aircraft test
- **H** Hang gliding
- **M** Military operations
- **P** Parachute dropping
- **S** Soaring
- **T** Training

Altitudes are inclusive unless otherwise indicated. For example, "CYA 125(A) to 5000′" indicates an aerobatic flight alert area, active from the surface to 5000 feet MSL.

Navigational information consists of isogonic lines and values, local magnetic disturbance notes, aeronautical lights, airway intersection depictions, and VFR checkpoints. Most symbols are similar counterparts on NOS charts. Finally, Canadian charts cost more than double American charts.

14

Publications

Effective application of charts during preflight planning and while navigating en route is dependent upon an understanding of supplemental publications. Simple charts published in the early days of aviation (see Chap. 9) naturally evolved hand in hand with the complexities of flying. Before too long, charts could no longer reasonably depict all the data. Charted information changed more rapidly than it was possible to update, reprint, and distribute charts in a cost-efficient manner (see chart currency in Chap. 10). Most aeronautical publications are merely extensions of aeronautical charts.

Aeronautical publications are most often produced and supplemented by the same agency that publishes the respective charts; NOS charts are supplemented by the *Airport/Facility Directory* (A/FD) published by NOS, and the NOTAM publication and NOTAM system are administered by the FAA. DMA supports its charts through flight information publications (FLIPs) and the aeronautical chart updating manual (CHUM). Canada and other countries that produce charts have similar supplementary publications, such as the *Canada Flight Supplement*.

This chapter focuses on publications that supplement NOS visual and instrument aeronautical charts: A/FD, Alaska supplement, Pacific chart supplement (Fig. 14-1), and the NOTAM publication. Chap. 15 discusses the FLIP, CHUM, and other publications.

Some pilots subscribe to the volume of the A/FD where they do most of their flying. Often kept in a flight case, along with other charts, the A/FD's carry immense value. Although charts provide essential data, the directory provides the details. For long trips out of a directory's coverage area, visit the FSS and use the directories that cover the route. Directories are also available on a one-time-sale basis from many chart suppliers. Pilots planning long trips would be well advised to obtain the directories that cover their routes. By reviewing

Fig. 14-1. The Airport/Facility Directory, Alaska Supplement, *and* Pacific Chart Supplement *support NOS visual and instrument charts for the respective regions.*

the directory, all pertinent data can be obtained and noted or logged on a flight planning form.

Airport/facility directory

The A/FD is published in paperback books, measuring $5^3/_8 \times 8^1/_4$ inches, on the standard 56-day revision cycle. The directory is an alphabetical listing of data on record with the FAA for all airports that are open to the public, associated terminal control facilities, air route traffic control centers (ARTCCs), and radio aids to navigation within the contiguous states, Puerto Rico, and the Virgin Islands. Radio aids and airports are listed alphabetically. Airports and associated cities are cross-referenced when necessary. The information directly supports visual charts through the airport listing and aeronautical chart bulletin. Instrument charts are supported through NAVAID restrictions and detailed airport services and information not available on charts.

The A/FD is divided into seven booklets; coverage is depicted in Fig. 14-2. Each directory contains the following:

- General information
- Abbreviations
- Legend, A/FD
- A/FD
- Seaplane bases
- Notices
- Land and hold short operations
- Simultaneous operations on intersecting runways
- FAA and National Weather Service telephone numbers
- Air route traffic control centers

Fig. 14-2. *The seven booklets of the* Airport/Facility Directory *cover the United States, including Puerto Rico and the Virgin Islands.*

- FSDO addresses and telephone numbers
- Preferred IFR routes
- VOR receiver checkpoints
- Parachute jumping areas
- Aeronautical chart bulletin
- Tower en route control
- National Weather Service upper-air observing stations
- En-route flight advisory service outlets and frequencies

The inside front cover of the directory provides general information. This consists of information about corrections, comments, and procurement of the directory, with addresses and telephone numbers. General information contains the directory publishing schedule, along with airport and airspace information publication cutoff dates.

Next is a table of contents and list of abbreviations commonly used in the directory. Other abbreviations are contained in the legend and not duplicated in the list. Abbreviations are also contained in the appendix. This section is followed by the directory legend.

The A/FD lists public-use airports and NAVAIDs that are part of the National Airspace System (NAS). These are listed alphabetically by city or facility name within the state and cross-referenced when necessary. This section is followed by a listing of public-use seaplane bases.

The notices section contains additional information within the coverage of the directory. Other data in this section consists of advance flight plan filing requirements, special flight procedures, temporary closure of facilities, and other general information.

Figure 14-3 is an example of the notices section. It advertises laser-light demonstrations, controlled firing areas, flight test operations, border crossing procedures, and aerobatic operations. Keep in mind that information contained in the publications will normally not be provided during an FSS or DUAT briefing.

A listing of land and hold-short operations (LAHSO) and simultaneous operations on intersecting runways (SOIR) is provided. The *Aeronautical Information Manual* contains specific details on hold-short operations and markings.

Telephone numbers are provided for FAA and NWS facilities within the directory area of coverage, listed alphabetically within the state. The

Fort Collins, Colorado
Christman Airport, Fort Collins, Colorado
(Until December 31, 1995)

A Laser Light Demonstration will be conducted at Christman Airport (CO55), Fort Collins, Colorado, (LAT 40°56′N/LON 105°09′W) sunset to sunrise, FL 230 and below until December 31, 1995. The Laser light beam may be injurious to Pilot's and/or Passenger's eyes within 3,600 feet of light source. Cockpit illumination--flash blindness may occur beyond these distances.

Las Vegas, Nevada

A laser light demonstration will be conducted nightly between sundown and dawn at the Las Vegas Hilton, 3000 Paradise Road, Las Vegas, Nevada (LAS VORTAC 354 degree radial at 2.5NM LAT 36° 08′10″N/LONG 115°09′04″W). The beam may be injurious to eyes if viewed within 700 feet vertically and 2,000 feet laterally of the light source. Cockpit illumination--flash blindness may occur beyond these distances.

Luxor Hotel and Casino, Las Vegas, Nevada
(Until Further Notice)

A laser light demonstration will be conducted nightly between sundown and 2:00 AM at the Luxor Hotel and Casino, 3922 Las Vegas Blvd. South, Las Vegas, Nevada (LAS VORTAC 280 degree radial at 1 Nautical Mile LAT 36°05′44″N/LONG 115°10′34″W). The beam may be injurious to eyes if viewed within 800 feet vertically and 1,300 feet laterally of the light source. Cockpit illumination--flash blindness may occur beyond these distances.

Rio Hotel and Casino, Las Vegas, Nevada
(Until Further Notice)

A laser light demonstration will be conducted nightly between sundown and dawn at the Rio Hotel and Casino, 3700 West Flamingo Road, Las Vegas, Nevada (LAS VORTAC 315 degree radial at 3 Nautical Mile Lat 36°06′59″N/LONG 115°11′12″W). The beam may be injurious to eyes if viewed within 800 feet vertically and 1,300 feet laterally of the light source. Cockpit illumination--flash blindness may occur beyond these distances.

CONTROLLED FIRING AREA (CFA) EAST OF YUMA, AZ

The military has established a controlled firing area (CFA) east of Yuma, AZ. The CFA is bordered by the following fixes: BZA058015 - BZA068035 - BZA072034 - BZA075030 - BZA075015 - BZA058015. Operations will be conducted at or below 3000′AGL. The hours of operation are Monday through Saturday from sunrise to sunset.

NASA FLIGHT TEST OPERATIONS—CROWS LANDING FLIGHT FACILITY

CAUTION: NASA flight test operations conducted immediately south of the Crows Landing Flight Facility CLASS D airspace below 2500 MSL. Pilots should monitor NALF Crows Landing tower in this area.

SAN DIEGO, CALIFORNIA SOUTHBOUND INTERNATIONAL BORDER CROSSING.

Pilots crossing the International border southbound into Mexican airspace, in the vicinity of San Diego, are encouraged to cross Tijuana International Airport at midfield to avoid arriving and departing aircraft. Pilots requesting transition through the Brown Field CLASS D airspace should contact Brown Tower on frequency 126.5. All others should contact Tijuana Approach Control on frequency 119.5 prior to crossing the border. Southbound aircraft are requested to squawk 1260 prior to crossing the border unless otherwise advised by ATC.

AEROBATIC OPERATIONS EAST OF SANTA PAULA, CA

Practice and competitive aerobatic maneuvers are regularly scheduled in the vicinity of the Fillmore VORTAC, sunrise to sunset, from 1,500′ AGL to 5,500′ MSL. Aerobatic area is defined by the following fixed radial distances: FIM220004 through FIM260008 through FIM285009 through FIM360005 through FIM055014 through FIM070013 and on the FIM247013 within a 3 NM radius.

SALT LAKE CITY INTERNATIONAL AIRPORT
Salt Lake City, Utah

Salt Lake City Center, Tower, and Approach are authorized to eliminate the word "city" as part of the facility identification in communications over air traffic control frequencies.

Fig. 14-3. *Special notices supplement charts with additional information to support the safety of flight.*

availability of special services is noted—for example, recorded aviation weather and fast-file flight plan filing services. Flight standards district office (FSDO) addresses and telephone numbers are listed.

Air route traffic control center sector frequencies are provided. These are listed alphabetically by location and altitude stratum (low or high), within the individual ARTCC, for the area of coverage of the directory.

Preferred IFR routes are listed. This system has been established to help pilots in planning their routes, to minimize route changes during flight, and aid in the efficient, orderly management of air traffic.

Approved VOR receiver checkpoints and VOR test facilities (VOTs) are listed alphabetically within the states. Type of check, ground or airborne, and checkpoint description are provided. VOT facilities are listed separately.

Parachute jumping areas, depicted by a small parachute symbol on sectional and TAC charts, are tabulated alphabetically, within the states. This is where a pilot would obtain details on charted parachute jumping areas. Unless otherwise indicated, all activities are conducted during daylight hours and in VFR weather conditions. This section also outlines procedures for parachute jumping areas to qualify for inclusion on charts.

The aeronautical chart bulletin provides major changes in aeronautical information that have occurred since the last publication date of each sectional, terminal area, and helicopter chart. Additionally, users of world aeronautical and U.S. Gulf coast VFR aeronautical charts should make appropriate revisions to their charts from this bulletin.

Figure 14-4 is an example of the aeronautical chart bulletin. It deletes three airports from the San Francisco Sectional: Atwater, Pearce, and Schurz. Class E airspace in the Red Bluff and Redding areas has changed. These changes could be significant, depending on the operations planned.

Tower en-route control (TEC) refers to IFR operations conducted entirely within approach control airspace. These listings are similar to preferred routes. Pilots can use this section to determine preferred routes, altitudes, and type of operations allowed (jet, turboprops, nonjet, etc.).

232 **AERONAUTICAL CHART BULLETIN**

SAN FRANCISCO SECTIONAL
54th Edition, March 30, 1995

Delete ATWATER arpt 37°20'05"N, 120°36'19"W. Delete PEARCE arpt 38°56'05"N, 122°37'24"W. Delete SCHURZ arpt 38°56'10"N, 118°48'24"W.

Revise RED BLUFF, CA Class E Airspace: Within a 6.5-mile radius of the Red Bluff Municipal Airport and within 2.6 miles either side of the 161 deg. bearing from the Red Bluff Municipal Airport extending from the 6.5-mile radius to 10 miles south of the Red Bluff Airport. This Class E airspace area is effective during the specific dates and times established in advance by a Notice of Airmen. The effective date and time will thereafter be continuously published in the Airport/Facility Directory; That airspace extending upward from 700 feet above the surface within a 6.5-mile radius of the Red Bluff Municipal Airport and within 8 miles east and 4 miles west of the 161 deg. bearing from the Red Bluff Municipal Airport extending from 2 miles south to 17 miles south of the Red Bluff Municipal Airport. That airspace extending upward from 1200 feet above the surface within a 17.4-mile radius of the Red Bluff VORTAC and within 7.8 miles each side of the Red Bluff VORTAC 291 deg. radial, extending from the 17.4-mile radius to 45.2 miles west of the Red Bluff VORTAC and within 26.1-mile radius of the Red Bluff VORTAC, extending from the north edge of V-195 to the west edge of V-23 and within 7.8 miles west of and 8.7 miles east of the Red Bluff VORTAC 342 deg. radial, extending from the 17.4-mile radius to 58.2 miles north of the Red Bluff VORTAC and within 8.7 miles west and 5.2 miles east of the Red Bluff VORTAC 015 deg. radial, extending from the 17.4-mile radius to 48.7 miles north of the Red Bluff VORTAC and within an area bounded by a line beginning at lat. 40 deg. 41'27"N, long. 121 deg. 54'42"W; to lat. 40 deg. 34'40"N, long. 121 deg. 52'34"W; to lat. 40 deg. 21'46"N, long. 121 deg. 56'49"W; to lat. 40 deg. 22'35"N, long. 122 deg. 01'04"W, to the point of beginning and that airspace within a 20.9-mile radius of the Red Bluff VORTAC, extending from the Red Bluff VORTAC 015 deg. radial clockwise via the 20.9-mile arc to lat. 40 deg. 00'00"N.

Revise Redding, CA Class E Airspace: That airspace extending upward from 700 feet above the surface within a 4.3-mile radius of the Redding Municipal Airport and within 1.8 miles each side of the Redding Instrument Landing System (ILS) localizer North course, extending from the 4.3-mile radius to 10 miles north of the threshold of Runway 16 and within 8 miles west and 5.5 miles east of the 179 deg./359 deg. bearing from/to the Lassn NDB extending from 9.5 miles north of the Lassn NDB to 16 miles south of the Lassn NDB and that airspace within a 5.5-mile arc of the Redding VOR/DME from the Redding VOR/DME 100 deg. radial clockwise to the Redding VOR/DME 152 deg. radial. That airspace extending upward from 1200 feet above the surface north of the Redding VOR/DME within an arc of a 20-mile radius of the Redding VOR/DME within an arc of the 20-mile radius of the Redding VOR/DME, extending from the east edge of V-23 clockwise to the west edge of V-25.

Military Training Routes
No Changes

SAN FRANCISCO TERMINAL AREA CHART
46th Edition, March 30, 1995
No major changes.

Military Training Routes
No Changes

Fig. 14-4. *The aeronautical chart bulletin revises visual charts as needed between publication cycles.*

The page opposite the inside back cover of the directory contains a graphic of National Weather Service (NWS) upper-air observing stations and weather radars. These are scheduled balloon releases at 1100 and 2300 coordinated universal time (UTC); therefore, pilots cannot expect to be notified by NOTAM unless the release is delayed beyond 1130 or 2330 UTC. NOTAMs are issued for unscheduled balloon releases.

The inside back cover of the directory provides the location and frequency of flight watch outlets within the area served by the volume. All low-altitude flight watch outlets are on the common frequency of 122.0; therefore, only high-altitude frequencies are listed. Each center's airspace is served by a discrete high-altitude flight watch frequency.

Listings in the airport/facility portion of each directory merely indicate the airport operator's willingness to accommodate transient aircraft. A listing does not mean that the facility conforms with any federal or local standards or that it has been approved for use on the part of the general public. Information on obstructions is taken from reports submitted to the FAA. This information has not been verified in all cases. Pilots are cautioned that obstructions not indicated in this tabulation or on charts might exist and create a hazard to flight operations. Detailed specifics concerning services and facilities in the directory are contained in the *Aeronautical Information Manual*.

Directory legend

Figure 14-5 is an example entry in the directory. Features include:

- City/airport name
- Location identifier
- Airport location
- Time conversion
- Geographic position of airport
- Charts on which the facility is located
- Elevation
- Rotating light beacon
- Servicing available
- FUEL AVAILABLE
- Oxygen available
- Traffic pattern altitude
- Airport of entry and landing rights airports
- FAR 139 crash, fire, and rescue availability
- FAA inspection data
- Runway data
- Airport remarks
- Weather data sources
- Communications
- Radio aids to navigation
- Bearing and distance from nearest usable VORTAC

OAKLAND

METROPOLITAN OAKLAND INTL   (OAK)   4S UTC-8(-7DT)   N37°43.28'W122°13.24'   **SAN FRANCISCO**

06   B   S4   FUEL 100LL, JET A   OX 1,2,3,4 TPA-See Remarks   LRA ARFF Indx D   **H-2A,L-2F,A**
RWY 11-29: H10000X150 (ASPH-PFC)   S-200, D-200, DT-400, DDT-900   HIRL CL   **IAP**
    RWY 11: MALSR. Rgt tfc.   **RWY 29:** ALSF2 TDZ.
RWY 09R-27L: H6212XI50 (ASPH-PFC)   S-75, D-200, DT-400, DDT-800   HIRL
    RWY 09R: VASI(V4L)-GA 3.0 TCH 46'. Tree.   **RWY 27L** VASI(V4L)-GA 3.0 TCH 55'.
RWY 09L-27R: H5453XI50 (ASPH)   S-75,D-115,DT-180   HIRL
    RWY 09L VASI (V4L)-GA 3.0 TCH 381. **RWY 27R:** MALSR. Building. Rgt tfc.
RWY 15-33: H3366X75 (ASPH)   S-12.5, D-65, DT-100   MIRL
    RWY 33: Rgt tfc.
AIRPORT REMARKS: Attended continuously. Fee Rwy 11-29 and tiedown. Birds on and in vicinity of
    arpt. Rwy 09L-27R and Rwy 15-33 CLOSED to air carrier acft, except air carrier acft may use Rwy 09L and
    27R for taxiing. Rwy 09L-27R and Rwy 09R-27L CLOSED to 4 engine wide body acft except Rwy 09R-27L
    operations avbl PPR call operations supervisor 510-577-4067. All turbo-jet/fan acft, all 4-engine acft and turbo-
    prop acft with certificated gross weight over 12,500 pounds are prohibited from tkf Rwys 27R/27L or ldg Rwy 09L
    and Rwy 09R. Preferential rwy use program in effect 0600-1400Z‡: All acft preferred north fld arrive Rwys
    27R/27L or Rwy 33; all acft preferred north fld dep Rwys 09R/09L or Rwy 15. It these rwys unacceptable for
    safety or ATC instructions then Rwy 11-29 must be used. Prohibitions not applicable in emerg or whenever Rwy
    11-29 is closed due to maintenance, construction or safety. For noise abatement information ctc noise abatement
    office at 510-577-4276. 400' blast pad Rwy 29 and 500' blast pad Rwy 11. Rwy 29 and Rwy 27L distance
    remaining signs left side. Acft with experimental or limited certification having over 1,000 horsepower or 4,000
    pounds are restricted to Rwy 11-29. Rwy 09R-27L FAA gross weight strength DC 10-10 350,000 pounds, DC 10-
    30 450,000 pounds, L-1011 350,000 pounds. Rwy 11-29 FAA gross weight strength DC 10-10 600,000 pounds,
    DC 10-30 700,000 pounds, L-1011 600,000 pounds. TPA-Rwy 27L 606(600), TPA-Rwy 27R 1006(1000). Rwy 29
    centerline lgts 6500'. Flight Notification Service (ADCUS) available.
COMMUNICATIONS. ATIS 128.5 (510) 635-5850 (N and S Complex)   **UNICOM** 122.95
    OAKLAND FSS (OAK) on arpt. 122.5 122.2. TF 1-800-WX-BRIEF. NOTAM FILE OAK.
Ⓡ **BAY APP CON** 135.65 133.95 (South) 135.4 134.5 (East) 135.1 (West) 127.0 (North) 120.9 (Northwest) 120.1
    (Southeast)
Ⓡ **BAY DEP CON** 135.4 (East) 135.1 (West) 127.0 (North) 120.9 (Northwest)
    OAKLAND TOWER 118.3 (N Complex) 127.2 (S Complex) 124.9
    GND CON 121 75 (S Complex) 121.9 (N Complex)   **CLNC DEL** 121.1
AIRSPACE: CLASS C svc ctc **APP CON**
RADIO AIDS TO NAVIGATION: NOTAM FILE OAK.
    OAKLAND (H) VORTACW 116.8   OAK   Chan 115   N37°143.55'W122°13.42'   at fld.10/17E.   **HIWAS.**
    RORAY NDB (LMM) 341   AK   N37°43.28'W122°11.65'   253° I.3 NM to fld.
    ILS 108.7   I-INB   Rwy29
    ILS 111.9   I-AAZ   Rwy11
    ILS 109.9   I-OAK   Rwy27R   LMM RORAY NDB.

Fig. 14-5. *The* Airport/Facility Directory *provides detailed
information that cannot be included on charts.*

City/airport name. Airports and facilities are listed alphabetically
by associated state and city. Where a city name is different from the
airport name, the city name appears on the line above the airport
name, as in Fig. 14-5; the city name is Oakland and the airport name
is Metropolitan Oakland International. Airports with the same associated city name are listed alphabetically by airport name and will be
separated by a dashed line. All others are separated by a solid line.

Location identifier. The official location identifier is a three- or
four-character alphanumeric code assigned to the airport. These
identifiers are used by ATC in lieu of the airport name for flight
plans and in computer systems. It's important to distinguish between the letter "O" and the number "0." In Fig. 14-5, the identifier

for the airport is Oakland (OAK). Location identifiers might appear as 4OH4, OW15, and 84XS. The first instance is "four oscar hotel four," which should not be confused with "four zero hotel four." Incorrect use of numbers and letters is particularly significant for pilots obtaining weather briefings and filing flight plans with DUATs.

Airport location. Airport location is expressed as distance and direction from the center of the associated city in nautical miles and cardinal points: 4 S (four nautical miles south of the city).

Time conversion. Hours of operation of all facilities are expressed in coordinated universal time (UTC) and shown as "Z" or zulu time. The directory indicates the number of hours to be subtracted from UTC to obtain local standard time and local daylight savings time: UTC-8-7DT (subtract 8 hours from UTC to obtain local standard or 7 hours to obtain local daylight time). The symbol "‡" indicates that during periods of daylight savings time, effective hours will be one hour earlier than shown. For example in the eastern time zone tower: 1100–2300‡. This tower operates from 6 a.m. until 6 p.m. local, during standard and daylight savings time.

Geographic position of airport. The location of the airport is expressed in degrees, minutes, and tenths of minutes of latitude and longitude. For example: N37°43.28′W122°13.24′.

Charts. The sectional and en-route low- and high-altitude chart, and panel, on which the airport or facility can be found is depicted. Helicopter chart locations are indicated as COPTER. In Fig. 14-5, our airport can be found on the San Francisco sectional chart, panel A of the H-2 en-route high-altitude chart, panel F of the L-2 en route low-altitude chart, and the IFR area chart A.

Instrument approach procedures. Instrument approach procedure (IAP) indicates that a public-use FAA instrument approach procedure has been published for the airport.

Elevation. Elevation is given in feet above mean sea level, is the highest point on the landing surface, and is never abbreviated. When elevation is sea level, it will be indicated as (00), below sea level a minus (−) will precede the figure. In Fig. 14-5, the airport elevation is 6 feet MSL.

Rotating light beacon. The letter "B" indicates the availability of a rotating beacon. These beacons operate dusk to dawn unless otherwise indicated in airport remarks.

Servicing. Available services are represented by code:
- S1—Minor airframe repairs
- S2—Minor airframe and minor powerplant repairs
- S3—Major airframe and minor powerplant repairs
- S4—Major airframe and major powerplant repairs

Fuel. Availability and grade of fuel are also coded:
- 80—Grade 80 gasoline (red)
- 100—Grade 100 gasoline (green)
- 100LL—100LL gasoline, low-lead (blue)
- A—Jet A kerosene, freeze point $-40°C$
- A1—Jet A-1 kerosene, freeze point $-50°C$
- A1+—Kerosene with icing inhibitor, freeze point $-50°C$
- B—Jet B wide-cut turbine fuel, freeze point $-50°C$
- B+—Jet B wide-cut turbine fuel with icing inhibitor, freeze point $-50°C$
- MOGAS—Automobile gasoline used as an aircraft fuel (Automobile gasoline may be used in specific aircraft engines that are FAA certified. MOGAS indicates automobile gasoline, but grade, type, and octane rating are not published. Due to a variety of factors, the fuel listed might not always be obtainable to transient pilots. Confirmation of availability should be made directly with fuel vendors at planned refueling locations.)

Oxygen. The availability of oxygen is indicated by one of the following:
- OX 1—High pressure
- OX 2—Low pressure
- OX 3—High pressure (replacement bottles)
- OX 4—Low pressure (replacement bottles)

Traffic pattern altitude. The first figure shown is traffic pattern altitude (TPA) above MSL; the second figure, in parentheses is TPA above airport elevation. In Fig. 14-5, the pilot is referred to the remarks section for TPA.

Airport of entry and landing rights airports. *Airport of entry* (AOE): A customs airport of entry where permission from U.S. Customs is not required, but at least one hour advance notice of arrival must be furnished. *Landing rights airport* (LRA): Application for permission to land must be submitted in advance to U.S. Customs, and at least one hour advance notice of arrival must be furnished. Advance notice of arrival at AOE and LRA airports can be included in the flight plan when filed in Canada or Mexico where flight notification service (ADCUS) is available. Airport remarks will indicate this service. This notice will also be treated as an application for permission to land in the case of an LRA. Although advance notice of arrival *might* be relayed to customs through Mexican, Canadian, and American communications facilities via the flight plan, *the aircraft operator is solely responsible for ensuring that customs receives the notification.* In the example, Oakland is a landing rights airport, and notice that ADCUS is available.

Certificated airport (FAR 139). Airports serving Department of Transportation certified carriers and certified under FAR Part 139 are indicated by the ARFF index, which relates to the availability of crash, fire, and rescue equipment. Index definitions are listed in the directory and FAR 139, "Certification and Operations: Land Airports Serving Certain Air Carriers." When the ARFF index changes, due to temporary equipment failure or other reasons, a NOTAM D will be issued advertising the condition.

FAA inspection. All airports not inspected by the FAA will be identified by the note "Not insp." This indicates that airport information has been provided by the owner or operator of the field.

Runway data. Runway information is shown on two lines. Information common to the entire runway is shown on the first line, while information concerning the runway ends is shown on the second or following line. Lengthy information will be placed in airport remarks. Runway directions, surface, length, width, weight-bearing capacity, lighting, gradient, and remarks are shown for each runway. Direction, length, width, lighting, and remarks are shown for seaplanes. The full dimensions of helipads are shown. Runway lengths prefixed by the letter "H" indicate that the runways are hard-surface concrete or asphalt. If the runway length is not prefixed, the surface is sod, clay, and the like. Runway surface composition is indicated in parentheses after runway length:

- AFSC—Aggregate friction seal coat
- ASPH—Asphalt
- CONC—Concrete

- DIRT—Dirt
- GRVD—Grooved
- GRVL—Gravel or cinders
- PFC—Porous friction courses
- RFSC—Rubberized friction seal coat
- TURF—Turf
- TRTD—Treated
- WC—Wire combed

Runway strength data is derived from available information and is a realistic estimate of capability at an average level of activity. It is not intended as a maximum allowable weight or as an operating limitation. Many airport pavements are capable of supporting limited operations with gross weights of 25 to 50 percent in excess of the published figures. Permissible operating weights, insofar as runway strengths are concerned, are a matter of agreement between the owner and user. When desiring to operate into any airport at weights in excess of those published, users should contact the airport management for permission. Runway weight-bearing capacity is indicated by code:

- S—Single-wheel type landing gear (DC-3)
- D—Dual-wheel type landing gear (DC-6)
- T—Twin-wheel type landing gear (DC-6, DC9)
- SBTT—Single-belly twin tandem landing gear (DC-10)
- DT—Dual-tandem type landing gear (B707)
- TT—Twin-tandem type landing gear (B-52)
- DDT—Double dual-tandem type landing gear (B747)
- TDT—Twin delta-tandem landing gear (C-5, Concorde)
- AUW—Maximum weight-bearing capacity for any aircraft irrespective of landing gear configuration
- SWL—Single-wheel loading
- PSI—Pounds per square inch, maximum PSI runway will support

Quadricycle and dual-tandem are considered virtually equal for runway weight-bearing consideration, as are single-tandem and dual-wheel. The omission of weight-bearing capacity indicates information is unknown. Three zeros are added to the figures for gross weight capacity. For example, S-90 single-wheel type landing gear, weight 90,000 pounds.

Lighting information is found underneath airport remarks: prior arrangement only, operating part of the night only, pilot controlled, specific operating hours, and the like. Because obstructions are usually lighted, obstruction lighting is not included in the lighting code. Unlighted obstructions on or surrounding an airport will be noted in airport remarks. Nonstandard (NSTD) runway lighting indicates that the light fixture systems are not FAA approved because the color, intensity, or spacing does not meet FAA standards. Nonstandard lighting will be shown in airport remarks. Types of lighting are shown with the runway or runway end they serve.

The type of visual approach slope indicator (VASI) and its location are described by a three-digit alphanumeric code. The code begins with the letter V, indicating a VASI system. The next figure indicates the number of boxes utilized: 2, 2-box; 4, 4-box; and 6, 6-box. The last letter indicates which side of the runway has the unit when it is a single-side installation: L is left, and R is right. For example, V6R would be a 6-box VASI on the right side of the runway; V16 is a 16-box VASI on both sides of the runway.

Runway gradient will be shown only when it is 0.3 percent or more. When available, the direction of upward slope will be indicated. Lighting systems, obstructions, and displaced thresholds will be shown on the specific runway end. Right-hand traffic patterns for specific runways are indicated by "Rgt tfc."

In Fig. 14-5, RWY 11-29 is hard surface, 10,000 × 150 feet. The surface is asphalt with porous friction courses. Wheel-bearing capacity is single-wheel 200,000, dual-wheel 200,000, dual-tandem 300,000 pounds, and double dual-tandem 900,000. High-intensity runway lights and runway centerline lights are available. Runway 11 has an MALSR approach lighting system and a right-hand traffic pattern. Runway 29 is equipped with ALSF2 approach and touch-down-zone lights.

Airport remarks. Airport remarks provide supplemental information on data already shown or additional airport information. Data is confined to operational items affecting the status and usability of the airport.

Weather data sources. This section indicates the availability of weather data or an automated weather observing system (AWOS). AWOS is available in one of four systems:

- AWOS-A reports altimeter setting only
- AWOS-1 reports altimeter setting, wind, and usually temperature, dewpoint, and density altitude
- AWOS-2 reports AWOS-1 data plus visibility
- AWOS-3 reports AWOS-2 data plus cloud/ceiling information

Other types of weather information are denoted by one of the following contractions:

- SAWRS—Supplemental aviation weather reporting station for current weather information
- LAWRS—Limited aviation weather reporting station for current weather information
- LLWAS—Low-level wind shear alert system
- HIWAS—Hazardous in-flight weather advisory service, which is a continuous broadcast of SIGMETs, AIRMETs, and urgent pilot reports (HIWAS is noted in the radio aids to navigation section)

Communications. Communications information is listed in the following order along with the frequency:

- CTAF—Common traffic advisory frequency
- ATIS—Automatic terminal information service
- UNICOM—Aeronautical advisory station
- FSS—Flight service station
- APP CON—Approach control, "R" indicates the availability of radar
- Tower—Control tower
- GND CON—Ground control
- DEP CON—Departure control
- CLNC DEL—Clearance delivery
- PRE TAXI CLNC—Pretaxi clearance

Pretaxi clearance procedures have been established at certain airports to allow pilots of departing IFR aircraft to receive the IFR clearance before taxiing for takeoff. The availability of radar is indicated by the circled letter "R."

Airspace. The section indicates the type of airspace (Class B, C, D, or E) and who provides the service. In the example, Class C service is provided by "Bay approach." For airports served by part-time towers, this section will give the type of airspace (and its base) in effect when the tower is closed.

Radio aids to navigation. The directory lists all NAVAIDs, except military TACANs, that appear on NOS visual or IFR charts, and those upon which the FAA has approved instrument approach procedures. NAVAIDs within the National Airspace System have an automatic monitoring and shutdown feature in the event of malfunction. Unmonitored (UNMON), means that an ATC facility cannot observe the malfunction or shutdown if the facility fails. NAVAID NOTAM files are listed on the radio aids to navigation line. At times, this NOTAM file will be different from the airport NOTAM file—for example, the Watsonville localizer/NDB and Salinas (SNS) VOR are listed in its radio aids to navigation section. The localizer and NDB NOTAM file is OAK. The Salinas VOR NOTAM file is SNS.

Radio class designators and standard service volume (SSV) classifications, discussed in Chap. 6, are listed: (T) terminal, (L) low altitude, and (H) high altitude. In addition to SSVs, restrictions within the normal altitude or range of a NAVAID are published. For example, "VOR unusable 030-090 beyond 30 nautical miles below 5000'" indicates that the VOR cannot be relied upon for navigation between the 030 and 090 radials beyond 30 nautical miles, below an altitude of 5000 feet.

Latitude and longitude coordinates, relation of the NAVAID to the airport, facility elevation, and magnetic variation are provided. In Fig. 14-5, "at fld.10/17E" means that the facility is at the field, has an elevation of 10 feet, and a magnetic variation of 17°E.

ASR/PAR indicates that surveillance (ASR) or precision (PAR) radar instrument approach minimums are published. The availability of automatic weather broadcast (AB), direction finding service (DF), and HIWAS are also listed.

Pertinent remarks concerning communications and NAVAIDs are included in this section. For example, possible interference to approach aids due to aircraft taxiing in the vicinity of the antenna, nonavailability of the emergency frequency at the tower, or tower local control sectorization will be listed in this section.

Refer to Fig. 14-5. Let's translate some of the airport data beginning with information for Runway 09R-27L. This runway is 6212 by 150 feet, hard-surfaced, and consists of asphalt with porous friction courses. It has high-intensity runway lights. The runways are served with a visual approach slope indicator (VASI), a 4-box VASI on the left side of the runway (V4L), the glide angle is 3° that crosses the 09R threshold at 46 feet and 27L threshold at 55 feet. Trees are close to the final approach of 09R.

From the airport remarks section, 09R-27L is closed to four-engine wide-body aircraft except by prior approval (PPR). All turbo-jet/fan, four-engine, and turboprop aircraft with weight over 12,500 pounds are prohibited from taking off from 27L or landing on 09R between 10 p.m. and 4 a.m. local time (0600–1400‡). Runway 27L has distance-remaining signs on the left side. The traffic pattern altitude 27L is 606 feet MSL or 600 feet AGL.

Alaska Supplement

The Alaska supplement (AK) provides an A/FD for the state of Alaska—it's a joint civil/military flight information publication—and a FLIP-A/FD for Alaskan civil and military visual and instrument charts. The Alaska supplement contains the following sections:

- General information
- A/FD legend
- A/FD
- Notices
- Associated data
- Procedures
- Emergency procedures
- Airport diagrams
- Position reports

General information, legend, and directory sections contain generally the same information as the A/FD, except that they include military and private airports. The legend and directory also contain information on jet-aircraft starting units, military specifications for aviation fuels and oils, military oxygen specifications, and arresting gear.

Facilities covered by the FAA and DOD NOTAM system are indicated by a diamond symbol for FAA/DOD NOTAMs or the section symbol for civil NOTAMs only. Pilots flying to airports not covered by the

NOTAM system should contact the nearest flight service station or the airport operator for applicable NOTAM information.

Airports in the supplement are classified into two categories: military/federal government and civil airports open to the general public plus selected private airports. Airports are identified using an abbreviation:

- A—U.S. Army
- AF—U.S. Air Force
- ANG—U.S. Air National Guard
- AR—U.S. Army Reserve
- CG—U.S. Coast Guard
- DND—Canadian Department of National Defense
- FAA—Federal Aviation Administration
- MC—U.S. Marine Corps
- MOT—Canadian Ministry of Transport
- N—U.S. Navy
- NG—U.S. Army National Guard
- PVT—Private use only, closed to the public
- NMFS—National Marine Fisheries Service
- USFS—U.S. Forest Service

No classification indicates an airport open to the general public.

Airport lighting is indicated by number:

1. Portable lights—electrical
2. Boundary lights
3. Runway floods
4. Runway or strip
5. Approach lights
6. High-intensity runway lights
7. High-intensity approach lights
8. Sequenced flashing lights (SFL)
9. Visual approach slope indicator system (VASI)
10. Runway end identifier lights (REIL)
11. Runway centerline lights (RCL)

An "L" by itself indicates temporary lighting such as flares, smudge pots, or lanterns. An asterisk preceding an element indicates that it operates on request only, by phone, telegram, radio, or letter; otherwise, lights operate sunset to sunrise, except where pilot controlled lighting (PCL) is indicated.

In addition to the information contained in the A/FD, the Alaska supplement provides unique data for the area it serves. In Alaska, some FAA flight service stations provide long-distance communications: air/ground and a weather broadcast via VOLMET on high-frequency radio. These are published in the supplement. The supplement also provides military air refueling and military training route data.

The procedures section of the supplement contains weather/NO-TAM procedures, ARTCC communications, military and civilian air defense identification zone (ADIZ) information, and other general data. A separate section provides emergency procedures. This section contains air intercept signals, air/ground emergency signals, and search and rescue procedures. The final section provides airport diagrams that are similar to those found on instrument approach procedures charts. The back cover contains position report, flight plan, and change of flight plan sequences for in-flight operations.

Pacific chart supplement

The Pacific chart supplement (PAC) is a civil flight information publication. It serves as an A/FD for the state of Hawaii and those areas of the Pacific served by U.S. facilities: American Samoa and Kiribati-Christmas Island; Tern, Kure, and Wake Island; and, the Caroline, Mariana, and Marshall Islands. The supplement contains ATC procedures for operating in the Pacific, including the same information found in the domestic terminal procedures publication, for its area of coverage. The Pacific chart supplement contains the following:

- General information
- A/FD legend
- Airport/FD
- Notices
- Associated data
- Procedures
- Emergency procedures

- Airport diagrams
- Terminal procedures
- Position reports

General information, legend, and directory are similar in content and format to the domestic directory and the Alaskan supplement. Notices are divided in special, general, and area categories. Special notices include information of a permanent or temporary nature and sectional chart corrections. General notices include navigational warning areas, preferred routes, and general information on flying to Hawaii. Area notices provide general information for operations in the covered areas, including terminal area graphics and Hawaiian island reporting service. Associated data contains NAVAID and communications information, VOR receiver checkpoints, parachute jumping areas, special-use airspace, visual navigation chart bulletin, and military training routes.

Procedures provide information on oceanic navigation and communications requirements, oceanic position reports, routes to the U.S. mainland, Scatana procedures (security control of air traffic and air navigation aids), and ADIZ procedures. Emergency procedures and airport diagrams contain the same information as the Alaska supplement. The final section of the supplement contains terminal procedures. This section contains the same information in the same format as the domestic terminal procedures publication. SIDs, STARs, and instrument approach procedures are contained in this section.

NOTAM publication

Notices to Airmen are published every 14 days (Fig. 14-6). Published NOTAMs are sometimes referred to as Class II. The international term Class II refers to the fact that the data appears in printed form for mail distribution rather than distributed on the FAA's telecommunications systems (FDC NOTAMs and NOTAM Ds). As of June 22, 1995, the publication is divided into four sections. Section one contains information of a general nature, such as airways, flight restrictions, airports, facilities, and procedural NOTAMs. Section two contains FAR Part 95 revisions to minimum en-route IFR altitude and changeover points. Section three contains international NOTAMs. Section 4 has special notices too long for section one that concern a wide or unspecified geographical area, or the fourth section includes items that do not meet the criteria for section one. Information in section four varies widely, but is included because of its impact on

U.S. Department
of Transportation

**Federal Aviation
Administration**

*NOTICES TO
AIRMEN*

Domestic/International

May 11, 1995

Next Issue
May 25, 1995

INTERNATIONAL NOTAMs

Effective March 30, 1995, International NOTAMs are now included in this
publication as Part 3.

Notices to Airmen included in this publication are NOT given during pilot briefings
unless specifically requested by the Pilot.

Airspace–Rules and Aeronautical Information Division (ATP–200)

Fig. 14-6. *Pilots are responsible for information contained in the*
Notices to Airmen *publication.*

flight safety, such as controlled airspace, terminal area graphics, loran-C status information, fly-ins, and the like. As stated on the cover
page: "*Notices to Airmen* included in this publication are NOT given
during pilot briefings unless specifically requested by the Pilot." NO-
TAMs contained in the publication are not available from DUATs.

The NOTAM publication provides NOTAM information current to approximately three weeks prior to its issuance date. Because the publication must be assembled and mailed, it cannot be more timely. Information that is not known soon enough for publication or is temporary is distributed as NOTAM Ds and FDC NOTAMs on the FAA's telecommunications systems. The publication advises users of the last FDC NOTAM number contained in the booklet: "FDC NOTAMs listed thru 5/2576" (the most recent FDC NOTAM is 5/2576). FDC NOTAMs issued after this number are obtained from the FSS and through a DUAT vendor. FDC NOTAMs for temporary flight restrictions are not published in the NOTAM publication; however, they are almost always issued for a short duration.

NOTAMs are issued for the opening, closing, or any change in the operational status of the following facilities or services:

- Airports
- Runway data
- Airport operating restrictions (ARFF)
- Approach lighting systems
- Control zones
- Displaced thresholds
- Runway lighting
- Navigational facilities
- Airport traffic control towers
- Flight service stations
- Weather (AWOS)

Obtaining charts and publications

Additional sources of useful information for pilots are available from NOS and other agencies. NOS suggests two publications for basic reference and supplementary data: NOAA catalog of aeronautical charts and related products plus the NOAA subscription order brochure for aeronautical charts and related products.

The catalog of aeronautical charts and related products describes each IFR and VFR aeronautical chart and chart-related publication, including digital products. Selected related products available from the FAA are also described. Lists of chart agents, prices, chart coverage, and other information needed to select and order charts and publications are provided.

The subscription order brochure for aeronautical charts and related products contains complete ordering information and order forms for all NOS aeronautical products available on a subscription basis. Both publications are available free, upon request from:

NOAA Distribution Branch, N/CG33
National Ocean Service
Riverdale, MD 20737-1199
301-436-6990

By the way, earlier in the chapter, three abbreviations that might be location identifiers were presented. They actually decode as follows:

- 4OH4—Millertime
- OW15—Crash in international
- 84XS—You asked for it, you got it (Toyota)

15

Supplemental and international publications

Additional publications that support charts are published by the public and private sector: DMA's flight information publications, Canada flight supplement, *Publicacion de Informacion Aeronautica* for Mexico, and NOS supplemental documents (Fig. 15-1). Private vendor publications are most often airport directories that offer supplemental airport information, such as an airport diagram plus names and telephone numbers of airport, restaurant, lodging, and transportation services.

Flight information publications

The DMA's products that are equivalent to the *Airport/Facility Directory* are flight information publications (FLIP) planning documents that are intended primarily for use in ground planning at base, squadron, and unit operations offices. They are revised between publication dates by issuing replacement pages or a planning change notice (PCN) on a schedule or as-required basis. Separate documents are designated general planning and area planning.

General planning documents contain general information on FLIPs, divisions of airspace, aviation weather codes, aircraft categories and codes, loran/omega chart coverage, and information on operations and firings over the high seas. They also include information on flight plans and pilot procedures that have common worldwide applications, and information on International Civil Aviation Organization (ICAO) procedures.

Area planning documents contain planning and procedural data for specific areas of the world. They include those theater, regional, and

Fig. 15-1. *Various chart publishers also issue flight information booklets to supplement the charts.*

national procedures that differ from the standard procedures. Additionally, area planning military training routes are available.

Area planning documents are available for North and South America; Europe, Africa, and the Middle East; and Pacific, Australia, and the Antarctic (Fig. 15-2). These publications supplement the visual, en-

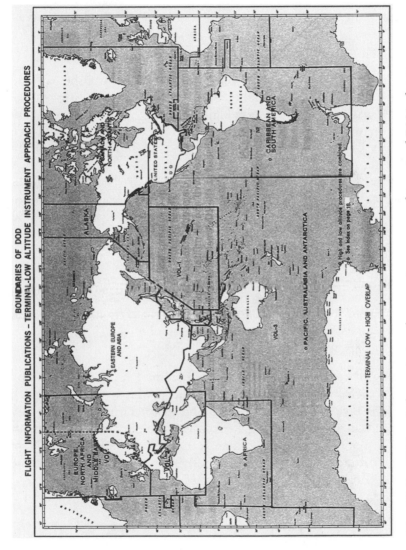

Fig. 15-2. *FLIPs supplement visual, en route, area, and terminal DMA charts.*

route, area, and terminal DMA publications discussed in previous chapters.

Special-use airspace FLIPs are published in three books that contain tabulations of all prohibited, restricted, danger, warning, and alert areas. They also include intensive student jet training areas, military training areas, known parachute jumping areas, and military operating areas. FLIP specials use airspace documents that are available for the same areas of coverage as the area planning documents.

In addition to FLIP charts and publications, DMA publishes a flight information handbook, plus the aeronautical chart updating manual (CHUM). The flight information handbook is a bound book containing aeronautical information required by DOD aircrews in flight, but is not subject to frequent change. Sections include information on emergency procedures, international flight data and procedures, meteorological information, conversion tables, standard time signals, and ICAO and NOTAM codes. The handbook is designed for worldwide use in conjunction with DOD FLIP en-route supplements.

DMA publishes the CHUM semiannually with monthly supplements. The CHUM contains loran and miscellaneous notices and all known discrepancies to DMA and most NOS charts affecting flight safety. Current chart edition numbers and dates for all DMA charts are listed in the CHUM. The publication is intended for U.S. military use. DMA also produces VFR and IFR en-route supplements. These serve the same purpose as the A/FD for the military.

Another publication is devoted to area planning and military training routes for North and South America; the material provides textual and graphic descriptions and operating instructions for all military training routes and refueling tracks. This publication supplements the area planning/military training routes chart discussed in Chap. 13. Each route and track is described by location (radial/distance from the nearest NAVAID) and altitude. Normal use times are also provided. This publication is available at any flight service station.

DMA maintains a public sales program administered by NOS. Through NOS, DMA has a free catalog that contains product descriptions, availability, prices, and order procedures for DMA-produced aeronautical products. Although charts and publications primarily cover areas outside the United States, many domestic charts covering the United States are available for purchase by the general public.

NOS sales agents are located at or near principal civil airports worldwide. DMA products may also be ordered directly by mail.

Use of any obsolete charts or publications for navigation is dangerous. Aeronautical information changes rapidly. It is crucial that pilots have current charts and publications.

Aeronautical Information Manual

The *Aeronautical Information Manual* (AIM) contains basic information needed for safe flight in the U.S. National Airspace System. It includes chapters describing navigation aids, airspace, and the air traffic control system. AIM also provides information on flight safety and safe operating practices. It also includes the FAA's pilot/controller glossary. The AIM is designed to provide pilots with basic flight information and ATC procedures, and it contains items of interest to pilots concerning health and medical factors.

The AIM is revised on a 112-day cycle, approximately three times a year. Pilots insert revised pages into a loose-leaf document. A completely new document is published approximately every two years at the time of a normal revision.

AIM is the FAA's resource for pilots to understand operating in the air traffic control system. With the exception of most regulations, the AIM is the pilot's handbook to operating in the airspace. All of the navigational aids, lighting aids, and procedural descriptions that are not within the scope of this book are defined and explained in the AIM. I recommend that any serious pilot subscribe to, or purchase from one of the many private sources, a copy of this document. It is invaluable to the flight instructor as well as the student pilot. Features include:

- Navigation aids
- Aeronautical lighting and other airport visual aids
- Airspace
- Air traffic control
- Air traffic procedures
- Emergency procedures
- Safety of flight
- Medical facts for pilots
- Pilot/controller glossary

Obtain AIM from the Government Printing Office (GPO). GPO will also provide, free of charge, subject bibliographies on a variety of aviation-related topics, which will guide the reader to government publications available through the Superintendent of Documents. Related subject bibliographies include: aircraft, airports, and airways; aviation information and training; and weather. These bibliographies and a free catalog are available from:

Superintendent of Documents
U.S. Government Printing Office
Washington, D.C. 20402
202-783-3238

Other supplemental products

Various additional supplemental products are available. Among these are a digital aeronautical chart supplement and a digital obstacle file.

The *Pilot's Handbook of Aeronautical Knowledge* contains essential information used in training. Subjects include the principles of flight, airplane performance, flight instruments, basic weather, navigation and charts, and excerpts from other flight information publications. This is a basic text and is ideal for the person getting started in aviation or interested in obtaining a general overview of aviation-related topics.

A Guide to Federal Aviation Administration Publications is a 60-page document that contains information on the wide range of FAA documents and publications and how they can be obtained. It lists available publications by category and gives the various sources. Listed also are civil aviation related publications issued by other federal agencies. Obtain a free copy by ordering FAA-APA-PG-9 from:

U.S. Department of Transportation
M-494.3
Washington, D.C. 20590

The *Location Identifiers Handbook* 7350.5 lists the location identifiers authorized by the Federal Aviation Administration, Department of the Navy, and Transport Canada. It lists U.S. airspace fixes with latitude, longitude, Rho Theta descriptions, and procedure codes. The handbook also includes guidelines for requesting identifiers and procedures for making assignments.

The *Contractions Handbook* 7340.1 lists contractions for general aeronautical, National Weather Service, air traffic control, and aeronautical weather usage. The handbook provides encode and decode sections, plus air carrier, air taxi, and nationality identifiers: N (USA), C (Canadian), G (United Kingdom), and the like. The last sections provide a list of aircraft type contractions: C150, BE35, PA28, and the like.

Digital supplement

The digital aeronautical chart supplement (DACS) is designed to provide digital airspace data that is not otherwise readily available to the public. This publication was originally used only by air traffic controllers but is now available to pilots for use in flight planning and might become available on diskettes and tape. The following sections of the DACS are available separately or as a set:

- Section 1: High-altitude airways in the contiguous United States
- Section 2: Low-altitude airways in the contiguous United States
- Section 3: Reserved
- Section 4: Military training routes
- Section 5: Alaska, Hawaii, Puerto Rico, Bahamas, and selected oceanic routes
- Section 6: STARs and profile descent procedures
- Section 7: SIDs
- Section 8: Preferred IFR routes
- Section 9: Air route and airport surveillance radar facilities

Features include:

- Routes listed numerically by official designation
- NAVAIDs and fixes listed by official location identifier
- Fixes without official location identifiers (airway intersections and ARTCC boundary crossing points) listed by five-digit FAA computer code
- Latitude and longitude for each fix listed to tenths of seconds
- Magnetic variations at NAVAIDs
- Controlling ARTCC
- Military training route descriptions (scheduling activity, altitude data, and route width)

- Preferred IFR route (includes departure or arrival airport name and effective times)
- Radar facilities (ground elevation, radar tower height, and type of radar facility)
- Data that is new or deleted since the last edition is clearly marked or listed

The NAVAID digital data file contains the geographic position, type, and unique identifier for every navigational aid in the United States, Puerto Rico, and the Virgin Islands. These data are chart-independent and can be applied to an NOS chart for which the data are required. Loran and RNAV avionics can use these data without modification. The data is government certified and is compatible with the ARTCC system. This information is made available to the public, including avionics manufacturers, software developers, flight planning services, pilots, navigators, and other chart producers. Features include:

- NAVAID identifier
- Type of NAVAID
- NAVAID status (commissioned or not commissioned)
- Latitude and longitude to tenths of a second
- Name of NAVAID
- NAVAID service volume category
- Frequency of NAVAID
- NAVAID elevation
- Magnetic variation
- ARTCC code where NAVAID is located
- State or country where NAVAID is located

These documents are used by chart producers and programmers of navigation systems. This information allows loran equipment to alert pilots of military and restricted areas and to assist pilots in navigation.

Digital obstacle file

This quarterly file contains a complete listing of verified obstacles for the United States, Puerto Rico, and the Virgin Islands with limited coverage of the Pacific, Caribbean, Canada, and Mexico. Each obstacle is assigned a unique NOS numerical identifier. The obstacles are listed in ascending order of latitude within each state. A monthly revision file contains all changes made to verified obstacles during the previous four-week period. The old record, as it appeared before the change, and the new record are shown. Features include:

- Unique NOS obstacle identifier
- Verification status
- State
- Associated city
- Latitude and longitude
- Obstacle type
- Number of obstacles
- Height above ground level
- Height above mean sea level
- Lighting
- Horizontal and vertical accuracy code
- MARKING, IF KNOWN
- FAA study number
- Julian date of last change

Commercial products

In addition to supplementary products from government sources, private vendors produce a variety of publications. These cover the entire spectrum from the copier-quality reproductions of local ATC and UNI-COM frequencies to high-quality airport diagram and data publications.

Jeppesen supplements its chart services with Federal Aviation Regulations and an airport and information directory, known as the *J-AID*. Features include:

- Radio aids to navigation, including coordinates, variation, and elevation
- Weather information sources
- Sunrise and twilight tables plus other common conversions
- A pilot/controller glossary, airport and NAVAID lighting and markings, and services available to the pilot
- Federal Aviation Regulations
- International entry requirements
- Emergency procedures
- Airport diagrams and other airport information

A free catalog is available:

Jeppesen Sanderson
55 Inverness Drive East
Englewood, CO 80112-5498
303-799-9090

The Aircraft Owners and Pilots Association (AOPA) publishes *Aviation USA* for sale to members and nonmembers. The manual is a guide to AOPA member services as well as an airport directory. It also has a section containing much of the material in the AIM. Publications of this type provide information on airport operators, transportation, lodging, and food services.

A number of commercially available publications are similar to *Aviation USA,* but with a subscription update service. These manuals provide a comprehensive listing of airports with airport diagrams and communication and NAVAID frequencies. Phone numbers for services, restaurants, and hotels and motels are included. The Oakland FSS subscribes to a publication that covers the state of California. Its telephone numbers for FBOs, hotels, and restaurants have been tremendous time savers when locating an overdue aircraft and a pilot who has inadvertently forgotten to close a flight plan. Pilots should not overlook the value of these publications. Most airport pilot shops carry these products, or they can be ordered from an aviation catalog.

Previous chapters have mentioned Howie Keefe's Air Chart System. Additional information on this product can be obtained from:

Howie Keefe's Air Chart System
12061 Jefferson Blvd., Suite A
Culver City, CA 90230-6219
310-822-1996

International publications

Information for international flights is available from various public and private sources. Among these are the U.S. government's *International Flight Information Manual* and international NOTAMs. In addition to these documents, the U.S. Customs Service publishes the *Guide for Private Flyers and Customs Hints for Returning Residents.* Pamphlets are available for sale by the U.S.

Government Printing Office and provide detailed customs requirements and procedures for pilots departing and returning to the United States.

The International Civil Aviation Organization (ICAO) provides extensive information for private and commercial international aviation operations. Several private companies provide information, flight planning, meteorological, customs service, and other international services to pilots.

International Flight Information Manual

The *International Flight Information Manual* is designed primarily as a preflight and planning guide for use by U.S. nonscheduled operators, business, and private aviators contemplating flights outside of America. This manual, which is complemented by the international NOTAM publication, contains foreign entry requirements, a directory of airports of entry, including data that is rarely amended, and pertinent regulations and restrictions. Information of a rapidly changing nature is not included—for example, hours of operation, communications frequencies, and danger-area boundaries including restricted and prohibited areas. The pilot assumes the responsibility for securing that information from other sources: charts, NOTAMs, and en-route supplements. The basic manual is revised every April with changes issued in July, October, and January. This manual is available at most flight service stations.

International NOTAMs, which are now a section in the Notice to Airmen publication, provides NOTAM service on a worldwide basis. International NOTAMs cover temporary hazardous conditions plus changes in facility operational data and foreign entry procedures and regulations. This publication supplements the international flight information manual and is also available at most flight service stations. The respective international NOTAM publication and manual are available on a one-year subscription service.

The U.S. *Aeronautical Information Publication* (AIP) is issued every two years with changes every 16 weeks. This is a comprehensive aeronautical publication containing regulations and data for safe operations in the National Airspace System. It is produced in accordance with the recommended standard of the International Civil Aviation Organization (ICAO).

Canada flight supplement

The Canada flight supplement is a joint civil/military publication issued every 56 days that contains information on airports. This supplement serves the same purpose as the U.S. A/FD. It is published
under the authority of Transport Canada, Aeronautical Information
Services, and of the Department of National Defense. The flight supplement contains the following sections:

- Special notices
- General section
- Aerodrome/facility directory
- Planning
- Radio navigation and communications
- Military flight data and procedures
- Emergency

Special notices direct the pilot's attention to new or revised procedures.

The general section contains information of a universal nature, such
as procurement, abbreviations, and a cross-reference airport location
identifier and name section. This section provides an airport and facilities legend for the supplement.

The aerodrome/facility directory serves the same purpose as its U.S.
directory counterpart. This section might include an aerodrome diagram. The diagram depicts the airport and its immediate surrounding area. An obstacle-clearance circle is provided to assist VFR pilots
operating within close proximity to the airport, but should not be
considered minimum descent altitudes. Altitudes shown are the
highest obstacle plus 1,000 feet within a 5-nautical-mile radius. The
last portion of this section contains a directory of North Atlantic airports and facilities: Azores, Bermuda, Greenland, and Iceland.

The planning section contains flight plan filing, position reporting,
and supplementary data. It provides definitions of airspace classes,
VFR chart updating data (aeronautical chart bulletin), and preferred
IFR routes. A list of designated airway and oceanic control boundary
intersection coordinates is provided.

Radio navigation and communications lists NAVAIDs by location and
identifier. This section provides lists of marine radio beacons, commercial broadcasting stations, and other commonly used frequencies

and services. The military flight data and procedures section contains procedures and flight data for military operations in Canada and the North Atlantic.

The emergency section provides information similar to that contained in the Alaska supplement and Pacific chart supplement. Features include:

- Transponder operation
- Unlawful interference (hijack)
- Traffic-control light signals
- Fuel dumping
- Search and rescue
- Recommended procedures to assist in search
- Procedures when spotting someone in distress
- Small craft distress signals
- Avoidance of search and rescue areas
- Emergency radar assistance
- Emergency communication procedures
- Two-way communications failure
- Information signals
- Military visual signals
- Interception of civil aircraft
- Interception signals
- Signals for use in the event of interception

Canada produces a water aerodrome supplement that provides tabulated and textual information to supplement Canadian VFR charts. This bound booklet contains an aerodrome/facilities directory of all water landing areas shown on Canadian VFR charts. Communications, radio aids, and associated data are also listed.

Pilots planning flights into Canada may obtain a pamphlet, *Air Tourist Information Canada,* free of charge. This pamphlet describes procedures when entering Canada and also lists pertinent aeronautical information publications. This publication is available from:

Transport Canada
AANDHD
Ottawa, Ontario, Canada K1A 0N8

Canadian charts and publications and a free Canadian aeronautical chart catalog are available from:

Canada Map Office
Energy, Mines and Resources Canada
615 Booth Street
Ottawa, Ontario, Canada K1A 0E9
613-952-7000

Pilots planning flights to Canada would be well advised to obtain the tourist pamphlet and chart and publications catalog. Publications are available on a one-time-sale basis. When you are airborne is no time to realize how helpful a particular chart or publication would be.

Publicacion de Informacion Aeronautica

The Mexican government publishes the Publicacion de Informacion Aeronautica (PIA)—in Spanish, of course. This document is similar to its U.S. and Canadian counterparts. (Mexico publishes a series of VFR charts at the scale of 1:250,000, which is equivalent to the TAC series in America.) The PIA also contains coverage of the ONC series at a 1:1,000,000 scale. The PIA contains both low- and high-altitude en-route instrument charts.

The PIA contains an airport/facility directory. Features include:
- Abbreviations
- Time zones
- Sunrise-sunset tables
- Location identifiers
- Radio communications facilities
- Meteorological services
- Flight information region* boundaries
- Flight information region* frequencies
- Airspace information
- Search and rescue information
- Description of available charts
- Glossary of terms
- Aeronautical circulars

*A flight information region (FIR) is equivalent to a U.S. ARTCC.

The PIA, like the U.S. Pacific Supplement, contains instrument approach charts. IAPs are similar in format to their U.S., Canadian, and DOD equivalents, except in Spanish. The PIA also provides SIDs, STARs, graphic depiction of certain areas of controlled airspace, and airport diagrams.

Mexican charts and publications are available from:

SENEAM
Oficina Divulgación y Reproducción Gráfica
Boulevard Puerto Aereo nr. 485
Delegación Venustiano Carranze
c.p. 15500 Mexico, D.F.

ICAO publications

The International Civil Aviation Organization publishes a catalogue of publications and audio-visual training aids. Most ICAO publications are available in English, French, Russian, and Spanish. Arabic is being introduced on a gradual basis. The catalogue indicates which language versions are available. Here is a list of ICAO publications:

- Conventions and related acts
- Agreements and arrangements
- ICAO rules of procedure and administrative regulations
- Annexes to the convention on international civil aviation
- Procedures of air navigation services
- Regional supplementary procedures
- Assembly
- Council
- Air navigation
- Air transport
- Legal
- Miscellaneous publications
- Indexes of ICAO publications
- Audio-visual training aids

ICAO publications cover all aspects of the organization, from the convention and related acts to procedures for air navigation service, air navigation, and air transport rules. Many of the more pertinent documents are retained at FAA flight service stations for reference. A free catalogue is available by writing:

Catalogue of ICAO Publications
International Civil Aviation Organization
1000 Sherbrooke Street West, Suite 400
Montreal, Quebec
Canada H3A, 2R2
514-285-8219
514-288-4772 (fax)

Commercial international products

Various commercial vendors provide international flight planning services. Among them are the Aircraft Owners and Pilots Association (AOPA) and Jeppesen. AOPA provides international services to its membership. Jeppesen provides its services per a set fee.

AOPA has developed flight planning guides for Alaska, Canada, Mexico, and the Bahamas. These booklets include:

- Preflight planning and preparation
- Flying in the country
- Flight rules
- Returning to the United States
- Local information
- Airports of entry
- Supplemental information
- Special flight considerations

AOPA has information on training, services, and requirements for pilots planning to fly the North Atlantic. This includes addresses and telephone numbers for weather and flight planning vendors. Documents provided include various circulars, major changes in procedures and requirements, and the latest information available. AOPA is a resource that should not be overlooked.

Pilots who are not members of AOPA should consider its services. To contact AOPA, write or call:

Aircraft Owners and Pilots Association
Frederick Municipal Airport
421 Aviation Way
Frederick, MD 21701-4798
301-695-2000

As well as aeronautical charts for the world, Jeppesen's international weather and flight planning services consist of:

- Jetplan
- Metplan
- NOTAM services
- International flight services
- OPSDATA service

Jetplan provides optimized flight plans that use the most fuel-efficient routes and altitudes. The computerized database contains the world's low- and high-altitude routes, airports, SIDs, and STARs. Metplan is Jeppesen's weather service. It provides all required weather products and weather information to meet flight operational needs. Jeppesen's weather information complies with FAA/FAR and ICAO requirements. Jeppesen's NOTAM services provide pilot and flight operation personnel with regulatory flight information for the world. Selected information is designed to meet specific user needs. International flight services provide the user with information regarding permits, ground handling, flight plans, weather, ground transportation, and accommodations. Jeppesen's Opsdata service provides airport analysis and airport data services. This provides the user with maximum allowable weights and other related information for high-performance aircraft.

Part Three

Communicate

There are a lot of planes in the air. Sometimes there are traffic jams—even a few accidents. Communication does a great deal to facilitate the crowds and assist the individual pilots in getting around. So you know how to fly; you know how to get there; now you audition for the show...

16

A simulated flight and the ATC contacts

This will be just an introduction to the principal facilities with which you would typically communicate or monitor, but it will include those you must contact, those you should contact, and the sequence in which the communications should be established. As such, it will hopefully give those unfamiliar with the air traffic system a feeling for what it is and the roles its various facilities play.

A simulated flight and the ATC contacts

To set the scene:

- The flight will be nonstop from the Kansas City Downtown Airport to the Memphis, Tennessee, International Airport, via the Victor 159 Airway, which runs generally southeast/northwest over Springfield, Missouri, and Walnut Ridge, Arkansas.

- You're going to file a flight plan with the Columbia, Missouri, Flight Service Station (FSS) that serves Kansas City, and close it on arrival with the Jackson, Tennessee, AFSS, serving Memphis. (I should note here that Flight Service Stations are a part of the air traffic system but do not control traffic in any respect, at any time.)

- Kansas City Downtown Airport lies under, not in, the Kansas City Class B airspace. The floor of the Class B airspace is 3000 feet over the airport and 5000 feet a few miles to the south.

- Kansas City Departure Control is located in the Kansas City International Airport tower, 15 miles northwest of Downtown Airport.

- You intend to request en-route VFR traffic advisories from the Kansas City Air Route Traffic Control Center and Memphis Center.

- Memphis International is in a Class B airspace, which means that radar contact with Memphis Approach Control and specific radio clearance into the airspace is mandatory.

- As the Downtown Airport and Memphis International are both under or in Class B airspaces, all aircraft operating within 30 nautical miles of the primary airports must be equipped with altitude-reporting transponders (Mode C). The transponder must be on, and the switch must be in the ALT (altitude) position.

Filing the flight plan

The first facility to contact is the Columbia FSS for a weather briefing. Unless you happen to be at an airport that has a Flight Service Station (FSS) on the property, or have computer access to a briefing service, obtaining the weather briefing is done by telephone. For the most complete briefing, the FAA recommends a certain sequence of initial information to give to the specialist. The sequence is summarized in the Flight Service Station chapter.

Satisfied that the weather is no problem, you give the briefer the balance of the information necessary to enter the flight plan. All done through a computer, the flight plan is then stored in the FSS until you radio the facility just before takeoff and ask that the flight plan be opened.

Pretakeoff: Automatic terminal information service

You've completed the preflight check and started the engine. What's next?

With radios on and the transponder in the Stand By (SBY) position (warming up), the first job is to tune in the frequency for the Automatic Terminal Information Service (ATIS). The ATIS is a locally recorded message, updated hourly, that provides in sequence the following local airport information:

- Airport name
- Information code (phonetic alphabet: Alpha, Bravo, Charlie, and so forth) of the current report

- UTC time (24-hour coordinated universal time) of weather observations, such as "1355 Zulu weather"
- Sky conditions
- Visibility
- Temperature and dew point
- Altimeter setting
- Instrument approach in use
- Current runway(s) in use
- Information code repeated (phonetic alphabet)

Additional information might be included, such as reports of severe weather, Notices to Airmen (NOTAMs), warnings of construction work near the runway (obstructions, such as cranes), or other conditions of concern to the pilot. The fact that you have monitored the current ATIS must be communicated to Ground Control before you taxi out and in the initial contact with the tower or Approach Control when you land.

Pretakeoff: Clearance delivery

At those airports where the service is provided, your initial call should go to Clearance Delivery (CD) on its published frequency. After you advise CD of your location on the airport, the fact that you have the current ATIS report, that you're operating VFR, your destination, and intended cruising altitude, CD will coordinate that information with the local tower and, if applicable, with Departure Control. Once the coordination has been completed, CD will call back to confirm your clearance into the Class B or C airspace, the altitude(s) to fly in the airspace, the Departure Control frequency to use, and the transponder code to enter in your transponder. These are brief radio exchanges, but this preliminary service to VFR as well as IFR operations saves radio communicating time and expedites departures from the busy Class B and C airport environments.

Pretakeoff: Ground control

If Clearance Delivery, as such, does not exist at the departure airport, Ground Control (GC) may well provide the same service. At any rate, GC represents the next or, if there is no CD, your first radio contact with ATC.

The ground controller operates from the tower itself and is responsible for directing all aircraft and ground equipment movements on the taxiways or when crossing active runways. If you're just shuttling your plane around the ramp, you can do that all day long without anyone's permission. Before your wheels touch a taxi strip, however, you must contact Ground Control for clearance to proceed.

The departure call to Ground Control is usually very brief. Basically, after establishing contact with Ground Control, you simply give your aircraft type and N number; where you are on the airport; that you have Alpha, Bravo, or whatever the current ATIS information happens to be; and that you have received clearance from Clearance Delivery. GC will then authorize you to taxi to the active runway.

Even though cleared to taxi out, it's important to stay tuned to the GC frequency. You're still under the controller's jurisdiction and will be until you are at the runway hold line and ready to contact the tower. In a word, Ground Control is to be monitored from the ramp until at the runway hold line, prior to contacting the tower, and after landing, from a point past the hold line to the ramp. Those are the rules.

Pretakeoff: Flight service station

You've taxied to the run-up area, completed the pretakeoff checks, and are ready to go. First, though, it's time to open the flight plan. This involves a change to one of the published FSS frequencies, and the call is both simple and brief. One thing to keep in mind, though: The call is addressed to "—Radio," not "—Flight Service." Radio is the standard term for all radio communications with any FSS. The one exception arises when you want to contact an FSS flight watch specialist for a weather update while in flight. That call is addressed to "—Flight Watch."

Something else is important: Include in this and all other FSS radio calls the frequency you're calling on and, if different, the frequency you're receiving on, as well as your location. In the automated FSSs, the in-flight specialist staffing the radio might have up to 48 frequencies to monitor, with certain frequencies duplicated. Unless you announce which frequency you're on and where you are, the specialist might find it impossible to respond.

With the flight plan opened, taxi to the hold line.

Control tower

At the hold line—and not before—change frequencies and contact the tower for takeoff clearance. Include in the call the intended direction of flight. Just don't make this call when you're back in the run-up area, maybe 100 feet away. You're not ready to go that far from the runway. Get up to the yellow hold line so that when the tower clears you, you're in a position to move without delay.

Prior to the beginning of the takeoff roll, change the transponder from the SBY position to ON or ALT. The latter puts the unit in the altitude-reporting mode (a Mode C transponder). There's no choice. FAA regulations say that if the aircraft is transponder-equipped, the set must be turned to the ON or ALT position. And from here on, like it or not, when you're within radar range of any controlling agency, an image representing your aircraft will be on the radar screen.

Airborne: Departure control

Since this is a Class D airport underlying a Class B airspace, the next service, if you want it as a VFR pilot and had not requested same from the tower before takeoff, involves traffic advisories from Kansas City Departure Control. Assuming that you do want the service, when airborne and in an established climb, ask the tower to approve a frequency change. The controller might tell you, "Stay with me for a while," if the local traffic in the vicinity is heavy; but otherwise, he'll come back with, "Frequency change approved."

When you hear that, switch to the Departure Control frequency; establish contact; identify yourself, your present heading, and destination; and request traffic advisories. If the controller's workload permits, perhaps he'll give you altitude or heading directions and advise you of other traffic in your line of flight. Regardless of what you're told or not told, your continuing responsibility is simple: Keep your head out of the cockpit, watch for other traffic, listen, acknowledge advisories or instructions, and do what the controller says.

In a short while, Departure will tell you that you're leaving the Class B airspace, and he might or might not turn you over to the Kansas City Center for flight following. A *handoff*, as it is called, however, is not an automatic procedure; the controller may simply tell you, "Radar service terminated. Squawk one two zero zero. Frequency change approved. Resume own navigation."

Airborne: Air Route Traffic Control Center (ARTCC, or Center)

Because you do want flight following, your first responsibility, after tuning to the Center frequency, is to listen to be sure the air is clear before you make the initial radio contact. If it is clear, merely address the call to such and such Center and give your aircraft and N number. No more "Kansas City Center, Cherokee 1234 Alpha." After Center has responded, give your present position, altitude, destination, the numerical code in your transponder, such as "1-2-0-0," and your request for flight following. The Center controller, if his or her workload permits, will give you a different code to enter in the transponder, after which the controller will respond, "Radar contact."

Once you hear that, your progress will be monitored by a facility, which, in the case of Kansas City, is physically located in Olathe, Kansas, about 25 miles southwest of the Downtown Airport. This is one of 21 Centers in the contiguous United States. Each Center is responsible for the control of IFR traffic over its assigned area and, when conditions permit, provides traffic advisories to VFR aircraft.

As you move down the airway toward Springfield, along with some possible advisories of traffic, the controller tells you to "Contact Kansas City Center on—frequency." This means that you're leaving that controller's sector of responsibility and are being turned over to another controller, perhaps sitting only a few feet away, who will be monitoring your progress in his or her sector and on a different frequency. All you do now is to acknowledge the instruction, repeat back the new frequency (to be sure you've copied it correctly), enter it in the radio, and contact the new controller: "Kansas City Center, Cherokee Eight Five One Five November with you. Level at seven thousand five hundred."

As you near Springfield, Center will tell you to contact Springfield Approach Control on a certain frequency. Why Springfield? You're not going to land there. No, but you'll be passing through the airspace for which that Approach Control is responsible and will thus be under the facility's surveillance. Once you are out of that airspace, Springfield will probably turn you over to Center again, assuming Center's workload in that sector will be able to handle you.

Depending upon the length of the flight, this process of transfer could happen several times, until, in the example we're using, you get about 50 miles southeast of Springfield. At that point, Kansas City will come on the air and tell you to contact Memphis Center on a certain frequency. You're now in Memphis's territory, and future advisories will come from them.

Regardless of the Center, you might hear at any time something like this: "Cherokee Eight Five One Five November, traffic at one o'clock, three miles, also southeast, altitude unknown." The controller has spotted another aircraft on the screen at your one o'clock position and is alerting you to its presence. First acknowledge that you've received the call, and if you see the aircraft, tell the controller. "Roger, Center, Cherokee One Five November has traffic." If you haven't spotted it, the response goes, "Roger, Center, Cherokee One Five November no contact. Looking." If you sight the traffic in a minute or so, advise the controller: "Cherokee One Five November has traffic." Controllers appreciate that information.

Airborne: Memphis ATIS

About 40 or so miles out of Memphis, tune to the Memphis ATIS for the latest weather information and the runway(s) in use. Knowing the active runway this far out gives you time to visualize and plan the probable traffic pattern you'll fly.

Airborne: Approach Control

About 30 miles out, Center turns you over to Memphis Approach Control for vectors (headings) and advisories into the Class B airspace surrounding the airport. Remember that you must contact Approach before entering any Class B airspace, whether you've been in contact with Center or not, and you must receive specific clearance into it. Once entrance is approved, if you're unfamiliar with the location of an airport, don't be afraid to tell Approach. The controller will give you the vectors that will lead you straight to it.

Airborne: Control Tower

Finally, when you are within a few miles of the airport, Approach tells you to contact Memphis International Tower on a certain frequency. This is the last lap as the tower sequences you into the traffic pattern and gives the clearance to land.

Postlanding: Ground Control

Once you are on the ground and clear of the runway, the next call is to Ground Control for taxi clearance to whatever location or fixed-base operator you've selected. In real-life situations, if you're not sure where to go or how to get there, ask Ground Control for progressive taxi instructions.

Postlanding: Flight Service Station

When you are parked at the ramp, the final radio call is to Jackson Flight Service to close out the flight plan. You can do this by dialing 1-800-WXBRIEF in the operator's facility; but it's a good idea to make the call as soon as you've shut down the engine. It's an easy call to forget, and if you do forget, the FSS will start looking for you 30 minutes after your estimated arrival. A needless search does not make for happy authorities.

In that same vein, if you find in a real-life situation that headwinds or other factors are slowing you down and that you're going to miss your estimated time of arrival (ETA), contact the nearest FSS in flight and revise the ETA. Or perhaps a favorable tailwind is pushing you along faster than expected, and you determine that you've got enough fuel to reach another airport farther down the airway. The same principle applies. Radio the nearest FSS; advise it of the destination change, your new ETA, and the hours of fuel remaining. In a word, don't keep Flight Service in the dark if you've filed a flight plan. Search procedures will be started unless those folks know what's going on.

By way of summary

In sequence, the 14 FAA service and controlling facilities involved in this example are

- Flight Service (telephone)
- ATIS
- Ground Control
- Flight Service
- Tower
- Departure Control
- Kansas City Center

- Springfield Approach
- Memphis Center
- ATIS
- Approach Control
- Tower
- Ground Control
- Flight Service

With an absolute minimum of one telephone call and 13 radio-frequency changes, you might get the impression that you have been a little busy. Were they all mandatory? No. Going VFR, you are required to listen to the Kansas City ATIS, contact Ground Control and the tower, monitor the Memphis ATIS, contact Memphis Approach (because Memphis is in a Class B airspace), Memphis tower, and Memphis Ground Control. Otherwise, filing a flight plan and requesting traffic advisories from Departure (when below the floor of a Class B airspace) and the two Centers is your choice.

If you do request advisory services, keep in mind that the ATC system will be tracking you, as in this simulated flight, across 3001 miles of geography. Somebody will know who you are and where you are at all times. Changing radio frequencies and listening for your aircraft N number are small prices to pay for the added security the system can offer the VFR pilot, when, of course, its workload permits it to handle VFR traffic in addition to IFR operations. Nothing will happen, though, until you pick up the mike and make the first call.

The objective for the preceding scenario is merely to introduce the basic elements of the airspace system and the various FAA agencies that would or could enter the picture on a typical VFR cross-country trip. It is a sketchy overview, but greater thoroughness is reserved for the next chapters and what goes on behind the scenes as we file flight plans, get advisories, and the rest.

What I've outlined, however, is part of the National Airspace System (NAS), which includes operations in controlled as well as uncontrolled environments. So with this as a start, the next logical step is to take a closer look at the system and the regulations pertaining to it.

17

A closer look at
the airspace system

Reviewing or describing the airspace system in the twenty-first century has certain risks. What is valid today might be invalid tomorrow. Changes are in the wind or are already blowing across the pilot's horizon. The rash of midair collisions and reported near-misses has stirred the Department of Transportation and its Federal Aviation Administration into a welter of activity uncommon in the bureaucracy. Things are moving rapidly. Regulations are getting tougher; airspace violators are being tracked and punished more severely; and general aviation is finding freedom of the air a diminishing privilege. In the process, it is the general-aviation VFR pilot who is bearing the brunt of the increasing restrictions. Some of the prophecies of the 1930s are coming true.

To a certain degree, however, perhaps the VFR pilot population has brought at least a portion of the restrictiveness on itself. While the volume of incidents is relatively small, there have been too many instances of illegal penetration of controlled airspaces and violations of VFR flight regulations, as the cases indicate. Whether they are products of carelessness, inattention, or ignorance matters little. The fact remains that federal authorities are taking action to reduce the hazards of the "crowded skies."

That said, pilots still have considerable freedom to go almost anywhere with at least a private license and an aircraft that has the required avionics. They say the freedom is theirs, but freedoms or rights are not always honored because denial or delay of clearances into controlled areas is not a rarity these days. The cause is not all regulatory, although that is certainly a contributor. Until the ATC facilities are fully staffed and the equipment matches today's state of the art, additional regulations might further impinge on the freedoms that remain.

Whatever the future, the basic elements of the airspace system will undoubtedly survive. If that's a reasonable prophecy, let's begin with a general review of the system, followed by a discussion of the special-use airspaces in Chap. 18. Then, in succeeding chapters, we'll look at tower-operated airports more closely, including the control areas that surround an increasing number of the busier terminals.

Sources for identifying airspace

Before one ventures forth, it's only smart to determine what airspaces will or might be penetrated: controlled, uncontrolled, prohibited, restricted, military operation areas (MOAs), or whatever. Four basic determining sources exist that can be used singly or in combination, depending on the nature of the flight: the sectional aeronautical chart (or just plain sectional), the en-route low-altitude chart (ELAC), the Airport/Facilities Directory (A/FD), and the terminal area chart (TAC) for operating in Class B airspaces.

Primarily designed for instrument operations, the ELAC is an excellent reference for VFR cross-country operations, if it is used in conjunction with the sectional. Unlike the sectional, ground details such as rivers, highways, obstructions, terrain altitudes, towns, and the like are omitted. Instead, the ELAC focuses on radio navigation facilities and includes crucial data not found on the sectional. Combined, the two charts provide about all the information one could want on a cross-country journey. Updated about every 2 months by NOS, 28 individual charts cover the 48 states.

The Airport/Facilities Directory, as the name implies, contains data about airport facilities and services, including runways, types of fuel available, communication facilities, frequencies, radio aids to navigation, and other information not readily available elsewhere.

The terminal area chart is an enlargement of the Class B surface and airport information depicted on the sectional charts. As an exploded and less cluttered "picture" of the area that lies within the large blue square on the sectional identifying the existence of a Class B airspace, the TAC provides the sort of visual image of the airspace so necessary to operate in a busy airport environment. With any of these sources, be sure to refer to the current issue. Because of changes, what was fact on the last chart or A/FD might not be valid today.

How the airspace is depicted

Simply said, the U.S. airspace basically conforms to the worldwide ICAO system in terms of structure and designation. The U.S. system, then, consists of six types of airspace, each identified alphabetically from A through E and then G. (The omission of an F airspace is not a typographic error. That airspace is common overseas, but the United States has no counterpart.) Conceptually, the system is easy to understand, as perhaps Fig. 17-1 illustrates, but more explanation than just a figure is necessary:

The airspaces—a general description

In essence, Class A (Alpha) is the high-altitude (18,000 to 60,000 feet MSL) en-route positive control airspace for IFR operations only; Classes B (Bravo), C (Charlie), and D (Delta) are airport terminal airspaces; Class E (Echo), except where these B, C, and D airspaces exist, comprises the en-route low-altitude airspace, generally from 1200 feet AGL to 18,000 feet MSL. At the same time and under specified conditions, certain non-B, C, or D airports are identified as Class E; Class G is uncontrolled airspace, which typically rises from the surface to 1200 feet AGL and includes all other airports not justifying a B, C, D, or E classification. Adding a little more meat to

Airspace classifications effective September 16, 1993

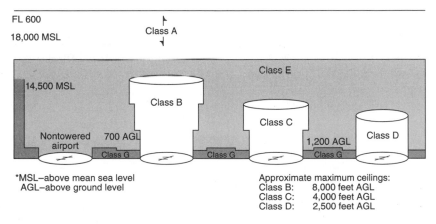

Fig. 17-1. *This commonly seen FAA drawing shows the six types of airspaces from Class A to Class G, with Class A being the most restrictive and Class G the least.*

these skeletal descriptions, the following summaries, coupled with reference to Fig. 17-2, may help you to understand the system.

Class A

These are positive control areas from 18,000 to 60,000 feet. All operations are conducted under instrument flight rules (IFR) and are subject to ATC clearances and instructions. Aircraft separation and safety advisories are provided by ATC.

Class B

These are the largest and busiest U.S. airports and their surrounding areas, based on the number of instrument flight operations and passenger enplanements. Operations may be IFR or VFR, but all aircraft are subject to ATC clearances and instructions. ATC provides aircraft separation and safety advisories.

Class C

These are the medium to large airports, again based on the number of instrument operations and passenger enplanements, and their surrounding areas. Operations may be either IFR or VFR, and all aircraft are subject to radio contact with ATC and instructions. ATC provides aircraft separation between IFR/IFR and IFR/Special VFR (SVFR) aircraft. VFR operations are given traffic advisories and, on request, collision resolution instructions.

Class D

These are tower-controlled airports that have fewer instrument flight operations and passenger enplanements than Class C or B airports and the immediate surrounding areas. Operations may be either IFR or VFR, with all aircraft subject to radio contact with ATC and subsequent instructions. ATC separation service is provided for IFR aircraft only, but all aircraft are given traffic advisories.

Class E

These are typically nontowered airports with weather-reporting sources and approved for Part 135 commuter and on-demand flight operations. Also, Class E represents all the controlled airspace aloft, rising typically from 700 or 1200 feet AGL to 18,000 feet MSL but excluding Class A, B, C, and D airspaces.

Airspace features	Class A airspace	Class B airspace	Class C airspace	Class D airspace	Class E airspace	Class G airspace
Operations permitted	IFR	IFR and VFR	IFR and VFR	IFR and VFR	IFR and VFR	IFR and VFR
Entry requirements	ATC clearance	ATC clearance	ATC clearance for IFR. All require radio contact.	ATC clearance for IFR. All require radio contact.	ATC clearance for IFR. All require radio contact.	None
Minimum pilot qualifications	Instrument rating	Private or student certificate	Student certificate	Student certificate	Student certificate	Student certificate
Two-way radio communications	Yes	Yes	Yes	Yes	Yes for IFR	No
VFR minimum visibility	NA	3 statute miles	3 statute miles	3 statute miles	3 statute miles[1]	1 statute mile[2]
VFR minimum distance from clouds	NA	Clear of clouds	500 feet below, 1000 feet above, and 2000 feet horizontal	500 feet below, 1000 feet above, and 2000 feet horizontal	500 feet below, 1000 feet above, and 2000 feet horizontal[1]	Clear of clouds
Aircraft separation	All	All	IFR, SVFR, and runway operations	IFR, SVFR and runway operations	IFR, SVFR	None
Conflict resolution	NA	NA	Between IFR and VFR operations	No	No	No
Traffic advisories	NA	NA	Yes	Workload permitting	Workload permitting	Workload permitting
Safety advisories	Yes	Yes	Yes	Yes	Yes	Yes

Fig. 17-2. *This abbreviated chart summarizes the operating regulations and ATC services available within the various classifications.*

On a sectional chart, the Class E airport is surrounded by a segmented magenta circle or cookie-cutter figure (representing what is called a transition area) which establishes the airspace designed to protect arriving or departing IFR aircraft during IMC (instrument meteorological conditions) weather. Within the "design," VFR operations are excluded or limited when IMC prevail. In all other Class E airspace outside the airport environment, operations may be IFR or VFR. Separation service is provided to IFR aircraft only, but to the extent practical, traffic advisories to VFR aircraft are provided, if so requested.

Class G

These are nontowered airports without weather-reporting sources and not approved for Part 135 commuter or on-demand operations, and most airspace outside the airport vicinity up to 1200 feet AGL. Operations may be IFR or VFR, but no ATC service is available. Again, Fig. 17-2 provides additional information, in summary form, about the features and operating requirements within each of the airspaces.

"Subject to air traffic control"

Before proceeding, and perhaps to remove present or future confusion, the meaning of a pilot being "controlled by ATC" or "subject to ATC clearances and instructions" warrants clarification. Under instrument flight rules, either expression implies that the IFR pilot, among other things, files a flight plan, receives clearances or amended clearances from ATC, is subject to ATC instructions while in flight, makes required position reports, varies aircraft airspeed if so directed (within safe operational limits), maintains approved altitudes and flight routes, requests altitude or heading deviations, and, barring extenuating circumstances, adheres to and coordinates with ATC all flight operations or instructions.

These overall responsibilities do not imply that the IFR pilot has no control over the management of the flight. The pilot always has the ultimate control. The ATC system, directly administered to the pilot by the controller, is charged with the orderly flow and separation of all traffic in the controlled airspaces. If that is to be achieved, pilot adherence to a planned operation, to approved deviations, and to ATC instructions is essential.

The VFR pilot, on the other hand, whether having filed a flight plan or not, is under no such constraints—with one very important exception. That exception exists when he or she is operating within the confines of a Class B, Class C, or Class D environment.

Controlled airspace

Logic would have it that the airspace either is controlled or isn't. In one sense, however, that isn't quite true, because if you're flying VFR, you may well be in a controlled area but not literally under the control of any ATC facility. On the surface, that statement doesn't make a lot of sense, but Part 1.1 of the Federal Aviation Regulations tends to clarify the issue—at least partially:

> *Controlled airspace means an airspace of defined dimensions within which air traffic control service is provided to IFR flights and to VFR flights in accordance with the airspace classification.*

Note: Controlled airspace is a generic term that covers Class A, Class B, Class C, Class D, and Class E airspace. What this means, in essence, is that IFR aircraft are subject to air traffic control in all five airspaces. VFR aircraft cannot operate in Class A airspace, but they are subject to control in Classes B, C, and D (the terminal environments that have operating control towers).

As to Class E operations, control in that airspace is a different matter. For example, you're on an VFR cross-country trip and have contacted an Air Route Traffic Control Center for traffic advisories. If the workload permits, the Center controller will provide this service and also alert you to potentially hazardous conflict situations. You're still flying VFR, however, and are thus free to change altitudes or directions, as long as you adhere to VFR flight regulations. (Be sure, though, to advise the controller before making such changes.) You are thus being helped by ATC, but are not under its control.

Here is another VFR cross-country situation: You're flying an airway in the Class E airspace, but you haven't called Center for traffic advisories. Despite being in a controlled environment, you can climb, descend, change headings, or do whatever you want as long as you adhere to the VFR altitude, operational, and cloud separation regulations.

So, to make it simple, you're never under ATC control when flying VFR in the controlled Class E airspace.

Uncontrolled airspace (Class G)

As far as uncontrolled airspace is concerned, Class G is the airspace in which no ATC service to either IFR or VFR aircraft will be provided (other than possible traffic advisories when the ATC workload permits and radio communications can be established).

Class G is quite limited, though. As the shaded area in Fig. 17-1 illustrates, the ceilings of the uncontrolled airspace are 700 feet AGL, 1200 feet AGL, and 14,500 MSL. That's well and good, but it still leaves a fair number of questions unanswered. Finding the answers is where the sectional chart comes into play.

If you check any sectional, as mentioned earlier, you'll see a lot of airports surrounded by magenta designs that identify "transition areas." Most of these designs are circular; others are irregular in structure. The sectional legend flap implies, but doesn't state as such, that within the areas bounded by the magenta, the ceiling of the uncontrolled airspace is 700 feet AGL. It further implies, but doesn't state, that outward from the magenta design, the ceiling of the uncontrolled airspace is 1200 feet AGL.

The fading or fuzziness indicates that everywhere inside the magenta, the 700-foot ceiling prevails, except at those airports where a segmented circle or design in magenta defines this Class E airspace. By the same token, the airspace outside the magenta identifies the geography where the 1200-foot ceiling prevails. Accordingly (other than for any intervening prohibited or restricted areas), in the open country between any airport with magenta and any other airport with magenta, the uncontrolled Class G ceiling is 1200 feet AGL.

Fine, but what about the uncontrolled airspace that goes up to 14,500 feet, per the shaded area at the extreme left in Fig. 17-1? As a general rule, the 1200-foot AGL ceiling or floor exists over most of the United States. It's a different story, however, in the mountainous regions of the west.

For example, look at the excerpt from the Phoenix sectional (Fig. 17-3). This is just a segment of a Victor airway, but you'll notice that the outside lines of the airway (shaded on the sectional here; in blue in full-color versions) are sharp, while the shading fades inward toward the airway center line. This means that the controlled airspace starts at 1200 feet above the ground everywhere within the 8-nautical-mile width of the airway. Outside the airway, however, and beyond the

Fig. 17-3. *In this excerpt from the Phoenix sectional, controlled airspace begins at 1200 feet above ground level between the boundaries of the airway. Outside those boundaries, the airspace is uncontrolled from the surface up to 14,500 feet MSL.*

sharp demarcation lines, it's a different story. In these areas, the airspace is uncontrolled up to the 14,500-foot level. Remember, though, that this symbology is peculiar only to the mountainous areas in the western United States. Elsewhere, the Victor airways are simply narrow blue lines on a sectional.

Controlled airspace, classes A and E

Disregarding airports and their immediate surroundings for the moment, Fig. 17-1 depicts the other two controlled airspaces: Classes A and E. To operate in the Class A airspace, regardless of the weather, you must file an IFR flight plan and adhere to all IFR regulations and procedures. Control by ATC from 18,000 to 60,000 feet is, indeed, positive.

As to Class E except in Hawaii, where there are no vertical limits, the normally 8-nautical-mile width of airspace along federal airways, from 1200 up to 18,000 feet, is identified as controlled. To repeat, though, if you are flying an airway on a VFR flight plan, or no flight plan at all, you're in a controlled airspace but are not under ATC's influence. You might have established contact with a Center for traffic advisories, and if so, you'll probably have several communication exchanges with Center or Approach Control personnel as you proceed along the airway; however, you're still VFR in VFR conditions. Accordingly, if you have been receiving advisories, your only responsibilities are to adhere to the VFR regulations, to advise Center if you are going to change altitudes or deviate from your announced route of flight, and to keep your eyes open.

Until you near a Class B, Class C, or Class D airspace, you can fly the airways all day and never call or be in contact with any controlling agency. That's not a necessarily good idea, but it's still the VFR pilot's privilege.

VFR regulations

With the basic structure of the airspace out of the way (except for the Class B, C, and D airports), a brief review of the altitude, visibility, and cloud separation regulations affecting VFR pilots is in order. These might be well known, but no discussion of operating within the airspace system would be complete without at least a short summary.

Flying VFR places that old and oft-repeated responsibility on the pilot: See and avoid, whether the airspace is controlled or uncontrolled. That's as applicable in the open country as in the vicinity of a tower-operated airport. In a controlled environment, such as around an airport, you might be given advisories that you "... have traffic at two o'clock," but the ultimate responsibility for conflict avoidance is still yours.

Despite that obvious fact, the skies would be chaotic and infinitely more dangerous if pilots were allowed to roam about completely at will, relying solely on the hope that the other pilot will see and avoid. Consequently, and to minimize the potential of conflict, regulations have been established governing altitudes to be flown as well as visibility and cloud separation minimums that must be observed.

Altitudes.

First, let's look at the altitudes, again from AIM, as illustrated in Fig. 17-4. The chart is self-explanatory, but an easy way to remember whether the VFR altitude should be in odd or even thousands of feet is to use the acronym WEEO: westbound even, plus 500; eastbound odd, plus 500. Note, though, that those altitude requirements don't apply when you're flying at 3000 feet AGL or below.

If everyone adhered to the prescribed altitudes, we'd have a lot fewer near-misses. Evasive action is too frequently necessary when you're at the correct altitude, and some sightseer barrels toward you head-on at the same altitude. This often happens above 3000 feet

VFR flight altitudes

If your magnetic course (ground track) is	More than 3,000' above the surface but below 18,000' MSL, fly
0° to 179° 180° to 359°	Odd thousands MSL, plus 500' (3,500, 5,500, 7,500, etc.) Even thousands MSL plus 500' (4,500, 6,500, 8,500, etc.)

West Even—East Odd : plus 500'

Fig. 17-4. *The required VFR flight altitudes above 3000 feet AGL.*

AGL within 30 or 40 miles of an airport where, for some reason, altitude adherence seems less important.

And then there are cases when both aircraft are going by the book, but one is headed southeast and the other northeast. Unless at least one pilot has alert and roving eyes, a disaster is in the making. If you have your head in the cockpit much more than 10 seconds, instead of scanning the surrounding atmosphere, you ought to be getting a little nervous. The rate of closure, even between light aircraft, will not tolerate inattention to the outside world.

Visibility and cloud separation.

Now let's look at the visibility and cloud separation minimums. Figure 17-5 tells the story, and there's not much to add. Okay, you're at the proper altitude, but the visibility is decreasing or there's a bunch of puffy clouds up ahead. Again, to see and avoid, the FARs are very specific about visibility minimums and separation from clouds or overcasts.

Have trouble remembering the numbers? Follow this sequence: feet below, feet above, feet laterally. Then double the figures as you go: 500 feet below, 1000 feet above, 2000 feet laterally. Just don't forget that 2000 feet from one cloud also means 2000 feet from the next, for a 4000-foot separation between those billowing whites.

Despite the regulations, for some the desire to skim along a few feet above or below a cloud layer is irresistible. It's a great way to get the sensation of speed, while others have trouble overcoming the

Basic VFR Weather Minimums

Airspace	Flight Visibility	Distance from Clouds
Class A	Not Applicable	Not Applicable
Class B	3 statute miles	Clear of Clouds
Class C	3 statute miles	500 feet below 1,000 feet above 2,000 feet horizontal
Class D	3 statute miles	500 feet below 1,000 feet above 2,000 feet horizontal
Class E Less than 10,000 feet MSL	3 statute miles	500 feet below 1,000 feet above 2,000 feet horizontal
At or above 10,000 feet MSL	5 statute miles	1,000 feet below 1,000 feet above 1 statute mile horizontal
Class G 1,200 feet or less above the surface (regardless of MSL altitude). Day, except as provided in section 91.155(b)	1 statute mile	Clear of clouds
Night, except as provided in section 91.155(b)	3 statute miles	500 feet below 1,000 feet above 2,000 feet horizontal
More than 1,200 feet above the surface but less than 10,000 feet MSL. Day	1 statute mile	500 feet below 1,000 feet above 2,000 feet horizontal
Night	3 statute miles	500 feet below 1,000 feet above 2,000 feet horizontal
More than 1,200 feet above the surface and at or above 10,000 feet MSL.	5 statute miles	1,000 feet below 1,000 feet above 1 statute mile horizontal

Fig. 17-5. *This excerpt from the Aeronautical Information Manual specifies the VFR visibility and cloud separation minimums for flight within the various airspaces.*

temptation to burst into a nice puffy cloud just for the fun of it or to see if they really can fly on instruments.

Are you a gambler or a Russian roulette devotee? That's what it's all about. Some other character might also be testing his new-found instrument skills in the same cloud. Or a perfectly legitimate IFR aircraft may be climbing or letting down through the cloud layer. Suddenly there's the IFR aircraft and there you are. The other pilot is legal, you're not, but that won't make much difference if the dreaded midair collision becomes reality.

The old saying that "There are old pilots and there are bold pilots, but there are no old bold pilots" is a time-worn adage, but age hasn't dimmed its validity. Violating the rules might give some people a sense of bravado and make for good hangar talk. In those cases, though, bravado is named stupidity.

Uncontrolled airports

Fundamentally, the existence or nonexistence of an operating tower and a qualified weather-reporting source on an airport property determines whether an airport is classified as controlled or uncontrolled. If there is no tower at all, the airport is uncontrolled in the sense that no one is directing the routine flow of traffic in the traffic pattern or within the general 5-statute-mile radius of the airport proper. Also, many airports have control towers, but the towers operate only part-time, say, from 7:00 a.m. to 11:00 p.m. When closed, the airport reverts to an uncontrolled status, again in the sense that there is no external control of traffic within the immediate airport environment. All told, more than 87 percent, or roughly 4450 out of 5132, airports in the United States available for unrestricted public use, excluding heliports and seaplane bases, fall into this description of uncontrolled.

There's another aspect, however, to this matter of controlled versus uncontrolled. If you check almost any sectional, you'll see a fair number of magenta nontowered airports surrounded by a segmented magenta circle or design. This design indicates that in IMC weather either a Center or a nearby Approach Control facility will provide control of arriving IFR aircraft down to the surface and of departing IFR aircraft from the surface to the aircraft's eventual assigned en-route altitude. These generally are the Class E airports which, in the FAA context, are thus controlled.

At many of the other magenta airports, as pointed out earlier, where there is no segmented circle, you'll note a solid magenta design— again, basically circular or cookie-cutter in form—surrounding the airport. This symbol, which becomes fuzzy as it fades inward, indicates that the floor of the Class E airspace, normally 1200 feet AGL, is 700 feet AGL within the confines of the symbol, and that ATC control of IFR aircraft in IMC will be provided down to or up from that AGL altitude. Keep in mind, though, that these designs and altitudes apply only in IMC. In VMC, you can figuratively erase them from the sectional.

A caution about operating VFR in these so-called uncontrolled airports: Being uncontrolled does not mean that you have total freedom to operate as you please in or around these many airports. The FAA visibility and cloud separation minimums still apply, as do radio communication responsibilities and the traffic pattern procedures.

Class G airports

The Class G airports are typically the least complex in terms of operating and communicating requirements. As a general rule, but certainly not exclusively, these are the small, low-volume airports that are unattended and probably with no air-to-ground radio facilities at all. Once again, such airports are easily identified on the sectional chart by the magenta coloring of the airport symbol and the information about the airport, its elevation, runway data, and radio frequency. To be more specific, refer to the Warren, Arkansas, example in Fig. 17-6.

The airport data block says that the field elevation is 235 feet, the runway length is "38" (3800 feet), and the frequency is the italicized 122.9, followed by the C in the solid circle. This symbol, whether in magenta or blue (blue for controlled airports), always identifies the CTAF (common traffic advisory frequency) for that airport. Incidentally, the L does not indicate runway length, but implies that at single- or multiple-runway airports at least one runway can be lighted for night operations. An asterisk before the L would mean that the lights are pilot-activated by depressing the microphone button a certain number of times. (Refer to the A/FD to determine how many times to press the mike button at a given airport.)

By inference, then, Warren has no tower, no Flight Service Station, no unicom. It is uncontrolled in every sense, which means that

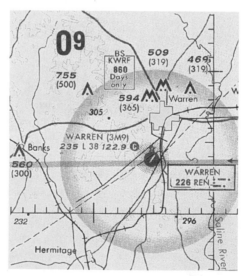

Fig. 17-6. *Warren airport, as shown on the sectional, is a Class G multicom airport with the aircraft-to-aircraft 122.9 CTAF.*

- You can operate in and around the airport at will, observing only the basic rules of traffic pattern flying.
- If you maintain an altitude of 1200 feet AGL or less, the visibility must be at least 1 statute mile and you must be able to remain clear of any clouds.
- To operate above 1200 feet, the visibility must be at least 3 miles, with the standard cloud separation of 500 feet below, 1000 feet above, and 2000 feet horizontally maintained.
- Radio position reports and intentions are always communicated on the 122.9 CTAF, and all calls are addressed to "(Airport name) traffic"—or, in this case, to "Warren traffic."

Class E airports

A Class E airport is something like a chameleon in that it can change "colors" depending on the existence or nonexistence of certain conditions. For example:

- An airport has no tower, but a qualified weather-reporting source, either human or automated, such as ASOS or AWOS, exists on the property. As long as that source is on duty or functioning, the airport is a Class E. When weather information is not available (the human source goes home for the night and there is no ASOS or AWOS), the airport reverts to a Class G.
- Another airport has a tower plus a weather-reporting source on duty. This, then, is a Class D airport. The tower, however, closes, say, from 1900 hours local to 0700 hours. If the weather-reporting source is still available, the field becomes Class E during those 12 hours of downtime. Should both the tower and the weather source go off duty simultaneously, the airport drops to a Class G.

As is apparent, the existence or nonexistence of an operating tower and a weather-reporting source is the determining factor. Depending on what is available, the airport is a Class E or D; without either, it is a Class G. To simplify the discussion here, however, let's focus solely on the typical Class E airport, which has weather reporting but no tower.

Identifying a Class E airport

To identify a Class E, do as you did with a Class G: First find an airport on the sectional that is depicted in magenta with a segmented magenta circle surrounding the airport. One of many examples is

Minnesota's Grand Rapids Gordon Newstrom Field, illustrated in Fig. 17-7. Also, as part of the airport data block, check the italicized radio frequency. If the frequency is 122.7, 122.8, or 123.0, as is so commonly the case, you have further evidence that the airport is a Class E and provides unicom radio service. Then for additional details about the airport, refer to the Airport/Facilities Directory (Fig. 17-8). This establishes the fact that the airspace is indeed Class E and provides service from 1100 to 0300Z Monday through Saturday and 1400 to 0300Z Sunday. At other times, it reverts to Class G.

From an operating and communicating point of view, the advantage of unicom is that when contacted for a field advisory, a unicom operator (typically an employee of a fixed-base operator) will give you the current wind direction and velocity, the favored runway in use, and the existence of any reported traffic in the pattern or area. In no way, however, is unicom a controlling agency. It is merely a field advisory service.

In addition to field advisories, the fixed-base operator (FBO), upon request, can call a taxi, make a phone call for you, relay messages to people waiting for you, alert a mechanic for repairs, and so on.

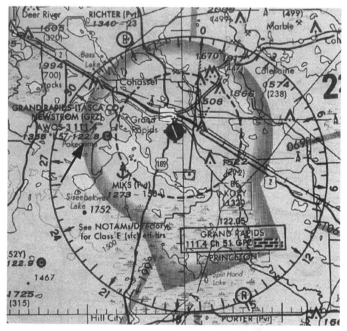

Fig. 17-7. *Grand Rapids, Minnesota, is a nontowered Class E airport, identified in part as such by the 122.8 CTAF.*

```
GRAND RAPIDS/ITASCA CO GORDON NEWSTROM FLD   (GPZ)   2 SE   UTC-6(-5DT)          TWIN CITIES
        N47°12.67' W93°30.59'                                                    H-1E, 3G, L-10G
        1355   B   S4   FUEL 100LL, JET A   OX 2   ARFF Index A                          IAP
        RWY 16-34: H5755X100 (ASPH-PFC)   S-23, D-38, DT-78   HIRL   1.0% up SE
           RWY 16: REIL. VASI(V4L)—GA 3.0° TCH 50'. Thld dsplcd 423'. Tree.
           RWY 34: MALSR. VASI(V4L)—GA 3.0° TCH 50'.
        RWY 04-22: 2968X135 (TURF)
           RWY 04: Trees.        RWY 22: Trees.
        RWY 10-28: 2470X100 (TURF)
           RWY 10: Trees.        RWY 28: Trees.
        AIRPORT REMARKS: Attended dalgt hrs. Rwy 04-22 and Rwy 10-28 CLOSED to wheel acft winter months. Deer and
           gulls on and invof arpt. Rwy 16-34 has a 582' paved area north end of rwy. Rwy 16 REIL omni directional.
           ACTIVATE HIRL Rwy 16-34, VASI Rwy 16 and Rwy 34, REIL Rwy 16 and MALSR Rwy 34—122.8.
        WEATHER DATA SOURCES: AWOS-3 111.4 (218) 326-8337.
        COMMUNICATIONS: CTAF/UNICOM 122.8
           PRINCETON FSS (PNM) TF 1-800-WX-BRIEF. NOTAM FILE GPZ.
           RCO 122.05 (PRINCETON FSS)
      (R) MINNEAPOLIS CENTER APP/DEP CON 127.9
    ——▶ AIRSPACE: CLASS E svc Mon-Sat 1100-0300Z‡, Sun 1400-0300Z‡ other times CLASS G.
        RADIO AIDS TO NAVIGATION: NOTAM FILE GPZ.
           (L) VORW/DME 111.4    GPZ     Chan 51    N47°09.82' W93°29.31'    337° 3.0 NM to fld. 1400/6E.
           VOR portion unusable 120°-192° byd 12 NM blo 5000'.
           DME portion unusable 140°-215° byd 20 NM blo 3000'.
                              215°-045° byd 20 NM blo 3500'.
           GALEX NDB (LOM) 272    GP    N47°07.82' W93°28.88'    344° 5.0 NM to fld.
           ILS 110.1    I-GPZ    Rwy 34.    LOM GALEX NDB.
```

Fig. 17-8. *The Airport/Facilities Directory provides more information about the Grand Rapids airport and the services available.*

These services are available wherever unicom exists, whether the airport is controlled or uncontrolled.

To obtain either those services or a field advisory at any nontower airport, the initial radio contact is addressed to "(Airport name) unicom." All subsequent calls reporting your position or intentions are then addressed to "(Airport name) traffic" on the airport's same published unicom frequency.

Airports with closed towers

The next possibility is an airport with unicom and a tower that operates part-time, say, from 0700 to 1900. When the tower is open, the airport is controlled and the rules for operating within at least a 5-mile radius apply (subsequently explained in greater detail). When the tower is closed, though, the airport essentially becomes Class E or G airspace. The only difference, in radio use, is that field advisories would be obtained by first contacting unicom on its frequency (assuming the FBO is open) and then addressing position reports to "(Airport name) traffic" on the tower frequency, even though the tower is closed.

How do you know whether a tower is open full- or part-time? The sectional will tell you, as will the A/FD. First, find a blue airport on any sectional. If there is a star or asterisk immediately following the tower frequency, the tower is part-time. For example, look at Fig. 17-9 and Sioux City's Sioux Gateway Airport. The star to the right of the frequency—CT 118.7*—is the clue. Now turn to the inside flap of the sectional and the table data under the control tower frequencies (Fig. 17-10). Along with other information, you'll see that the tower is in operation from 0600 to 2400 local time. During those hours, the airport is controlled. At all other times, it is Class E and, as far as local traffic is concerned, uncontrolled. Unicom, however, can be reached on 122.95, as indicated by the italicized frequency on the sectional.

Here is another variation. A part-time tower and a Flight Service Station are on the airport (although this is becoming a matter of history as more and more FSSs have been consolidated into regional locations). With the tower closed, the FSS provides the airport advisories on the tower frequency, not 123.6 or any of the published FSS frequencies. For emphasis, however, at no time is a Flight Service Station a controlling agency. Other than being capable of providing more accurate and complete weather information, an on-field FSS functions basically as unicom, as far as flight operations are concerned, but it will not provide personal message services. Whom do you call when, and what frequency do you use? There are a lot of possible combinations; Fig. 17-11 might help clarify the situation.

Fig. 17-9. *A star to the immediate right of the tower frequency and just before CTAF symbol identifies the tower as part-time.*

CONTROL TOWER FREQUENCIES ON OMAHA SECTIONAL CHART

Airports which have control towers are indicated on this chart by the letters CT followed by the primary VHF local control frequency. Selected transmitting frequencies for each control tower are tabulated in the adjoining spaces, the low or medium transmitting frequency is listed first followed by a VHF local control frequency, and the primary VHF and UHF military frequencies, when these frequencies are available. An asterisk (*) follows the part-time tower frequency remoted to the collocated full-time FSS for use as Local Airport Advisory (LAA) during hours tower is closed. Hours shown are local time. Ground control frequencies listed are the primary ground control frequencies.

Automatic Terminal Information Service (ATIS) frequencies, shown on the face of the chart are primary arrival VHF/UHF frequencies. All ATIS frequencies are listed below. ATIS operational hours may differ from control tower operational hours.

ASR and/or PAR indicates Radar Instrument Approach available.

"MON-FRI" indicates Monday thru Friday.

CONTROL TOWER	OPERATES	TWR FREQ	GND CON	ATIS	ASR/PAR
CENTRAL NEBRASKA REGIONAL	0700-1900	118.2 388.2	121.9 388.2	127.4	
DES MOINES INTL	CONTINUOUS	118.3 257.8	121.9 348.6	119.55 283.0	ASR
EPPLEY	CONTINUOUS	132.1 256.9	121.9	120.4	
FOSS	0500-2300	118.3 257.8	121.9 348.6	126.6	ASR
LINCOLN	0530-2400	118.5 253.5	121.9 275.8	118.05 302.2	
OFFUTT AFB	CONTINUOUS	123.7 348.4	121.7 275.8	126.025 273.5	ASR/PAR
SIOUX GATEWAY	0600-2400	118.7 254.3	121.9 348.6	119.45 277.2	

Fig. 17-10. *Another source for determining a tower's period of operation is the reverse side of the sectional's legend panel.*

Controlled airports, Class D

A step up from the Class G and E uncontrolled airports are those airports that are controlled and fall under the Class D airspace category. Such an airport has an operating control tower which, besides controlling traffic within the airspace, typically has Ground Control, perhaps what is called Clearance Delivery, unicom for nonoperational requests, and a weather-reporting source, whether human or automated, such as ASOS or AWOS.

To repeat just for emphasis: If an airport has an operating tower, then the airport is controlled and every pilot is under the jurisdiction of ATC. If there is no tower or if the tower is closed, then the airport becomes Class E or G where the VFR weather and operating regulations apply and there is no ATC control of local traffic pattern operations.

Radiating out from Class D airports is an airspace with a 4.3-nautical-mile radius that defines the outer limits of that airport's control. If Fig. 17-12 were in color, you could easily spot Nantucket Island Memorial as a towered airport by the blue airport symbol, the airport data block, and the segmented circle. Additional information in the data block includes "CT" (obviously meaning Control Tower) and the tower's CTAF frequency of 118.3, which is followed by an asterisk, indicating that the tower is part-time. The data block further tells us that ATIS can be monitored on 126.5, the field elevation is 48 feet MSL; the longest runway is 6300 feet and is lighted. Finally, 122.95 is the Nantucket unicom frequency. One other bit of information

Tower status	On-site FSS status	On-site RCO	FBO Status	Field advisories type/radio frequency	Radio frequency for self-announce position reports (CTAF)	ATC radio frequency
Tower Open						Tower
Tower Closed	FSS Open			AAS/Tower[5]	Tower	
Tower Closed	FSS Closed	RCO	FBO Open	UFA/UNICOM[1]	Tower	
Tower Closed	FSS Closed	RCO	FBO Closed	FFA/RCO	Tower	
Tower Closed	FSS Closed	RCO	No FBO	FFA/RCO	Tower	
Tower Closed	FSS Closed	No RCO	FBO Open	UFA/UNICOM	Tower	
Tower Closed	FSS Closed	No RCO	FBO Closed	Not Available	Tower	
Tower Closed	FSS Closed	No RCO	No FBO	Not Available	Tower	
Tower Closed	No FSS	RCO	FBO Open	UFA/UNICOM[1]	Tower	
Tower Closed	No FSS	RCO	FBO Closed	FFA/RCO	Tower	
Tower Closed	No FSS	RCO	No FBO	FFA/RCO	Tower	
Tower Closed	No FSS	No RCO	FBO Open	UFA/UNICOM	Tower	
Tower Closed	No FSS	No RCO	FBO Closed	Not Available	Tower	
Tower Closed	No FSS	No RCO	No FBO	Not Available	Tower	
No Tower	FSS Open			AAS/123.6[3]	123.6[2]	
No Tower	FSS Closed	RCO	FBO Open	UFA/UNICOM[1]	123.6[2]	

No Tower	FSS Closed	RCO	FBO Closed	FFA/RCO	123.6[2]	Tower
No Tower	FSS Closed	RCO	No FBO	FFA/RCO	123.6[2]	
No Tower	FSS Closed	No RCO	FBO Open	UFA/UNICOM	123.6[2]	
No Tower	FSS Closed	No RCO	FBO Closed	Not Available	123.6[2]	
No Tower	FSS Closed	No RCO	No FBO	Not Available	123.6[2]	
No Tower	No FSS	RCO	FBO Open	UFA/UNICOM[1]	UNICOM	
No Tower	No FSS	RCO	FBO Closed	FFA/RCO	UNICOM	
No Tower	No FSS	RCO	No FBO	FFA/RCO	122.9[4]	
No Tower	No FSS	No RCO	FBO Open	UFA/UNICOM	UNICOM	
No Tower	No FSS	No RCO	FBO Closed	Not Available	UNICOM	
No Tower	No FSS	No RCO	No FBO	Not Available	122.9[4]	

[1] Last hour's official weather observation available from FSS over RCO if weather observer is on duty.
[2] Or as listed in A/FD.
[3] Where available. Some AFSSs may not offer this service.
[4] MULTICOM.
[5] FSS will reply on tower frequency.

AFSS - Automated Flight Service Station
ATC - Air Traffic Control
FBO - Fixed - Base Operator with UNICOM
FSS - Flight Service Station (All AFSSs are open 24 hours.)
RCO - Remote Communications Outlet
AAS - FSS Airport Advisory Service (winds, weather, favored runway, altimeter setting, reported traffic
     within 10 miles of airport)
FFA - FSS field advisories (last hour's winds, weather, and altimeter setting, if observer is on duty at airport)
UFA - UNICOM field advisories (winds, favored runway, known traffic, altimeter setting [at some locations])

Fig. 17-11. *This chart summarizes the facilities and frequencies to use for field advisories (winds, reported traffic, favored runway, and, in some cases, weather observations) and for self-announce position reports. Check the A/FD, though, for exceptions.*

443

about the airspace itself is its ceiling. At about the four o'clock position southeast of the airport symbol, you'll see the number 25 inside a segmented box. This represents the AGL ceiling of Class D at Nantucket (these ceilings do vary slightly from Class D to Class D, but 2500 feet AGL is the standard across the country).

For additional information about the airport, the A/FD covering the northeastern United States is the logical source (Fig. 17-13). As the Nantucket example illustrates, many more details are provided, including fuel available, runway and lighting data, airport physical and operating information, communications, the type of airspace, and navigational radio aids. With this further wealth of information available, one going into Nantucket, or indeed any airport, would be foolish not to refer to the appropriate A/FD as part of the flight planning process.

It's undoubtedly evident by now that any pilot intending to take off, land, or transit this cylinder of controlled airspace has certain re-

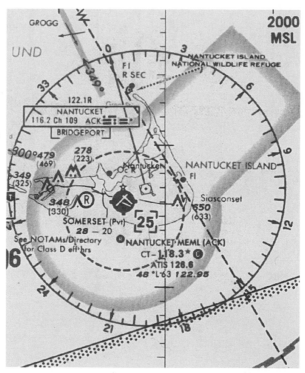

Fig. 17-12. *The segmented circle around Nantucket's airport identifies the Class D area of control.*

```
NANTUCKET MEM   (ACK)   3 SE   UTC-5(-4DT)   N41°15.18' W70°03.61'                    NEW YORK
   48   B   S4   FUEL 100LL, JET A   ARFF Index A                                     N-3I, L-25D
   RWY 06-24: H6303X150 (ASPH)   S-75, D-170, DT-280   HIRL CL   0.3%up NE.               IAP
     RWY 06: MALSF. VASI(V4L)—GA 3.0°. Thld dsplcd 539'.        RWY 24: SSALR. TDZL.
   RWY 15-33: H3999X100 (ASPH)   S-60, D-85, DT-155   MIRL
     RWY 15: REIL. Building.         RWY 33: REIL. VASI(V4R)—GA 3.0°TCH 43'.
   RWY 12-30: H3125X50 (ASPH)   S-12
     RWY 12: Trees.       RWY 30: Trees.
   AIRPORT REMARKS: Attended continuously. Be aware of hi-speed military jet and heavy helicopter tfc vicinity of Otis
      ANGB. Deer and birds on and invof arpt. Rwy 12-30 VFR/Day use only aircraft under 12,500 lbs. Arpt has noise
      abatement procedures ctc Noise Officer 508-325-6136 for automated facsimile back information. PPR 2 hours
      for unscheduled air carrier ops with more than 30 passenger seat, call arpt manager 508-325-5300. When twr
      clsd ACTIVATE MALSF Rwy 06; SSALR Rwy 24; HIRL Rwy 06-24; MIRL Rwy 15-33 and twy lgts—CTAF. Rwy 24
      SSALR unmonitored when arpt unattended. Twy F prohibited to air carrier acft with more than 30 passenger
      seats. Fee for non-commercial acft parking over 2 hrs or over 6000 lbs. NOTE: See Land and Hold Short
      Operations Section.
   WEATHER DATA SOURCES: ASOS (508) 325-6082. LAWRS.
   COMMUNICATIONS: CTAF 118.3   ATIS 126.6 (508-228-5375) (1100-0200Z‡) Oct 1-May 14, (1100-0400Z‡) May
      15-Sept 30.   UNICOM 122.95
      BRIDGEPORT FSS (BDR) TF 1-800-WX-BRIEF. NOTAM FILE ACK
      RCO 122.1R 116.2T (BRIDGEPORT FSS)
   Ⓡ CAPE APP/DEP CON 126.1 (1100-0400Z‡) May 15-Sept 30, (1100-0300Z‡) Oct 1-May 14.
      BOSTON CENTER APP/DEP CON 128.75 (0400-1100Z‡) May 15-Sept 30, (0300-1100Z‡) Oct 1-May 14.
      TOWER 118.3 May 15-Sep 30 (1100-0300Z‡), Oct 1-May 14 (1130-0130Z‡)
      GND CON 121.7   CLNC DEL 128.25
   AIRSPACE: CLASS D svc May 15-Sep 30 1100-0300Z‡, Oct 1-May 14 1130-0130Z‡ other times CLASS G.
   RADIO AIDS TO NAVIGATION: NOTAM FILE ACK.
      (N) VOR/DME 116.2   ACK   Chan 109   N41°16.91' W70°01.60'   236°2.3 NM to fld. 100/15W.
      WAIVS NDB (LOM) 248   AC   N41°18.68' W69°59.21'   240° 4.8 NM to fld.
      NDB (HH-ABW) 194   TUK   N41°16.12' W70°10.80'   115° 5.5 NM to fld.
      ILS/DME 109.1   I-ACK   Chan 28   Rwy 24.   LOM WAIVS NDB. ILS unmonitored when twr clsd.
```

Fig. 17-13. *The A/FD for Nantucket provides much more information about the airport and the facilities and services available.*

sponsibilities to fulfill before penetrating it or operating within it— a few of which include the following:

- Two-way radio contact with the tower must be established before you operate in any part of the Class D airspace, and radio contact must be maintained while in this airspace.

- You are expected to comply with the control tower instructions, unless doing so would cause you to violate a VFR regulation or create a potentially hazardous situation. In such instances, you must immediately so advise the tower.

- Controller instructions are to be tersely acknowledged in a few words that tell the controller you have received the message and understand what is expected of you. Merely "Rogering" an instruction does not necessarily communicate understanding.

- Permission to transit the airspace at altitudes below the 2500-foot ceiling of the airspace must be obtained from the tower controller before you enter the area. Above that ceiling, you're out of Class D, and no radio contact is required.

- You do not report on downwind, base, or final, as at an uncontrolled field, unless the tower requests you to do so. The controller knows where you are, and those calls only needlessly consume airtime and the controller's attention.

It's obvious that if a tower exists, the airport has enough activity to warrant such a traffic controlling agency. Consequently, permission to take off is required, and the initial contact before entering the airspace, for either landing or transiting, should be established at least 10 or 15 miles from the field. In the latter cases, merely identify your aircraft (type and N number), position, altitude, the fact that you have monitored the current ATIS (if this service exists on the airport), and your intention (to land or to transit). For instance: "Billard Tower, Cherokee 8515 November over Perry Dam at four thousand five hundred with Charlie (the current ATIS report), landing Billard."

If you're below the ATA's ceiling and merely want to fly through the area, the call goes: "Billard Tower, Cherokee 8515 November over Perry Dam at two thousand three hundred. Request approval to transit your area to the west." From then on, it's just a matter of acknowledging and following the controller's instructions.

Now let's assume that instead of specifically authorizing you to enter the airspace for landing or transiting, the controller calls you back with this: "Cherokee eight five one five November, stand by." What should you do? Answer: Keep right on going. The fact that the controller has included your N number in his or her response to your call means that radio communications with Billard tower have been established and despite the instructions to stand by, you are clear to enter the airspace.

But suppose the response to your initial call is this: "Aircraft calling Billard, stand by." Or "Aircraft calling Billard, remain outside the Class D airspace and stand by." In these cases, your N number was not included in the response. Therefore, radio communications have not been established, and you must remain clear of the airspace until you have reestablished contact.

By the same token, ground movements around the airport that would put you on any taxiway or runway require approval of Ground Control. In sum, whether transiting, landing, taxiing, departing, or shooting touch-and-goes at a controlled airport, radio contact with the tower personnel, including Ground Control, is mandatory.

That's all well and good, but what happens if the tower is part-time and closed when you want to take off or land? In that case, the airspace reverts to a Class E or Class G and radio calls are just like those at a unicom-only uncontrolled airport.

To be more specific about that, refer again to Fig. 17-11 and take the example with the arrow where there is no FSS on the field, no remote communications outlet (RCO) to an FSS, and the FBO is open. For field advisories, you'd call the FBO on the unicom frequency, and for position reports, you'd transmit and listen on the tower frequency. Or if there is no FSS, no RCO, and the FBO is closed or none exists, you wouldn't be able to get a field advisory unless you heard other traffic in the pattern making routine position reports and stating the runway in use: "Beetown Traffic, Cessna zero zero zero alpha left downwind, landing two four Beetown."

Regardless of what facility is open, is closed, or doesn't exist at a tower airport, the only operational position-reporting frequency you would use is the tower's. This is illustrated in Fig. 17-11, along with the several potential tower-closed variables you might encounter.

More on Class D and E airspace purposes and designs

Some further discussion is in order relative to the blue and magenta designs, segmented or otherwise, that surround Class D and E airport symbols. While partially explained elsewhere, all questions that a curious reader might ask about these designs have not been answered. Should that be the case, even while recognizing the possibility of some repetition, we dig a little more deeply. What do the various segmented designs mean or represent?

In addition to establishing the 4.3-nautical-mile radius of Class D airspace, the blue segmented design around Class D airports indicates that all traffic within that design is controlled from the surface to the ceiling of the airspace—approximately 2500 feet AGL. At the same time, when IMC or marginal VFR weather exists, ATC has the authority to limit or prohibit entirely VFR operations within that segmented airspace.

While most Class D airports have perfectly cylindrical airspace designs, many have built-in or added-on segmented extensions. The purpose of these extensions is simply to expand the areas of ATC of

IFR aircraft arriving during periods of IMC weather as they make the transition from Class E airspace to the Class D airport environment. And, quite logically, the same principle applies in reverse to departing IFR aircraft in IMC weather.

If a given extension reaches out 2 nautical miles or less from the core segmented circle, the extension appears in blue on the sectional as an integral part of the circle. It is thus part of the Class D controlled airspace that exists from the surface to the Class D ceiling. Operationally, this means that pilots must contact the tower before entering one of those extensions—just as they would prior to penetrating any other portion of Class D airspace. The Pendleton, Oregon, airport (Fig. 17-14) illustrates this extension that lies just west of the airport symbol on the sectional.

Often, though, a longer extension than just 2 nautical miles of controlled airspace is needed for instrument approaches or departures. In those cases, the extension is added to the Class D circle but appears in magenta rather than blue. That color change identifies the extension as Class E airspace, signifying that control within the area

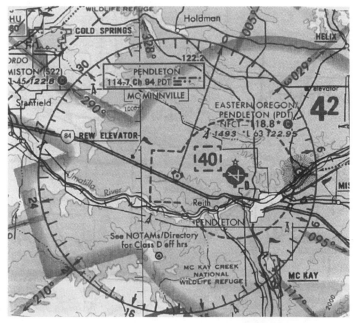

Fig. 17-14. *This Pendleton, Oregon, excerpt from the Seattle sectional shows how a Class D airspace extension— the keyholelike design to the west of the airport symbol—is charted.*

runs from the surface up to the floor of the next-higher overriding airspace, be it Class A, B, or C. Tennessee's McKeller-Sipes Airport (Fig. 17-15) illustrates this addition, although, of course, it shows up in black here instead of the sectional's magenta. Since this transition area is Class E airspace, pilots flying VFR in VFR weather are not required to contact the control tower prior to entering or operating in that portion of the airspace. Radio contact with the Class D tower is, of course, still required before the pilot penetrates any portion of the blue segmented Class D airspace itself.

Keep in mind that the purpose of these Class E magenta designs and symbols in the Class D airport environment is to define the areas in which ATC of IFR aircraft exists during periods of IMC weather. In VFR conditions, you can dismiss those magenta designs as though they faded off the sectional. That doesn't mean, however, that VFR

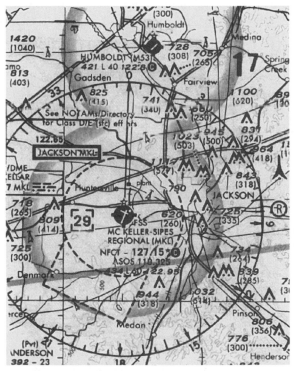

Fig. 17-15. *A Class E extension (southwest of the airport) to a Class D airspace is illustrated in this McKeller-Sipes (Tennessee) airport. Note that the Class D segmented circle is closed and the Class E extension is added on, whereas in Fig. 17-14, the circle is not closed and the entire area within the segmented design is Class D airspace.*

flight regulations have also faded away. Far from it. The visibility and cloud separation regulations continue in place, as do those affecting radio communications and traffic pattern operations.

How can there be control at nontower Class E airports?

That's a valid question. To a large extent it depends on geography. At Class E airports located near a Class B or C airspace, the B or C Approach Control facility, through its radar facilities, would assume the controlling responsibility. On the other hand, should the non-tower airport be beyond the radar coverage of any Approach Control facility but near a Center remote outlet, then the Center would provide the necessary radar coverage and instructions, probably down to the final approach. To determine if a Class E airport has radar Approach Control, check the appropriate A/FD. It will tell you one of three things:

1. There is the symbol ® in the left margin, followed by "(blank) APP/DEP CON" or "(blank) Center APP/DEP CON." This tells you that radar service is provided to the airport in question by either an Approach Control or a Center facility.

2. Only one of the two facilities—"...APP/CON" or "...Center"— is listed, but no ® symbol precedes the listing. The implication here is that that facility will provide radar service to IFR aircraft down to a certain altitude. Because of lack of radar coverage below that altitude, however, the facility will apply nonradar, or "manual," control of landing or departing aircraft. To do so requires ongoing radio communications between controller and pilot relative to the aircraft's position and altitude, as well as pilot compliance with the controller's instructions. To ensure separation of participating aircraft, the Air Traffic Control manual 7110.65J for controllers contains many procedures and instructions that controllers are to follow to maintain the necessary lateral, longitudinal, vertical, and visual separations of aircraft for safe operations in IMC weather.

3. If neither "(blank) APP/DEP CON" nor "(blank) Center APP/DEP CON" appears in the A/FD legend, you can be sure that no controlling service is available at the airport in question.

What is required for an airport to be Class E?

There may be several factors, but at least one of the following must be located on the airport:

- An operating control tower with a qualified weather observer on duty
- A Flight Service Station
- A federally designated qualified weather observer
- A National Weather Service office

The basic requirement is the on-field resource qualified to report accurate and current weather data. If such a source is part-time and the airport is without any weather-reporting capability, it then becomes Class G.

What are the solid magenta designs that surround so many airports, large and small?

You may recall the earlier discussion of these designs and that whatever their shape—circular, keyhole, or a cookie-cutter pattern—the inside of the design becomes fuzzy as the magenta fades inward toward the airport. These structures, called *transition areas,* exist to contain and protect IFR aircraft during IMC weather as they make the transition from en-route flight to arrival and, conversely, departure to an in-flight status. The inward fading of the magenta indicates that the floor of the Class E airspace within the design is now 700 feet AGL rather than the usual 1200-foot floor that generally prevails outside the design. However, when the airport is Class D or E, with the segmented blue or magenta circle around the airport proper, the 700-foot floor exists only to that segmented design. At that point and inward, the floor of the controlled airspace drops from 700 feet down to the surface.

Outside those solid magenta shapes around so many airports, however, and beyond the design's sharp external edge, the floor of the Class E airspace returns to the normal 1200-foot AGL altitude that prevails throughout so much of the United States. In other words, barring interruptions by any Class B, C, or D airspaces, the open-country geography between all solid magenta designs around airports is Class E airspace, wherein all Class E operating rules and regulations in terms of flight altitudes, visibility, and cloud avoidance minimums apply.

How much are VFR pilots affected by all this?

In general terms, only marginally. To operate VFR, you must have at least 3-mile visibility and a 1000-foot ceiling. In addition to meeting those standard minimums, you are required to make contact with the Class D tower well before you penetrate any portion of the airspace identified by the blue segmented circle or design on the sectional. This requirement applies whether you plan (1) to land at the primary airport; (2) to enter the airspace below the published ceiling of 2500 feet AGL just to transit a portion of it; or (3) to land at a satellite airport that lies within the Class D airspace. Otherwise, in VFR weather, you can disregard any segmented magenta add-ons to the Class D structure as well as the solid magenta transition area(s) that surround the Class D airport environment.

Similarly, if you're operating into or out of a Class E nontowered airport and the weather is VFR, don't worry about the segmented magenta design around the airport, the transition areas, or the magenta 700-foot floor symbol. These have no meaning for VFR pilots when VFR visibility and ceiling conditions prevail. They exist solely to provide protection and control of IFR aircraft in IMC weather.

Special VFRs (SVFRS)

Special VFRs have been mentioned in passing but have never been explained as to what they are and how they can be obtained. So now is the time to be more specific. Let's assume that you are not IFR-qualified and that you want to depart from an airport where the current weather is below the 3-mile, 1000-foot visibility and ceiling VFR minimums. From various weather-reporting sources, however, you do know that once you get out of the immediate airport area, conditions are VFR to your intended destination. Now is the time to request an SVFR, with the hope that ATC will approve same.

The SVFR weather minimums and general regulations

If the visibility is at least 1 mile and you are able to remain clear of clouds, you can request an SVFR clearance from the controlling ATC facility. As a rule, however, only one fixed-wing (FW/SVFR) aircraft at a time is permitted within the horizontal and vertical boundaries of the surface-based controlled airport airspace. Approval will depend on the existing traffic and whether granting an SVFR would interfere with IFR operations. Also, as I just implied, the SVFR is valid only within the established limits of the controlled airspace. Beyond those limits, the pilot must be able to adhere to the basic VFR minimum ceiling and visibility regulations, both below and above the 700- and 1200-foot floors.

Regardless of IFR traffic, no SVFR would be approved unless the following minimum safe altitudes stated in FAR 91.119 could be met:

- No pilot may operate an aircraft over any congested area of city, town, or settlement, or over any open-air assembly of persons, below 1000 feet above the highest obstacle within a radius of 2000 feet of the aircraft.
- The minimum altitude for fixed-wing aircraft, while flying over other than congested areas, is 500 feet AGL, except over open water or sparsely populated terrain.

Thus, if the ceilings are such that these portions of the FARs would be violated, a request for an SVFR would be denied. By the same token, however, these minimums must not be violated by fixed-wing aircraft, no matter what the weather. (The FARs are a little more lenient for helicopters.)

What about night SVFRs? These are prohibited between sunset and sunrise unless the pilot is instrument-rated and the aircraft is equipped for IFR flight.

Can SVFRs be obtained at all airports?

No. Some of the busy, high-density Class B and C airports will not approve SVFRs. The ban, where it exists, is indicated on the sectional chart by the notation "No SVFR." Typically, the notation is located just above the airport data block, as Fig. 17-16 illustrates. Also, the specific airports imposing the ban are listed in FAR Part 91, Appendix D, Section 3.

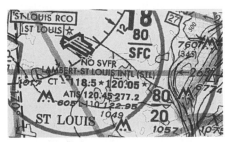

Fig. 17-16. *The "No SVR" directly above the name of the airport indicates just that—no special VFR clearances are permitted at St. Louis Lambert Field.*

How is an SVFR obtained at a Class E airport?

In nontower locations, the SVFR is requested through the Flight Service Station, operating tower, Approach or Departure Control, or Air Route Traffic Control Center responsible for serving the area.

Why contact an FSS? It's not a controlling agency.

True, but the FSS will relay the SVFR request to the appropriate ATC facility, such as Approach Control or Center. That facility will approve or disapprove the request, based on existing traffic, and will inform the FSS accordingly, which, in turn, will inform the pilot. Although it's an abbreviated example of the radio call, the phraseology, after establishing contact with the FSS, would go something like this:

> *Pilot: Jonesville Radio, Cherokee one two three four echo requests Special VFR out of the Jonesville Class E surface area for VFR northwest to Smithtown via Victor four zero zero. Departure time immediate upon approval of request. Three four echo.*

There will be a period of silence while the FSS specialist contacts the responsible controlling facility and then reports back:

> *FSS: Cherokee 34 echo, ATC clears Cherokee one two three four echo out of the Jonesville Class E surface area for northwest departure on Victor four zero zero for Smithtown. Maintain Special VFR conditions at or below three thousand while in the Class E surface area. Report leaving the area.*

What facility would you contact in other situations?

That depends on where you are geographically and the facilities that are on or off the airport. Generally speaking, however, here are some broad guidelines on whom to call for an SVFR clearance:

Departing:

- Class E airport with no facility physically on the airport: Call the nearest FSS, the Approach or Departure Control, or Center responsible for traffic control in the Class E airport airspace.

- Class B, C, or D airports: AIM says call the tower. Alternatives are Ground Control or Clearance Delivery. If uncertain, make the first call to Ground Control and ask if it is the facility that initiates the SVFR request. Ground Control will then either tell you whom to contact or will handle the request itself.

Arriving or Transiting: (You are flying VFR but have learned from ATIS or an FSS that your destination airport is below the VFR minimums of 3000 feet and 1 mile. Just be sure to call 15 to 20 miles out so that there will be time to receive the SVFR approval before you inadvertently penetrate a controlled airport airspace.) Your destination airport is a

- Class E. Call the FSS, Center, or Approach Control responsible for the Class E traffic control.
- Class D with tower but no Approach Control. Call the tower.
- Class D, C, or B with tower and Approach Control. Call Approach Control.
- Satellite airport lying within the surface-configured Class D airspace of the primary airport or the core area of a Class C or B airspace. Call the Class D Approach Control, if such exists, or the Class D tower. Call Approach Control if the primary airport is Class C or B.

Some of this talk about the airspaces and SVFRs undoubtedly gets confusing at times, and perhaps I've belabored the subject(s) to the extreme. They're all part of the system, though, and thus must be understood by those who use the airspaces. I would suggest, however, whether you are a student or VFR, IFR, private, or commercial pilot, that you master the airspace concepts and stay on top of whatever changes may be forthcoming. Currency here, as well as in all other aspects of flying, not only could be, but is, essential to your well-being.

Temporary flight restrictions

Curiosity and/or morbidity is rather common in humans. A major fire is raging; an explosion has devastated a wide area; a tornado has wiped out a portion of a town; a major sporting event is underway. Whatever the incident, many of us are tempted to get airborne and see what's going on from above. Before we do so, it would be wise to do a little checking.

If there is an incident that warrants temporary flight restrictions, the restrictions are established by the area manager of the Air Route Traffic Control Center that has jurisdiction over the area in which the incident has taken, or is taking, place. Then, through FAA procedures, a

NOTAM is issued stipulating what restrictions are in effect. Largely, these are designed to bar sightseers from the area and to prevent interference with emergency, rescue, or disaster relief measures.

Normally, the restricted airspace extends to 2000 feet AGL within a 2-nautical-mile radius of the event. These dimensions can vary, however, with the seriousness of the occurrence and the rescue or relief operations involved. The NOTAM fully explains the restrictions, how long they will be in effect, what aircraft are permitted within the area, and the military or federal agency responsible for coordinating emergency activities.

FARs state the restriction regulations, and AIM provides further explanation. Recognizing the occasional tendency to be ambulance chasers or just plain gawkers, we should review current regulations, plus check for NOTAMs that might have been issued, before giving into that tendency, should it exist; it could keep us out of trouble.

Automated Weather-Observing System (AWOS) and Automated Surface Observation System (ASOS)

If you glance at the airport data blocks on almost any sectional chart, you'll find a goodly number that are equipped with AWOS (automated weather-observing system) or ASOS (automated surface observation system). A "goodly" number is probably not very descriptive, however, because estimates, including those from the National Weather Service ASOS Program Office in Silver Spring, Maryland, put the current installations in excess of 1000, with many additional systems underway or projected.

The advantages of an automated system are many, ranging from minute-to-minute real-time observations, to more consistent information, undiluted by human judgment, interpretation, vision-blocking obstructions, or darkness, to greater economic use around the clock of airport personnel. In addition to these advantages is the obvious safety factor. Particularly with ASOS's more extensive observations and reports, pilots, when approaching an unattended Class E or G airport, will have a much clearer picture of the current conditions than would otherwise be possible or probable. And, in a similar vein, AWOS or ASOS on the field opens the airport to increased local and transient traffic, ultimately resulting in a more profitable operation for the community.

Much has been written describing and justifying both AWOS and ASOS. For our purposes here, however, a brief summary of the two systems and the data they transmit would seem more to the point. Consequently, what follows focuses on what the systems are, what they do, and what they don't do.

AWOS and ASOS: General

As the National Weather Service's *Tool Box* (a summary of automated surface observations for ASOS trainers) puts it, both AWOS and ASOS are composed of "electronic sensors, connected to a computer, that measure, process, and create surface observations every minute. These systems provide 1-minute, 5-minute, hourly, and special observations 24 hours a day." Through computer-generated voice subsystems, the minute-by-minute observations are transmitted over discrete VHF radio frequencies and can be received up to 25 nautical miles from the airport as well as up to 10,000 feet AGL. Transmissions may also be over the voice portion of a local navigational aid. Whichever the vehicle, the weather message is 20 to 30 seconds long and is updated each minute. Consequently, if you monitor the reports from a given airport over a reasonable period of time, you can develop a picture of the prevailing conditions and whether they are degenerating, improving, or remaining static.

Although AWOS and ASOS would seem to have a close-cousin relationship, there are differences that make ASOS the more complete system and the system that is in the gradual process of replacing AWOS. Neither system, however, is likely to eliminate entirely the need for human observations, if only because the electronic systems are limited in their horizontal coverage and cannot forecast weather trends. They only report what is happening—not what will happen. Of course, as I said, if you listen to enough consecutive reports of what is going on now at an airport, you'll be able to detect or establish a weather pattern trend. That trend, though, is a product of deduction, not prognostication.

There is another factor related to human involvement: The automated systems report only the weather that exists directly above the sensors. They have no ability to accumulate and digest conditions that might be surrounding or approaching the airport. For example, the observing system reports highly favorable conditions at a given airport in terms of ceiling, visibility, winds, temperature, density altitude, dew point, and the like—in essence, perfect flying weather. But only a couple of miles or so off the airport property, a raging thunderstorm has blackened the sky. The automated observing systems will not report that storm because they're not capable of processing horizontal or diagonal weather. Despite all their data-collecting capabilities, the current systems have a built-in vertical tunnel vision which, at times, requires human intervention, or "augmentation," to provide a more complete report of present and/or anticipated conditions. Whenever this augmentation by a qualified observer is considered necessary, it is, as AIM puts it, "identified in the observation as observer weather."

Location of the sensors on the airport is important. Generally speaking, the chosen site is near the touchdown point of the principal instrument runway; but also to be considered in that decision are local conditions such as nearby lakes, rivers, an ocean, or terrain that could adversely affect accurate observations. If such potentially distorting elements exist, additional sensors could be installed to produce more accurate reports.

Of the two systems, AWOS is the older, having been around in one form or another for more than 30 years. Despite a continuing pattern of adding information to the system, AWOS does not produce the variety of observations provided by ASOS. Consequently, ASOS is gradually replacing AWOS, and AWOS will presumably disappear from the scene within the next few years.

AWOS, in brief

Four levels make up the AWOS:

1. AWOS-A: Only reports altimeter setting.
2. AWOS-1: Reports altimeter setting, winds, temperature, dew point, and density altitude.
3. AWOS-2: Reports the information provided by AWOS-1, plus visibility.
4. AWOS-3: Reports the information provided by AWOS-2, plus cloud and ceiling data.

In addition to the radio transmissions, most AWOS messages can be monitored on the ground via telephone. For the level of an AWOS at a given airport, plus the radio frequency and telephone number, consult the appropriate A/FD. If AWOS or ASOS is on the field, you'll see in bold print an A/FD entry similar to this: Weather Data Sources: AWOS-3 133.8 (814) 443-2114.

ASOS, in brief

The ASOS program, which is the primary surface weather-observing system, is a joint effort of the National Weather Service, the Department of Defense, and the FAA. The system, as AIM puts it, provides "continuous minute-by-minute observations and performs the basic observing functions necessary to generate an aviation routine weather report (METAR) and other aviation weather information." Equipped with at least one sensor for each unit, a given ASOS, as the National Oceanic and Atmospheric Administration (NOAA) describes it, will be able to "observe" and report these conditions:

- Cloud height and amount (clear, scattered, broken, overcast) up to 12,000 feet
- Visibility (to at least 10 statute miles)
- Precipitation identification (type and intensity for rain, snow, and freezing rain)
- Barometric pressure and altimeter setting
- Ambient temperature and dew point temperature
- Wind direction, speed, and character (gusts, squalls)
- Rainfall accumulation
- Selected significant remarks such as variable cloud heights, variable visibility, precipitation beginning and ending times, rapid pressure changes, pressure change tendency, wind shift, and peak wind

Figure 17-17 is an NOAA illustration of the principal elements and sensors of an ASOS installation.

With all its positive features, however, ASOS is not designed to report:

- Clouds above 12,000 feet
- Tornadoes
- Funnel clouds
- Ice crystals
- Drizzle
- Freezing drizzle
- Blowing obstructions (snow, sand, or dust, snowfall and snow depth)

To paraphrase the *NWS Tool Box,* data related to many of these latter conditions are, or can be, provided by other sources. Meanwhile, NWS adds that "…new sensors will be added to measure some of these weather elements."

As a reminder again, these automated systems provide 24-hour observations of weather conditions within 2 to 3 miles of the sensor site. Beyond those limits, lateral coverage is not possible. Keep in mind, though, that observing systems are really in their infancy, despite the fact that AWOS has been around since the 1960s. As new sensors are developed and computer technology advances, improvements in the systems are inevitable.

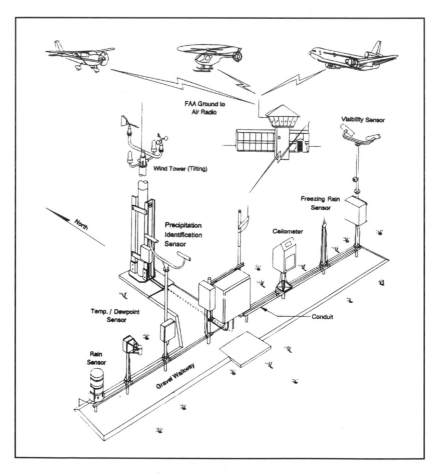

Fig. 17-17. *The elements of an ASOS airport installation.*

These, then, are some of the AWOS and ASOS basics, although there's more that could be said about this electronic weather data processing system. An overview of what the systems do and can't do would be more meaningful than a technical discussion of their components, the application of mathematical logic (algorithms), or why human augmentation is often an essential supplement to the automated report. But like so many other aspects of aviation, this is just one more subject or area in which proficient pilots keeps themselves informed. If you tend to do most of your flying in Class E or G airport environments where 24-hour automated weather broadcasts may be available, knowledge of the system, what it's telling you, and its limitations could be especially important. There will be changes— mostly for the good—as time moves along, so staying on top of these safety-enhancing systems should be a high-priority project for every pilot, whatever his or her rating or experience level.

18

Special-use airspace

Glancing at almost any sectional reveals a variety of lines and designs that depict airspaces set aside for special use. In all cases, these are blocks of space established for purposes of national security, welfare, and environmental protection, plus military training, research, development, testing, and evaluation (military training/RDT&E). Although the airspace reserved for security, welfare, and the environment could require flight detours or altitude changes, their size and sparseness make them relatively minor obstacles to VFR and IFR operations. Not so, though, with the areas designed for military training/RDT&E, which provide space for all sorts of flight-training maneuvers, bombing runs, missile launching, aerial gunnery, or artillery practice.

It's thus apparent that unauthorized or careless penetration of active special-use airspace (SUA) could be somewhat hazardous to the unwary pilot. That being the case, let's start with a summary of the policy behind this airspace, followed by the various types of SUA and the VFR pilot's responsibilities relative to them.

Military operations requirements

The FAA has a controller's handbook titled *Special Military Operations* (7610.4G) that summarizes the basic FAA policy relative to military operations requirements. In essence, the policy recognizes that the military has a continuing need to conduct certain training as well as research, development, testing, and evaluation activities, and that these activities should take place in airspaces large enough to contain the planned activity and should be as free from nonparticipating aircraft as is practical. Accordingly, four types of special-use airspace exist primarily, but not exclusively, for military purposes: restricted areas (R), military operations areas (MOAs), alert areas (A), and warning areas (W). In addition, but not considered special-use airspace, there are the Air Defense Identification Zones (ADIZs) and cross-country low-level VFR and IFR military training routes (MTRs).

The one other type of SUA, classified as prohibited areas (P), is established for national security, national welfare, or environmental protection. As mentioned, however, these are relatively few in number and small enough in area to be of only infrequent concern to the pilot. Prohibited areas, along with MTRs, ADIZs, and the varieties of military SUA, are depicted on sectional and other aeronautical charts.

As a further matter of policy, the government agency, organization, or military command responsible for establishing SUA must specify the proposed area's

- Vertical limits
- Horizontal limits
- Hours of use, plus notations indicating whether the use is continuous, intermittent, by NOTAM, or otherwise

The policy also states that

- The SUA will be limited to the minimum area(s) necessary to support operational requirements.
- The area will be designed to conduct the maximum number of different types of military activity in the same airspace area.
- The military shall provide procedures for joint-use scheduling in the area (joint use meaning use by both participating and nonparticipating aircraft).

Does the military just grab a chunk of airspace that it decides it needs? Hardly. Take the case of an MOA. The only aspect of the process occasionally subject to whimsy is the MOA's name. From the time the Air Force determines the need for the airspace until the FAA headquarters in Washington issues the final approval, eight or nine steps have intervened. These include reviews by, and coordination with, the military's regional office, the Air Route Traffic Control Center responsible for the area in which the SUA will be established, the FAA's regional office, interested entities on whom the airspace might have impact (such as airport managers, pilot organizations, and pilots themselves), and if military activities are proposed below 3000 feet AGL, the Environmental Protection Agency (EPA). Above 3000 feet, a categorical exclusion exists, and the EPA does not become involved. What with consultations, negotiations, and the need to follow the military and government chain of command, it can easily take up to two years from start to finish to establish a military SUA.

Moreover, once the SUA comes into being, each using agency (as subsequently defined) must submit annual reports on the previous 12 months' activity. In the case of restricted areas, this includes the usage by daily hours, days of the week, and the number of weeks during the year that the space was released to the controlling agency for public use. In effect, the report justifies the existence of the space as it stands or reflects the need for revisions. Some definitions might help in understanding the basic policy:

Using agency

This refers to the agency or military unit that is the primary user and scheduler of the particular airspace.

Controlling agency

This is the FAA facility, almost always the ARTCC, in which the airspace is located. Depending on a variety of factors, Center might exert direct control of military traffic operating within that airspace.

Participating aircraft

These aircraft, usually but not necessarily military, are involved in the training/RDT&E.

Nonparticipating aircraft

These are aircraft for which a Center has separation responsibility (meaning IFR aircraft) but which have not been authorized by the using agency to operate through or in the SUA. Before we discuss SUA in a little greater detail, let's examine the three basic areas that fall under the broad heading of SUA.

Prohibited

Throughout the country, certain geographic areas have been set aside to protect wildlife and recreational properties. In these prohibited areas, either flight is banned completely or aircraft must observe the established minimum altitude. If you check the sectional charts or the ELAC (en-route low-altitude chart), you'll find a few such areas, one example being located in the arrowhead portion of Minnesota on the U.S.-Canadian border. The minimum altitude over the areas identified as P-204, P-205, and P-206 is 4000 feet AGL.

Other types of prohibited territory include those that are considered important in terms of national security or history. Examples are the White House and government buildings, mostly along the Mall; Mount Vernon, Virginia; Kennebunkport, Maine (former President Bush's family residence); Camp David; the Naval Support Facility near Thurmont, Maryland; and a Department of Energy nuclear facility in Amarillo, Texas.

Restricted

Approximately 180 restricted areas exist across the country, with many composed of subsegments, such as A, B, and C. The subsegments, however, are not counted as part of the 180 total.

Military operation areas

There are also approximately 180, with each specific MOA counted as a single area, regardless of the number of subsegments. Finally, you who fly exclusively in Connecticut or Rhode Island are lucky. Those are the only two states of the 50 that have no special-use airspace within their borders.

Special-use airspace types

Now for a closer look at the types of SUA, why they exist, and the restrictions, if any, they place on the VFR pilot.

Prohibited areas (P)

These areas are identified in sectional charts as illustrated in Fig. 18-1. The Camp David area is depicted on the Washington sectional by the inward-pointing shaded lines (in blue in the full-color originals) and the "P-40" identification. The SUA table on the sectional (Fig. 18-2) states that flight below 5000 feet is prohibited, the ban is continuous, and there are no air-ground communications in the area.

The same principles apply in other such areas. From the surface to whatever altitude is specified, the area is prohibited, and that means

Fig. 18-1. *How the sectional chart depicts a prohibited area.*

SPECIAL USE AIRSPACE ON WASHINGTON SECTIONAL CHART

Unless otherwise noted altitudes are
MSL and in feet; time is local.
Contact nearest FSS for information.
†Other time by NOTAM contact FSS

The word "TO" an altitude means "To and including."
"MON-FRI" indicates "Monday thru Friday"
FL – Flight Level
NO A/G – No air to ground communications

U.S. P–PROHIBITED, R–RESTRICTED, A–ALERT, W–WARNING, MOA–MILITARY OPERATIONS AREA

NUMBER	LOCATION	ALTITUDE	TIME OF USE	CONTROLLING AGENCY**
P-40	THURMONT, MD	TO BUT NOT INCL 5000 (UNDERLIES R-4009)	CONTINUOUS	NO A/G
P-56	WASHINGTON, DC	TO 18,000	CONTINUOUS	NO A/G WARNING - AVOID THIS AREA
P-73	MOUNT VERNON, VA	TO BUT NOT INCL 1500	CONTINUOUS	NO A/G
R-4001 A	ABERDEEN, MD	(1) UNLIMITED (2) TO 10,000	(1) 0700-2400 (2) 0000-0700 HIGHER ALTITUDES BY NOTAM 24 HRS IN ADVANCE	ZDC CNTR
R-4001 B	ABERDEEN, MD	TO 10,000	INTERMITTENT BY NOTAM 24 HRS IN ADVANCE HIGHER ALTITUDES BY NOTAM 24 HRS IN ADVANCE	ZDC CNTR

Fig. 18-2. *The special-use airspace box on the sectional summarizes the SUA flight restrictions and times of use.*

prohibited: no ifs, ands, or buts. Prior approval is required and must be properly obtained to penetrate any of this space.

Restricted areas (R)

More common, and larger in territory, are the restricted areas. These blocks of space, when active, pose serious and often invisible hazards to nonparticipating aircraft and those not specifically authorized to enter the area. What sorts of hazards? Well, how about artillery fire, missiles, aerial gunnery, or bombing as examples? That should be enough to discourage violation of the area by any VFR pilot when it's active.

The areas are easy to spot on the sectional because of the vertical (formerly diagonal) blue lines that establish the horizontal perimeters.

Figure 18-3 illustrates two areas located near Brookfield, Kansas, identified as R-3601A and R-3601B. The table on the Wichita sectional (Fig. 18-4) provides further information about the altitudes within which military activity can be conducted, the time(s) of use, and the controlling agency. In this case, it is ZKC, which is the Kansas City ARTCC.

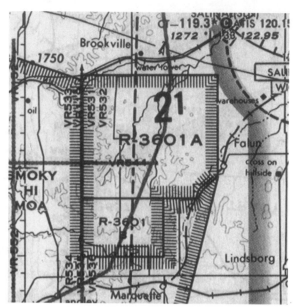

Fig. 18-3. *Restricted areas on the sectional are identified by the R number and the border lines (in blue on the original).*

SPECIAL USE AIRSPACE ON WICHITA SECTIONAL CHART

Unless otherwise noted altitudes are MSL and in feet; time is local.
Contact nearest FSS for information.
†Other time by NOTAM contact FSS

The word "TO" an altitude means "To and including."
"MON-FRI" indicates "Monday thru Friday"
FL – Flight Level
NO A/G – No air to ground communications

U.S. P-PROHIBITED, R-RESTRICTED, A-ALERT, W-WARNING, MOA-MILITARY OPERATIONS AREA

NUMBER	LOCATION	ALTITUDE	TIME OF USE	CONTROLLING AGENCY**
R-3601 A	BROOKVILLE, KS	TO FL 180	0800-1800 MON, WED, FRI, SAT, 0800-2230 TUES, THURS, †24 HRS IN ADV	ZKC CNTR
R-3601 B	BROOKVILLE, KS	TO 6500	0800-1800 MON, WED, FRI, SAT, 0800-2230 TUES, THURS, †24 HRS IN ADV	ZKC CNTR
R-3602 A	MANHATTAN, KS	TO 29,000	CONTINUOUS	ZKC CNTR
R-3602 B	MANHATTAN, KS	TO 29,000	CONTINUOUS	ZKC CNTR
A-562 A	ENID, OK	TO 10,000	SR-3 HR AFTER SS MON-FRI	NO A/G
A-562 B	ENID, OK	TO 10,000	SR-SS MON-FRI	NO A/G
A-683	WICHITA McCONNELL AFB, KS	TO 4500	0800-1900 MON-FRI	NO A/G

**ZKC-Kansas City

Fig. 18-4. *Another example of the SUA data summary in the sectional chart.*

You're flying VFR at 7500 feet and would like to cross through R-3601B going west to east. Can you do it without permission or clearance? Yes. The area's altitude of use is from the surface to and including 6500 feet MSL. Above that altitude, R-3601B doesn't exist; thus, no approval to transit the airspace is required, and there is no threat from ground or airborne military activities.

Transiting R-3601A is a different story. This airspace extends from the ground up to FL180, and during the hours of use, ZKC (Kansas City Center) would route even IFR aircraft around or above the area. Obviously, VFR flight should not even be contemplated.

During the published hours of use, the using agency is responsible for controlling all military activity within the R area and determining that its perimeters are not violated. When scheduled to be inactive, the using agency releases the airspace back to the controlling agency (Center), and, in effect, the airspace is no longer restricted.

By the same token, it's entirely possible that no activity will be scheduled during some of the published hours of use. In those instances, the using agency again releases the space to the controlling agency for nonmilitary operations during that period of inactivity.

From a VFR point of view, then, the pilot's responsibility is rather apparent:

- When you are plotting a VFR flight and the route crosses a restricted area, determine from a current sectional the altitudes of activity and the days and hours of use. If the flight would penetrate the area when it is active, there's only one admonition: Stay out. An active R area is a land for no man. Furthermore, if you should wander into such an area intentionally or carelessly, anticipate a chat with the FAA and a resulting violation filed against you.

- If the flight is planned during a period when the area is published as inactive, don't plod ahead in blind confidence. Note the likely statement "O/T by NOTAM," meaning possible use at times other than stated. When this occurs, the using agency notifies the controlling ARTCC (16 hours in advance, in the cases of R-3601A and B). Center then advises all Flight Service Stations within 100 miles of the area, and a NOTAM is issued by those FSSs reflecting the nonpublished activation of the area. So, even though you're planning to head out on a Sunday morning or any other time when an R area is apparently not in use, don't do so until you have contacted the appropriate FSS. Then specifically ask if there is a NOTAM stating that the R area will be active at the time you will pass through it. If there is such a NOTAM, you again have only one choice: Reroute and stay out; an R area is not to be fooled with.

- Even though a chart might indicate that a military-related R area is effective 8 a.m. until noon, Monday through Friday, it doesn't mean that the area is always active during that period. If the military is not using the airspace and has released it to the controlling agency, you might get permission from the controlling agency (usually Center) to pass through the R area. All it takes is a telephone or radio call, and it can save you time and money. Be sure you call, though, within 2 hours of your estimated transit through the area, to make sure that the military does not intend to reactivate it on short notice.

Military operations areas

Military operations areas pose the largest potential obstacle to direct VFR flight between two points because of their relative size and number. This is not always the case, of course, but it happens frequently enough to cause diversions or detours when an MOA is active.

What is a MOA? It's the vertical and horizontal chunk of airspace established to segregate certain military flight training exercises, such as combat maneuvers, aerobatics, and air intercept training, from nonparticipating IFR traffic. While activity in a restricted area is described as potentially hazardous to nonparticipating aircraft, the same adjective is not employed in describing an MOA. Perhaps that's a matter of semantics, however. An F-16 barreling straight up or straight down could be considered somewhat hazardous to an aircraft that had penetrated an active MOA.

To draw a clearer distinction between a restricted area and an MOA, restricted areas are established primarily for artillery, missiles, lasers, ground-to-ground and air-to-air gunnery, bombing practice, and similar training/RDT&E exercises. Because of the nature of the activities, the ground and airspace above a restricted area are, in effect, "deeded" to the military by the government and made a matter of record in the *Federal Register*.

An MOA, on the other hand, is designed for the flight training exercises cited above. Furthermore, the airspace might be requested by the military (Air Force, Navy, or Air National Guard), but it is the FAA in the final analysis that agrees to establish an MOA. In other words, one might say that an MOA comes into being only at the discretion of the FAA.

Identifying MOAs

Undoubtedly familiar to every pilot are the vertically striped magenta bands on the sectional that block off those large chunks of terrain and the airspace above them. Designed as they are with consideration for both military and nonmilitary operations, the length and width of an active MOA could add miles and certainly minutes to a VFR flight. (As an aside, hot is the usual pilot jargon to refer to an active MOA or restricted area, as in "Demo 2 MOA is hot.")

As just one example of size, take the Ada West MOA in central Kansas (Fig. 18-5). It's not unusually large, but it does stretch 40 nautical miles east to west and 18 to 30 nautical miles north to south. That could be a fair piece of geography to circumnavigate if the MOA were hot.

Should you or should you not avoid an MOA? First, check the sectional table (Fig. 18-6). Printed in magenta on the original is the MOA name, altitudes of use, time(s) of use, and controlling agency. Note particularly the Ada West altitudes of use: 7000 feet. This means that the MOA can be active from 7000 to flight level 180 (18,000 feet), as per the asterisk footnote, and the time of use is from sunrise (SR) to sunset (SS) Monday through Friday. As a matter of general information, MOAs may extend vertically above FL180 through the designation of the airspace as an ATC assigned airspace (ATCAA).

The time of use raises a point. These are the "published" periods during which the MOA is most likely, but not necessarily, active. Said another way, these are the times that the using agency has identified as the "probable" periods of activity based on operations schedules, availability of aircraft, pilots, and similar considerations. The area's using agency, let's say an Air Force unit, has, in each case, a scheduling office that is responsible for establishing a real-time activity schedule for the MOA and forwarding it, as well as any subsequent changes, to the controlling ARTCC. That office is also responsible for developing procedures with other Air Force user units to ensure that those units notify the scheduler as soon as possible of any periods

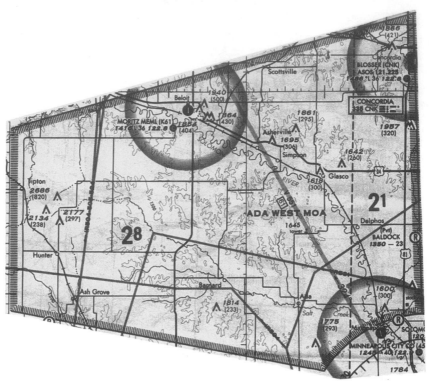

Fig. 18-5. *An excerpt from the Wichita sectional shows the expanse of the Ada West MOA.*

MOA NAME	ALTITUDE OF USE*	TIME OF USE†	CONTROLLING AGENCY**
ADA EAST	7000	SR-SS MON-FRI	ZKC CNTR
ADA WEST	7000	SR-SS MON-FRI	ZKC CNTR
BISON	1000 AGL	0830-1130 & 1330-1600 MON-FRI & ONE WEEKEND PER MONTH	ZKC CNTR
KIT CARSON A	100 AGL TO BUT NOT INCL 9000	INTERMITTENT BY NOTAM	ZDV CNTR
KIT CARSON B	9000	INTERMITTENT BY NOTAM	ZDV CNTR
LINCOLN	8000	BY NOTAM (NORMALLY 0900-1600 TUE-SUN)	ZMP CNTR
MT. DORA EAST, NORTH, WEST HIGH	11,000	BY NOTAM	ZAB CNTR
MT. DORA EAST, NORTH, WEST LOW	1500 AGL TO BUT NOT INCL 11,000	BY NOTAM	ZAB CNTR
PINON CANYON	100 AGL TO 10,000	INTERMITTENT BY NOTAM	ZDV CNTR
SMOKY	500 AGL TO BUT NOT INCL 5000	0900-1700 MON-SAT	ZKC CNTR
SMOKY HIGH	5000	0900-1700 MON-SAT	ZKC CNTR
VANCE 1A	10,000	1 HR BEFORE SR- 1 HR AFTER SS MON-FRI	ZKC CNTR
VANCE 1B	7000	1 HR BEFORE SR- 1 HR AFTER SS MON-FRI	ZKC CNTR

*Altitudes indicate floor of MOA. All MOAs extend to but do not include FL 180 unless otherwise indicated in tabulation or on chart.
†Other time by NOTAM contact FSS.
**ZAB-Albuquerque, ZDV-Denver, ZKC-Kansas City, ZMP-Minneapolis.

Fig. 18-6. *The MOA data in the sectional detail the altitudes, times of use, and controlling agency—in this case, ZKC (Kansas City Center).*

(1 hour or longer) of MOA nonuse after the initial schedule has been established. The purpose, of course, is to permit the scheduling office to return complete control of the airspace to the responsible ARTCC for nonmilitary use. As with restricted areas, however, "O/T by NOTAM" (other times by NOTAM) also applies to the MOAs.

Consequently, keep in mind that the time of use only establishes the hours during which the using agency is free to schedule activity. Through its scheduling office, however, that agency must advise the ARTCC on a daily basis of the planned schedule, as well as changes in that schedule that might occur. Otherwise, the MOA cannot be used for its designated purposes.

MOAs and the VFR pilot

Can you, as a VFR pilot, enter any MOA at any time, whether active or not? Yes. It is not a prohibited or restricted area. Should you enter? If the area is hot at your intended altitude, the answer is an absolute no. It's like cloud-busting—you're playing Russian roulette.

For one thing, an active MOA might be full of fighters zipping around you at close to Mach 1 speeds. Also, military training involves acrobatics and abrupt maneuvers. To permit such training, participating aircraft in a MOA are exempt from FAR 91, which states:

No person may operate an aircraft in acrobatic flight...within a control zone or a Federal airway...[or] below an altitude of 1,500 feet above the surface....

Freed of these regulations, you could have aircraft going straight up, straight down, inverted in flight, or whatever, while you're plodding along in their training airspace. To compound the risk, despite the fact that military pilots are probably far better trained in the see-and-avoid concept than the typical civilian, they are often very busy while maneuvering in an MOA and can't devote much time or attention to looking for nonparticipating aircraft.

Here is a point of clarification: Participating aircraft are exempt from FAR 91 only when they are operating between the published MOA floor and ceiling. In the Ada example, from 7000 feet MSL to FL180, the FAR doesn't apply; however, below 7000 feet, participating aircraft must abide by all FARs. The rules for them are the same as for any pilot, VFR, IFR, airline, or general aviation.

So, once again, how do you know whether you should or shouldn't penetrate an MOA? Let's go back to the Ada West example (Fig. 18-5). If you planned a daylight trip during the week and wanted to cruise at any altitude above 7000 feet MSL through the area, you'd be wise to forget it. Make a flight plan for a lower altitude or detour the whole MOA. Not that you are prohibited from cruising at 7500 or 8500 feet, but you're likely to encounter activity anywhere between 7000 and 18,000 feet. Unless, within a few hours of your departure, you can determine from a Flight Service Station that no operations have been scheduled, despite the published time of use, it's just not worth the risk.

However, suppose you want to venture forth on a weekend. Presumably the MOA will be quiet, per the sectional, and a flight through it at any altitude would pose no problem. But don't be too sure. A call to your Flight Service Station is in order. During the briefing, if the information is not volunteered, specifically ask if any NOTAMs have been issued about activity in the Ada MOAs.

Assuming no NOTAMs have been issued, you're probably in good shape. But let's say that your last phone call to the FSS was 4 or 5 hours ago. Now you're airborne, and what was fact then might not be fact now. The using agency could have scheduled an activity in the interim. If so, you ought to know about it. So what do you do? As you're flying along and approaching the MOA, get on the radio and call the controlling Air Route Traffic Control Center—in the Ada case, Kansas City Center. If you have not been in contact with Center for traffic advisories up to now, the call is simple:

> **You:** *Kansas City Center, Cherokee Eight Five One Five November.*

> **Center:** *Cherokee Eight Five One Five November, Kansas City Center, go ahead.*

> **You:** *Center, Cherokee One Five November is about twenty north of Ada West, level at seven thousand five hundred. Can you advise if the area is hot?*

> **Center:** *Cherokee One Five November, negative. There is no reported activity at this time and no NOTAMs issued.*

> **You:** *Roger, center. Thank you. Cherokee One Five November.*

If the area is hot, Center will confirm that fact and probably tell you the type of activity that's going on.

Instead of contacting Center, why not call the nearest FSS? Although both Center and the FSS will have the same schedule of activity, only Center has the most accurate reading of what participating aircraft are actually in the area and where they are; thus, Center is the best source for determining the current, real-time MOA activity.

Let's say the MOA is hot. Who provides any control of the participating aircraft? (*Participating* means the aircraft authorized and scheduled to be in the MOA.) An ATC facility, such as Center, might assume the responsibility when requested by the military. Otherwise, when certain conditions are met and a letter of agreement exists, control of an MOA, or any other ATC Assigned Airspace (ATCAA), can be transferred to a military radar unit (MRU). It is now the MRU's responsibility to keep its aircraft within the altitudes and boundaries of the airspace, to provide traffic advisories to participating aircraft, to separate participating aircraft, and to advise Center immediately when participating aircraft cannot remain within the allocated airspace. So, in this case, there is control, but it is of the military, by the military.

Suppose, going VFR, you have been warned by a Center that an MOA is hot, but you choose to enter it anyway. What can you expect from Center? First, Center would probably advise the military radar facility that a VFR aircraft had penetrated the MOA. Center then might give you advisories of potential conflicting military traffic, but don't count on it! If the controller has told you that the MOA is hot but you decide to drive on into it anyway, much more likely is a call saying that "Radar service is terminated. Frequency change approved. If you want later flight following, contact Center when clear of the MOA." The wording may differ somewhat, but that's the gist of what the controller will probably transmit to you.

The common concern of controllers about committing to giving advisories in this situation is the very good possibility that they won't be able to maintain continuous radar contact with the military aircraft during their maneuvering and constant altitude changes. And yet the controllers potentially could be held responsible, should a midair incident occur. It's a logical concern. Traffic advisories are meaningful only when the traffic can be seen on the radarscope and its altitude verified.

So don't rely on much, if any, help from Center in an active MOA. Center will provide separation between nonparticipating IFR aircraft cleared into all MOAs or ATCAAs, but it's a different matter when VFR aircraft are involved. You've been warned that the MOA is hot. You're within your rights to enter the area, but you have to assume the risk and the responsibility to see and avoid. Is the risk really worth the miles and minutes you're saving? You be the judge.

MARSA

Here's an acronym with which many pilots might not be familiar: MARSA, or *military assumes responsibility for separation of aircraft*. This means that when an agreement between the military and the FAA controlling agency has been reached (via letter or otherwise), the military using agency has the right to invoke MARSA. Depending upon the conditions or purposes for doing so, separation and control of participating aircraft become the responsibility of the military radar units, *airborne radar units* (ARUs); or it might be nothing more than visual separation and the pilots' responsibility to see, avoid, and stay within the confines of the assigned airspace.

One example cited by Lt. Col. John Williams, formerly Air Force representative to the FAA's Central Region, was the Williams MOA, just east of Phoenix, Arizona. Designed for extensive pilot training, the MOA was divided into several horizontal and vertical internal segments, with one aircraft assigned to a specific segment. Albuquerque Center, through radar coverage, was responsible for keeping each aircraft within its assigned segment to ensure proper separation. The problem was that the Center controllers, trying to do their jobs, were on the air so much with the participating aircraft that the flight instructors found it difficult to communicate with their students. Consequently, an agreement was engineered with Center that MARSA would automatically be in effect during student training exercises.

Another example is aerial refueling operations. MARSA begins when a tanker and receiver are in the air refueling airspace and the tanker advises ATC that it is accepting MARSA. From then on, until MARSA is terminated, the tanker and receivers are responsible for their own separation.

This is from the FAA's Special Military Operations handbook, paragraph 1-33:

The application of MARSA is a military service prerogative and will not be invoked indiscriminately by individual units or pilots....ATC facilities do not invoke or deny MARSA. Their sole responsibility concerning the use of MARSA is to provide separation between military aircraft engaged in the MARSA operations and other nonparticipating IFR aircraft. (Author's italics)

This is perhaps a matter of little concern to the VFR pilot, but if you should be considering entering an MOA or flying along a military training route (we'll discuss those in a moment), recognize that MARSA might have been invoked, and that the military is assuming responsibility for the separation of its aircraft and its aircraft alone. ATC is now out of the picture entirely, other than separating its non-participating IFR aircraft from those of the military.

That's pretty much the MOA story. Of all special-use airspace, MOAs typically offer the largest obstacle to a direct flight between two points; however, the VFR pilot can penetrate them safely if he chooses between two safe courses of action:

- Fly above or below the altitudes of scheduled activity.
- Determine from the FSS the extent of activity, if any, within the assigned altitudes by asking for MOA NOTAMs, and then establish contact with the controlling Center for current activity updates.

Otherwise, be wise. Stay out of MOAs if you don't know what's going on within those magenta boundaries.

Alert areas

Now we come to a slightly different breed of SUA. Except for prohibited areas and air defense identification zones, alert areas are the only type of special-use airspace that exists for other than just military operations. Alert areas might contain a high volume of pilot training or unusual types of aerial activity, neither of which is classified as hazardous to aircraft. While most SUAs do indicate military pilot training, those near Miami, Wichita, and along the Texas and Louisiana Gulf Coast, for example, denote heavy civilian flying activity. In essence, they exist to "alert" pilots to areas of high-density air traffic.

As with the other areas discussed, the alerts are depicted on the sectional by the familiar vertically striped bands (in blue on the original) that define the perimeters (Fig. 18-7), and the areas are further detailed on the sectional table (Fig. 18-8), outlining location, altitudes of use, and time(s) of use. Underneath the Controlling Agency section is a "No A/G" notation (*no air-ground communications*), which means that no controlling agency is dedicated specifically to the alert area, and no special frequency is assigned for operations within the area.

That might require a little more explanation. Take the case of alert area A-562B, located just a few miles north of Enid, Oklahoma's, Woodring Airport and Vance Air Force Base. If you were coming from the north for landing at Woodring, it might be faster and perhaps more logical to fly straight ahead through the alert area, despite the density of student training. Once in the area, you'd undoubtedly hear all sorts of radio communications, especially to and from the Woodring Class D tower—including the tower's response to your call requesting landing instructions. So, yes, there are air-ground communications in these alert areas, but none of it from the ground relates to control of the operations or traffic within the area itself. So in that context, "No A/G" is absolutely correct.

Unlike the other military-use airspace, the type of activity in the alert area is stated on the sectional. An example is "high-density stu-

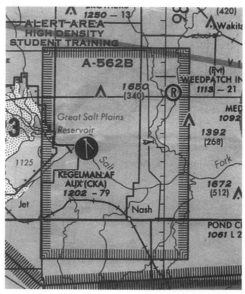

Fig. 18-7. *The alert area at Enid, Oklahoma, is identified by A-562B and the inward-directed lines.*

SPECIAL USE AIRSPACE ON WICHITA SECTIONAL CHART

Unless otherwise noted altitudes are
MSL and in feet; time is local.
Contact nearest FSS for information.
†Other time by NOTAM contact FSS

The word "TO" an altitude means "To and including."
"MON-FRI" indicates "Monday thru Friday"
FL – Flight Level
NO A/G – No air to ground communications

U.S. P–PROHIBITED, R–RESTRICTED, A–ALERT, W–WARNING, MOA–MILITARY OPERATIONS AREA

NUMBER	LOCATION	ALTITUDE	TIME OF USE	CONTROLLING AGENCY**
R-3601 A	BROOKVILLE, KS	TO FL 180	0900-1700 MON-SAT †6 HRS IN ADV	ZKC CNTR
R-3601 B	BROOKVILLE, KS	TO 6500	0900-1700 MON-SAT †6 HRS IN ADV	ZKC CNTR
R-3602 A	MANHATTAN, KS	TO 29,000	CONTINUOUS	ZKC CNTR
R-3602 B	MANHATTAN, KS	TO 29,000	CONTINUOUS	ZKC CNTR
A-562 A	ENID, OK	TO 10,000	SR TO 3 HRS AFTER SS MON-FRI	NO A/G
A-562 B	ENID, OK	TO 10,000	SR-SS MON-FRI	NO A/G
A-639 A	USAF ACADEMY, CO	3000 AGL TO 12,000	SR-SS MON-FRI EXC HOL †DAYLIGHT ONLY	NO A/G
A-639 B	USAF ACADEMY, CO	3000 AGL TO 12,000	SR-SS MON-FRI EXC HOL †DAYLIGHT ONLY	NO A/G
A-683	WICHITA MC CONNELL AFB, KS	TO 4500	0800-1900 MON-FRI	NO A/G

** ZKC-Kansas City

Fig. 18-8. *Another example of how the sectional denotes altitudes and times of use of an alert area.*

dent training" in A-562B, or perhaps "high volume of helicopter and seaplane traffic," or whatever the reason for existence of the area might be.

Okay, but what's the difference between this and the other areas described? First, participating aircraft in an alert area are governed by all FARs, including the ban on acrobatics on a federal airway and below 1500 feet AGL. Second, no permission is required to enter an alert area. Third, participating as well as nonparticipating aircraft are equally responsible for collision avoidance. Because no agency (other than routine FAA or military air traffic control) is issuing traffic advisories or alerts, or providing any sort of separation between aircraft within the area, whether IFR or VFR, alert is thus a good adjective to describe the area.

Warning areas (W)

Another chunk of special-use airspace are the warning areas (Figs. 18-9 and 18-10); however, these areas should be of little concern to the typical VFR pilot because they're located offshore. Actually, there is almost no difference between warning and restricted areas in terms of the types of activity and the hazards to nonparticipating aircraft.

Fig. 18-9. *The W-50 warning areas off the Atlantic coast at Norfolk, Virginia.*

R-6611 A	DAHLGREN COMPLEX, VA	TO 40,000	0800-1700 MON-FRI †48 HRS IN ADVANCE	ZDC CNTR
R-6612	DAHLGREN COMPLEX, VA	TO 7000	0800-1700 MON-FRI †48 HRS IN ADVANCE	ZDC CNTR
R-6613 A	DAHLGREN COMPLEX, VA	TO 40,000	0800-1700 MON-FRI †48 HRS IN ADVANCE	ZDC CNTR
A-220	MC GUIRE AFB, NJ	TO 4500	0800-2200	MC GUIRE RAPCON
W-50 A, B, C	DAM NECK, VA	TO FL 750	INTERMITTENT BY NOTAM	ZDC CNTR
W-72 A	NORTH CAROLINA	E OF 75°30'00" UNLIMITED, W OF 75°30'00" TO BUT NOT INCL 2000	INTERMITTENT	ZDC CNTR
W-72 B	NORTH CAROLINA	UNLIMITED	INTERMITTENT	ZDC CNTR
W-105 B	NARRAGANSETT, RI	TO BUT NOT INCL FL 180	INTERMITTENT	ZBW CNTR

Fig. 18-10. *More information about W-50 from the sectional.*

Nonregulatory warning areas

These areas are designated over international waters in international airspace beyond 12 nautical miles from the U.S. coast, and thus they cannot legally be regulated by the FAA. However, for any nonparticipating pilot, the admonitions about penetrating a restricted area apply equally when the area is active: Said simply: Stay out.

Regulatory warning areas

These areas extend from 3 to 12 nautical miles from the U.S. coast (over areas now considered U.S. territorial waters) and contain the same form of hazardous activity as nonregulatory warning areas and restricted areas. They serve to warn nonparticipating pilots of the potential dangers. Within regulatory warning areas, pilots must abide by the operating rules of FAR Part 91.

Air Defense Identification Zones (ADIZs)

Finally, we have these offshore areas called Air Defense Identification Zones (ADIZs). Unlike the airspace reserved for training, research, development, testing, and evaluation, these areas fit better under the national security category. More specifically, an Air Defense Identification Zone (ADIZ) is an area of airspace over land or water in which the ready identification, location, and control of civil aircraft is required in the interest of national security.

Flight regulations

Considering the purpose of an ADIZ, specific regulations pertain to operations into and within one of these security areas (excerpts), In brief,

- Aircraft must have a functioning two-way radio and a transponder with altitude-reporting equipment. The only exception to the radio requirement pertains to procedures for aircraft without two-way radios and operated on a defense VFR (DVFR) flight plan.
- An IFR or DVFR (defense VFR) flight plan must be filed.
- IFR and DVFR position reports are required.
- Flight plan deviations by IFR aircraft in uncontrolled airspace, and by DVFR aircraft, are prohibited unless the appropriate aeronautical facility has been notified prior to the deviation.
- Radio failures must be reported to the proper facility as soon as possible.

These few regulations make it rather clear that unauthorized penetration of an ADIZ could be a serious matter. The whole ADIZ concept, however, shouldn't concern most of us. With only a few exceptions, the areas begin 20 or more nautical miles off the coasts, and other than for those pilots who fly aircraft equipped for extensive overwater flight, the ADIZs lie a little far offshore for the average general aviation plane or pilot. Despite that, no discussion of the special-use airspace would be complete without at least a brief mention of ADIZs and some of the rules pertaining to them.

Controlled firing areas

This area is one I haven't mentioned for a couple of reasons. First, although the activities within the controlled firing area could be hazardous, those activities are suspended immediately when radar, spotter aircraft, or ground lookout positions detect an approaching aircraft. Second, because of this feature—and the fact that they do not cause nonparticipating aircraft to alter their flight route—the areas are not identified on any aeronautical chart. They pose no problem to either VFR or IFR flight, but just be aware that such things exist. If some of what I've said about the SUA is a little confusing, perhaps Fig. 18-11 will provide a more succinct summary.

Military training routes

No review of the airspace system would be complete without a brief discussion of those thin gray lines on the sectional or the brown lines on the en-route low-altitude charts that identify the military training routes (Fig. 18-12). (They're shown in pink on VFR wall planning charts.)

What they are

Similar to special-use airspace, MTRs exist because of the recognized need for high-speed, low-altitude military pilot training in the interest of national security. A Department of Defense and FAA joint venture, MTRs come in two forms: IFR, which is charted as IR, and VFR, which is charted as VR.

How they are identified

All routes flown exclusively below 1500 feet AGL are assigned a four-digit number, such as IR 1221 or VR 1756. Routes with one or more segments flown above 1500 feet have three numbers: IR 804. Thus, a route with, say, three segments below 1500 feet and only one above will have a three-digit identification. As Fig. 18-13 indicates, the numbers are allocated in blocks and identify the FAA region in which the route's entry point is located. Three-digit numbers are not used in the Southern Region.

SPECIAL USE AIRSPACE MATRIX

Type of airspace	Purpose/activity	Dimensions	Designated hours of operation	Nonparticipating aircraft permitted during designated hours?	Chart / publication
Prohibited Area	To prohibit flight over a surface area in the interests of national security or national welfare.	VERT: Min. required FLOOR: Surface HORZ: As required	Continuous	No	Sectional/WAC ELAC IFR Planning Federal Reg.
Restricted Area	To confine or segregate activities considered hazardous to nonparticipating aircraft. Guns and bombs	VERT: As required FLOOR: As charted HORZ: As required	Charted times	No[1]	Sectional/WAC ELAC IFR Planning Federal Reg.
Warning Area (Nonregulatory)	To contain activity that may be hazardous to nonparticipating aircraft. Guns and bombs	Defined dimensions over international water outside 12-nautical mile limit	Charted times	Yes[2]	Sectional/WAC ELAC IFR Planning Federal Reg.
Warning Area (Regulatory)	To contain activity that may be hazardous to nonparticipating aircraft. Guns and bombs	Defined dimensions between 3 and 12 nautical miles offshore	Charted times	Yes[2]	

Fig. 18-11. *This matrix summarizes the types, purposes, and pertinent operating data related to special-use airspaces.*

SPECIAL USE AIRSPACE MATRIX

Type of airspace	Purpose/activity	Dimensions	Designated hours of operation	Nonparticipating aircraft permitted during designated hours?	Chart/ publication
Military Operations Area	To contain nonhazardous training activity in airspace as free as possible of nonparticipating aircraft Acrobatics, maneuvers	VERT: As required to FL180 FLOOR: Normally 1200' HORZ: As required	Charted times	Yes[2]	Sectional/WAC ELAC IFR Planning
Alert Area	To alert nonparticipating pilots of high volume nonhazardous activity. Fixed-wing, oil rigs, helicopter training, etc.	VERT: To FL180 FLOOR: Surface HORZ: Avoid airways airports	Charted times	Yes[3]	Sectional/WAC ELAC
Controlled Firing Area	Hazardous to nonparticipating aircraft. Rockets, blasting, field artillery	VERT: 1000' above highest altitude activity FLOOR: Surface HORZ: 5 statute miles visibility - 360	By NOTAM	Yes	Not charted

[1] unless airspace has been released to ATC and pilot obtains ATC permission

[2] but not recommended when active

[3] use caution

Fig. 18-11. *Continued.*

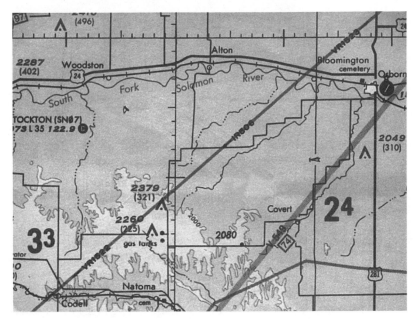

Fig. 18-12. *The thin lines on the sectional identify the VR and IR military training routes. Note the small arrows that indicate the direction of the routes. VR 1522 and IR 506 both are southwest routes, while VR 1523 goes northeast.*

	Route Numbers	
FAA Region	**One or More Segments above 1500 Feet**	**All Segments at or below 1500 Feet**
Southern	1 thru 99*	1001 thru 1099
Southwest	100 thru 199	1100 thru 1199
Western-Pacific	200 thru 299	1200 thru 1299
	980 thru 999	1980 thru 1999
Northwest Mountain	300 thru 499	1300 thru 1499
Central	500 thru 599	1500 thru 1599
Great Lakes	600 thru 699	1600 thru 1699
Eastern	700 thru 799	1700 thru 1799
New England	800 thru 899	1800 thru 1899
Alaska	900 thru 979	1900 thru 1979

*Leading zeros are dropped

Fig. 18-13. *The MTR numbering system is based on the region in which the route originates.*

Route structures

Another aspect of the structure is the small arrows adjacent to the route numbers, indicating the direction of flight along the route (Fig. 18-12). And related to this is the fact that a given numbered route always has one-way traffic. As Fig. 18-12 illustrates, VR 1522 is a southwest route flown under 1500 feet; IR 506, also southwest, is flown above 1500 feet; and VR 1523 is a northeast route under 1500 feet—three different numbers for the same track over the ground. The example further illustrates that a given route can have two-way traffic flowing on it, but when it does, the reciprocal is always given a different route number. This is unlike the Victor airways where the airway number is the same regardless of the direction in which it is flown.

When an MTR is indeed one-way traffic throughout, it does have a couple of advantages for the nonparticipating pilot. If you're crossing or paralleling the route, you at least know the flow of the military traffic and in which direction to keep your eyes peeled. Also, in planning a cross-country trip, if your route coincides with a one-way MTR, you'd be wise to plot your course well to the right or left of the MTR centerline, especially if you're traveling in the same direction as the potential traffic. A jet coming up behind you at 250+ knots might find you a little hard to spot if you're in its direct line of flight—and unless you've got a rear-view mirror and a swivel neck, you'd probably never see him at all. This is just another reason for getting a thorough FSS briefing and maintaining contact with the ARTCC for en-route traffic advisories. Those are the folks who can help keep you out of trouble.

Whether IR or VR, the routes below 1500 feet are structured to skirt uncontrolled but charted airports by at least 3 nautical miles or 1500-foot altitude. Similarly, the routes are designed to avoid populated areas and controlled airport areas, and to cause as little disturbance as possible to people and property on the ground.

Although the sectional and the ELAC depict an MTR with the thin colored line, do not let the thinness deceive you. The actual width of an MTR can be considerably greater than the charts intimate. If there is any standard at all, it is probably 5/5 (meaning 5 miles either side of the centerline).

Don't take that as a rule, though. The variations are considerable. For example, IR 514, originating at Lincoln, Nebraska, varies from

4/4 to 16/25; IR 608, Pensacola, Florida, is 10/10 throughout; VR 1128, Tinker Air Force Base, Oklahoma, is 2/2; and VR 1180, Cannon Air Force Base, New Mexico, fluctuates from 5/5 to 7.5/7.5.

The point for the nonmilitary VFR pilot is that just because you might be a little to the right or left of an MTR centerline doesn't mean the potential for conflict no longer exists. Those B-52s, or what-have-you, could be anywhere within the route's established limits. Once again, vigilance is the key.

Military pilot rules on an MTR

Whether VFR or IFR, the military pilot is responsible for remaining within the confines of the published MTR width and altitude. Flights are to be conducted at the minimum speed compatible with the mission requirements. However, while on the MTR and if below 10,000 feet MSL, military aircraft are not bound by FAR 91, which limits aircraft speed to 250 knots indicated (288 miles per hour) below that altitude. When exiting or before entering an MTR, that FAR does apply, unless the aircraft manual recommends a higher speed for safe maneuverability.

Visual flight rule operations on a VR route are conducted only when weather conditions are better than standard VFR minimums (1000 and 3). More specifically, flight visibility must be 5 miles or better, and flights are not conducted when the ceiling is less than 3000 feet AGL.

If operating IFR, on an IR route, will the military pilot get the standard services and aircraft separation from ATC? Perhaps, but mission requirements, the altitudes flown, or the inability of Center's radar to pick up the target could make those services impossible. In such cases, through a letter of agreement between the scheduling unit and the appropriate ATC facility, the route or routes might be designated for MARSA. Then the military assumes sole responsibility for separation of its aircraft.

Route scheduling

Each MTR has a designated military unit responsible for scheduling all military flights intending to use that route. When it is practical to do so, the scheduling unit, each day and prior to 2400 hours, confirms with the appropriate FSS (the tie-in FSS) the next day's planned route utilization. When that much advance notice is not possible, and barring any other agreement, the scheduling must be accomplished and communicated to the FSS at least 2 hours before use.

The schedule confirmed with the tie-in FSS is the hourly schedule for each IR and VR route, and it includes the route number, aircraft type, number of aircraft on the mission, proposed times when the aircraft will enter and exit the MTR, and altitude(s) to be flown. If a given route is closed or a scheduled aircraft cancels, the scheduling unit is required to relay any changes to the tie-in FSS as soon as possible.

With this information on hand, the Flight Service Station tonight would be able to give you a reasonably accurate briefing on tomorrow's activity. Changes might occur, though, so the closer the briefing to your actual departure time, the more accurate the MTR's status will be.

MTRs and the VFR pilot

A few precautions will minimize the risks of MTRs to the VFR pilot:

- In planning a cross-country trip, note what MTRs will cross or parallel your route of flight.

- When you call the Flight Service Station for a briefing, specifically ask for the scheduled military activity at the approximate time you would be on or crossing an MTR.

- Get an updated activity report from the nearest FSS when you are within 100 miles of the MTR.

- Establish and maintain contact in flight with the appropriate Air Route Traffic Control Center for routine traffic advisories as well as the actual, real-time military activity on the MTRs in your line of flight.

- Keep your transponder on. (If you have one, that's an FAA requirement anyway.) Many military aircraft have airborne radar and could spot you as a target and take evasive action before an emergency arose.

- Stay above 1500 feet AGL. That sounds like an obvious admonition, but remember that there could be a fair volume of high-speed, low-level operations anywhere from 100 feet on up.

- Finally, keep your head out of the cockpit and your eyes open. Military aircraft aren't easy to see. They've been camouflaged to blend in with the sky or the terrain.

MTRs are rarely discussed or considered as potential hazards to the VFR pilot, but they can be. And what perhaps adds to the potential is the MTR's apparent innocence. Unlike the distinct definition of a MOA or a restricted area, there are just those grayish lines on the sectional—nothing to cause us to sit up and take notice. But take notice we should. That empty sky out there could soon be darkened by a stream of B-52s or some other sample of airborne military machinery. A paragraph in the *Aeronautical Information Manual* seems to sum up MTRs:

> *Nonparticipating aircraft are not prohibited from flying within an MTR; however, extreme vigilance should be exercised when conducting flight through or near these routes. Pilots should contact FSSs within 100 nm of a particular MTR to obtain current information of route usage in their vicinity. Information (available) includes times of scheduled activity, altitudes in use on each route segment, and actual route width. Route width varies for each MTR and can extend several miles on either side of the charted MTR centerline.... When requesting MTR information, pilots should give the FSS their position, route of flight, and destination in order to reduce frequency congestion and permit the FSS specialist to identify the MTR routes that could be a factor.*

SUAS: A small price to pay

Discussions with many pilots, experienced as well as student, tend to point to two principal areas in which initial and refresher training are deficient: radio communications and special use airspace. The obvious intent of this chapter is to fill in some of the possible gaps in the latter.

SUAs do consume a fair amount of airspace, but it is airspace set aside with the mutual consent of several agencies of the government—and for the best possible purpose: our national defense and security. That purpose can be achieved only through the training, research, development, testing, and evaluation of our defense resources, human and materiel. MOAs, restricted areas, and the like could indeed present barriers to direct-line flight or require altitude deviations, and we might complain silently or loudly about those infringements on our freedom, but they are small prices to pay for the reasons this airspace exists.

19

Transponders

Upcoming discussions of class B and class C controlled airport areas will focus, in part, on the crucial role of the transponder. Without a transponder, or without an understanding of its proper use and the associated terminology, operating in many of the controlled airspaces would be next to impossible. Accordingly, this chapter outlines the radar beacon system, then the transponder and what it does, and finally transponder terminology.

To avoid confusion, it's important to note that the system and the radarscope images discussed here pertain only to terminal Approach Control facilities. This is called an *automated radar terminal system* (ARTS) and is a different system from that used in the en-route Centers. ARTS II and ARTS III identify advanced models of the basic system.

Air Traffic Control Radar Beacon System (ATCRBS)

ATCRBS permits more positive control of airborne traffic and consists of two basic sky-scanning features. The first is called *primary radar,* the other *secondary radar.* Primary radar sweeps its area of coverage and transmits back to the radarscope images of obstacles it encounters, such as buildings, radio towers, mountains, heavy precipitation, or aircraft. The size of the return depends on the reflective surface encountered by the radar sweep. A small, fabric-covered aircraft would produce an almost negligible reflection on the radarscope compared to that of an all-metal Boeing 747. Whatever its construction, an aircraft appears on the controller's scope as merely a small blip, showing the aircraft's range (distance) and direction (azimuth) from the radar site.

These limited returns produce a very imprecise means of identification; they only indicate to the controller that there's an airplane out there on a certain bearing and a certain number of miles from the radar site. With such scant data, a controller would be hard-pressed to track, control, separate, or identify traffic in the aircraft's vicinity with any degree of accuracy.

To overcome those limitations, a secondary system exists that incorporates a ground-based transmitter/receiver, called an *interrogator,* along with an operating aircraft transponder. The two radar systems, primary and secondary, then function in unison and constitute the Air Traffic Control Radar Beacon System.

As the two systems make their 360° sweep, the secondary beacon antenna transmits a signal that "interrogates" each transponder-equipped aircraft (or *target,* in controller parlance), and "asks" the transponder to "reply." As the transponder replies, the synchronized primary and secondary signals produce a distinctly shaped image on the radarscope. The image, however, only indicates that the transponder is on and, unless the pilot has been otherwise directed, the transponder has been set to the standard VFR transmitting code of 1200. As the terminology goes, it is "squawking one two zero zero," or "squawking twelve hundred," or "squawking VFR."

But this, too, is limited data—particularly in a heavy-traffic area. For more specific identification, each transponder has a small button that, when the pilot activates it, more definitively identifies that particular aircraft. Thus, when a controller tells you to "ident" and you push the Ident button, the radarscope image changes again, permitting the controller to positively identify your aircraft among all the targets on the screen.

Very superficially, that's the basic principle of the radar beacon system. Figures 19-1 and 19-2, taken directly from AIM, illustrate and identify the various ground and airborne images produced by ARTS III. The transponder, however, is the airborne unit that maximizes the system's value in the traffic control process.

Transponder modes and features

Undoubtedly, most general-aviation pilots have at least a speaking acquaintance with the transponder. That acquaintanceship is particularly likely if they have been flying in or around any type of high-density terminal airspace. For other pilots not entirely conversant

ARTS III Radar Scope With Alphanumeric Data

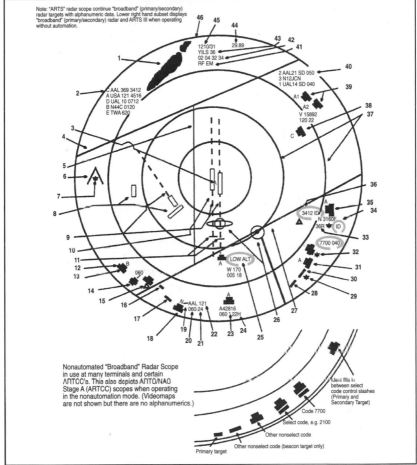

Fig. 19-1. *These are the radarscope symbols and images produced by ARTS III.*

with the transponder or the regulations pertaining to its use, a few words might be in order.

Transponder modes

Transponders are usually referred to in terms of types or *modes,* of which seven are either in use or available (general aviation interest is currently centered on Modes 3/A and C):

Example–

1. AREAS OF PRECIPITATION (CAN BE REDUCED BY CP)

2. ARRIVAL/DEPARTURE TABULAR LIST

3. TRACKBALL (CONTROL) POSITION SYMBOL (A)

4. AIRWAY (LINES ARE SOMETIMES DELETED IN PART)

5. RADAR LIMIT LINE FOR CONTROL

6. OBSTRUCTION (VIDEO MAP)

7. PRIMARY RADAR RETURNS OF OBSTACLES OR TERRAIN (CAN BE REMOVED BY MTI)

8. SATELLITE AIRPORTS

9. RUNWAY CENTERLINES (MARKS AND SPACES INDICATE MILES)

10. PRIMARY AIRPORT WITH PARALLEL RUNWAYS

11. APPROACH GATES

12. TRACKED TARGET (PRIMARY AND BEACON TARGET)

13. CONTROL POSITION SYMBOL

14. UNTRACKED TARGET SELECT CODE (MONITORED) WITH MODE C READOUT OF 5000'

15. UNTRACKED TARGET WITHOUT MODE C

16. PRIMARY TARGET

17. BEACON TARGET ONLY (SECONDARY RADAR) (TRANSPONDER)

18. PRIMARY AND BEACON TARGET

19. LEADER LINE

20. ALTITUDE MODE C READOUT IS 6000' (NOTE: READOUTS MAY NOT BE DISPLAYED BECAUSE OF NONRECEIPT OF BEACON INFORMATION, GARBLED BEACON SIGNALS, AND FLIGHT PLAN DATA WHICH IS DISPLAYED ALTERNATELY WITH THE ALTITUDE READOUT)

21. GROUND SPEED READOUT IS 240 KNOTS (NOTE: READOUTS MAY NOT BE DISPLAYED BECAUSE OF A LOSS OF BEACON SIGNAL, A CONTROLLER ALERT THAT A PILOT WAS SQUAWKING EMERGENCY, RADIO FAILURE, ETC)

22. AIRCRAFT ID

23. ASTERISK INDICATES A CONTROLLER ENTRY IN MODE C BLOCK. IN THIS CASE 5000' IS ENTERED AND "05" WOULD ALTERNATE WITH MODE C READOUT

24. INDICATES HEAVY

25. "LOW ALT" FLASHES TO INDICATE WHEN AN AIRCRAFT'S PREDICTED DESCENT PLACES THE AIRCRAFT IN AN UNSAFE PROXIMITY TO TERRAIN. (NOTE: THIS FEATURE DOES NOT FUNCTION IF THE AIRCRAFT IS NOT SQUAWKING MODE C. WHEN A HELICOPTER OR AIRCRAFT IS KNOWN TO BE OPERATING BELOW THE LOWER SAFE LIMIT, THE "LOW ALT" CAN BE CHANGED TO "INHIBIT" AND FLASHING CEASES)

26. NAVAIDS

27. AIRWAYS

28. PRIMARY TARGET ONLY

29. NONMONITORED. NO MODE C (AN ASTERISK WOULD INDICATE NONMONITORED WITH MODE C)

30. BEACON TARGET ONLY (SECONDARY RADAR BASED ON AIRCRAFT TRANSPONDER)

31. TRACKED TARGET (PRIMARY AND BEACON TARGET) CONTROL POSITION A

32. AIRCRAFT IS SQUAWKING EMERGENCY CODE 7700 AND IS NONMONITORED, UNTRACKED, MODE C

33. CONTROLLER ASSIGNED RUNWAY 36 RIGHT ALTERNATES WITH MODE C READOUT (NOTE: A THREE LETTER IDENTIFIER COULD ALSO INDICATE THE ARRIVAL IS AT SPECIFIC AIRPORT)

34. IDENT FLASHES

35. IDENTING TARGET BLOSSOMS

36. UNTRACKED TARGET IDENTING ON A SELECTED CODE

37. RANGE MARKS (10 AND 15 MILES) (CAN BE CHANGED/OFFSET)

38. AIRCRAFT CONTROLLED BY CENTER

39. TARGETS IN SUSPEND STATUS

40. COAST/SUSPEND LIST (AIRCRAFT HOLDING, TEMPORARY LOSS OF BEACON/TARGET, ETC.)

41. RADIO FAILURE (EMERGENCY INFORMATION)

42. SELECT BEACON CODES (BEING MONITORED)

43. GENERAL INFORMATION (ATIS, RUNWAY, APPROACH IN USE)

44. ALTIMETER SETTING

45. TIME

46. SYSTEM DATA AREA

Fig. 19-2. *The meanings of the various symbols as numbered in Fig. 19-1.*

- Modes 1 and 2: assigned to the military
- Mode 3/A: used by both military and civilian aircraft
- Mode B: reserved for use in foreign countries
- Mode C: Mode 3/A transponder equipped with altitude-reporting capabilities
- Mode D: not presently in use
- Mode S: a relatively new mode that may come into use at a later time (S means select, or selective address)

The big advantage of a Mode S-equipped aircraft is that it will permit ATC to selectively interrogate and address that aircraft, even in high-traffic-density situations. In the same vein, it can function in concert with the FAA's TCAS (Traffic alert and Collision Avoidance System). Despite these and other safety-oriented features, however, this element of the FAA's National Airspace System Plan (NASP) has badly missed its intended 1995 implementation date and, at the time of writing, appears to be at a standstill. Even should such a system be mandated, the costs to convert to a Mode S transponder would likely be beyond the reach of those who own or rent the typical general-aviation aircraft. The objectives and the value of the overall program are unquestioned; the potential costs, however, raise barriers that may be hard to overcome.

Features

Mode 3/A is the standard transponder that does a reliable job of establishing radarscope identification of a given aircraft. The unit pictured in Fig. 19-3 happens to be an Allied Signal Bendix/King KT 76A, but all makes are basically the same.

The unit is activated by the function selector at the extreme left of the set. The selector positions, as the picture shows, range from OFF to SBY (standby), ON (or NORMAL on some sets), ALT (altitude), and TST (test).

When the engine is started, the selector should be turned to SBY. This allows the set to warm up, which normally takes 2 to 3 minutes, but it does not transmit any signal. At the same time, check to be sure that the transponder has been set to 1200, unless ATC has given you a different code. When you have been cleared for take-off, turn the selector to ON, if you don't have altitude-reporting capabilities (Mode C). The set is now in the Mode 3/A posture, and after you're airborne, this symbol would appear on the radarscope. (For purposes here, the size of the symbol is considerably exaggerated.)

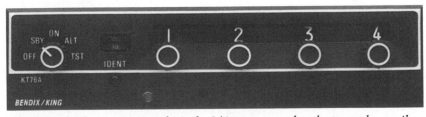

Fig. 19-3. *This is a typical Mode 3/A transponder that can be easily converted to a Mode C altitude-recording unit.*

The symbol, however, will reflect only your relative geographical position in the airspace and the fact that you're transmitting the 1200 code. If the set has altitude-reporting capabilities (Mode C), the switch should be turned to ALT. Then this symbol will appear on the controller's scope, once the aircraft is airborne. Now the controller can determine your azimuth, range, and current altitude—in this case, 4500 feet.

A firm FAA ruling is that the transponder must be in the ON position, or in the ALT position if it has altitude-reporting capabilities. And it must be left on from takeoff until landing, unless a controller instructs otherwise.

Skipping the ident feature for the moment and moving to the right of the switch positions in Fig. 19-3, we see the next prominent feature is the four CODE windows and the 1-2-0-0 numeric display, plus knobs below each window which change the numbers to whatever code ATC requests. The 1200 code is the standard VFR code and, under normal operating conditions, should never be changed unless ATC so advises. However, a change is likely to be requested if you are in a controlled area, such as Class B or Class C airspace, or in most cases, when you're receiving en-route traffic advisories from an Air Route Traffic Control Center on a cross-country excursion. In any of these situations, you may still be flying VFR, but ATC will assign you a specific *discrete* code so that your aircraft can be distinguished from all others in the airspace. That discrete code will be yours alone as long as you are in that ATC facility's airspace and are being controlled or given advisories.

Relative to these codes, reference is often made to a *4096 transponder.* What this means is that each of the four knobs can dial in eight digits from 0 to 7. Combined and multiplied, the four knobs thus produce a total of 4096 different codes. Anticipating what discrete code ATC might assign you is impossible. With only a few exceptions, which I'll review in a moment, the code is selected by a computer according to the *National Beacon Allocation Plan.* It thus could be anything within the block of codes assigned to a given facility.

If ATC asks you to *squawk* (meaning to dial in) a certain code, first write down the numbers so you won't forget them (it's easy to confuse 0465 with 0456 or any other potential combination). Next, repeat the code back to the controller to confirm that you have copied it correctly. Then immediately enter the code in the transponder. When you do, an image similar to this will appear on the radarscope:

The line is called the leader line, and 045 is the aircraft's altitude (based on Mode C's altitude-reporting capabilities). The 13 represents the aircraft's ground speed of 130 knots, while W indicates that the west sector controller is handling the aircraft.

Going back to the unit itself, located between the function selector and the four code selector knobs are two features, both of which are difficult to see in this photograph—one the reply light and the other the Ident push button.

As to the light, every time the transponder is interrogated by a radar beacon in its 360° sweep, the light blinks momentarily, thus indicating that the set is working properly and is transmitting a signal that represents your aircraft back to the ground unit. In effect, the transponder is "replying" to the interrogator. If the light appears to be blinking almost constantly, it simply means that the transponder is replying to several interrogators that are located within radar range of your present position.

The Ident button comes into play when a controller asks you to "Ident." In effect, he or she wants you to push the button once—and only once. When you do, the four-number code assigned to you "blossoms" on the scope, as the following drawing attempts to illustrate, thus further distinguishing your aircraft from the others in the area. (Also see example 35 in Fig. 19-1.) It's important to stress, however, that you push this Ident button only when instructed to do so. Don't touch it unless you hear something like "Cherokee, Eight Five One Five November, ident." Then push it, but, again, only once.

Transponder codes

As I've said, the standard VFR code is 1200 and should always be entered in the transponder during normal flight operations, unless ATC directs otherwise. You should, however, be familiar with other select codes.

For instance, if you have a bona fide emergency—a fire, engine failure, loss of control, whatever—and if time permits, immediately enter code 7700. When you do, as the controllers describe it, "Lights light up and bells ring" (example 32, Fig. 19-1). All radarscopes within range will signal an emergency, and controllers at a nearby radar-equipped tower, Center, or Approach Control will be trying to contact you by radio. That won't do much good if you're not on (or monitoring) one of their frequencies or the 121.5 emergency frequency, but at least ATC will be able to follow the track of your air-craft and be in a better position to spot your approximate position, should you go down. That 7700 code, however, is strictly an emer-gency code and should never be entered, even momentarily, in the transponder window.

Until recently, the transponder code for radio failure was 7700 for 1 minute, followed by 7600 for 15 minutes, and then this sequence re-peated. Now, however, the code is simply 7600, and the pilot should leave it there unless advised otherwise by ATC. This change is cited in the *Aeronautical Information Manual,* paragraph 6-4-2, which reads as follows:

> *Transponder operation during two-way communications failure*
>
> **a.** If an aircraft with a coded radar beacon transponder ex-periences a loss of two-way radio capability, the pilot should adjust the transponder to reply on MODE A/3, Code 7600.
>
> **b.** The pilot should understand that the aircraft may not be in an area of radar coverage.

Figure 19-4 recaps these and other standard codes. The starred codes should never be used by civilian pilots.

A few words about Mode C

In light of the FAA regulations effective in 1989 and 1990, some ad-ditional comments about Mode C are essential. Until fairly recently, Mode C for strictly VFR operations was more of a voluntary addition to the Mode 3/A transponder than a necessity. True, it was manda-tory for flight above 12,500 feet MSL and in the Class B airspaces, but otherwise it was not a requirement. Times have changed, however.

The rash of midair collisions such as the Cerritos, California, accident cited in App. A, in high-density traffic areas during the latter part of the 1980s triggered that change. Coupled with more and more re-ported close calls, the public, the media, and congressional pressure forced the FAA to take action. Particularly in terminal areas, the FAA responded by tightening pilot and equipment requirements in con-trolled airspaces. Mostly, but not exclusively, the tightening related to the transponder and its ability to report aircraft altitudes in those terminal areas. Enter, then, Mode C.

Transponder Code	Type of Flight	When Used
0000*	Military	North American Air Defense
1200	VFR	All altitudes, unless instructed otherwise by ATC
4000*	Military VFR/IFR	In Warning and Restricted areas
7500	VFR/IFR	Hijacking
7600	VFR/IFR	Loss of radio communications
7700	VFR/IFR	In an emergency—"Mayday"
7777*	Military	Intercept operations
Any code	VFR/IFR	When using center or Approach Control and ATC assigns a specific, or "Discrete" code

Note: The starred (*) codes are for military operations only, and never to be used by civilian pilots.

Fig. 19-4. *A summary of the standard military and civilian transponder codes, including the starred codes that civilian pilots should never use.*

What makes a transponder Mode C?

A regular Mode 3/A transponder becomes Mode C by one of two conversions. The first is to remove the present altimeter and replace it with an *encoding* altimeter. Externally, this looks like any ordinary altimeter; internally, however, there are some modifications. Electronically connected to the transponder, the altimeter aneroid bellows expand and contract with altitude changes, and the changes are converted to coded responses by the transponder. When interrogated by the sweeping radar beacon, the transponder replies by transmitting those responses, which are then decoded on the ground. The process concludes when the target's altitude appears on the radarscope.

The second conversion option is to leave the old altimeter right where it is and to install a *blind encoder*. This unit does just what the altimeter does, differing only in that it is a small black box, usually physically mounted on the firewall or elsewhere under the instrument panel and out of the pilot's sight.

Which is preferable? Operationally, both are equally effective. The blind encoder, however, can be removed for repairs and the airplane is still flyable, except, of course, where Mode C is required.

On the other hand, if the encoding altimeter goes on the fritz, it has to be removed, you'll have no altimeter, and the aircraft is grounded pending repairs. Plus, you've probably replaced a perfectly good altimeter. If you're lucky, you might be able to sell it; otherwise, you're stuck with a reliable but useless instrument.

Transponder terminology

To round out the transponder picture, the terminology associated with transponder use comes into play. These are the most common terms or expressions (some of which I've already touched on), whether the equipment is Mode 3/A or C.

Squawk

Back in the early World War II days, the Allies had a radar beacon system known as IFF (*identification friend or foe*). Our aircraft were equipped with transmitters that, when queried by the radar sweeps, sent back a sound similar to that of a parrot's squawk, identifying themselves as friends. That squawk has carried over and is used by controllers and pilots alike to mean that the transponder has been turned on and that a certain code has been, or should be, dialed in. When a controller tells you to "Squawk four three two one," he's telling you to enter those digits in the unit. In essence, squawk means transmit. The two words could be used interchangeably, but squawk has hung on, reminiscent of those days 50-plus years ago.

Ident

When you hear "Cherokee One Five November, ident," the controller wants you to push the Ident button one time so that your aircraft can be more positively identified on the radarscope. At this point, it's appropriate, but not required, to acknowledge the call with a simple "Roger." Say no more. Just push the button. Saying

"Roger, Cherokee One Five November identing" is unnecessary. As the ID on the scope blossoms, the controller will know that you followed the instruction.

Squawk (number) and ident

This instruction means to enter a certain four-digit code in the transponder and then ident. For example, "Cherokee Eight Five One Five November, squawk one two zero zero and ident." If you've already got that standard VFR code in the unit, nothing more needs to be done except the act of identing. If you've been given a different code, say, 4321, then it's proper to read back the code to be sure you understood it. "Roger, four three two one." Now enter it and ident, if the controller told you to. If you don't hear "Ident," don't touch the button.

Stop squawk

This is an instruction to turn the transponder off entirely.

Squawk standby

For whatever reason, the controller says this to have you discontinue transponder replies, but not turn the transponder off. So merely switch from ON or ALT to the SBY position. Replies cease, but the set will still be warm for immediate future use.

Stop altitude squawk

If you hear this, turn the switch to the ON (or in some sets, NORMAL) position, not to OFF. ATC wants you to continue transmitting your position on Mode 3/A, but not your Mode C altitude.

Recycle or reset

Occasionally, a squawk will transmit a questionable signal or one that is the same code given to another aircraft. Also, one of the code numbers might not quite slot into place. Whatever the case, ATC is asking you to reset the four digits so that you are transmitting the assigned code.

Squawk mayday

If you've reported an emergency of some sort, the controller with whom you've been in contact might tell you to "squawk Mayday," meaning that you are to dial in the 7700 emergency code. (Mayday comes from the French word *M'aidez,* pronounced mayday, meaning "Help me.")

A Passport to safety

Transponders, Modes 3/A and C, are things of the present and immediate future. More and more aircraft have to have them to operate in any environment other than uncontrolled airports, away from terminal control areas, and below 10,000 feet MSL. The specific requirements, especially for Mode C, are reviewed in Chap. 20.

Requirements aside, the transponders are one more piece of equipment that enhances safety in the air. Without them, ATC would be hard-pressed to direct and separate terminal or en-route traffic with any degree of accuracy.

Despite the furor raised by many general-aviation interests over some of the FAA's Mode C rulings, the cost to equip an aircraft with altitude-reporting capability is relatively minor when the increased safety factor is considered. It's also a minor price to pay for the privilege of entering and operating in the most strictly controlled airspaces: the airspaces near and surrounding the busy air carrier airports in some 150 major U.S. cities. Those terminal airspaces, along with the associated transponder requirements, are reviewed in Chap. 20.

20

Operating in Class B, Class C, and TRSA airport environments

Some aspects of the controlled airport environments, including Class D airports, have already been discussed, along with references to Class B and C airspaces. Now, however, it's time to be more definitive about those latter airspace structures and the facilities that manage air traffic in the terminal area. Specifically, I'm talking about the control tower and Approach/Departure Control.

The controlled airport environment: Be prepared!

When the figures are rounded, we have more than 18,000 airports in the United States, including heliports, STOL ports, and seaplane bases. Of that total, some 13,000 are *limited-use* airports, meaning that they are privately owned and access to them requires prior owner approval, and slightly more than 5000 are for public use. Among those 5000, roughly 680 have control towers, 400 of which are FAA-administered; 90 have towers staffed by private contractors, identified on sectional charts as NFCT (nonfederal control tower) in the airport data block (Fig. 20-1); and 175 airports have military towers.

So, again with rounding, more than 5000 airports are for your use and mine, while 680, or 13 percent, are classified as controlled. The remaining 87 percent are Class E and Class G uncontrolled fields requiring only self-announce communications. That gives us a lot of room to roam about, unfettered by the rules, procedures, or requirements involved in operating in the controlled airport environment.

The only problem, if there is a problem, is that these 680 or so aerodromes are located, for the most part, in the more populated areas and the larger cities. Consequently, access to them demands certain

Fig. 20-1. *The "NFCT" located to the left of the tower frequency identifies that tower as nonfederal and thus operated by private contractors.*

pilot qualifications as well as aircraft radio and avionics equipment. It further requires pilot knowledge of what the controlled airspaces are and how to operate within them.

If there is a knowledge deficiency, one of two things will happen: The pilot will go miles out of the way to avoid a Class C or Class B airspace, or he or she will fly unannounced into a controlled terminal area, potentially causing all sorts of havoc. The first alternative is needlessly costly in time and fuel; the second is both illegal and dangerous.

In interviews with many air traffic controllers, one common plea to pilots emerged: Be prepared. Do some homework before you sally forth into a controlled environment. Know the procedures and requirements. Know how and when to use the radio. Know what to say and what you can expect to hear. If you haven't understood an instruction, don't be afraid to ask the controller to "Say again" or "Speak more slowly." Don't just "Roger" an unclear instruction simply because you're reluctant to reveal your uncertainty to the listening audience.

And there are other elements in this matter of preparation, such as knowing the tower and ground frequencies of the airport you're entering. One controller cited this as a frequent example of unpreparedness, of which even air carrier pilots are occasionally guilty. As the controller said, "There are three people up there in the front end of a Boeing 727, making pretty good money, and they haven't even taken the time to look up the correct tower frequency."

In another case, a pilot called in for landing instructions and then asked the tower if the airport "was north or south of the river." The controller admitted the temptation to retort that the airport was north of the river, "where it's been for 50 years." The controller overcame the temptation, but I wonder how many times controllers have bitten their tongues when questions born of inadequate preparation have come through their headsets. A mere glance at the sectional before departing, or even en route, would have told this pilot exactly where the airport was located.

Still another example: Many airports have navigation aids, such as VOR/DME or nondirectional beacons (NDBs), right on the field; however, controllers say it is not unusual for pilots to contact the tower and ask for vectors or headings to the airport. Of course, if the necessary avionics aren't on board or aren't working properly, the request can be justified. Otherwise, just dialing in the appropriate frequency and centering the needle would lead these pilots directly to their destination. But these same folks probably didn't take a few extra minutes to check the sectional or the *Airport/Facility Directory* before venturing forth.

The lack of preparation arose many times in discussions with controllers, whether based in a tower, Approach Control, or Center, and with specialists at Flight Service Stations. Of course, pilots have to know what they're preparing for, so knowledge becomes the obvious cornerstone. Flying into or out of a controlled environment requires no rare or elusive skill, but it does demand those two basics: knowledge and preparation.

The remainder of this chapter thus briefly reviews the roles of two controlling facilities: the tower and Approach/Departure Control. Then the chapter summarizes the areas that comprise Class C and Class B airspaces including how they're identified, their structure, the pilot and equipment requirements, and finally the operating regulations, particularly the VFR regulations, pertaining to each.

Two terminal-controlling agencies

Two facilities are responsible for the orderly terminal arrivals and departures of both IFR and VFR aircraft. One is the air traffic control tower (ATCT), and the other is Approach/Departure Control.

The tower exercises control over only that traffic within the approximate 5-mile radius of the airport. Where it exists, Approach/Departure is responsible for separating aircraft, particularly IFR operations, outward from the airport area to about a 30-mile radius around the primary airport. Vertically, Approach/Departure's airspace rises to the upward limits of its radar coverage responsibility, which might vary from approximately 10,000 feet to 15,000 feet AGL, depending upon the terminal.

Two acronyms identify the most common Approach/Departure facilities: first, is *TRACON,* which means Terminal Radar Control. The other is *RAPCON,* Radar Approach Control. What's the difference? The first is FAA-operated, while the second is a military operation.

TRACON means that the actual facility, with its radar equipment, computers, phones, and personnel, is housed in the control tower structure but not in the glass-enclosed tower cab itself. All the larger airports have TRACON because of the space needed for equipment and personnel.

RAPCON is the Air Force/FAA acronym for Radar Approach Control, and where it exists, it is entirely removed from the civilian airport tower structure. It does, however, provide radar service to nearby civilian airports and civil aircraft, but it can be physically located at a military airport some miles away.

The control tower and its responsibilities

The control tower (Fig. 20-2) has four primary responsibilities:

- To separate and sequence aircraft transiting the area or in the traffic pattern
- To expedite arrivals and departures
- To control ground movements of aircraft and motorized vehicles
- To provide clearances as well as local weather and airport information to pilots, the last two primarily, but not exclusively, via the tape-recorded ATIS messages.

The basics of operating within the airport airspace are covered previously. But a few additional points related particularly to Class D airports are in order.

- The tower has direct and sole control over only the traffic that is within the limits of its assigned airspace.

Fig. 20-2. *The Clearance Delivery position is in the foreground with ground controllers and local traffic controllers in the background.*

- Outside that airspace, the tower at a Class D airport may issue traffic advisories to aircraft with which it has been in contact, if its workload permits. That, however, is secondary to the controller's primary responsibility, which is to keep track of the types and N numbers of aircraft that have called in and to scan the skies within the area to sequence or separate landing, departing, and transiting traffic. Consequently, don't be lulled into a false sense of security just because you're in the vicinity of a Class D control tower. The VFR pilot's neverending responsibility is to keep both eyes open and be the best conflict resolver. That's true anywhere, but especially so near or around a tower-controlled airport where the volume of traffic is sufficient to warrant the very existence of that facility.

- The sole purpose of your first inbound radio call is to inform the controller of your presence, that you intend to enter Class D airspace and, if the tower is radar-equipped, to help identify your particular aircraft. So informed, the controller can now better plan the flow and sequencing of traffic into the airspace and the traffic pattern.

All Class B and C primary airport towers have the same radar display that appears on the Approach/Departure Control scopes. Called BRITE [or *bright radar tower equipment* (or the more advanced D-BRITE, the D standing for digital)], the equipment reproduces the displays received in the TRACON (Fig. 20-3). What appears, though, is a video compression transmitted to the tower by a phone line. If a Class D satellite tower is equipped with BRITE, as more and more are, it, too, is receiving its images via phone lines from the primary airport TRACON. (The BRITE acronym, by the way, also describes the purpose of the equipment, which is to produce images that can be easily seen in the bright environment of the tower cab.)

Despite its presence, however, not all tower controllers are radar-qualified. This doesn't mean that they have had no training and are unable to interpret what they see on the scope; it just says that they have not gone through the extensive FAA training as well as current approach control experience that is required to classify them as radar-certificated. Without that certification, the controller can only suggest that a pilot take certain actions beyond the 5-mile radius, if the radar image so indicates; but the controller cannot issue directives that would constitute an order or a command. Instead

Fig. 20-3. *The enhanced D-BRITE radar in the tower displays the same data as those on the Approach Control TRACON scopes.*

of saying, "Turn left zero three zero heading," it would be, "Suggest you turn left zero three zero heading."

Class C airspace: Structure, requirements, and regulations

Class C airspaces are the fastest-growing addition to the controlled airspace system and are replacing the terminal radar service areas (TRSAs) throughout the country. The Class C concept was developed to provide more positive control of aircraft in the terminal area than was, or is, the case under the TRSA system. Pilot participation, meaning establishing radio contact with Approach Control and receiving traffic advisories, is optional in a TRSA. It's the pilot's choice, assuming the flight is VFR. In a Class C airspace, participation is mandatory.

Qualification criteria

To qualify as a Class C airspace, the airport must have an operating control tower served by radar Approach Control and the airport must meet at least one of the following criteria: 75,000 annual instrument operations at the primary airport; or 100,000 annual instrument operations at the primary and secondary airports in the terminal hub area; or 350,000 annual enplaned passengers at the primary airport.

As of the date of writing, 122 Class C airspaces exist. For those interested in the specific locations, check the current edition of the *Aeronautical Information Manual.*

Identifying Class C airspaces

Class C (or "Charlie") airspaces are easy to spot on the sectional, primarily because of the two solid magenta circles surrounding the primary airport. Figure 20-4 illustrates the design, with the circles that encompass the Jackson, Mississippi, International airport. Also, along with the usual frequency and basic runway information found in conjunction with the airport name, the sectional identifies the floors and ceilings of the airspace. In the Jackson example, 44/SFC means that the Class C controlled airspace—the airspace in which contact with ATC is mandatory—extends from the surface to 4400 feet AGL inside the inner circle. The 44/17 (located at approximately the two o'clock position in the outer ring) indicates that the Class C airspace starts at 1700 feet AGL in that ring and rises to the same 4400-foot ceiling.

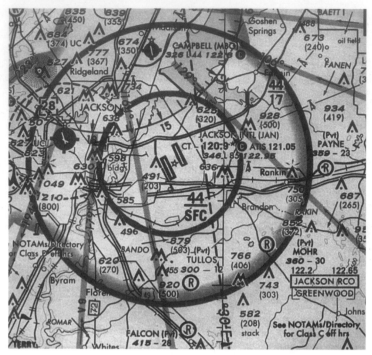

Fig. 20-4. *How the sectional depicts Class C airspaces.*

Another Class CD–identifying source is the *Airport/Facility Directory,* as in Fig. 20-5. The fact that Jackson International is Class C is clearly established by the entry AIRSPACE: CLASS C. The A/FD also tells you what Approach or Departure frequency to use, depending on your location relative to the airport. As you can see in the ®APP/DEP CON entry, if your position or intended flight path is anywhere between 333° and 152° from the airport (generally north and east), you should contact Approach on the 123.9 frequency. Should your flight path be anywhere between 153° and 332°, the frequency is 125.25. When the tower and the Approach/Departure Control facility are closed, which they are from 0500 to 1200Z, Memphis Center handles the approaching and departing traffic on 132.5 (the next line down in the A/FD).

One other source for frequencies is the reverse side of the sectional's legend flap (Fig. 20-6). Here, you'll find some of, but not all, the same basic information.

The Class C airspace structure

The sectional outlines the circular plan view of Class C airspace, but there's more to it than that. Figure 20-7, a cross section, illustrates the

```
- - - - - - - - - - - - - - - - - - - - - - - - - - - - - - - - - - - - - - - - -
JACKSON INTL  (JAN)  5 E   UTC-6(-5DT)  N32°18.67' W90°04.55'                    MEMPHIS
    346   B   S4   FUEL 100, JET A,   OX 2   LRA   ARFF Index C              H-4G, 5C, L-17D
    RWY 16R-34L: H8501X150 (ASPH-CONC-GRVD)    S-130, D-165, DT-300   HIRL   CL       IAP
       RWY 16R: REIL. VASI(V6L)—Upper GA 3.25° TCH 86.3'. Lower GA 3.0° TCH 52.4'. Trees. 0.4% up.
       RWY 34L: MALSR. TDZL. Trees.
    RWY 16L-34R: H8500X150 (ASPH-GRVD)    S-75, D-200, DT-358   HIRL   CL
       RWY 16L: ALSF2. TDZL. Trees.    RWY 34R: REIL. VASI(V4L)—GA 3.0° TCH 52'. 0.7% down.
    AIRPORT REMARKS: Attended continuously. Heavy bird activity and invof arpt. Ldg fee for non-commercial acft over
       25,500 pounds, fee waived for larger non-scheduled acft with sufficient fuel purchase. Noise abatement
       procedures—acft over 12500 pounds on Rwy 16L and 16R climb rwy heading to 15.6 DME or as directed by
       ATC. When twr clsd ACTIVATE REIL Rwy 16R-120.7. U.S. Customs user fee arpt.
    WEATHER DATA SOURCES: ASOS (601)932-2822. LLWAS.
    COMMUNICATIONS: CTAF 120.9   ATIS 121.05   UNICOM 122.95
       GREENWOOD FSS (GWO) TF 1-800-WX-BRIEF. NOTAM FILE JAN.
       RCO 122.65 122.2 (GREENWOOD FSS)    RCO 122.1R 112.6T (GREENWOOD FSS)
——▶ ®APP/DEP CON 123.9 (333°-152°) 125.25 (153°-332°)(1200-0500Z‡)
    ®MEMPHIS CENTER APP/DEP CON 132.5 (0500-1200Z‡)
       TOWER 120.9 (1200-0500Z‡)   GND CON 121.7
——▶ AIRSPACE: CLASS C svc 1200-0500Z‡ ctc APP CON other times CLASS E.
    RADIO AIDS TO NAVIGATION: NOTAM FILE JAN.
       (H) VORTAC 112.6   JAN   Chan 73   N32°30.45' W90°10.06'   153° 12.6 NM to fld. 360/05E.
       ALLEN NDB (LOM) 365   JA   N32°24.75' W90°07.17'   157° 6.5 NM to fld.
       ILS 109.3   I-FRL   Rwy 34L.   (Unmonitored when twr closed)
       ILS 110.5   I-JAN   Rwy 16L.   LOM ALLEN NDB. (Unmonitored when twr closed)
       ASR (1200-0500Z‡)
```

Fig. 20-5. *The A/FD further identifies a Class C and its Approach Control frequencies.*

CLASS B, CLASS C, TRSA, AND SELECTED RADAR APPROACH CONTROL FREQUENCIES

FACILITY	FREQUENCIES	SERVICE AVAILABILITY
MEMPHIS CLASS B	125.8 338.3 (356°-175°) 119.1 291.6 (176°-355°)	CONTINUOUS
COLUMBUS AFB CLASS C	127.95 389.8 (090°-165°) 135.6 226.0 (165°-310°) 120.4 229.15 (310°-090°) O/T 127.1 269.4 ZME CNTR	0700-1900 MON-THU 0700-2100 FRI 1000-1700 SAT-SUN 0930-1700 HOL EXC CHRISTMAS & NEW YEARS DAY O/T CLASS G; E 700' AGL & ABOVE
JACKSON CLASS C	123.9 317.7 (333°-152°) 125.25 319.2 (153°-332°) O/T 132.5 259.1 ZME CNTR	0600-2300
LITTLE ROCK CLASS C	135.4 353.6 (042°-221°) 119.5 306.2 (222°-041°)	CONTINUOUS
SHREVEPORT CLASS C	119.9 351.1 (153°-319°) 118.6 350.2 (320°-152°)	CONTINUOUS
FORT SMITH TRSA	125.4 393.0 (075°-255°) 120.9 228.4 (256°-074°) O/T 119.25 398.9 ZME CNTR	0530-2300
LONGVIEW TRSA	118.25 270.3 (EAST AT OR BELOW 4000') 124.67 128.75 379.15 385.4 (WEST AT OR BELOW 5000') 133.1 (AT OR ABOVE 4500') O/T 135.1 269.2 ZFW CNTR	0600-2200
MONROE TRSA	126.9 (180°-359°) 118.15 (360°-179°) 388.0 O/T 135.1 346.25 ZFW CNTR	0545-2300
LITTLE ROCK AFB RADAR	119.5 306.2	CONTINUOUS
NAS MERIDIAN RADAR (MC CAIN)	119.2 374.9 (E) 120.5 269.6 (S) 120.95 269.6 (W) 314.8 (N) O/T 124.4 323.0 ZME CNTR	0700-2300
NOLF WILLIAMS RADAR	276.4	0700-2300

ZFW – Fort Worth, ZME – Memphis
O/T indicates other times

Fig. 20-6. *Another reference source for class identification is the reverse side of the sectional legend flap.*

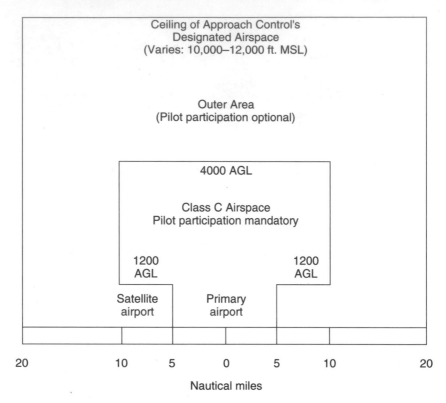

Fig. 20-7. *This drawing represents the basic structure of the Class C airspaces.*

typical lateral and vertical structure. Surrounding the primary airport is the *inner circle,* extending 5 nautical miles out from the airport and rising from the surface to approximately 4000 feet AGL. This might vary slightly, but it is the typical Class C ceiling.

Next comes the *outer circle,* with a 10-nautical-mile radius and a 1200-foot AGL floor. The floor, too, might vary slightly in altitude from Class C to Class C, but the outer circle's ceiling is the same as that for the inner circle.

Finally, the *outer area* is not depicted on any aeronautical chart, but it's there nonetheless. The perimeter of the area is generally circular, with a 20-nautical-mile radius from the primary airport. This dimension, however, is not entirely standard because individual facilities, in the interest of safety and local traffic count, have certain latitudes in defining the shape and size of the respective outer areas. Consequently, the limits can vary from a radius of 20 miles up to 30 or 40 miles, depending on the airport. Whatever its radius, the outer area still has the same common vertical limits of responsibility delegated to that particular facility, approximately 10,000 feet MSL.

Pilot qualifications and aircraft equipment

Class C airspaces have no minimum pilot requirements. From student on up, any pilot has access to this Class C airspace. For equipment, a two-way radio capable of communicating with Approach/Departure Control and the tower is required. An operable Mode C transponder is also required within the inner and outer circles (not the outer area) up to and including 10,000 feet MSL.

Operating regulations

To approach or depart a Class C airspace, the following regulations apply.

- When approaching a Class C airspace with intentions to land at any airport within the inner circle or to transit the airspace (lower than the 4000-foot ceiling), two-way radio communication with Approach Control must be established before you enter the outer circle—not the outer area. In the outer area, the pilot has the option of establishing contact for any traffic advisories. It's up to him or her. If contact is made, however, Approach Control has no option: It must provide the advisory service, whatever its workload might be inside the Class C airspace itself.

- While a literal "clearance" into a Class C airspace is not required, establishment of two-way radio communication is. To be more specific, if you call Approach and all you hear is "Stand by," you have not established two-way radio communication (by FAA's definition) and you may not enter the outer circle; that "Stand by" might have been addressed to another aircraft, unbeknownst to you. If, however, the controller responds with your aircraft N number "Cessna 1234, stand by," you have met the requirement and may proceed into the airspace. The mere fact that Approach specifically acknowledged your call sign is sufficient to allow you to enter the airspace, unless directed otherwise by the controller.

- Once you are in the airspace, Approach will issue instructions, vectors, or altitude changes. These should be acknowledged and followed unless doing so would place you in jeopardy or cause you to violate a VFR regulation. In such cases, you must advise the controller of the situation so that alternate instructions can be issued.

- Prior to your entering the inner circle, Approach will turn you over to the tower for final landing instructions and landing clearance.

- In departing the primary Class C airport, and after monitoring the current ATIS, the first step is to call Clearance Delivery (CD), which is usually located in the tower, next to the Ground Control position. At airports without CD, contact Ground Control. In either case, state your intentions and record the instructions you are given. For example:

You: *Tallahassee Clearance, Cherokee Eight Five One Five November.*

CD: *Cherokee Eight Five One Five November, Tallahassee Clearance.*

You: *Clearance, Cherokee One Five November at Coastal Aviation with Hotel. VFR to Birmingham direct, requesting six thousand five hundred.*

CD: *Cherokee Eight Five One Five November, maintain four thousand. Departure frequency will be one two eight point seven. Squawk three three four two.*

You: *Roger, Cherokee Eight Five One Five November maintain four thousand, one two eight point seven, and three three four two.*

CD: *Readback correct, One Five November, contact Ground.*

You: *Will do. One Five November.*

Next is the usual call to Ground Control and then the tower for take-off clearance.

- If you are departing a satellite airport within the inner circle, contact the tower as soon as possible after takeoff and follow the normal traffic patterns.

- For departures, arrivals, or touch-and-goes at a satellite airport under the outer circle, if you plan to stay lower than the published outer circle floor and will not be entering the Class C airspace at any time, contact with Approach Control is unnecessary and shouldn't be made at all.

Services in the Class C airspace

Once two-way radio communication is established, Approach or Departure Control will give you vectors and approve altitude changes. Within the 10-mile radius, Approach will

- Sequence all arriving traffic into the airspace
- Maintain standard separation between IFR aircraft of 1000 feet vertically and 3 miles horizontally
- Separate IFR and known VFR aircraft through traffic advisories or a 500-foot vertical separation
- Issue advisories and, when necessary, safety alerts to VFR aircraft

Relative to the last point, ATC, by the book, is not required to separate VFR aircraft from other VFR aircraft in a Class C airspace; however, in traffic-intense areas, controllers might provide the service out of a sense of obligation and to help prevent the possible later necessity of issuing traffic or safety alerts. Regardless, ATC will advise VFR pilots of the bearing (usually by the "clock" position: "Traffic at 2 o'clock"), the approximate distance, and the altitude of other traffic that might pose a potential problem.

A *safety alert* is issued when conflict between two aircraft appears imminent. The pilot to whom the alert is directed would be wise to follow ATC's instructions immediately: no arguing, no questioning, no delaying, unless you are aware of a safer alternative. An emergency is in the offing. Can ATC refuse a pilot entry into the Class C airspace? Theoretically, no. In reality, though, controllers have occasionally been forced to deny entrance because of traffic saturation within the airspace and therefore have the prerogative of telling you, "Cessna 1234, remain outside the Class C (or Charlie) airspace and stand by." If you are told this, comply without argument. ATC is temporarily just too busy to accommodate any more activity.

Control compromise

Class C airport areas were designed to replace the TRSAs, in which pilot participation is strictly voluntary. The Class C area removes that option and thus increases the safety factor.

Has it met with popular acclaim? Not entirely. Pilots have been denied entry because of traffic saturation and ATC's resultant inability to sequence, separate, vector, and issue advisories. Early complaints have diminished, however, as pilots and controllers have mastered the system and polished local procedures. Perhaps it's not the ideal solution to traffic control in high-density areas, but it is certainly a better approach to safety than the looseness of the TRSA. In effect, Class C is a compromise, a balanced medium between the TRSA and the stricter controls of Class B.

Class B airspace: Structure, requirements, and regulations

Now, and with a bit of history, let's look at those chunks of airspace that were subjects of considerable controversy during the latter years of the 1980s: terminal control areas (TCAs). The source of controversy was not so much the existence of the TCAs but rather their design and the related FAA pilot and equipment proposals. With at least four midair collisions within or near a TCA and an increasing number of reported near misses, the combined public and congressional pressure on the FAA to take action was considerable.

In response, the FAA issued Notice of Proposed Rulemaking 88-2, which contained some 40 proposed airspace and rule changes. As hearings on 88-2 progressed, several of the changes were dropped and others modified as the result of strong general-aviation objections. The TCA resolution was thus the product of give and take by the FAA and those representing general-aviation interests, particularly the Aircraft Owners and Pilots Association (AOPA).

Why TCAs existed is best summed up by quoting FAA Handbook 7400.2, issued July 11, 1988:

> *The TCA program was developed to reduce the midair collision potential in the congested airspace surrounding an airport with high density air traffic by providing an area in which all aircraft will be subject to certain operating rules and equipment regulations. The TCA operating rules afford a level of protection that is appropriate for the large number of aircraft and people served by this type of airport. The TCA equipment requirements provide the air traffic control system with an increased capability to provide aircraft separa-*

tion service within the TCA. The criteria for considering a given terminal as a TCA candidate are based on factors which include the number of aircraft and people using the airspace, the traffic density, and the type or nature of operations being conducted.

Boiling down the factors used for TCA classification, the minimum criteria an airport must meet are at least 650,000 passengers enplaned annually, or the primary airport must have at least 150,000 annual instrument operations.

The current Class B airspaces

As of late 1998, the 31 cities and 35 airports listed in Table 20-1 are classified as Class B (Bravo) airspaces. The asterisks by certain locations indicate that special student VFR regulations exist, which I'll explain later in this chapter.

Identifying Class B airspace

If you refer to any sectional that has Bravo airspace, one of the first features that will catch your eye is a large blue-band rectangle or square surrounding the area— in this case (Fig. 20-8), the Seattle-Tacoma International Airport. The square represents the geographic area depicted on another chart, a *VFR terminal area chart* (TAC), which provides an enlarged and more detailed depiction of the area outlined

Table 20-1

Atlanta*	Kansas City	Philadelphia
Baltimore	Las Vegas	Phoenix
Boston*	Los Angeles	Pittsburgh
Charlotte	Memphis	St. Louis
Chicago*	Miami*	Salt Lake City
Cleveland	Minneapolis	San Diego
Dallas/Ft. Worth*	Newark*	San Francisco*
Denver	New Orleans	Seattle
Detroit	New York: La Guardia*	Tampa
Honolulu	John F. Kennedy*	Washington: Dulles, National*
Houston	Orlando	and Andrews Air Force Base*

Fig. 20-8. *The large band-box helps identify a Class B airspace and also represents the area covered by the exploded and more detailed terminal area chart (TAC).*

by the square. Anyone nearing or intending to enter a Class B airspace should definitely have a current issue of that airport's TAC on board.

The Bravo airspace lies within the square and is easily spotted by the series of blue circles surrounding the primary airport. These might be exclusively circles, but more likely than not, there will be some irregularities, so structured to accommodate traffic at other nearby airports and/or IFR operations at the primary field.

The sectional has another identifying feature: Outside the blue square is a small rectangular box with the name of the terminal on the top

Fig. 20-9. *The small rectangular box near the band—in blue on the original—further identifies Class B.*

line, plus recommendations that pilots use the VFR terminal area chart for "greater detail and clarity of information" (Fig. 20-9). And there is the always reliable A/FD as another source of data and reference.

So, identifying a Bravo airspace should be no problem. As with Class C, however, the sectional plan view does not give a clear picture of the Class B structure or configuration. The following might help those not familiar with that airspace.

Class B airspace structure

The typical description of a Class B airspace is that it resembles an upside-down wedding cake, with the core surrounding the primary airport from the surface to the airspace's ceiling. The various levels, layers, or shelves (they're all synonymous) then extend laterally from the core at 3- to 10-nautical-mile increments (typically), with each layer having a prescribed altitude floor. All levels, however, rise to the same common ceiling of the particular airspace.

Figure 20-10 illustrates the basic structure. This Class B rises from the surface to 8000 feet (80/Surface) at the core. Moving up one level, the floor of the next-higher level is 2000 feet MSL with the same 8000-foot MSL ceiling (80/20). Farther out, the altitudes are from 3000 to 8000 feet (80/30), and so on. Thus, when operating below any of the floors, you're out of the airspace and out of any positive control by ATC (unless, of course, you happen to be within a satellite airport's Class C or D airspace). It's another story, though, within the Bravo airspace itself. Here is where the rules and regulations enter the picture.

Pilot requirements

To land or take off at the 13 busiest Class B airports (those identified by asterisks in Table 20-1), the pilot must have at least a private license. Student pilots, however, are allowed to enter the Class B

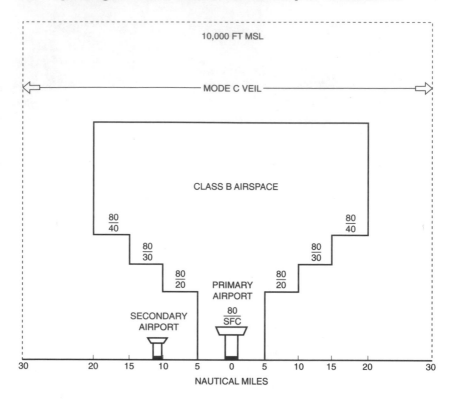

Fig. 20-10. *The structure of the basic Class B airspace is similar to this, but dimensions and configurations will vary considerably from one Class B to another.*

airspace surrounding those airports and may land on and depart from the primary airport in all other Class B airspaces, if an authorized flight instructor has endorsed the student's logbook within the past 90 days. The endorsement must certify that operating instructions within the specific Class B airspace have been given and that, as the result of actual flight in the area, the student has demonstrated competency to make a safe solo flight in the Class B environment.

Equipment requirements

Unless otherwise authorized by ATC, the aircraft must be equipped with an operable VOR or *tactical air navigation* (TACAN) receiver, an operable two-way radio capable of communicating with ATC on the terminal frequencies, and an operable transponder with altitude-reporting capabilities (Mode C or Mode S).

Exceptions to these regulations are aircraft not originally certificated with an engine-driven electrical system, balloons, and gliders, as long as they remain outside any Class A, Class B, or Class C airspace and operate below the ceiling of the airspace area or 10,000 feet MSL, whichever is lower.

The Mode C veil

Another feature of this Class B airspace is the *Mode C veil*. If you check any sectional that has a Class B airport, such as the Seattle example in Fig. 20-8, you'll see a thin blue circle at the very outer extremes of the airspace. This represents the 30-nautical-mile "veil" that surrounds the primary airport (Fig. 20-10). The initial FAA regulation, which was effective back in July 1989, stipulated that all aircraft operating anywhere within that 30-mile radius had to have a Mode C transponder and that it had to be turned to the ALT position. Whether you were going into the actual Bravo airspace, were flying above its ceiling (but below 10,000 feet MSL), or were just shooting touch-and-goes at a grass strip 25 miles out from the core of the airspace, it made no difference. A turned-on Mode C was mandatory.

As you might imagine, this edict created a storm of protest, particularly from those aircraft owners or renters who operated out of the many small and uncontrolled airports near the outer fringe of the 30-mile circle. Reacting to the protests, the FAA issued Special Federal Aviation Regulation (SFAR) No. 62, which you can find either at the beginning or toward the end of FAR Part 91. This SFAR modified the regulation by excluding from the Mode C requirement aircraft that were based at or operated into or out of those so-called fringe airports. At the same time, in conjunction with Part 91.215, it also excluded balloons, gliders, and any aircraft "...not originally certificated with an engine-driven electrical system or which has not subsequently been certified with such a system installed.... "

As part of the modification, however, the FAA stipulated that all non–Mode C aircraft were restricted to a 2-nautical-mile operating radius of the excluded airport or a 5-nautical-mile radius, if so directed by ATC. A second restriction stated that pilots wanting to leave the local airport area must do so by the most direct routing possible. *Direct,* in this case, means following as straight a line as possible out of the veil while avoiding natural obstacles, towns, or populated areas in the flight path that would cause pilots to violate the minimum safe altitudes stated in FAR 91.119. Said another way, once you go beyond the 2-nautical-mile airport radius, you must get out of the Mode C veil as rapidly as possible. There is no wandering around inside the veil or putting down at some other Mode C–excluded field that might be several miles away. By the same token, an aircraft intending to land at one of these Mode C–excluded airports must enter the veil at a point as close to the airport as possible and proceed directly to it—again, without wandering or delaying.

There's one other piece of this SFAR No. 62: the maximum operating altitudes for non–Mode C aircraft in the veil area. Listed in the SFAR under the various Class B airports are those many smaller airports that are excluded from the Mode C requirement and the maximum AGL altitudes at which these aircraft can fly within the veil area. The altitudes vary from one Class B to another, but they are consistent within each. For example, the ceiling for all the veil airports around the Atlanta Class B airspace is 1500 feet AGL; at Boston, it's 2500 feet; at Houston, 1200 feet, and so on. Across the country, the lowest ceiling is 1000 feet, and the highest is 2500 feet.

And still there is a bit more about SFAR No. 62. First, it really is a temporary ruling, because it was supposed to expire on December 30, 1993. In effect, it did expire officially, since nothing has been committed to writing to extend it. The Mode C exclusion still lives, though, and probably will remain alive but under a different SFAR number, name, FAR, or what-have-you. In fact, according to the FAA's Washington, D.C., office, a study of the subject is under way (as of early 1998) and a final ruling or decision should be forthcoming in the near future. (I'm not sure what *near future* means in this case, but I was given reasonable assurance that the exclusionary aspect of the SFAR would be continued.)

All told, approximately 300 of these are Mode C–excluded airports, so if you're planning a trip to one that is in the 30-mile radius, check the sectional first and then the *Aeronautical Information Manual.* AIM has a complete listing of the exclusions under the Bravo airspace in which the airport is located, along with the maximum operating altitude at that airport.

To summarize, and for reemphasis, the fact that an airport is exempt from the Mode C requirement does not mean that non–Mode C aircraft can operate inward toward the primary airport. They can land, take off, or practice touch-and-goes all day long, as long as they remain below the maximum altitude and stay within the 2- or 5-nautical-mile radius of the excluded airport. Otherwise, after takeoff, they must leave the veil as directly and expeditiously as possible. And the same principle applies to landing without Mode C: Enter the veil below the maximum SFAR No. 62 published altitude, proceed directly to the airport, join the traffic pattern, and land.

Yes, these regulations do limit the airspace for non–Mode C aircraft, especially near urban areas and high-density airports, but Mode C has a lot of advantages, not the least of which is the added safety it provides. Plus, it should keep pilots more alert to what they're doing and where they are.

Relative to that, if you accidentally or intentionally penetrate a Class B airspace without approval, you can be sure that someone on the ground knows it. Your position, combined with your Mode C altitude readout, gives you away. Oh, you might be able to outrun the radar coverage and get home unidentified, but the FAA is continually developing more sophisticated means of tracking airspace violators. And once you're caught, it could mean a license suspension, plus a black mark on your record. With Mode C in operation, though, you know you're probably being monitored, which alone ought to have a powerful effect on regulation abidance and attention to what you are doing.

Entering Class B airspace

If you want to land at a Class B primary airport or transit a portion of Bravo airspace to save time and fuel, one basic rule applies: You must be specifically and verbally cleared into the airspace by Approach Control before entering the airspace at any level.

Approach controllers stress this point again and again. What frequently happens is that the pilot calls Approach well outside the airspace. Approach responds. The pilot gives the aircraft position and makes his or her intentions known. Approach tells the pilot to squawk a certain transponder code and then to ident. The pilot squawks and idents. Approach comes back with "Radar contact." Now the problem: The pilot interprets "Radar contact" as approval to enter the Class B airspace—which isn't the case at all!

Perhaps a portion of the misunderstanding stems from *AIM,* where two definitions of radar contact appear in the Pilot/Controller Glossary, the second portion of which states:

The term (radar contact) used to inform the controller that the aircraft is identified and approval is granted for the aircraft to enter the receiving controller's airspace.

This quote refers to a controller-to-controller communication, not controller-to-pilot. Pilots reading the definition might interpret it to mean that they are cleared into the airspace as soon as a controller calls them and announces "Radar contact." Not so. Until you distinctly hear "Cleared into the Class B (or Bravo) airspace," "Cleared to enter the Class B airspace," or "Cleared through the Class B airspace," you must remain outside that controlled area. Circle, slow down, do a 180, do whatever you have to do. Just don't go where you haven't been invited. It's illegal and can result in your being fined, grounded, and cited for the violation. The offense will also be on your record for insurance companies to see.

But forget the punishment for the moment. Whether intentional or inadvertent, violating the airspace is just plain dangerous. You're in a controlled traffic environment with high-speed turbine aircraft buzzing around you in arrival, transiting, or departure postures. It's Approach Control's job to see that all aircraft are properly separated; but if you barge in unannounced or without clearance, you could be the source of all sorts of traffic problems, up to, and potentially including, the dreaded midair.

Admittedly, though, willful violation of a Bravo airspace is rare. The primary cause is lack of knowledge or lack of preparation. Legally entering this area under VFR is no trick, if you've done the necessary homework, including answering at least these questions:

- Do I have at least a private pilot license? If not, and I'm flying solo as a student, have I received the proper Class B airspace training and logbook endorsement(s)?
- Do I have a VOR receiver that works?
- What are the Bravo lateral and vertical limits?
- How far out and at what checkpoint should I make the initial call to Approach Control?
- What are the radio frequencies for ATIS, Approach, Departure, tower, and Ground Control? And is my radio capable of using those frequencies?
- How should the initial call be worded? What should I say and in what sequence?

- What am I going to do if I'm not immediately cleared into the airspace?
- What will I do if I haven't understood an instruction?
- Have I familiarized myself with the airport layout, including the runways and the upwind, downwind, and base-leg headings for each runway?
- What action will I take if I can't get into the airspace at all? Will I have enough fuel to go around it or get to an alternate airport?
- Is my transponder working? Am I familiar with the terminology associated with its use? Has it been turned to the ALT position?

Don't wait until you can taste the icing on the proverbial Bravo cake. If you want to enter the airspace, give yourself plenty of time to make contact and receive the controller's instructions. Approach controllers recommend an absolute minimum of 10 miles from the airspace shelf you're planning to enter. Better still, make it 20 or 30 miles out. Approach's radar will still easily pick you up. Even at a relatively modest 110 knots of airspeed, it won't take you long to eat up those few miles.

How should this initial call go, and what could you expect to hear? Perhaps something like this:

You: Kansas City Approach, Cherokee Eight Five One Five November.

App: Cherokee Eight Five One Five November, Kansas City Approach, go ahead.

You: Approach, Cherokee Eight Five One Five November is over Lake Perry at five thousand five hundred, landing International with India.

App: Cherokee One Five November, squawk zero two six three and ident.

You: Roger, zero two six three. Cherokee One Five November. (Now push the Ident button just once.)

In a moment, Approach might come back with

App: Cherokee One Five November, radar contact.

Are you cleared into the Bravo airspace? Again, no. Wait until you hear

> **App:** *Cherokee One Five November, cleared into the Bravo airspace.*

Included in this message might be instructions on headings to turn toward, altitude changes, or advisories of other aircraft — or these might come later. Whatever the case, from this point on, Approach Control will guide you into and within the Class B airspace until you near the tower controlled area. At that point, Approach will tell you to contact the tower for landing instructions.

Just one more thing: Do you have to make that first call to Approach before you enter the 30-mile Mode C veil? No. You're not in the Bravo airspace yet. As long as you have Mode C and it's switched to the ALT position, you're perfectly legal. Depending on your altitude, though, a call 30 miles or so out is about right because you could be as close as 10 miles (or maybe closer) to one of the higher Class B shelves — and those 10 miles can evaporate quickly before clearance into the airspace is approved. Give yourself plenty of time so that you won't have to involve yourself in aerial gyrations while you go through the contact, squawk, ident, and entrance approval procedures.

Departing a Class B airport

With only a couple of exceptions, departing a Class B primary airport is the reverse process and the same as leaving a Class C.

First, monitor ATIS.

Second, call clearance delivery, stating your intentions. You'll receive initial departure headings, altitude(s), squawk code, and Departure Control frequency.

Third, call Ground Control for taxi clearance.

Fourth, call the tower for takeoff permission.

Fifth, contact Departure when the tower so advises.

Sixth, follow Departure's instructions until you are clear of the airspace and Departure tells you to contact Center on such and such frequency or advises you that "radar service is terminated," that you are free to dial in whatever frequency you wish, and to "squawk one two zero zero." Or the controller might say, "Squawk VFR," which means the same thing.

Alternately, you might hear this:

> **Dep:** *Cherokee One Five November, two seven miles north of International. Resume own navigation. Remain this frequency for advisories.*

What's Departure telling you? First, that you've left the lateral or vertical limits of the Bravo airspace and are free to change headings or altitudes as you wish. Next, the controller is saying, without saying it, that you're still in Departure's airspace and will be for another 15 miles or so. The controller thus wants you to continue to monitor that frequency, to "stay with me," in the event that advisories of other traffic in your line of flight become necessary.

Finally, when you're clear of Departure's area of responsibility, if the controller hasn't automatically handed you off to Center, he or she will call you once more and tell you to change to whatever radio frequency you wish. It will be a short call, no more than this:

> **Dep:** *Cherokee One Five November, radar service terminated. Squawk one two zero zero. Frequency change approved. Good Day.*

Now you can tune to another tower, a Flight Service Station, call Center to request VFR advisories, or, presumably, but not wisely, turn the radio off completely. The choice at this point is yours, as Departure will not provide traffic advisories once radar service is terminated.

Transiting a Class B airspace

The procedures for transiting a Class B airspace are the same as those for entering for landing. There's nothing new to be said, other than the fact that you will be with Approach or Departure Control the entire time.

Whether you'll receive transiting clearance will depend on the volume of traffic in the airspace at the time. Every controller I talked with, however, said he would gladly help a VFR pilot get through the area, workload permitting, but it might not be a straight line. For reasons of IFR Approach and Departure procedures and the runway(s) in use, you might have to be routed a few miles out of the way. Or you could be asked to change altitudes. Regardless of the deviations, however, your journey will be shorter than if you avoided the airspace altogether because of procedural uncertainty or fear of entering a territory presumably reserved for the "big" boys. If the airspace is crowded, Approach will tell you. Then circumnavigation may be your only choice. Otherwise, you'll get all the help you need. So saith those responsible for traffic control in this airspace.

Class B VFR routes

Despite—or maybe in tune with—those comments about transiting a Bravo airspace, there are a couple of alternatives at some, but not all, Class B airports. Where the alternatives do exist, they fall under the general heading of published VFR routes. As stated in *AIM,* "Published VFR Routes for transitioning (sic) around, under, and through complex Class B airspaces were developed through a number of FAA and industry initiatives." *AIM* goes on to say, "All of the following terms, i.e., 'VFR Flyway,' 'VFR Corridor,' 'Class B Airspace Transition Route,' and 'Terminal Area VFR Route,' have been used when referring to some of the different types of routes or airspace." As of the date of writing, the Bravo airports that offer one or more of these VFR routes are as follows:

Andrews Air Force Base (Washington, D.C.)

Atlanta

Baltimore

Charlotte

Dallas/Ft. Worth

Denver

Detroit

Dulles (Washington, D.C.)

Houston

Los Angeles

Miami

National Airport (Washington, D.C.)

O'Hare (Chicago)

Phoenix

St. Louis

San Diego

Seattle

By way of clarification, the following summarizes the basic features of the two more common routes and when ATC clearance is or is not required.

- VFR flyways. Again per *AIM,* a flyway is a general path for use by pilots planning flights into, out of, or near a complex terminal to avoid the immediate Class B airport airspace. This really means following paths that allow you to skirt the

Bravo airport airspace that rises from the surface to its published ceiling—typically but not universally 8000 feet. These are not VFR pathways around the whole Class B airspace (you can do that any time without using flyways). Instead, they are routes to help VFR pilots avoid the areas of major traffic flows—meaning, of course, those surface-to-ceiling airspaces that surround the Bravo airports.

Where these flyways exist, they are usually depicted on the reverse side of the terminal area chart (TAC) of the airport in question. Take Seattle, for example. Figure 20-11 illustrates the flyways—the bands on all sides of the terminal area with the altitude limitations, such as "Below 3000" and "Below 5000" (lower left). These bands appear in blue on the actual TAC, as does the statement of the purpose of the VFR flyways in Fig. 20-12. Clearance by ATC to enter a flyway is not required, but pilots should set their transponders on 1201 while on the flyway. Before using one of these shortcuts at a Bravo airport, check the local operating regulations. They can vary from one locale to another. Los Angeles, for instance, limits the airspeed to 140 knots, requires that navigation lights be on, and excludes all turbine-powered aircraft from the routes. Other airports could have different regulations, so know what's required before you venture forth.

- Class B VFR transition routes. (See Fig. 20-11 and note the double-pointed arrows that cross the airport area east to west.) These routes, which are depicted on the face of the TAC chart as well as the reverse side (but not on the sectional), make it possible for VFR aircraft to be cleared through the Class B airport airspace at certain airports. On both sides of the TAC, the routes appear as double-pointed arrows in magenta (Fig. 20-11 again). Their purpose, as *AIM* puts it, is this: "A Class B Airspace Transition Route is defined as a specific flight course depicted on a Terminal Area Chart (TAC) for transiting a specific Class B airspace. These routes include specific ATC-assigned altitudes, and pilots must obtain an ATC clearance prior to entering Class B airspace on this route." (*AIM* is quite "specific" here, isn't it?) Additional instructions (Fig. 20-13) are included on the TAC's reverse side, plus those in boxes at either end of the route (Fig. 20-11).

Note that this route takes the pilot directly over the Class B airport at 2500 feet or lower. The flyway, on the other hand, directs the pilot away from any part of the Bravo airspace.

Fig. 20-11. *This chart illustrates the VFR flyways and transition routes in the Seattle Class B airspace.*

THIS CHART IDENTIFIES VFR FLYWAYS DESIGNED TO HELP VFR PILOTS AVOID MAJOR CONTROLLED TRAFFIC FLOWS. IT DEPICTS MULTIPLE VFR ROUTINGS THROUGHOUT THE SEATTLE AREA WHICH MAY BE USED AS ALTERNATES TO FLIGHT WITHIN THE ESTABLISHED CLASS B AIRSPACE. ITS GROUND REFERENCES PROVIDE A GUIDE FOR IMPROVED VISUAL NAVIGATION. THIS IS NOT INTENDED TO DISCOURAGE REQUESTS FOR VFR OPERATIONS WITHIN THE CLASS B AIRSPACE BUT IS DESIGNED SOLELY FOR INFORMATION AND PLANNING PURPOSES.

Fig. 20-12. *Another excerpt from the Seattle terminal area chart states the purpose of the VFR flyways.*

Because of their rarity, or even nonexistence, I'm intentionally skipping a discussion of the uncontrolled VFR corridors and the terminal area VFR routes. In essence, a *corridor* is the equivalent of a hole, or series of holes, drilled in one or more levels of the Bravo airspace. No ATC clearance or communication with ATC is required. The terminal

VFR TRANSITION ROUTES

THIS CHART ALSO IDENTIFIES VFR TRANSITION ROUTES IN THE SEATTLE CLASS B AIRSPACE. OPERATION ON THESE ROUTES REQUIRES ATC AUTHORIZATION FROM SEATTLE TOWER. UNTIL AUTHORIZATION IS RECEIVED, REMAIN OUTSIDE CLASS B AIRSPACE. DEPICTION OF THESE ROUTES IS TO ASSIST PILOTS IN POSITIONING THE AIRCRAFT IN AN AREA OUTSIDE THE CLASS B AIRSPACE WHERE ATC CLEARANCE CAN NORMALLY BE EXPECTED WITH MINIMAL OR NO DELAY. ON INITIAL CONTACT, ADVISE ATC OF POSITION, ALTITUDE, ROUTE NAME DESIRED, AND DIRECTION OF FLIGHT. REFER TO CURRENT SEATTLE VFR TERMINAL AREA CHART FOR USER REQUIREMENTS.

Fig. 20-13. *Additional instructions to those in AIM outline the operating regulations in VFR Class B transition routes.*

area routes, on the other hand, are based on reference to prominent surface features and/or electronic navigation aids. For anyone interested in more details, *AIM* has brief descriptions of both.

Remaining VFR

Whatever the reason for entering a Class B airspace, assuming you're VFR, you have the responsibility to adhere to VFR regulations. If clouds are around, Approach might give you vectors or altitude changes that would cause you to penetrate a cloud or cloud bank, all the while advising you to "remain VFR." If that happens, it's your responsibility to advise the controller accordingly. Yes, you're under ATC, but that doesn't authorize you to violate the VFR regulations.

Remember, though, that the standard VFR cloud separation of 500 feet below, 1000 above, and 2000 horizontally does not apply in the Class B airspace. Your only requirement is to remain clear of clouds, which, you recall, should reduce cloud avoidance altitude or direction changes and thus minimize disruptions to the flow of other traffic in the area. But do tell the controller immediately if a given instruction would cause you to violate that regulation. Your safety and the safety of others, plus your legality, could be at stake.

Related to the VFR issue is the possibility, in either Class B or Class C, that ATC will assign you a non-VFR altitude, perhaps 4000 feet going eastbound or 5500 feet westbound. That's ATCOs prerogative while you're in one of the controlled airspaces; however, when you hear "Resume own navigation and altitudes. Maintain VFR," it means that you are to climb or descend to the appropriate VFR altitude.

Know before you go

Class B airspaces do scare off the ill-prepared or the unknowledge-able, which is understandable. If you're not ready to cope with the airspace, it's better to be scared off than to drive into the area, re-plete with unjustified self-confidence. The consequences, to put it mildly, could be serious.

Do pilots do this? Unfortunately, yes. In one example noted by an Approach controller, a Cessna 210 departed a satellite airport under-neath, but not in, a Bravo airspace. The aircraft showed up on Ap-proach's radar just east of the primary airport and well within the airspace. It had never contacted Approach and was not Mode C–equipped; thus, Approach had no idea at what altitude it was flying.

Meanwhile, the controller had vectored a descending Boeing 727 toward its downwind leg, and as the controller put it, "For some reason—I don't know why—I told the 727 to level off at 5000 feet, advising the pilot that I had traffic out there, wasn't talking to it, and didn't know its altitude. The 727 captain came back in a second, advising that the traffic was dead ahead at 4500 feet."

Intuition? Premonition? The controller didn't know. But the con-troller did know that if the 727 had been allowed to continue its de-scent, the likelihood of a 727-210 conflict was almost inevitable. To track down the 210, the controller called the satellite airport tower, explained what had happened, and wanted to know the aircraft's N number. The tower had the information, and, by chance, the 210 pi-lot just then reestablished radio contact with the satellite tower. The end result was that the pilot of this errant aircraft had a few not so pleasant chats with FAA personnel.

Yes, some pilots do barge in where only the prepared should tread.

On the other hand, do controllers really want to play cops and rob-bers? Perhaps a few, but of those I interviewed and talked with, many of whom were pilots themselves, not one wanted to nail a pi-lot for a really minor infraction. Playing the bad guy is not in the makeup of most of those folks who are dedicated to serving the fly-ing public, but with safety the overriding consideration.

Don't let these comments give the wrong impression. The FAA is very firm about violations. While the controller has the option of infor-mally warning an offender of a minor violation, the decision to issue a written warning is up to supervision. If a supervisor who is either monitoring the position or is made aware of what happened and subsequently so concludes, the controller has no alternative but to write up, and thus formalize, the incident. The fact is: If you are guilty of an offense, be prepared for the consequences, minor or major.

I say again, though, that most controllers want to establish a collaborative relationship with the pilot population, while, in the process, hoping to bury any remnant of what at one time might have been a dictatorial, police-type reputation. Simply said, they are not sadists who enjoy hurting the very people they exist to serve.

The TRSA—A few words about the terminal RADAR service area

This element of the original terminal radar program is a peculiar mix of airspaces with some rather unique features.

1. The TRSA has never been a controlled airspace from a regulatory aspect.

2. It has no counterpart in the ICAO airspace system and is thus identified by a name rather than a letter.

3. Because of its level of IFR operations and passenger enplanements, the overall airspace lies between a Class C and a Class D, but in most cases, the airports themselves are Class D. If you look closely at Fig. 20-14, you'll see the segmented circle (blue on the sectional) surrounding the airport itself, thus identifying the airport and the 4.3-nautical-mile radius around it as a Class D.

4. The TRSA provides radar approach control service to IFR and VFR aircraft, but VFR traffic has the option of using or not using the service.

5. There are only a relatively few TRSAs around, and sooner or later all will probably either move up to a Class C or drop down to a Class D tower-only airport. In essence, you might view the TRSA as a Class D towered airport surrounded by circles or segments of partially controlled airspaces.

Despite their relative rarity and their peculiarities, however, I can't overlook their existence or the procedures involved in entering, departing, or transiting one of these neither-nor entities.

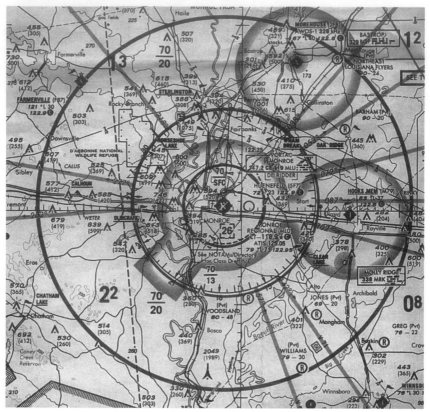

Fig. 20-14. *A typical TRSA resembles Class B in general appearance and has some physical characteristics of Class C, but its operating requirements really place it between Class C and Class D.*

Identifying TRSAs

This is easy to do. They're spotted on the sectional chart by the black bands (they look almost gray) that surround the primary airport. The bands at the Monroe, Louisiana, TRSA (Fig. 20-14) reflect a perfect circle pattern, which is generally, but not necessarily, the typical design. Some TRSAs, for instance, have odd-shaped extensions for IFR aircraft protection; others have cutouts; still others have three or four bands with varying floor levels; and there are those that have only one perfectly circular ring or band. So design uniformity is not a TRSA trademark.

Note, too, in Fig. 20-14, the ceilings and floors of the rings: in the outermost ring and in the next ring. But then we come to the Class D airspace, again identified by the segmented circle and the segmented box enclosing the 26, indicating that the Class D airspace extends from the surface to 2600 feet MSL. Along with what the sectional tells you, more information about a TRSA is available in the *A/FD*. Figure 20-15 illustrates the Monroe Regional Airport data and, as per the arrow, how the type of airspace is identified.

TRSA services provided

First, an important point for VFR pilots: Specific clearance into a TRSA by Approach Control is not required. If, for whatever reason, you don't want to use the available radar services, you can refuse them in your initial call to Approach, or, if departing, to Ground Control, by merely stating, "Negative TRSA (pronounced "tersa") service." A question arises, however, as to why someone would not take advantage of the assistance ATC is ready to provide. Personally, I've always found it rather comforting to have help through a perhaps busy terminal environment. But should you choose to refuse it, just continue on your inbound course until you're within 10 miles or

---

MONROE REGIONAL (MLU) 3 E UTC−6(−5DT) N32°30.65' W92°02.26'   **MEMPHIS**
  79 B S4 **FUEL** 100LL, JET A OX 1 ARFF Index C   **H−4G, 5C, L−17C**
RWY 04−22: H7507X150 (ASPH−CONC−GRVD) S−75, D−170, DT−290 HIRL   **IAP**
  RWY 04: MALSR. **RWY 22:** MALSR. VASI(V4L)—GA 3.0° TCH 52'.
RWY 18−36: H5001X150 (CONC) S−60, D−75, DT−130
  RWY 18: Trees. **RWY 36:** Trees.
RWY 14−32: H5000X150 (ASPH) S−75, D−170, DT−290 MIRL
  RWY 14: VASI(V4L)—GA 3.0° TCH 70'. Thld dspicd 301'. Trees.   **RWY 32:** REIL. VASI(V4L)—GA 3.0° TCH 48'.
RUNWAY DECLARED DISTANCE INFORMATION

RWY 04:	TORA−7507	TODA−7507	ASDA−7507	LDA−7507
RWY 22:	TORA−7507	TODA−7507	ASDA−7507	LDA−7507
RWY 14:	TORA−5000	TODA−5000	ASDA−5000	LDA−4699
RWY 32:	TORA−5000	TODA−5000	ASDA−5000	LDA−5000
RWY 18:	TORA−5001	TODA−5001	ASDA−5001	LDA−5001
RWY 36:	TORA−5001	TODA−5001	ASDA−5001	LDA−5001

AIRPORT REMARKS: Attended continuously. Rwy 18−36 bumpy with vegetation growing thru cracks. Rwy 18−36 not avbl for air carrier ops with over 30 passenger seats. Taxlway A clsd to acft over 12,500 lbs E of Rwy 18 to Taxiway C. Taxiway E clsd to aircraft with wing span over 90' East of Rwy 04−22. When twr closed HIRL Rwy 04−22 preset med ints. ALS Rwy 04 and Rwy 22 unmonitored. ACTIVATE ALS Rwy 04 and Rwy 22—CTAF. Rwy 04−22 NE 1500' CONC wire combed; SW 6000' grvd ASPH. NOTE: See Land and Hold Short Operations Section.
WEATHER DATA SOURCES: LLWAS.
COMMUNICATIONS: CTAF 118.9 **ATIS** 125.05 (1145−0500Z‡) **UNICOM** 122.95
  DE RIDDER FSS (DRI) TF 1−800−WX−BRIEF. NOTAM FILE MLU.
  RCO 122.25 (DE RIDDER FSS)
Ⓡ **APP/DEP CON** 126.9 (180°−359°) 118.15 (360°−179°) (1145−0500Z‡)
Ⓡ **FORT WORTH CENTER APP/DEP CON** 135.1 (0500−1145Z‡)
  TOWER 118.9 (1145−0500Z‡) **GND CON** 121.9 **CLNC DEL** 121.65
AIRSPACE: CLASS D svc 1145−0500Z‡ other times CLASS E.
➤ **TRSA** svc ctc APP CON within 25 NM below 7000'.
RADIO AIDS TO NAVIGATION: NOTAM FILE MLU.
  (L) VORTACW 117.2 MLU Chan 119 N32°31.01' W92°02.16' at fld. 80/3E. **HIWAS.**
  SABAR NDB (LOM) 219 ML N32°27.25' W92°06.25' 042° 4.8 NM to fld. Unmonitored when tower closed.
  ILS 109.5 I−MLU Rwy 04. LOM SABAR NDB Unmonitored when tower closed.
  ILS 109.5 I−MZR Rwy 22. Unmonitored when tower closed.
  ASR (1145−0500Z‡)

Fig. 20-15. *Further identification and information about a TRSA is, as always, in the A/FD.*

so of the airport Class D airspace and then contact the tower for landing instructions. Meanwhile, though, it would be smart to keep monitoring the Approach frequency to learn what you can about other traffic in the area, where it is, its altitude, and its intentions.

However, if you have decided that you do want the service, make that known in your initial call. For example, when arriving, and following the basic IPAI/DS, all you need to say is this:

> *Monroe (or Regional) Approach, Cherokee One Four Six One Tango over Calhoun at five thousand five hundred, landing Monroe, squawking one two zero zero (or "squawking twelve" or "squawking VFR") with Echo. Request radar traffic information.*

Approach will now provide the TRSA services until you near the control tower's airspace, at which point Approach will tell you to contact the tower on its assigned frequency. From then on, the tower controller takes over until you're safely down and clear of the runway. Ground Control enters the picture at that point for taxi clearance to wherever you want to park.

And when you're departing, the call goes like this:

> *Monroe Clearance, Cherokee One Four Six One Tango at Legacy Aviation, VFR west to Shreveport via Victor 94 at four thousand five hundred with Echo. Request radar traffic information.*

Once you are at or just beyond the TRSA's outer limits, Departure will normally, but not always, hand you off to Center for flight following and will advise you accordingly.

That's all there is to it. If you don't want the TRSA service, merely substitute the word *negative* for *request*.

Thus, when you are departing, the sequence of contacts or calls is the same as that in a Class B or C:

1. ATIS
2. Clearance Delivery
3. Ground Control
4. Tower
5. Departure Control

6. Center (if handed off or you initiate the request for en route traffic advisories)

The TRSA service is designed to provide separation between all IFR and all participating VFR aircraft. VFR aircraft will be separated from other VFR/IFRs by 500 feet vertically, or by visual separation, or by target resolution, meaning that space is maintained on the radar screen between targets when the broadband radar system is used. In other words, there is "green between" the targets on the screen.

AIM, under the heading of "Terminal Radar Services for VFR Aircraft," has considerably more details on the subject than are given in this brief overview. Consequently, should you be planning to fly into a TRSA, you would do well to consult that resource for further details.

21

Terminal radar approach and departure control facilities

Picture any major terminal—say, an Atlanta, a Chicago, or New York's Kennedy Airport—with only the basic glass-enclosed room, or cab, that sits atop the control tower structure. Then picture the scene if three or four controllers in that cab were solely responsible for separating and sequencing all the arriving, departing, and transiting traffic within a 35- to 45-mile radius of the airport. Conservatively stated, the scene at best would be chaotic.

Enter, then, the role of Approach and Departure Control—the function that does the separating, sequencing, vectoring, and advising of all aircraft within Class B and C airspaces, except for the approximately 5-mile radius of airspace surrounding the primary airport and any satellite fields in the area. Approach Control (for simplicity's sake, I'll refer to the function as Approach), with its computers and radar equipment, is the bridge between the outside world and the tower for all arriving and departing aircraft. In essence, it does what no busy tower could possibly do all by itself with any degree of order or safety.

Where is the facility located and who operates it?

First, despite the basic sameness of the radar service, the overall approach and departure function has five different names or titles, depending on which government agency or service has the operating responsibility. More specifically, their names, their acronyms, and the responsible agencies are as follows:

- Terminal Radar Approach Control (TRACON), FAA
- Radar Approach Control (RAPCON), FAA/Air Force
- Radar Air Traffic Control Facility (RATCF), Navy/FAA

- Army Radar Approach Control (ARAC), Army
- Air Traffic Control Tower (ATCT), only the towers to which Approach Control authority has been delegated

As to the physical location of the facilities (by that, I mean where the facility is located on the airport itself), five possibilities exist:

- It's located in the control tower structure but in a darkened room on a floor somewhere below the tower cab itself.
- It could be separate from the tower structure, but is still on the airport property and provides approach service to the airport as though it were located in the tower building.
- The function is not physically on the airport but rather at a larger nearby terminal.
- The service might be provided by the ARTCC in whose area the airport is located.
- It could be provided by an ARTCC down to the minimum altitude at which its radar can capture the image of the aircraft. Lower than that altitude, the controller uses manual, nonradar procedures to separate and sequence IFR traffic.

With these five possibilities, how do you know what airports have what? The *Airport/Facility Directory* is the best determining source (Figs. 21-1 through 21-5).

SANTA BARBARA MUNI (SBA)   7 W   UTC−8(−7DT)   N34°25.57′ W119°50.42′   **LOS ANGELES**
    10   B   S4   FUEL 100LL, JET A   OX 1, 2, 3, 4   TPA—See Remarks   LRA   ARFF Index C   **H−2A, L−3B**
    RWY 07−25: H6052X150 (ASPH−PFC)   S−110, D−160, DT−245   HIRL   IAP
      RWY 07: MALSR. Tree. Rgt tfc.   RWY 25: REIL. VASI(V4L)—GA 3.0° TCH 46′. Thld dsplcd 314′. Fence.
    RWY 15R−33L: H4183X100 (ASPH)   S−48, D−63, DT−100   MIRL
      RWY 15R: REIL. Tree.   RWY 33L: Road. Rgt tfc.
    RWY 15L−33R: H4179X75 (ASPH)   S−35, D−41, DT−63
      RWY 15L: Thld dsplcd 217′. Tree.   RWY 33R: Rgt tfc.
    AIRPORT REMARKS: Attended 1330−0600Z‡. Fee for fuel after hours call 805−964−6733 or 967−5608. Numerous
      flocks of birds on and in vicinity of arpt. Arpt has noise abatement procedures ctc arpt manager 805−967−7111.
      Due to limited ramp space at the airline terminal non−scheduled transport category acft with more than 30
      seats are required to ctc arpt manager 24 hour PPR to arrival. TPA—1000(990) small acft, 1500(1490) large
      acft. Pure jet touch/go or low approaches prohibited. When twr clsd ACTIVATE MIRL Rwy 15R−33L CTAF. NOTE:
      See Land and Hold Short Operations Section.
    WEATHER DATA SOURCES: LAWRS.
    COMMUNICATIONS: CTAF 119.7   ATIS 132.65 (805) 967−0283 (1400−0700Z‡)   UNICOM 122.95
    HAWTHORNE FSS (HHR) TF 1−800−WX−BRIEF. NOTAM FILE SBA.
→ Ⓡ APP/DEP CON 125.4 (330°−150°) 120.55 (151°−329°) (1400−0700Z‡)
→ Ⓡ L.A. CENTER APP/DEP CON 119.05 (0700−1400Z‡)
    TOWER 119.7 (1400−0700Z‡)   CLNC DEL 132.9   GND CON 121.7
    AIRSPACE: CLASS C svc 1400−0700Z‡ ctc APP CON other times CLASS E.
    RADIO AIDS TO NAVIGATION: NOTAM FILE HHR. VHF/DF ctc FSS.
    SAN MARCUS (H) VORTAC 114.9   RZS   Chan 96   N34°30.57′ W119°46.26′   201° 6.1 NM to fld.
      3620/14E. · HIWAS.
    GAVIOTA (L) VORTACW 113.8   GVO   Chan 85   N34°31.88′ W120°05.47′   101° 13.9 NM to fld. 2620/16E.
    ILS/DME 110.3   I−SBA   Chan 40   Rwy 07.   Unmonitored when twr clsd.

Fig. 21-1. *The symbol* Ⓡ *means that the Santa Barbara Approach/ Departure Control facility is located on the airport. When closed, Los Angeles Center Approach/Departure Control provides the service.*

PANAMA CITY–BAY CO INTL   (PFN)   3 NW   UTC–6(–5DT)   N30°12.73' W85°40.97'   **NEW ORLEANS**
    21   B   S4   FUEL 100LL, JET A   OX 1, 2   LRA   ARFF Index B   **H–5D, L–18F, 19A**
RWY 14–32: H6308X150 (ASPH–GRVD)   S–100, D–174, DT–300   HIRL   **IAP**
    RWY 14: MALSR.   **RWY 32:** REIL. VASI(V4L)—GA 3.0° TCH 50'. Trees.
RWY 05–23: H4884X150 (ASPH)   S–40, D–70, DT–120   MIRL
    RWY 05: VASI(V4L). Trees.   **RWY 23:** VASI(V4L)—GA 3.0° TCH 39'. Trees.
AIRPORT REMARKS: Attended continuously. Banner towing below 500' adjacent to rwys and coastline. ARFF equipment
    and personnel meet FAR 139 index C, call (904) 769–4791/6033. Extensive Helicopter operations from ramp.
    Acft arriving/departing S.E.–N.W. use caution due to intensive military jets transiting arpt tfc area 1500' and
    above on apch to Tyndall AFB. ACTIVATE HIRL Rwy 14–32; MIRL Rwy 05–23; MALSR Rwy 14; REIL Rwy 32 and
    taxiway lgts—CTAF. Flight Notification Service (ADCUS) avbl. NOTE: See Land and Hold Short Operations
    Section.
WEATHER DATA SOURCES: LAWRS.
COMMUNICATIONS: CTAF 120.5   **ATIS** 128.3 (1200–0400Z‡)   **UNICOM** 122.95
    GAINESVILLE FSS (GNV) TF 1–800–WX–BRIEF. NOTAM FILE PFN.
    RCO 122.1R 114.3T (GAINESVILLE FSS)
►(R) **TYNDALL APP/DEP CON** 119.1 (blo 5000') 119.75 (above 5000') (1300–0500Z‡)
►   **JAX CENTER APP/DEP CON** 124.77 (0500–1300Z‡)
    TOWER 120.5 FCT (1200–0400Z‡)   **GND CON** 121.6
AIRSPACE: CLASS D svc 1200–0400Z‡ other times CLASS G.
RADIO AIDS TO NAVIGATION: NOTAM FILE PFN.
    (L) VORTAC 114.3   PFN   Chan 90   N30°12.98' W85°40.86'   at fld. 10/00W.
    LYNNE NDB (LOM) 278   PF   N30°19.60' W85°46.94'   143° 8.6 NM to fld.
    ILS 110.5   I–PFN   Rwy 14.   BC unusable. LOM LYNNE NDB. (ILS unmonitored when twr closed).

Fig. 21-2. *This is an example of a RAPCON, where Tyndall Air Force Base provides service to the civilian Panama City–Bay County airport.*

- -

FULTON CO ARPT–BROWN FLD   (FTY)   6 W   UTC–5(–4DT)   N33°46.75' W84°31.28'   **ATLANTA**
    841   B   S4   FUEL 100LL. JET A,   OX 1, 2, 3, 4   LRA   **H–4H, 6F, L–20E, A**
RWY 08–26: H5796X100 (ASPH–GRVD)   S–105, D–121, DT–198   HIRL   **IAP**
    RWY 08: MALSR. Trees.   **RWY 26:** REIL. VASI(V4L)—GA 3.0° TCH 52'. Trees. Rgt tfc.
RWY 14–32: H4158X100 (ASPH)   S–30   MIRL
    RWY 14: REIL. VASI(V2L)—GA 4.0° TCH 52'. Trees.   **RWY 32:** REIL. Thld dspicd 200'. Towers.
RWY 09–27: H2801X60 (ASPH)   S–35 D–45 DT–72
    RWY 09: Trees.   **RWY 27:** Trees. Rgt tfc.
AIRPORT REMARKS: Attended continuously. Rwy 32 has three lgtd twr on centerline 32 ft AGL (873 ft MSL) 650 ft from
    thld. When twr clsd ACTIVATE MALSR Rwy 08   CTAF; HIRL Rwy 00–20 preset step 3 ints only. Flocks of birds on
    and invof arpt during dalgt hrs. Rwy 14–32 CLOSED 0200–1100Z‡ indef. Noise sensitive area all quadrants; no
    run ups authorized on any ramp. Flight Notification Service (ADCUS) avbl. NOTE: See Land and Hold Short
    Operations Section.
WEATHER DATA SOURCES: LAWRS.
COMMUNICATIONS: CTAF 118.5   **ATIS** 120.175 (0800–0759Z‡)   **UNICOM** 122.95
    MACON FSS (MCN) TF 1–800–WX–BRIEF. NOTAM FILE FTY.
    ATLANTA RCO 122.6 122.2 (MACON FSS)
►(R) **ATLANTA APP/DEP CON** 121.0
    COUNTY TOWER 118.5 120.7 (0800–0759Z‡)   **GND CON** 121.7   **CLNC DEL** 123.7
AIRSPACE: CLASS D svc 0800–0759Z‡ other times CLASS E.
RADIO AIDS TO NAVIGATION: NOTAM FILE PDK.
    PEACHTREE (L) VOR/DME 116.6   PDK   Chan 113   N33°52.54' W84°17.93'   245° 12.5 NM to fld. 970/02W.
    FLANC NDB (MHW/LOM) 344   FT   N33°45.74' W84°38.34'   082° 6 NM to fld. NOTAM FILE FTY.
    NDB unusable byd 12 NM.
    ILS 109.1   I–FTY   Rwy 08.   LOM FLANC NDB. (G.S. unusable when twr not operating). LOC unusable byd
    15 NM blo 3500 ft MSL.

Fig. 21-3. *The Atlanta TRACON provides radar Approach/Departure service to the Fulton County airport, which is about 15 miles from Atlanta's Hartsfield Airport.*

Do the TRACON controllers have any view of the outside world as they fulfill their responsibilities? No. Their job is to control all traffic in their airspace (beyond the roughly 5-mile radius around the airport itself) by radar and radar alone. The room, or work area, is dark, with only the radarscopes and lighted computer buttons providing most of the illumination.

WILLIAMSPORT REGIONAL (IPT)   4 E   UTC−5(−4DT)   N41°14.52′ W76°55.31′   **NEW YORK**
    529   B   S4   **FUEL** 100LL, JET A   OX 1   ARFF Index A   **H−31, 61, L−24G**
    RWY 09−27: H6449X150 (ASPH−GRVD)   S−65, D−100, DT−190   HIRL   **IAP**
      RWY 09: REIL. VASI(V4L)—GA 3.0°TCH 58′. (Unmonitored). Trees.   **RWY 27:** MALSR. Railroad. Rgt tfc.
    RWY 12−30: H4280X150 (ASPH)   S−200, D−200, DT−400   MIRL
      RWY 12: Trees.   **RWY 30:** Brush. Rgt tfc.
    RWY 15−33: H3502X100 (ASPH)   S−13
      RWY 15: Thld dspicd 1175′. Trees.   **RWY 33:** Brush. Rgt tfc.
    AIRPORT REMARKS: Attended 1100−0400Z‡. Fuel and svcs available 1100−0400Z‡ Mon−Fri; 1100−0300Z‡ Sat−Sun
      and holidays; after hours call 717−323−1717. Arpt CLOSED to ultralight and banner towing ops. Rwy 15−33
      CLOSED to acft with seating capacity in excess of 30 passenger seats. Twy J CLOSED to acft with wing span
      over 78 ft. PPR 12 hours for unscheduled air carrier ops with more than 30 passenger seats 0500−1100Z‡ call
      arpt manager 717−368−2444 or 717−368−2446. Weather report not avbl 0500−1100Z‡. Rwy 9, VASI skewed
      10°to N of Rwy 9 centerline & originating 1050 ft inboard of landing threshold. Deer and birds on and invof arpt
      especially apch end Rwy 27. When twr closed ACTIVATE MALSR Rwy 27, HIRL Rwy 09−27, REIL Rwy 09 and twy
      lgts—CTAF; MIRL Rwy 12−30 off. For landside access from apron when twr clsd ctc IPT AFSS 717−368−1545 or
      frequency 122.65/122.2. Rwy 15−33 cracked with weeds growing through cracks. Ldg fee. NOTE: See Land and
      Hold Short Operations Section.
    WEATHER DATA SOURCES: ASOS (717) 368−3420.
    COMMUNICATIONS: CTAF 119.1   **UNICOM** 122.95
     WILLIAMSPORT FSS (IPT) on arpt. 122.65 122.2 122.1R. TF 1−800−WX−BRIEF. NOTAM FILE IPT.
     RCO 122.1R 114.4T (WILLIAMSPORT FSS)
➤ Ⓡ **NEW YORK CENTER APP/DEP CON** 124.9
     TOWER 119.1 FCT (1130−0330Z‡)   **GND CON** 121.9
    AIRSPACE: CLASS D svc 1130−0330Z‡ other times CLASS E.
    RADIO AIDS TO NAVIGATION: NOTAM FILE IPT. VHF/DF ctc WILLIAMSPORT FSS
     (L) VOR/DME 114.4   FQM   Chan 91   N41°20.31′ W76°46.49′   238° 8.8 NM to fld. 2090/09W.
     PICTURE ROCKS NDB (MHW) 344   PIX   N41°16.61′ W76°42.61′   267° 9.8 NM to fld.
     ILS 110.1   I−IPT   RWY 27.   GS unusable blo 650′. LOC unusable 1 NM to AER abv 2200 ft.
    COMM/NAVAID REMARKS: VHF/DF unusable 150°−200° byd 20 NM blo 5000′.
• •
    HELIPAD H1: H125X160 (CONC)
    HELIPORT REMARKS: Helipad H1 located adjacent to Twy 'J'.

Fig. 21-4. *The New York Center Approach/Departure Control provides radar service to Williamsport, PA, through New York's remoted radar facilities.*

HAYS MUNI   (HYS)   3 SE   UTC−6(−5DT)   N38°50.70′ W99°16.44′   **WICHITA**
    1998   B   S4   **FUEL** 100LL, JET A   ARFF Index Ltd.   **H−2E, L−6G**
    RWY 16−34: H6300X100 (ASPH)   S−28, D−48, DT−86   MIRL   **IAP**
      RWY 16: REIL. VASI(V4L)—GA 3.0° TCH 38′. P-lines.   **RWY 34:** MALSR. VASI(V4L)—GA 3.0° TCH 27′.
    AIRPORT REMARKS: Attended dalgt hrs. After hrs for fuel call number posted. Arpt CLOSED to air carrier ops with more
      than 30 passenger seats except 24 hrs PPR call arpt manager 913−628−7370. Ultralight activity on and invof
      arpt. ACTIVATE MIRL Rwy 16−34, VASI Rwy 16 and Rwy 34, REIL Rwy 16 and MALSR Rwy 34—CTAF.
    WEATHER DATA SOURCES: AWOS−3 125.525 (785) 625−3562.
    COMMUNICATIONS: CTAF/UNICOM 122.8
     WICHITA FSS (ICT) TF 1−800−WX−BRIEF. NOTAM FILE HYS.
     RCO 122.3 (WICHITA FSS)
➤ **KANSAS CITY CENTER APP/DEP CON** 124.4
    RADIO AIDS TO NAVIGATION: NOTAM FILE HYS.
     (L) VORTACW 110.4   HYS   Chan 41   N38°50.86′ W99°16.61′   at fld. 2020/10E.   **HIWAS.**
      DME unusable 220°−030° byd 35 NM blo 3700′
      VOR unusable 220°−030° byd 35 NM blo 4100′   030°−220° byd 35 NM blo 3500′
     NETTE NDB (LOM) 374   HY   N38°46.16′ W99°15.09′   339   4 7 NM to fld.
     ILS 111.5   I−HYS   Rwy 34.   LOM NETTE NDB (LOC only)

Fig. 21-5. *The Hays, Kansas, airport is beyond the Kansas City Center's radar coverage, but the Center can provide that service down to about 4000 feet AGL. From there on, the handling is by "manual," or nonradar, methods. The absence of the ® indicates this type of service.*

Questions and answers

Now for more details. What follows are a few questions asked of various Approach Control specialists, primarily at the Kansas City Class B and the Wichita Class C airspace facilities. In considering their responses, keep in mind that while basic FAA control procedures apply everywhere, there might well be certain local variations from facility to facility, depending upon the size, traffic volume, and airspace complexity of an individual location. Recognizing that minor differences can exist, the controllers' responses to the questions will hopefully clarify some issues that VFR, and perhaps IFR, pilots might have wondered about.

Q. I understand that towers are rated by levels—level 1 the least busy, up to level 5, the busiest. What determines these levels?

A. It's based on counts of all traffic handled—VFR, IFR, and satellite activity, whether landing, taking off, or transiting our area. We're (Kansas City) currently a level 4 with a minimum of at least 60 handlings per hour. A level 5 must handle at least 100 per hour, and a level 3 is in the 40 range.

Q. Approximately how many miles out from the airport does Approach's area of responsibility extend?

A. About 55 miles, which is well beyond the limits of the actual Class B airspace, of course. That doesn't mean that a VFR aircraft, for example, is under our control out that far, but our radar coverage is that extensive.

Q. I'm VFR approaching a Class B airspace for landing and have been with Center for flight following. Will Center now hand me off to Approach without my requesting it?

A. Very possibly, but it's not automatic. Just be prepared to introduce yourself to Approach for the first time in case there is no handoff and Center has terminated radar coverage, telling you to contact Approach, change to the advisory frequency, or whatever.

Q. How frequently do you have to reject a VFR request to transit the Bravo airspace?

A. I don't know how frequently, but the only time we reject a VFR request is when our traffic volume requires it.

Q. On average, how many controllers—tower and Approach—do you have on duty at one time?

A. Usually 10 or 11, plus two supervisors. In the tower, there's one each assigned to Clearance Delivery, Ground Control, local control, and the cab coordinator position. The rest are downstairs in Approach Control.

Q. Is there any way a pilot can determine what frequency to use to establish initial contact with Approach, assuming, of course, that there's been no handoff or previous contact with a Center?

A. Yes. Check the frequencies published in the A/FD or on the back of the sectional chart's legend flap. What frequency to use depends on the airport and the direction of flight. Coming into the airspace from the east, you'd probably contact Approach on one frequency, another frequency if coming from the west, and maybe yet a third if from the south. If, by chance, the one you chose isn't the right one, Approach will so advise you.

Q. Do you have any advance information on a VFR flight plan aircraft that would be entering your airspace?

A. Not unless the pilot has been receiving traffic advisories from Center and had been handed off to Approach for transiting or landing purposes. We'd know about that aircraft because we had accepted the handoff. Otherwise, we have no prior information. IFR, of course, is a different matter.

Q. For VFR traffic transiting the airspace, do you have a preferred altitude?

A. At some terminals, there are preferred, or perhaps required, VFR flyways or transition routes, but that's not universally true. Much depends on the runways in use, the flow of IFR traffic, and IFR traffic at satellite airports for which Approach is responsible.

Q. Another question on transiting the airspace: I'm coming in VFR from the east and want to go through the airspace to the west. I've been monitoring the east frequency for several miles, and there seems to be little traffic or activity. When I make my first call and request clearance to the west, the request is denied "because of workload." How can that be? The controller obviously isn't that busy.

A. No, (the controller) might not be, but what you're not likely to know is the volume of activity the west controller is handling. At the time of your call, he could be too busy with IFR traffic to accommodate a VFR transit. The east controller knows what's going on and would have to refuse your entry request.

Comment: Another factor to consider is that the controller might be engaged in several off-frequency activities, such as landline coordination with nearby VFR towers, Center sectors, or other Approach positions; he or she might be in the process of generating VFR or IFR flight strips or perhaps be involved in any number of functions of which the pilot could not be aware. So silence on the air should not imply that the controller is sitting there with nothing to do. As a Wichita Approach controller put it, "Actual time talking on the frequency is not at all indicative of a controller's workload—though I completely understand that that is about the only method a pilot has to try to judge (the controller's workload)."

Q. The controllers are assigned to specific geographic sectors, right?

A. Right. Again, however, the volume of traffic at a terminal can make a difference, as do the runways in use. At this location, we have three sectors: east and west, north and south, and satellite. The last covers eight airports in our area, including three with VFR towers, one that's a part-time military tower, and four that are uncontrolled.

Q. I have departed either the primary airport or one of the satellites, VFR, and have received advisories from Approach. I request a hand-off to Center for en-route advisories. What do you do when I make the request?

A. You've already told us your route of flight, and we know what position, or sector, at Center handles IFR and VFR traffic in that geographic area. It's then just a matter of activating what we call an "override" phone that connects us directly with that sector controller. We'll ask if he can handle a VFR flight. His workload permitting, he'll say yes, so we'll give him your aircraft type, N number, present position, altitude, first point of landing, and squawk. Meanwhile, he's manually filling out a VFR flight strip on you. Once the information has been relayed, we'll tell you to contact Center on such and such frequency. All you do now is get on the air, identify yourself with aircraft type and N number, verify your present altitude, if climbing, and the desired altitude; or if you've reached your cruising altitude, merely verify what it is. The call is simple: "(Blank) Center, Chero-kee Eight Five One Five November is with you at five thousand, climbing to seven thousand five hundred." Don't say any more. Center already has the other pertinent information.

Q. I imagine it's the same for arrivals when Center hands me off to you folks.

A. Exactly the same.

Q. What if Center is too busy to give me advisories?

A. The sector controller will tell me, I'll tell you, advise you that radar service is terminated, ask you to squawk 1200, and bid you good day. You're on your own now to see and avoid.

Q. What if I'm VFR and intend to land at the primary airport in a Bravo airspace, whether I've been using Center or not. Can you refuse to clear me into the airspace?

A. Only if the weather is below VFR minimums or if local conditions make it impossible for us to handle any VFR operations. Otherwise, assuming you meet the pilot and equipment qualifications, I'll clear you, separate you from IFR traffic, advise you of other possible VFR traffic, and vector you to the airport. There could be a clearance delay, though, if the departing or arriving IFR traffic were heavy.

Q. What if I want to go through a portion of the airspace and land at a satellite that is under but not in the Class B airspace? Would you still clear me and give me the normal services?

A. That's strictly a matter of workload. If I can, I will. Otherwise, stay under Class B and contact the tower at the satellite. If it's an uncontrolled field, self-announce your position and intentions 10 to 15 miles out.

Q. On a cross-country trip, using Center for advisories, my route takes me near an Approach Control airport. Maybe it's a Class B or C, but I'm going to go around or above the airport, regardless of what it is, and I'm not going to land there. Why, then, does Center tell me to contact Approach? I might be 30 miles away from the primary airport.

A. Because you'll be entering that Approach Control's airspace, and it's now Approach's responsibility, not Center's, to give you advisories.

Again, though, if Approach is too busy, the sector controller might not be able to accept the handoff. Center will then tell you that radar service is terminated and to squawk 1200. If you want advisories down the road, you'll have to reestablish contact with Center.

Q. Are these handoffs also handled by override phones?

A. For VFRs, yes. On an IFR flight plan, however, everything is computerized, and the transfer of control is more or less automatic. That's not the best use of the term, but it's reasonably descriptive of what happens.

Q. Speaking of automatic, or automated, I've heard a lot about something called a *slewball*. Explain it to me, would you?

A. First, what you're referring to is called a trackball today and looks a little like a computer mouse, except that it doesn't have the left and right pushbuttons, or clickers. Instead, picture a tennis ball with all but the top 10 percent sliced off and discarded. That remaining 10 percent is then a rounded, movable dome set in a casing that generally resembles the computer mouse. Then picture a video game or a computer monitor where you can move a pointer, an arrow, or a symbol to a certain position by manipulating a control of some sort The trackball is the control we manipulate, and the symbol in our unit is simply a letter, such as a W for a west controller, E for east, C for center, H for handoff, and so on. When we want to make some computerized change to an identified aircraft in the area, we move the trackball until the letter is placed on the radarscope symbol of that aircraft. Then, by punching certain computer buttons, we can change the aircraft transponder code, initiate a handoff to Center, or do anything else we wish that involves the computer.

Q. A question about those things called flight strips. Would you decode this one for me (Fig. 21-6)?

A. Sure. The aircraft N number is obvious, as is the type, a PA-28. For those unfamiliar with it, one of a series of letters follows the slash to indicate the type or types of avionic equipment aboard the aircraft. In this case, the A means that One Five November has a DME and a transponder with altitude-reporting capabilities. The 613 is merely the computer identification number. Moving over a column, 4626 is the transponder code given the pilot to squawk. P1620 indicates the proposed departure time, and 40 is the requested 4000-foot altitude. MCI is the departure point, and TOPEK is the "gate" through which

N8515N	4626	MCI	+TOPEK+			
PA28/A	P1620		MCI MKC V4 TOP STJ MKC			
613	40					

Fig. 21-6. *This computer-generated IFR flight strip originated at an FSS and was automatically sent to the Center computer, which then transmitted it to Approach Control 30 minutes before the planned 1620 UTC departure.*

the aircraft will be vectored out of our airspace. The balance indicates the route of flight to its termination at MKC.

Q. A second question, then: Who originates these strips? How do they end up in Approach?

A. They start at a Flight Service Station. When the pilot files an IFR flight plan, the FSS enters the information in its computer. That computer then "talks" to the central core, or host computer, in the Center, which interprets the data, processes the flight strip, and sends it to the appropriate Approach Control facility. Approach receives the printed strip from its computer 30 minutes before the proposed departure time. That time is valid for 2 hours, unless the pilot is delayed and contacts Approach, the tower, or at an uncontrolled airport, the Flight Service Station to request an updated departure.

Q. Is there anything like this for a VFR flight plan?

A. No. A VFR flight plan is just between the FSS at which it was filed and the FSS in whose area the terminating airport is located.

Q. Do Approach controllers ever work the tower?

A. Yes, if they have been tower-certificated. Here, we rotate. We can do that because we're all tower- and radar-certificated. That's not true everywhere, though. Many tower controllers have not gone through the radar training and certification process, so they're limited to strictly tower duty.

Q. Which do you think is the more stressful, tower or Approach?

A. I'd say Approach. You're working a lot more traffic in a relatively confined airspace. Because of the volume, you have to be constantly alert to where the traffic is, whether it's following your instructions, the proximity of one aircraft to another, and so on. The same is true in the tower, but to a lesser extent. Yes, if I had to choose, I'd say there is more pressure in Approach. Being able to work both positions, though, varies the job and makes it a lot more interesting.

22

Air route traffic control center

We've left the immediate airport vicinity and started heading toward a distant destination—well, it might not be "distant," but it's a cross-country excursion of some sort—and we have at our disposal one more FAA facility to help us along the way: the Air Route Traffic Control Center (ARTCC).

In one respect, "at our disposal" is somewhat of a misstatement, because Center exists primarily for IFR, not VFR, operations. Its first responsibility is to control IFR flight plan aircraft, ensure proper separation, monitor the aircraft's fix-to-fix and point-to-point progress, issue advisories, and sequence the aircraft into the terminal environment. Thus, in marginal weather conditions or when the volume of IFR traffic is heavy, Center might have no time to provide services to pilots operating VFR. Those are considered to be additional services and are at the discretion of the individual controller, depending on the workload.

Barring such situations, Center can play an important role in enhancing the safety of VFR flight by providing a variety of assistance: advising us of other traffic in our vicinity, alerting us if we should inadvertently (or perhaps carelessly) venture into an active MOA, offering vectors around potentially severe weather conditions, or helping us in an emergency.

Unfortunately, though, many VFR pilots are reluctant to request the services that are available. The reasons are varied: the belief that Center is only for the big boys who drive the widebodies or the high-time IFR pros; lack of knowledge of what Center does and how it functions; fear of being a nuisance to controllers who have enough to do without worrying about a Cessna 152 or Cherokee 140; uncertainty of how to communicate with the facility; reluctance to get on the air and perhaps display ignorance, make mistakes, or sound sort of stupid.

Whatever the case, whatever the cause, it's too bad, because Center not only will help but wants to help the VFR pilot whenever its workload permits. Controllers across the country have so stated, and the FAA encourages that help. Accordingly, a review of what a Center is, what it does, and how the VFR pilot can avail himself of the available service might help diminish some of this pilot reluctance—if such reluctance exists.

Center locations and airspace responsibilities

Twenty Centers blanket the continental United States, plus one each in Alaska, San Juan, Honolulu, and Guam. Those within the 48 states are listed in Table 22-1, while Figure 22-1 depicts all 24 Centers and the airspace for which each is responsible. As the figure indicates, any given Center has a fair piece of airspace to control, the typical Center being responsible for approximately 100,000 square miles or more. To service such an expanse, the geography is first divided into areas, and each area then subdivided into sectors, with one controller and perhaps an assistant controller usually handling a sector.

A logical question is how a Center, physically located perhaps several hundred miles from an aircraft, can maintain radio contact and radar coverage. The answer, of course, is through many remote air-ground radio outlets and a fewer number of remote radar antennas that are connected to the Center via microwave links and landlines. The microwave signal is essentially the primary radar carrier, while landlines serve as backups in the event of a microwave outage.

Figure 22-2 illustrates the Kansas City Center airspace and the various sector boundaries, as well as the radio frequencies in the sectors. From left to right, the extreme western limit just touches the southeastern corner of Colorado, and the eastern limit reaches almost to Indiana. The square boxes designate the *remote communications air/ground* (RCAG) locations and their radio frequencies.

Figure 22-2 also illustrates the changing of frequencies as you move across the airspace. Going from sector to sector means that you're leaving one controller's area and entering that of another. As this occurs, the first controller will call you and advise you to change from the current frequency to the one monitored by the next controller. This is the *handoff* process, which is very similar to the steps Approach takes when it transfers the monitoring of an aircraft from the terminal area to a Center. The only major difference is that this hand-

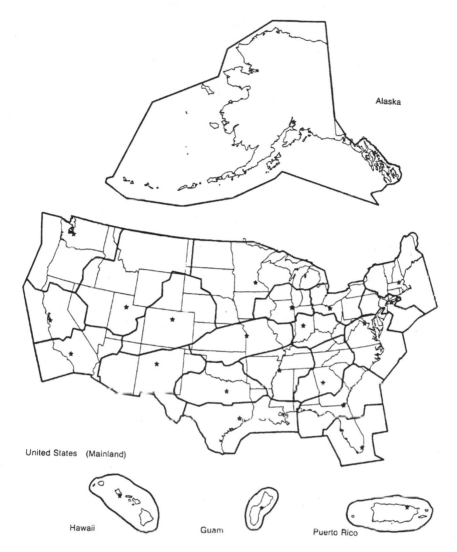

Fig. 22-1. *The locations of the 24 Centers and the general geographic areas for which each is responsible.*

off is between facilities, while the other is between sector controllers, separated, perhaps, by only a few feet.

Figure 22-2 illustrates something else: You'll note that several of the sectors have two or more frequencies back to Center. As the aircraft moves out of the range of one, the controller will tell the pilot "Change to my frequency, 134.9" or "Contact me now on 134.9." This simply means that the pilot will be talking to the same controller but

Table 22-1   The Air Route Traffic Control Centers in the continental United States.

Albuquerque	Houston	Minneapolis
Atlanta	Indianapolis	New York
Boston	Jacksonville	Oakland
Chicago	Kansas City	Salt Lake City
Cleveland	Los Angeles	Seattle
Denver	Memphis	Washington, D.C.
Fort Worth	Miami	

on another frequency. The key to what's happening is the use of *my* or *me*. Be alert to that, because it will have some effect on what you say when you make the change and reestablish radio contact.

How a Center is organized

Not all Centers are identical in physical layout, but the Kansas City Center (actually located in Olathe, Kansas, about 20 miles southwest of Kansas City) is typical in terms of size, staffing, and layout. Figure 22-3, which requires a bit of explanation, is a drawing of the facility's overall design, while Fig. 22-4 illustrates just one of the sector positions.

At the left and outside the boxed-in areas (Fig. 22-3), you'll see the letters A to D. These identify the four rows of controller positions and their related radar, radio, and telephone equipment. The rows are then divided into geographic airspace areas, arbitrarily designated here as Flint Hills, Trails, Ozark, Prairie, and Rivers/Gateway.

The next breakdown is the division of the areas into sectors, with each sector responsible for the control of high (HI) altitude or low (LO) altitude traffic. For ease of reference, Fig. 22-3 is an enlargement of the Trails area and its various sectors. Either above or below the position symbols is the coding for these sectors, such as SLN-HI (Salina High) and MKC-HI (Kansas City High). These are the sectors in Trails that handle flight operations from 24,000 up to 60,000 feet MSL. Meanwhile, the low-altitude operations (up through FL23,000 feet) are the responsibility of the sectors identified as IRK-LO (Kirksville), ANX-LO (Napolean, a VOR), BUM-LO (Butler), STJ-LO (St. Joseph), and FOE-LO (Forbes Air Force Base—low, 5000 feet

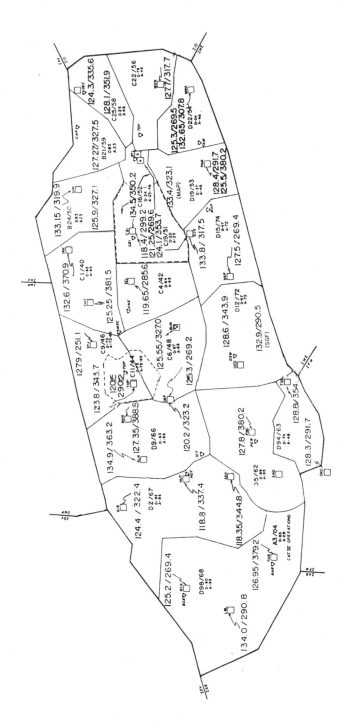

Fig. 22-2. *The sectors and respective frequencies within Kansas City Center's airspace. The squares identify remote communications outlets, and the triangles are VOR locations.*

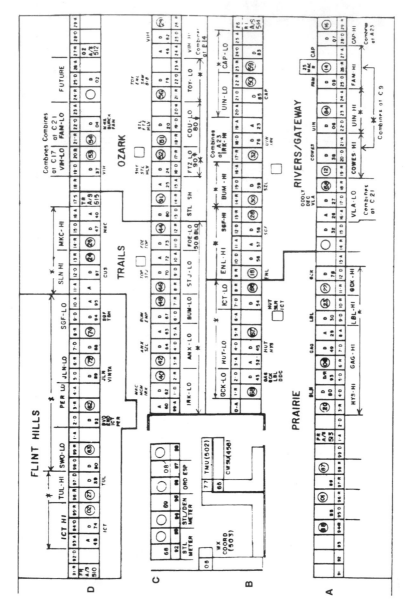

Fig. 22-3. *The floor plan of the Kansas City Center is typical of others throughout the country.*

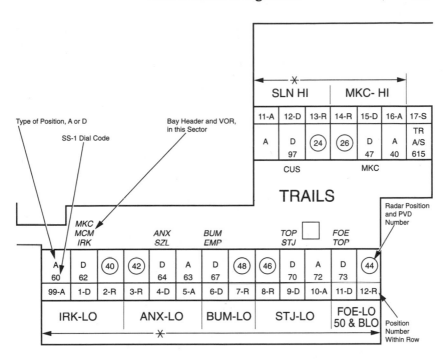

Fig. 22-4. *An enlargement of the Trails area and its sectors.*

and below). In small print, again, either above or below the position, you'll see the VORs in the sector, in this case, MKC, MCM (Macon, MO), and IRK in the IRK-LO sector. The encircled numbers (40, 42, 48, 46, etc.) identify the locations of the sector radarscopes, which are also called PVDs, or *planned visual displays.*

Adjacent to the PVD numbers are the letters A and D. What the A originally stood for is apparently lost in history, at least in the Kansas City ARTCC, but today it is commonly called the *A side,* to which new employees (meaning those who have completed the initial screening and training at the FAA's Oklahoma City Academy and are called *developmentals*) are assigned. It's not unusual, however, to find fully qualified and experienced personnel—personnel who have reached what is called the "full performance level," or FPL— working the position. Whatever the case, the A side is primarily responsible for removing the computerized flight strips from the printers and hand-delivering the strips to the appropriate sector controller. Additionally, for example, if a Flight Service Station wanted to correct or update a filed flight plan, the FSS would call the A side's telephone dial-code (such as 60, in the IRK-LO sector) and the A side would manually amend the flight plan data. His or her job is thus rather clerical in nature, but these tasks, basic as they may be, are essential to the FPL who is manning the PVD and to the smooth control of traffic about to enter the FPL's sector.

The letter D, in Fig. 22-5, represents the data, or D-side, position and also refers to the "developmental," still in a training capacity and not yet a qualified FPL. As with the A side, however, an FPL could be working this position, depending on staffing needs or staffing limitations. In either case, the D-side person is responsible for much of the paperwork, communicating information to the PVD FPL; racking, revising, or updating flight strips; handling landline phone calls; assisting in emergencies; and generally serving as the right hand to the FPL who is actually on the radio at a PVD and controlling sector traffic.

As part of the training process and when the workload permits, the developmental puts on a headset and observes as the FPL communicates with his or her sector traffic. Depending on his level of training, the developmental also could sit at the position and actually control the sector traffic, but (and it's an important but) he or she does so only under the eyes and ears of an instructor or an FPL.

Fig. 22-5. *As much as a darkened room will permit, this picture illustrates the elements that make up a sector suite in a typical Air Route Traffic Control Center.*

On the other hand, the developmental may have reached the stage of training and competency that enables him or her to function as a *manual,* or *nonradar,* controller. At that skill level, the developmental is qualified to handle IFR traffic in nonradar situations, such as an IFR aircraft that is preparing to land at a distant Class E or G airport and has descended to an altitude below Center's radar coverage. In cases like this, the only way to separate and control the aircraft in IMC weather is by radio, by following a strict set of procedures, and then by trusting in the pilot's flying skills and ability to follow instructions. Otherwise, it is not until the stage of radar certification is reached and the developmental is classified as an FPL that he or she mans the radar position free of immediate supervision or in-person monitoring. I should point out, however, that, regardless of the individual's level of experience, every radio communication is taped for whatever future need it might serve. So in a sense, even the most knowledgeable and long-tenured FPL can never escape some sort of moment-by-moment performance monitoring.

To round out the makeup of a sector, you'll note that either above or beneath the sector's PVD is a number and then the letter R, such as 2-R in the IRK-LO sector or 13-R in SLN-HI. The number is simply the position number within the row, and the R, quite logically, iden tifies the radar position.

The *selective signaling (SS-1) dial code* in Fig. 22-4 is the direct-line number that a controller or outside facility, such as Approach, would use to reach a specific sector and its A or D specialist. In the illustration, to contact the A or D in Sector 40, the caller would merely dial 60 or 62, respectively, and be directly connected to the position.

Position numbers in the row are identified 99-A and 1-D. For maintenance or a telephone outage, the problem would be reported as at Position 99-A or 1-D. Finally, the horizontal line with the asterisk defines the sectors that are combined during the midnight shift when traffic is at its ebb (Figs. 22-3 and 22-4).

One other set of functions not yet mentioned are those to the left between Rows B and C in Fig. 22-3. STL METER, STL/DEN METER, and ORD ESP (ESP means *en-route spacing*) are computerized displays of all aircraft currently within the Center's area of responsibility that are going to land at St. Louis, Denver, and O'Hare. Rather than allowing 20 or 30 planes to deluge a given terminal airspace at the same time, an orderly flow of arriving traffic is established by *metering.*

Based on terminal weather conditions, runways in use, or other local factors, the maximum number of arrivals per hour that the airport can accommodate is established. The Center computer then assigns each inbound flight a specific time by which it is to arrive over a given fix that is located well outside the terminal's Approach Control airspace. If the aircraft fails to arrive at the fix within 2 minutes before to 2 minutes after that assigned time, or if it doesn't appear that it will, the sector controller handling the flight will instruct the pilot to speed up, slow down, or take any other action necessary to slot the aircraft into the planned arrival pattern.

The metered displays of the aircraft for which that controller is responsible also appear on the controller's sector PVD, but only those aircraft. The controller is thus expected to take unilateral action to maintain the precise time schedule of flights into the terminal area.

In the case of O'Hare, the Chicago Center handles that traffic, but Kansas City tries to ensure the proper en-route spacing of ORD-bound aircraft in its airspace so that there won't be an unnecessary buildup once the traffic is handed over to Chicago. The same principle applies to traffic going into Denver.

Overall coordination is the responsibility of the *traffic management unit* (TMU). These specialists are constantly monitoring the meters to ensure that the computer-generated spacing and timing are being met. When discrepancies occur, it's up to the TMU, through the sector controllers, to ensure compliance with the designed arrival flow.

The other function in this complex is the *center weather service unit* (CWSU). The planned visual displays at the controllers' positions in a Center are all computer-generated; thus, from a weather point of view, what the controller sees on the scope is only a series of lines, some close-set, others more widely separated, superimposed with an H (heavy) to identify thunderstorm intensity. Unlike the terminal ARTS II or III radar that produces a much more realistic image of the weather, these lines alone are really of minimal help in guiding a pilot around or through conditions that might exist out there.

The CWSU, however, has a real radar picture of the weather, with the various degrees of thunderstorm activity in color. Using this and other available data, the CWSU briefs the controller supervisors twice a day on what's happening and what to expect. Then, if severe weather or turbulence develops between briefings, the CWSU issues a typed *general information* (GI) message that goes to all position printers, alerting controllers to current disturbances, where they are, and the degree of reported intensity.

That's the overall physical organization of a typical Center, a small portion of which is illustrated in Fig. 22-5. Other Centers might have different layouts, but the functions are the same, and basically so is the hardware. I say basically because advances are coming rapidly, and not all Centers receive the same new equipment at the same time.

Using Center's services

What happens when we contact a Center for advisories, and why is it to the VFR pilot's advantage to utilize the services available? Let's review a few of the procedures to follow, as well as certain controller recommendations, particularly for VFR pilots.

Determining the correct center frequency

Alternatives for obtaining ARTCC frequencies are offered because it's not always easy to determine the correct frequency for the area you're in. The *A/FD* and en-route low-altitude charts give a general idea, but as the Kansas City Center sector-frequency chart illustrated (Fig. 22-2), a given sector could have several frequencies. Furthermore, how many of us have a sector-frequency chart on board? It's not something routinely distributed to the public. These alternatives, however, are available:

- If you've filed a flight plan, ask the FSS specialist what the frequency is for your initial route of flight.
- If you forget or didn't ask the FSS, get the frequency from Clearance Delivery, Ground Control, or the tower.
- If departing a Class C or Class B airspace and you are not handed off to Center, ask Departure Control.
- If you're en route and want to establish contact with Center for the first time, merely radio the nearest FSS, giving your aircraft identification, position, and altitude, and then make the request.

Calling Center for advisories

Assume there has been no handoff from Approach or another Center. You decide en route that you want traffic advisories, so you get on the mike and call a Center for the first time. A point that controllers continually stress is this: Do nothing more than establish contact in that initial call. Don't volunteer any information other than the aircraft type and N number, such as: "Kansas City Center, Cherokee Eight Five One Five November." (Period!)

The reason for this is that the controller might be talking with an aircraft on another frequency and wouldn't hear your call or could be momentarily involved in other matters. Once the call is heard, though, the controller's probable first action would be to glance at the flight strips to see if there is one with your N number—which wouldn't be the case if you're VFR and this is your first contact with Center.

In the past, the procedure at this point was for the controller to begin hand writing a VFR strip. Now, however, the controller originates a computer flight strip and begins keying it in as soon as your call has been acknowledged and you start supplying the essential information. While talking or listening to you, the controller will be

1. Requesting a transponder code from the computer
2. Identifying your aircraft on radar
3. Typing in "VP" (meaning VFR flight plan), aircraft type, first landing point, and aircraft call sign.

The processed flight strip will then look just like an IFR strip, except that it will be identified VFR.

Keep in mind what is happening on the ground and the need to speak clearly. The dialogue after your initial call would go like this:

Ctr: *Cherokee Eight Five One Five November, Kansas City Center. Go ahead.*

You: *Center, Cherokee Eight Five One Five November is over Butler at seven thousand five hundred, VFR to West Memphis. Request advisories.*

Ctr: *Cherokee One Five November, Roger. Squawk one seven zero one and ident.*

You: *Roger, one seven zero one. Cherokee One Five November.*

Ctr: *Cherokee One Five November, radar contact. Altimeter two niner three five.*

You: *Two niner three five. Cherokee One Five November.*

From here on, Center will advise you of traffic that is or might be in your line of flight:

Ctr: *Cherokee One Five November, traffic nine o'clock, three miles, a Cessna 172, altitude seven thousand five hundred.*

You: *Negative contact. Looking. Cherokee One Five November.*

If, in a couple of minutes, you do spot the 172, tell Center:

You: *Center, Cherokee One Five November has the traffic.*

Ctr: *Cherokee One Five November, Roger. Thank you.*

Controllers like such information. It's one less thing they have to worry about.

Another reason for first establishing contact and waiting for the call to be acknowledged is that some pilots get on the mike and ramble on and on with unnecessary trivia, including the fact that they've got the family dog in the rear seat. As the trivia flows, the controller has two IFR aircraft on a merging pattern, both at the same altitude. Watching them come closer and closer, he feels his blood pressure rise, but the controller is helpless to do anything about the impending crisis. The rambler has the frequency tied up.

Also, when you request advisories, give Center only your next point of landing, not necessarily your final destination. If you're on the first leg of a four-leg flight, where you're ultimately going is useless information to the controller. Where you're going to land first is what matters.

It's the old admonition, again: Know what you need to say, say it, and get off the mike.

A word about frequencies

Perhaps you've experienced this: You've been using a Center for advisories, or were merely eavesdropping. You hear the controller call another aircraft, but you hear no response.

The controller might have as many as six frequencies to work with. You're talking on one, but a second pilot is transmitting on another. The controller's transmissions, however, go out on all frequencies assigned to that position.

IFR versus VFR flight strips

While controllers originate the VFR flight strips, IFRs are a different matter. When an IFR flight plan has been filed, the FSS computer sends the basic information to the Center computer, where it is automatically stored until 30 minutes before the planned departure. The computer then *spits out* (as controllers term it) the data on a flight strip. The A side tears off the strip, takes it to the appropriate sector controller position, inserts it in a plastic holder, and places the holder on a slanted rack, along with the other IFR and VFR strips (Figs. 22-6 and 22-7). Now when the aircraft is handed off from Approach, from sector to sector, or Center to Center, the controller responsible has the pertinent flight data right there (Fig. 22-8). Figure 22-9 is an example of an actual IFR strip, decoded in Table 22-2.

ARTCC radar system

Unlike Approach Control, which has the ARTS II or III radar that produces target symbols as well as alphanumeric data, Centers use the *National Airspace System Stage A,* or the Planned Visual Display we've already mentioned. With ARTS II or III, two systems function in unison: primary and secondary radar, or broadband and narrowband. It's the combination of the two that produces a particular aircraft symbol,

Fig. 22-6. *A typical controller's position in a Center, with the PVD and the flight strip racks.*

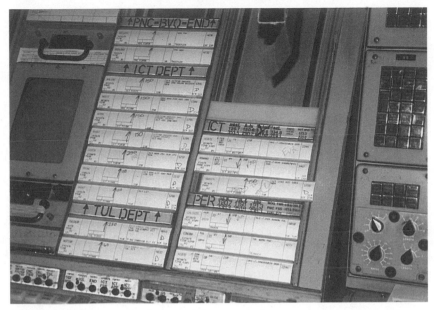

Fig. 22-7. *The IFR flight strips for yet-to-depart aircraft are on the left. Those en route are on the right.*

Fig. 22-8. *The controller has the racked strips immediately available as needed.*

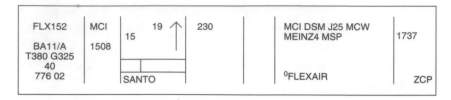

Fig. 22-9. *The only noncomputerized entry on this IFR strip is the penciled-in up arrow identifying the departing aircraft.*

Table 22-2   Figure 22-9 decoded.

FLX152	FlexAir Flight 152
BA11/A	Type of aircraft, BAC111, equipped with DME and Mode C transponder
T380	True airspeed of 380 knots
G325	Ground speed of 325 knots
40	Sector 40
776	Computer identification number
02	The second strip for this flight
MCI	Departed Kansas City International
1508	Departure time (UTC)
15 19	Time aircraft should be over the SANTO fix (1519 UTC)
230	Assigned altitude: flight level 230 (approximately 23,000 feet)
MCI DSM J25 MCW MEINZ4 MSP	Route of flight from MCI to Minneapolis
°FLEXAIR	The call sign when the flight has an unusual three-letter designation, such as FLX
1737	Transponder code
ZCP	Minneapolis Center, the facility receiving the handoff from Kansas City Center

(Note: ZMP is the normal code for Minneapolis Center, but ZCP is the computer coding.)

depending on whether it is just a primary target without an operating transponder, one with Mode 3/A, one with Mode C, and the like.

The Center's en-route NAS Stage A relies on just the narrowband radar. In this system, the signal goes out and hits the target, and then the "reply" is processed through the computer, which produces only a digital alphanumeric readout of the pertinent aircraft data. As Fig. 22-10 illustrates, there are no ARTS III-type symbols and no weather depictions, other than lines or solid areas, with the superimposed H indicating areas of heavy precipitation.

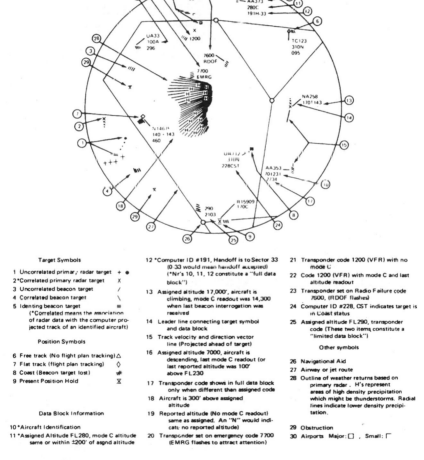

Target Symbols

1 Uncorrelated primary radar target   + ●
2 *Correlated primary radar target   X
3 Uncorrelated beacon target   /
4 Correlated beacon target   \
5 Identing beacon target   ≡
   (*Correlated means the association of radar data with the computer projected track of an identified aircraft)

Position Symbols

6 Free track (No flight plan tracking) △
7 Flat track (flight plan tracking) ◇
8 Coast (Beacon target lost) ⊹
9 Present Position Hold   X̄

Data Block Information

10 *Aircraft Identification
11 *Assigned Altitude FL280, mode C altitude same or within ±200' of asgnd altitude

12 *Computer ID #191, Handoff is to Sector 33 (0·33 would mean handoff accepted) (*Nr's 10, 11, 12 constitute a "full data block")

13 Assigned altitude 17,000', aircraft is climbing, mode C readout was 14,300 when last beacon interrogation was received

14 Leader line connecting target symbol and data block

15 Track velocity and direction vector line (Projected ahead of target)

16 Assigned altitude 7000, aircraft is descending, last mode C readout (or last reported altitude was 100' above FL230

17 Transponder code shows in full data block only when different than assigned code

18 Aircraft is 300' above assigned altitude

19 Reported altitude (No mode C readout) same as assigned. An "N" would indicate no reported altitude)

20 Transponder set on emergency code 7700 (EMRG flashes to attract attention)

21 Transponder code 1200 (VFR) with no mode C

22 Code 1200 (VFR) with mode C and last altitude readout

23 Transponder set on Radio Failure code 7600, (RDOF flashes)

24 Computer ID #228, CST indicates target is in Coast status

25 Assigned altitude FL290, transponder code (These two items constitute a "limited data block")

Other symbols

26 Navigational Aid

27 Airway or jet route

28 Outline of weather returns based on primary radar . H's represent areas of high density precipitation which might be thunderstorms. Radial lines indicate lower density precipitation.

29 Obstruction

30 Airports Major: □ , Small: ⌐

Fig. 22-10. *These are the alphanumeric data blocks and symbols produced by the NAS Stage A computer system and are quite unlike the ARTS III symbols.*

The Stage A radar does, however, produce various computer-generated images. These are briefly illustrated and described in Fig. 22-10, under the four main headings of Target Symbols, Position Symbols, Data Block Information, and Other Symbols.

Relative to the data block information in Fig. 22-10, note the possible variables from number 10 through number 25 that could be displayed. Of those, only the data in numbers 10, 11, and 12 (the starred items) constitute a full data block. The other alphanumerics convey additional information about the target when or if any condition listed between numbers 13 and 25 exists or materializes. For example, the assigned transponder code might have been 2345, but the pilot entered 2435 (number 17 in Fig. 22-10 at about the four o'clock position). Or the aircraft was assigned an altitude of 14,000 feet but is 300 feet above that altitude at 14,300 (number 18, Fig. 22-10 at the seven o'clock position). Or the pilot has radio failure and is squawking the RDOF (radio failure) 7600 code (number 23 at the twelve o'clock position, Fig. 22-10).

Figure 22-11 illustrates what basic data blocks would look like for an IFR and a VFR aircraft being tracked on Center's radar system. The IFR operation, when decoded, is explained in Table 22-3. The VFR data block is basically the same, except that VFR 55 identifies the aircraft as VFR and cruising at 5500 feet.

That, briefly, is the NAS Stage A system used in the various ARTCCs. Again, it differs from ARTS in that its symbols are computerized, while the ARTS images are generated by the combination of primary and secondary radar.

The handoff process

Let's first take the case of a VFR handoff. You're coming to the limits of a controller's sector—say, Sector 40—and the next sector on

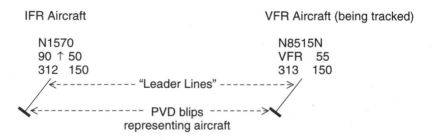

Fig. 22-11. *The design of IFR and VFR data blocks on a Center's scope with Stage A radar.*

**Table 22-3   The table explains the meanings
of the IFR data box in Fig. 22-11.**

N1570	: Aircraft ID
90	: Desired altitude—9000 feet
↑	: Climbing
50	: Present Mode C altitude—5000 feet
312	: Computer ID number, for computer storing or recall purposes
150	: Ground speed in knots

your route is Sector 48. To begin the handoff, Sector 40 rolls the trackball (just as in Approach Control) to the blip by your data block and punches the console Handoff button, with Sector 48.

Your data block appears on 48's scope with an intermittent time-sharing flashing of "H-48," meaning "Handoff to Sector 48." The flashing appears over the ground speed segment of the data block, showing first the ground speed and then the H-48. This continues until 48 accepts the handoff. To do this, the receiving controller rolls his or her trackball over to your blip and enters your call sign or computer identification into the computer. The computer accepts the handoff, the flashing stops on both scopes, and the 40 controller can now erase the data block from the screen. The A side then walks your flight strip to the Sector 48 scope and puts it on the slanted rack, and the handoff is complete.

At this point, or perhaps a moment or two before, Sector 40 will call you and tell you to change to one of Sector 48's frequencies, say, 125.3. The call sequences would then go like this:

> **Ctr:** *Cherokee One Five November, contact Kansas City Center on one two five point three.*

> **You:** *Roger, one two five point three. Cherokee One Five November.*

Make the frequency change and then reestablish contact:

> **You:** *Kansas City Center, Cherokee Eight Five One Five November is with you, level at seven thousand five hundred.*

> **Ctr:** *Cherokee One Five November, Roger. Altimeter three zero one five.*

You: *Three zero one five. Cherokee One Five November.*

That's all there is to it. From here on, keep the radio tuned to that frequency, listen for your call sign, and acknowledge all advisories.

There can be some variations in all this. For example, you're cruising along in the 3000- to 4000-foot range. As you move away from its remote outlet, that altitude could be too low for Center's radar to pick you up. You'd then be told that radar service was terminated but that you might try to establish contact again a few miles down the road.

Another variation: With a VFR flight strip in the computer, the hand-off from Center to Center is now computer-generated, much as an IFR strip. If Center A is handing a VFR off to Center B and B is too busy to provide advisories, the sector controller in B will call A on interphone and reject the handoff. The sector controller in A will then advise the pilot that "Radar service is terminated. Squawk 1200 (or VFR)."

Advantages of using Center for the VFR pilot

Yes, the ARTCCs do exist to monitor and separate IFR traffic. At the same time, though, knowing what their services are, plus being able to take advantage of them, can be of tremendous benefit to the VFR pilot. Let's consider just a few of the things Center can do.

Routine advisories

It should be comforting to realize that someone down there is aware of your existence, has you on radar, and is monitoring your progress in a specific parcel of airspace. It's also comforting, but should never be relaxing, to know that you're likely to be advised of some of the traffic that could impinge on your line of flight, or that questionable weather might lie ahead. And it's nice to know, if you are venturing into an active MOA, that you could be informed of the F-14s involved in aerobatic maneuvers, or a formation of B-52s is barreling along the military training route you happen to be paralleling.

Those are only samples of what a Center can do to enhance safety. That should not, however, nurture complacency. Lots of VFR aircraft are out there, with and without transponders, with and without Mode C, in contact or not in contact with Center. These targets might appear as only small blips on the radarscope, their altitudes un-known to all but their drivers. Whether you are VFR or IFR in clear skies, safety begins in the cockpit. Regardless of help from the ground, it's up to us to see and avoid.

Weather problems

A situation one controller cited is a good example of how a Center can be of considerable help when weather causes in-flight problems. It seems that a VFR pilot got a weather briefing early one morning for a flight from City A to City C. Conditions were reported favorable, so the pilot departed but did not file a flight plan. En route to City C, the pilot stopped at City B for about 3 hours and then headed out again.

Midway between B and C, snow showers, poor visibility, and the potential of icing were encountered. Becoming more than con-cerned, the pilot contacted a nearby FSS and explained the predica-ment. The FSS, in turn, called Center to see if it could help get the aircraft to safety. By this time, the pilot was flying at approximately 3100 feet, which was below Center's radar coverage. With instruc-tions to the pilot, Center was able to determine the aircraft's position by establishing a fix off two VORs.

Meanwhile, the weather at City C had deteriorated, making contin-ued flight in that direction even more hazardous. Getting the pilot to an alternative airport (we'll call that City D) near the aircraft's present location was the only option. The pilot could have retraced the flight back to clearer weather, but the potential of icing made it critical to get on the ground as soon as possible.

Another problem arose, however: The pilot was in the immediate vicinity of a restricted area, where the military was shooting off can-nons. Now there were two threats. Center called the military, how-ever, and had them shut down the range, and then successfully guided the pilot to the lighted runway threshold of City D's no-tower airport.

Center filed an *aircraft assist* report on the incident, and, in check-ing back, found that the only weather briefing the pilot had had was at the point of origin, some 7 or 8 hours earlier. None had been recorded at City B, where the aircraft was on the ground for 3 hours or more.

There are two obvious lessons here: First, get current weather briefings. Conditions can, and do, change rapidly, and forecasts do not always become reality. Second, if you get in trouble, don't wait until things become critical before contacting a Center. A crisis might well be avoided if Center is brought into the picture early enough.

As a Center area manager put it, "Give us a chance to help you before you really need help. Even if things look great right now, a short call is all that's necessary: 'Kansas City Center, Mooney One Two Three Four Alpha, over Emporia at six thousand five hundred to Dallas. What's the weather a hundred miles ahead?' We might be able to tell you that there are severe storms south of Wichita, or whatever. Don't be afraid to call us, but do it before an emergency develops."

What the manager said doesn't mean you should rely on Center as your primary source for weather information. The FSS is the first and best source, but Center can and will help, when help is needed.

Emergencies–Mechanical and otherwise

Here's another reason to use Center, or at least monitor its frequencies. As dramatized by an experienced controller:

> *Ginny up front suddenly breaks loose and spurts oil all over the windshield. You're frantically trying to read charts to find the nearest airport, keep the airplane level, see where you're going, and scour the terrain below for a place to put down the dying bird. And, if you still have time, squawk 7700.*

> *If you were in contact with me, I'd know exactly where you were, and I would know the closest airport to your position. I know my sector. That's my job. I could lead you to the airport and perhaps save your hide. At the very worst, had you been talking to me and couldn't reach the airport, I'd know where you had gone down and could start the rescue operations immediately.*

The benefits of at least tuning to a Center frequency are apparent. Even if you haven't been getting advisories, you can get on the air in a hurry with the Mayday call when the situation gets tight. That universal distress code gets immediate attention.

The same controller told about a student who was on his first solo cross-country, hopelessly lost, with fuel dwindling. The student knew a certain VOR had been crossed 30 or 40 minutes previously, but disorientation, coupled with growing panic, had set in.

The student had the presence of mind to call Center. The controller, who was also a pilot, in a reassuring tone vectored the student toward the nearest airport, flying along a four-lane highway that led to the town where the airport was located. As the controller said, "It was a narrow runway, but it was 20 miles long." The student landed with less than 15 minutes of fuel in the tank.

Before signing off, the controller told the student to call him on the phone as soon as possible. (You can picture the student's emotions when that edict was issued.) When the call came through, though, the controller told the student that he had done exactly the right thing: He had remained calm and asked Center for help in an emergency. The controller's only request was that the student tell the instructor what had happened.

The controller concluded the story by saying that handling the student took about 20 minutes, basically tying up the frequency during that time. Meanwhile, the controller had a bunch of air carrier operations to watch, contacting the airliners only when necessary. The air carrier pilots, however, were listening to the dialogue. As the controller put it, 25 pilots would have been howling if almost all attention had not been devoted to one lost neophyte in a Cessna 152.

Perhaps these incidents at least partly illustrate the potential value of Center to those of us who fly VFR. Perhaps they also hint at the sensitivity so common to the vast majority of those wearing the headsets in the Center complex. They want to help; they're there to help; and they will help, be it a routine flight or an emergency. The routine help is on a workload-permitting basis; the emergency help receives instant attention.

Position reporting and Center

You've filed a VFR flight plan and are using Center for en-route advisories. At some point, Center becomes too busy to handle you, you lose contact, or you voluntarily terminate contact yourself. Whatever the case, you subsequently have an emergency and go down. Ninety minutes after your flight plan has expired, the FSS alert notice (ALNOT) goes out, and Center receives a copy of it. In this situation, how much help will Center be?

Frankly, it's questionable. For one, Center keeps no written record of radio contacts with VFR aircraft. It has only the flight strip for those aircraft that had requested advisories. When it receives the ALNOT, one of Center's first actions is to go through the VFR flight strips to see if yours shows up, as it would in the case at hand. Another action is to ask the controller who last handled you if he or she remembers you and any communications that would shed light on your possible whereabouts or what might have happened. Finally, if necessary, a specialist in the quality assurance office will review the tape recording of all radio communications during the period you were in contact with Center to determine the time of your last transmission. This might take a lot of time, though, and meanwhile, you could be down and injured for hours before anyone finds you. Hopefully, your emergency locator transmitter (ELT) is working.

Relying solely on Center for position reports is better than contacting no one inflight, but it's not the best way to keep ATC advised of your whereabouts. Some pilots have even tried to make these reports when they haven't been in contact with Center at all—which is an almost totally futile gesture because the only record of the call would be just a brief moment on the tape, not even a flight strip.

No, for position reporting purposes, the facility is Flight Service, not Center. At the FSS, every position or flight progress report not only is tape recorded but also is written down. With the data almost immediately available, any FSS is far better equipped to initiate a rapid search-and-rescue mission—and over a smaller geographic area.

Recommendations for VFR pilots

I asked several Center controllers and managers what VFR pilots do, or don't do, that bugs them or causes them problems. The following are a few of their comments and recommendations.

Radio skills

We've discussed this enough, but knowledgeable use of the radio kept cropping up. The basic points, almost direct quotes, were as follows:

- Monitor the frequency and listen before you start talking. Don't pick up the mike and jump in until you're sure that the frequency is clear. Otherwise, only squeals and squawks will drift through the headset. Two people can't talk on the same frequency at the same time.

- Depending on the nature of the call, know what you should say, plan how you're going to say it, say it, and get off the air. Disconnected messages, rambling, and inconsequential trivia drive us nuts. (This, as you've probably gathered, was one of the most frequently voiced criticisms of VFR pilots, but even the IFRs were not immune.)

- Don't leave a frequency without telling us. You want to call an FSS for some purpose? Fine, but first advise us that "Cherokee One Five November is leaving you temporarily to go to Flight Service." (This assumes, of course, that you've been getting advisories from a Center.) When the call to the FSS is completed, reestablish contact with, "Center, Cherokee One Five November is back with you."

- In the same vein, a lot of pilots who receive advisories get near their destination and start dial-twisting to reach the tower or a unicom frequency, but they never tell us what they're doing. Meanwhile, we can't raise them and don't know whether they've gone down, have radio trouble, or what. If you're leaving our frequency for any reason, please tell us first, and then tell us when you're back with us.

- Call us before you get into trouble. We can probably help you. Otherwise, it might be too late.

- Don't be afraid to ask us for advisories (which means the same as *flight following*). We're here to help, if our workload will let us.

Stay VFR

One situation was cited: A VFR pilot who had been getting advisories called the controller at a certain point, saying, "Center, I won't need advisories any more. I'm out of the clouds now."

"Gulp," said the controller.

Remember that Center's radar won't pick up cloud layers or formations unless there is thunderstorm activity in them. In this instance, the controller couldn't know that the VFR character was, or had been, in the soup. VFR means always adhering to the visibility and cloud separation regulations, even when receiving advisories.

Report altitude changes

With or without Mode C, if you're getting advisories, Center has you pegged at your reported altitude. Being VFR, you have the freedom to climb or descend at your discretion, but don't do either until you have notified the controller. "Center, Cherokee One Five November is leaving six thousand five hundred for eight thousand five hundred due to turbulence." When you reach the new altitude, although it's not mandatory, it's a good practice to confirm the fact with another short call: "Center, Cherokee One Five November level at eight thousand five hundred."

The reason for the first call, particularly, is that Center has you at a certain altitude and is giving you and other aircraft advisories based on that altitude. If you move up or down at will (which is still your right), you could be venturing into the flight paths of other traffic. With Mode C, your variations will be spotted, but even then, the controller might have a question. Are you really changing altitudes, or is the Mode C malfunctioning? Things do go haywire, which is one reason controllers ask pilots to verify that they actually are at such and such an altitude. As one said, "I've seen Mode C report a Cessna 182 at 19,000 feet…a little unlikely for a 182."

En-route flight plan filing

Don't use Center for filing VFR or IFR flight plans. It's not designed to provide such a service, although I have heard it offered when an IFR-rated pilot has encountered, or was about to encounter, non-VFR weather conditions. It was a voluntary offer on the part of the controller, however, to help the pilot out of a potential predicament. Flight Service is the place to file, and even then, filing while in the air should be a last resort. It takes up a lot of the specialist's time.

Visits to a Center

The FAA and controllers alike highly recommend visits to a Center. Go as a group, or on your own, or attend one of the FAA's Operation Raincheck sessions. Whatever the case, just be sure to make arrangements in advance.

Nothing can replace a personal tour to get the real feel of what goes on. Even one trip through the facility will answer a lot of questions and alleviate some of the concerns bugging too many pilots about requesting the services a Center offers.

In lieu of a visit

While a personal visit might not be feasible for pilots who live hundreds of miles from the nearest Center, there are just as many pilots who are within an hour's drive of one but have never set foot on the property. Recognizing both realities, and the fact that a goodly number of pilots consider Center a somewhat mysterious institution, I've tried here to outline the organization of a typical facility, to explain how it functions, and to provide a few glimpses of what goes on behind the scenes. I hope, if you were previously uncertain, that you'll now have a little more confidence in your ability to use and profit from the services of a Center. There's an old adage: "What we're not up on, we're down on." What we don't understand, we avoid. But, the adage works in reverse: "What we're up on, we're not down on."

There's nothing mysterious about a Center, and the folks there aren't ogres. Many are pilots who know very well what goes on in the skies above them. Pilots or not, they're more than willing to help guide you, maybe even save you. But it's you who must make the first contact. No one in that darkened room can.

23

A case for communications skills

A few years ago, back in the days when the airspaces were called TCAs, ARSAs, ATAs, and the like, a CFII friend (Certificated Flight Instructor, Instrument) was returning to Kansas City's Downtown Airport (not the city's Class B primary airport) with a student from an instrument training flight. After Center had handed the CFII off to Approach Control and the CFII had established contact, another aircraft made its initial call, also to Approach:

Pilot: *Clay County TCA Approach Control, this is Cherokee November Four One Nine Six Six. Over.*

Approach: *Cherokee Four One Niner Six Six, Kansas City Approach.*

Pilot: *Clay County TCA Approach Control, November Four One Nine Six Six is over the interstate, and I want to land at the big airport. Over.*

Approach: *Cherokee Niner Six Six, squawk zero two five two, ident, and stand by.*

Pilot: *Clay County TCA Approach Control, November Four One Nine Six Six squawking zero two five two, identing, and standing by. Over.*

At this point the controller directed several other aircraft and lined the CFII up for the instrument approach to Downtown. The controller then returned to N41966.

Approach: *Cherokee Niner Six Six, I missed your ident. Please ident again.*

Pilot: Clay County TCA Approach Control, November Four One Nine Six Six squawking zero two five two and identing. Over.

Approach: [After a pause] Cherokee Niner Six Six, I'm still not receiving your ident. Remain clear of the TCA, and say your present position and altitude.

Pilot: Clay County TCA Approach Control, I'm still over the interstate at three thousand five hundred feet, and I want to land at the big Kansas City Airport. Over.

Approach: Cherokee Niner Six Six, which interstate are you over? There are several in this area.

Pilot: Clay County TCA Approach Control, November Four One Nine Six Six. I'm not sure which interstate, but it's near the city. I still want to land at the big airport. Over.

Approach: Cherokee Niner Six Six, I have not received your ident. Remain clear of the TCA and stand by.

Instead of doing what he was told, the pilot of 966 launched into an airwave-monopolizing discourse along these lines:

Pilot: Clay County TCA Approach Control, this is November Four One Nine Six Six. I don't know why you aren't receiving my ident. I just had it worked on, and the mechanic told me it was fine. I've got to land at the big airport because I told Agnes, my wife, I'd pick her and the kids up when they got in from Chicago. What will they think if I'm not there? Over.

Approach: Cherokee Niner Six Six, Kansas City International is a TCA, and I can't clear you to land unless your transponder is working. I am not receiving your ident, so remain clear of the TCA, and please stand by.

Continuing to ignore the explicit instructions, 966 rambled on:

Pilot: Clay County TCA Approach Control, November Four One Nine Six Six. I just had the transponder checked because the last time I was here the controller told me to stand by. I did, and the thing didn't work then. The guy at the radio shop said it worked fine, but I'm still having the same trouble. Can't you get me into the big airport? Over.

Throughout all of this, other aircraft were trying to get a word in to report positions, get clearances, and the like. But N41966 continued on and on as though he was the only one in the air.

After the last exchange, the controller saw the light. The pilot of 966, bathed in the glow of ignorance, did what he had been told. He entered "0252" in the transponder, pushed the IDENT button, and placed the switch in the STANDBY position. Of course he wasn't received! Once the mystery was solved, the pilot, quite unabashed by his display of incompetence, was cleared into Kansas City International—the "big airport."

This is about as accurate an account of the dialogue as is possible to recreate because, of course, the instructor didn't tape the real thing. It's only one incident, and, while unusual in some respects, it's not very different from what pilots and controllers hear every day. All a pilot has to do is listen with a critical ear. Some of the garbage that filters through speaker or headset from air to ground reflects an appalling lack of knowledge that is both unfunny and potentially hazardous to the ignorant pilot and those occupying the same general airspace.

Why the problem?

Who's to blame for the incompetence? Oh, we could probably point a finger at the instructor who eased over the whole subject of communications, but the main thrust of accusation must be directed at the pilot himself. The pilot of N41966 obviously had little interest in the subject. Otherwise, he would have been sure that he knew what he was doing before venturing into a controlled and congested traffic area. At the same time, we can blame him for a consummate egotism that allowed him to enter such an area with so little knowledge.

Pilots such as our friend in N41966 are dangerous because they don't know what they don't know. They are the airman's example of the Peter Principle. They've risen to their level of incompetence. If the flying abilities of this character were no better than his communicating skills, I wonder if he is still among the living—assuming he continued to exercise his private pilot privileges.

Yes, we can blame the pilot for incompetence, but others also share in the blame. Let's include the CFIs and CFIIs. And let's include the literature—or lack of it—that discusses the subject of radio communications.

Without exaggeration, it's a subject that probably receives the least attention and explanation of all of a pilot's flight training. Even the material currently in print offers little guidance on what to say and how to say it. *The Aeronautical Information Manual* takes a stab at a few examples, but the examples are limited; some conclude with the almost extinct "Over."

If flight instructors (certainly not all, but entirely too many) fail to teach more than the absolute rudiments of radio procedures, there are probably three reasons:

- The instructor isn't too sure about them himself. This should be an unlikely reason, but, as just an example, when getting back into flying after an absence of several years, I was told by a young CFI always to begin calls with "This is November 1461 Tango" and conclude with "Over." (So that I don't leave you wondering, "November" is usually used only by controllers when the pilot calling in hasn't identified his or her aircraft type. The controller then comes back with something akin to, "November 1461 Tango, say type of aircraft." The principle: Always include your make of aircraft in at least the initial call-up, as "Salem Approach, Cherokee One Four Six One Tango…" "Over" is rarely used, but it can be helpful to indicate the end of a lengthy transmission. Otherwise, it's not necessary, and you almost never hear it in normal pilot-controller dialogues.

If the instructor has accumulated most of his hours flying out of Cowslip Municipal, he probably isn't very confident of radio techniques. His own insecurity results in a superficial coverage of the subject as he preps his eager students for the FAA check rides.

- The instructor is a pro as a radio communicator, but teaching the subject takes time—mostly ground time, which is neither profitable nor exciting. So the student learns barely enough to get by.

- The airport is uncontrolled (meaning, no tower) and no controlled airport is within reasonable flight range. This is a logical reason for not teaching communications in depth, especially if the student plans to fly only on weekends and demonstrate his skills over Aunt Martha's vineyard. A good instructor emphasizes what is necessary to know, not what's nice to know.

That same instructor, however, must make it very clear that if the student (now private pilot) ever plans to fly to a controlled airport or through a Class B airspace or go on a cross-country, he must return for a thorough schooling in radio procedures. A fully equipped aircraft and a private pilot certificate entitle a pilot to land at any airport. However, the hardware and a piece of paper are hardly adequate to ensure the continued well-being of the pilot or the other airmen in the vicinity. It's dangerous to run out of knowledge—but it can happen easily and quickly to untrained pilots. The results may be devastating.

So okay—you've got a license, and you either own or rent a plane. You're a good pilot, confident of your ability. Now, like many of your counterparts, are you going to spend the rest of your flying days avoiding tower-controlled airports or being fearful of using Center (Air Route Traffic Control Center) on a VFR cross-country? If you've been well-trained in radio procedures, a busy airport or getting advisories from Center is neither a challenge nor a concern. Your radio skill makes flying just that much more fun. But if you're untrained or uncertain, you'll probably steer clear of the controlled areas and not bother Center because you think that's for the IFR pilots and the pros who wheel the wide-bodies.

This, of course, is nonsense. Admittedly, Center might not be able to help you on a busy day if you're VFR. A controller also has the right to refuse to give you routine en-route advisories or track you on radar if you come across as hesitant, uncertain, or lacking in knowledge. Center controllers might not do this very often, but requests in VFR conditions have been rejected when the pilot was obviously incompetent in the basics of radio communications. Otherwise, Center exists to serve all pilots—from the greenest student to the 30,000-hour airline captain. Besides, the FAA urges us to use this as well as all other facilities available to us.

Let's be careful not to oversimplify the matter of radio procedures. Mastering them takes time and practice. To underscore that point, the FAA's *Instrument Flying Handbook* makes these comments:

> *...Many students have no serious difficulty in learning basic aircraft control and radio navigation, but stumble through even the simplest radio communications. During the initial phase of training in Air Traffic Control procedures and radiotelephone techniques, some students experience difficulty....*

...Communication is a two-way effort, and the controller expects you to work toward the same level of competence that he strives to achieve. Tape recordings comparing transmissions by professional pilots and inexperienced or inadequately trained general aviation pilots illustrate the need for effective radiotelephone technique. In a typical instance, an airline pilot reported his position in 5 seconds whereas a private pilot reporting over the same fix took 4 minutes to transmit essentially the same information....The novice forgot to tune his radio properly before transmitting, interrupted other transmissions, repeated unnecessary data, forgot other essential information, requested instructions repeatedly, and created the general impression of cockpit disorganization....

Practicing for competence

Mastery of the technique starts with knowing what you want to say, what to listen for, how to respond, and when and how to use the mike that spreads your voice throughout the surrounding skies. As in any other field, the initial ingredient of proficiency is knowledge. The trick is to apply that knowledge in a logical sequence so that you can say what you want to say and get off the air. Once the knowledge is acquired, the next step is practice, followed by more practice, until what you know intellectually becomes an ingrained habit.

If you've ever been asked to make a speech, you know that you didn't just get up and talk. You either wrote the entire speech or outlined it, and then practiced it until you had the subject matter, sequence, body language, and voice inflections down pat. The first time around, you were probably a bit nervous. The second time was a little easier. Eventually, if you spoke or lectured enough, you became an old pro.

It's the same thing talking to the ground from an airplane. Knowledge coupled with practice will calm nerves and conquer whatever mike fright you might have. No matter how green you are, you'll come across as a professional.

You can practice in a couple of ways. One is to buy an inexpensive aircraft-band radio that picks up the various aviation frequencies. Then monitor the transmissions from your home. This method is less effective if you live far from a tower, but you can at least listen to the pilot's side of the communications exchange.

Another practice method is to use a tape recorder and do a little role-playing with yourself. You're the pilot and controller all in one. Make the initial call to Ground Control, and answer yourself as the controller would. Or pretend that you're in flight and want to land at X airport. Go through the same process. Using your knowledge of radio procedures, act out a series of scenarios on a mythical flight from the first contact with Ground Control until you have "landed," are off the active, and have called Ground Control again for taxi clearance.

Then play the tape back. Be your own worst critic. Be objective about the "dialogue." Ask yourself: "If I were a controller or another pilot listening to me, what would be my impressions of me?" If you're not satisfied, pick up the mike and go at it again.

If you practice this way enough, it won't take long to learn how to get the message across in the fewest possible words and with maximum clarity. Yes, you might feel a little silly sitting there talking to yourself, but that, too, shall pass. Even if it doesn't, it's a small price to pay for greater confidence and increased expertise.

Now you have the words down, but will you remember them, and in the proper sequence, when it comes to the real thing? If in doubt, write out what you want to say when you contact the various services. Put the notes on your knee pad and, if necessary, read from them as you make your calls. After all, people use checklists so they don't have to rely on a memory that might fail them, so why not adopt the same technique for radio contacts? (However, unlike checklists, experience makes the need for written notes unnecessary.)

A pause to regroup: I hope I am not exaggerating the case and the need for greater communication skills among the pilot population. Obviously, I don't think so. All you have to do is fly a few hours a week and keep an alert ear to what flows through the headset. On any given flight around a busy airport you'll hear everything from a terse "Okay," to a rambling recitation of superfluous trivia, to a series of mumbled incoherencies that no human or electronic decoder could decipher. If you question that statement, spend a few minutes with a controller and listen to what he or she has to say.

Controllers are human, too

Flight Service Stations, Towers, Approach and Departure Control, and the Air Route Traffic Control Centers are the pilot's valuable but unseen friends. They exist to serve the pilot and to make flying safer for all. Their services, however, aren't really free. You've paid for them through your taxes. The services are there to be used or not used, so why not take advantage of your annual donation to Uncle Sam and the taxes you find tagged onto your fuel bill?

Yes, you've paid for the service, but so has every other pilot, so it's not yours and yours alone to use or misuse. Any given service, particularly that offered by Center, can be denied you if you give an impression of incompetence. On occasion, those on the ground just don't have time to try to make sense out of nonsense and clarity out of obscurity. To do so might put someone else's life in jeopardy.

At times it might seem unlikely, but controllers happen to be human beings, too. They have good days; they have bad days. Even on their good days, though, they can quickly turn into vocal ogres when they encounter unmitigated stupidity over the air. On bad days, they can come across as halo-endowed saints when a knowledgeable pro solicits their assistance or advice.

While we're on the subject, the basic rules of courtesy over the air should always prevail. When either pilots or controllers resort to sarcasm or needless abuse, they are merely reflecting emotional immaturity—which is hardly a credit to the responsibility they bear. Controllers call it "chipping," a gentle term for "telling the other guy off."

Controllers recognize the humanity of man, and 99 percent never utter a word of recrimination when mistakes are made or ignorance shines brightly. A few don't have that level of patience, of course. They can chip with the best of them, as did one I overheard when he couldn't get a response from a pilot with whom he had just been in contact: "You gonna talk to me, boy? If you are, talk now."

To give them their due, controllers have to be models of tolerance and self-control to endure some of the things that go on in and over the air. Yes, some talk too rapidly, and some run their words together so that comprehension is nigh impossible. But the performance of the vast majority, even under pressure, sets a standard of excellence in their profession that pilots should strive to match in theirs.

If you have a problem with a controller, don't let anger overrule good judgment. The radio isn't the place for chipping. Wait until you're on the ground. Then call the facility and talk to the supervisor. Explain calmly what happened. Let the supervisor take it from there. Childish spleen-venting is out of place in the adult world, whether airborne or ground-bound.

While the controller is indeed the "controller," that doesn't mean he or she has to be obeyed at all costs. You are still the pilot in command. If the controller tells you to do something that you believe might endanger you, say so. Don't follow blindly into the path of possible destruction, but don't keep the controller in the dark about your concern or your alternate action.

In a very literal sense, a team is at work: you and those on the ground. They are there to ensure your safety and that of your fellow pilots. They can fulfill their responsibilities, however, only if you keep them informed and conduct yourself with the skill expected of a licensed pilot—private or ATP.

By the same token, if you help the controller when he or she asks you to lengthen your downwind leg, make a tight pattern, land long, speed up, slow down, make a high-speed landing run-out, or whatever, you'll be functioning as an effective team member. Remember: the controller can do without you, but you can't do without the controller. Whether you are new or experienced, it is entirely to your personal benefit to make it easy for the controller to do his or her job and thus help you do yours. Achieving that objective is a matter of communications—knowing what to say, how to say it, when to say it, and why it should be said. Knowledge plus skill equals professionalism.

All evidence that I have found indicates that a strong case can be made for greater pilot communication skills. The reason behind poor communication, whether it's the absence of literature on the subject or instructor reluctance to emphasize it, is secondary. The result is often a pilot's unnecessary fear of the microphone, which in turn tends to restrict his or her flying activities and limits the airborne adventures to which the license entitles him or her. The alternate result is unjustified confidence, as embodied by our friend in Cherokee November 41966.

What follows, then, will hopefully reduce any fears you might have and establish justified confidence in your ability to communicate as a professional. Whether flying is your vocation or avocation, that should be your objective.

A few words about phraseology

Because we'll soon begin illustrating the various radio calls and contacts, I want to be sure that some basic phraseology principles are understood. There's nothing difficult about it, but there is a certain standardization that is both accepted and expected. Reasonable variations are, of course, permissible. The examples that follow in this book, however, generally reflect the approved wording and structure as established by the *Aeronautical Information Manual* (AIM) and the FAA's *Air Traffic Control Manual*, 7110.65, for controllers.

As to wording, aircraft N-numbers are stated individually, preceded by the aircraft type. Cherokee 1461 Tango is announced as "Cherokee One Four Six One Tango," not "Cherokee Fourteen Sixty-One Tango." Land Runway 19 is "Land Runway One Niner," not "Nineteen." "Altimeter 29.65" is "Altimeter two niner six five," not "twenty-nine sixty-five." "Heading 270" is "Heading two seven zero," not "…two seventy."

In quoting altitudes, controllers state them in terms of thousands and hundreds: "maintain three thousand five hundred" or "expect seven thousand five hundred in ten minutes." Pilots can (and do) shorten altitude quotations by saying "level at three point five" or "leaving five point five for three point zero" or "over the field at two point three." This sort of verbal shorthand is acceptable from the pilot, but does not conform to FAA standards. Consequently, you won't hear controllers using that phraseology, and in the examples I cite from now on, I attempt to conform to FAA recommendations.

Accordingly, and to be sure that you understand and employ the correct phraseology, all numbers in the simulated dialogues that follow are spelled out. "Runway 19" will appear as "Runway One Niner" because that's the way it's pronounced. "Heading 240" is stated as "Heading two four zero" and so on.

Depending on the specific reference, decimal points might or might not be included in the quotation. For instance, when citing altimeter settings, the decimal is omitted. A setting of "30.08" is communicated as "altimeter three zero zero eight." On the other hand, the decimal is included in references to radio frequencies. "Contact Ground on 121.9" is stated as "Contact Ground on one two one point niner" or "Contact Ground, point niner."

One other explanation is apropos before we get into examples. You will note that at times I use the aircraft's type and full N-number, such as "Cherokee One Four Six One Tango." On other occasions, it's "Cherokee Six One Tango." Why the difference? When making the initial contact with each controller (Ground Control, Tower, each Center sector, etc.), the type of aircraft should be identified and its full N-number given (just in case the controller is handling another "Cherokee Six One Tango," a distinct possibility in congested areas). You can shorten the call sign to "Cherokee Six One Tango" after the controller does. Once the controller abbreviates your call sign, there's no point in giving the complete identification in subsequent calls to the same controller.

Even at uncontrolled (nontower) airports, the identification process should be the same. Make the type of aircraft you're flying known to others—and hope that they extend the same courtesy to you. There's a big difference between landing behind "Zero Zero Zero Zero Alpha" and "Learjet Zero Zero Zero Zero Alpha." It would be nice to know that you're trailing a jet rather than a Piper Cub. The wake turbulence of the former can be a bit more challenging.

Yes, a fair amount of verbal shorthand is acceptable, but not necessarily correct. Despite this phraseology latitude exercised by pilots (and perhaps even tolerated by the FAA), I have chosen not to take such liberties in the communication examples. The idea is to present the correct wording and phrasing here. Accepted but unapproved abbreviations can come later, if you so choose.

24

Unicom airport radio communications

The most modest air-to-ground communication (and also one that can be very helpful) is provided by unicom. In a sense, it's a step up from multicom and a step below the tower communications in a controlled airport traffic area.

What is unicom?

Simply stated, unicom permits radio contact with a ground facility on the airport. At many locations without a tower or Flight Service Station, the unicom operator fills the void by giving "field advisories" to pilots who call in and request them. The advisory consists of wind direction and velocity, possibly altimeter setting, the favored runway, and any reported traffic. Unlike a tower, however, unicom is not a controlling agency. The operator gives information, and that's all. The rest is up to the pilot.

Unicom also provides services of a nonflight nature. For example, if you want a fuel truck available for a quick turnaround, the unicom operator can make the arrangements. Maybe you need a taxi, or you'd like the operator to call your office or home to advise someone of your arrival time. Unicom is there to help you.

Of course, if an operating tower or FSS is on the field, unicom can't and won't give you runway, winds, or traffic information. That's the responsibility of the official facility. Unicom will, however, provide the nonflight services mentioned.

Who operates unicom?

At uncontrolled airports (the primary concern at the moment), the fixed-base operator (FBO) usually mans the unicom. The radio facility itself can be located anywhere at the airport, but it's generally in the lounge area, where a call-in can be handled by the FBO manager or a jack-of-all-trades employee who answers the phone, keeps the books, and sells candy and sectional charts. Typically, a barometer and wind speed/direction indicator are near the transceiver.

Keep in mind that the unicom operator is not a controller. Indeed, he might have only the most meager knowledge of what goes on in the air. He probably will give you the best information he has, but it's unwise to count on 100-percent reliability.

For example, the wind direction, its velocity, and the favored runway might be completely accurate. The operator then concludes with "no reported traffic." In reality, half a dozen planes could be in the pattern, but none has been using the unicom frequency for that particular airport. Admittedly, if six airplanes are flying around him, it's a bit far-fetched to believe that he doesn't know that traffic is in the area; the point, however, is that there has been no radio contact with him or on his frequency. In effect, there's "no reported traffic."

Don't disbelieve the unicom operator, but learn not to depend on his every word. He simply might not be in a position to know everything that's going on outside. Perhaps he's not even a pilot. Or, as in every walk of life, he could be one of the few who just doesn't care. The moral is to use the service, but be vigilant as you enter the airport area. Your own eyes are the best instruments you have to tell you what's happening in the real world.

How do you know if an airport has unicom?

The easiest way is to check the sectional chart. Just to be sure there is no confusion about what that chart tells you in that regard, let's look at a couple of examples.

Figure 24-1 shows an uncontrolled airport (so identified on the chart by the magenta coloring of the airport symbol and the related data). Starting at the twelve o'clock position, the airport is Fine Memorial in Missouri's Lake of the Ozarks and is equipped with AWOS-3, an Automated Weather Observing System, available on the 135.325

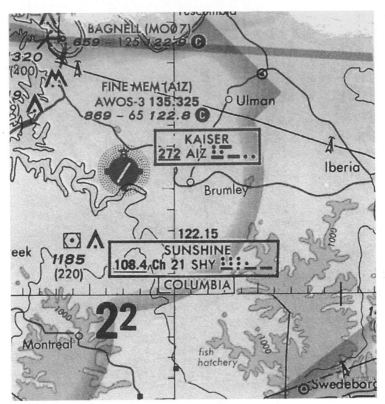

Fig. 24-1. *Fine Memorial (Kaiser) is a typical unicom airport, as depicted on the sectional chart.*

frequency. Elevation of the airport is 869 feet above sea level, the runway is 6500 feet long, and the CTAF unicom frequency is 122.8.

The white diagonal line in the airport symbol identifies the one runway and its general northeast-southwest direction. Specifically, the headings are 30 and 210 degrees, but that information is obtained by reference to the government-published *Airport/Facility Directory,* or *A/FD.* The series of small dots surrounding the airport symbol means that a nondirectional beacon (NDB) is located on the field proper. The rectangular box further identifies the NDB as "Kaiser" and, assuming the aircraft is equipped with Automatic Direction Finding equipment (ADF), the NDB can be accessed on frequency 272. Assurance that you have tuned in the correct NDB is determined by listening to the constantly repeated "AIZ" Morse code signal.

Finally, the star at the top of the airport symbol means that a rotating airport beacon is in operation from sunset to sunrise. The little ticks at the three, six, and nine o'clock positions signify that services are provided during normal working hours.

The sectional chart depiction, plus some additional symbology information on the chart's legend flap, provide a fair amount of information about a given airport, but it can't tell the whole story. The real details are best obtained from the appropriate *A/FD*.

Going a step further, but still on the subject of unicom, Fig. 24-2 illustrates how the sectional chart depicts the Columbia Regional Airport, which is tower-controlled. (All symbols and data relating to such an airport are always colored blue on the sectional chart.) With or without coloring, the fact that a tower is on the field is established by the "CT" (Control Tower) and the tower frequency, in this case, "119.3." Next comes the small star reflecting that the tower operation is part time, followed by the "C," or CTAF symbol. That symbol means that transmissions should be made on the tower frequency of 119.3 even when the tower is closed.

Unicom service is also available at this airport, but the only indication of that is the italicized 122.95 frequency. In this instance, unicom could be contacted for whatever nonoperational services you

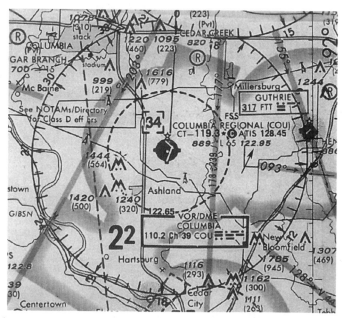

Fig. 24-2. *Columbia Regional is an example of a tower-controlled airport with a unicom facility for nonoperational services.*

might require, such as the need for a mechanic, a taxi, a call to the office, or the like. Otherwise, there's no reason to contact unicom because all flight-related information or instructions come from the tower controllers. When the tower is closed, the locally based FSS— not unicom—provides weather-related data, the favored runway, and reported traffic on the 119.3 CTAF frequency.

But back to the uncontrolled airport with unicom. Unicom's major drawback is that it might not be in operation when you need it. Maybe it's closed; maybe the person in charge has gone to lunch, is gassing an airplane, is on the phone, or just doesn't hear your call over the blare of rock music and the hangar talk of gathered pilots, or maybe he just doesn't want to be bothered. The vast majority of unicom operators are conscientious businesspeople who want to render a service; a few, however, merely consider unicom an interruption of more rewarding pursuits. The service can thus be limited. As a vehicle for keeping others informed and aware of your presence in the vicinity, however, unicom provides a safety measure that every pilot can and should use.

Contacting unicom

At airports with unicom but no tower or FSS, the unicom frequency is almost always 122.7, 122.8, or 123.0. These are the frequencies for airport advisories at uncontrolled airports. Let's say, then, that you're going into the Lee C. Fine Memorial Airport illustrated in Fig. 24-1. The call sign, or correct radio address, for Fine is "Kaiser," so 10 or 15 miles out, you tune to 122.8 and listen. Just as you did with multicom, you monitor the frequency to see what you can learn from aircraft that might be in the pattern or what information Kaiser unicom might be relaying to others that have already called in. If you can pick up the winds, favored runway, and so on merely by eavesdropping, all the better. You can spare the airwaves that one transmission.

Let's assume, though, that all is silent as you approach Kaiser. At least 10 miles out, the initial call to obtain the field advisory should go like this:

You: Kaiser unicom, Cherokee One Four Six One Tango.

Unicom: Cherokee One Four Six One Tango, Kaiser unicom.

You: Kaiser, Cherokee Six One Tango is ten miles south at four thousand, landing Kaiser. Request field advisory.

Unicom: *Cherokee Six One Tango, wind two three zero at one zero, variable. Favored runway is Two One. Reported traffic two Cessnas in the pattern.*

You: *Roger, Kaiser. Thank you. Six One Tango.*

A word of caution here: If you're monitoring unicom or listening to other aircraft as they report their positions in the pattern, be sure you're getting the right information from the right field. To illustrate what we mean: In one area of Kansas, at least four uncontrolled airports are within 25 miles of each other, all with the same 122.8 CTAF. Obviously, you don't have to have much altitude to hear all four (plus a few others more distant), so you could be picking up the winds and traffic at Airports B, C, or D when you were intending to land at Airport A, or vice versa. That's why it's essential to begin and end your transmissions with the name of the airport, just as you did with multicom.

At this point, several options are open to you. Your only intention is to land and tie down. Then your response is simply "Roger, Kaiser. Thank you. Six One Tango."

But maybe you want a taxi. Then it's "Roger, Kaiser. Would you call us a taxi to take us to Zandu Products?"

Or you need a mechanic: "Roger, Kaiser. Would you have a mechanic available? Our oil temperature is running high."

Or you'd like unicom to make a phone call: "Roger, Kaiser. Would you call 555-5678 and advise Mr. Schwartz that his party will be landing in about 15 minutes?"

But what if unicom doesn't respond to your first call? Try to rouse the operator a couple of more times. If you still get no response, make a blind call on 122.8 to any aircraft that might be in the pattern:

"Any aircraft at Kaiser, this is Cherokee One Four Six One Tango. Can you give me a field advisory?"

If someone answers, fine. You can then proceed to enter the pattern. Otherwise, it would be smart to fly over the field for a wind tee or sock check. The call, then, is the same as with multicom:

Kaiser Traffic, *Cherokee One Four Six One Tango is ten south at four thousand. Will cross the field at two thousand five hundred for wind check, landing Kaiser.*

Note that "Traffic" in this example is Roman. That's for emphasis, because once you have a field advisory from unicom or by other means, the unicom operator is out of the picture. Consequently, all position reports and flight intentions are now directed to other aircraft in the pattern or in the vicinity. Thus the calls are addressed to "(Blank) Traffic," not "(Blank) unicom."

Before entering the pattern

Case 1: You received no field advisory from any source, so you fly over the field to determine the wind direction:

Kaiser Traffic, Cherokee One Four Six One Tango over the field at two thousand five hundred. Will enter left downwind for Runway Two One, full stop, Kaiser.

Case 2: You received the field advisory from unicom or another aircraft:

Kaiser Traffic, Cherokee One Four Six One Tango entering left downwind for Runway Two One, full stop, Kaiser.

An admonition I must repeat: Keep your eyes open. Yes, unicom might have said that two Cessnas were reported in the pattern, but do you know that they constitute the only traffic? Has someone else shown up with no radio, a radio that hasn't been turned on, or who hasn't bothered to report his presence? This is an uncontrolled airport, so your most effective life preserver might be healthy skepticism liberally sprinkled with vigilance.

Turning base and final

Kaiser Traffic, Cherokee Six One Tango turning left base for landing Two One, Kaiser.

Kaiser Traffic, Cherokee Six One Tango turning final, landing Two One, Kaiser.

Down and clear of the runway

Kaiser Traffic, Cherokee Six One Tango clear of Two One, Kaiser.

Predeparture

Leaving a unicom airport is essentially the same as with multicom, except that a call to unicom can give you the wind, favored runway, and reported traffic. This call assumes, of course, that you haven't obtained the information from a personal visit with the unicom operator—which might not always be practical because you could be at one location on the field while the FBO is somewhere else, perhaps hundreds of yards away. Assuming you haven't talked with the operator, the call is simply:

> *Kaiser unicom, Cherokee One Four Six One Tango at (state your location on the field). Request airport advisory.*

Once you have the advisory, and while still stationary or taxiing slowly, get on the air again and tell the local traffic what you're doing—or going to do:

> *Kaiser Traffic, Cherokee One Four Six One Tango taxiing to (or back-taxiing on) Runway Two One, Kaiser.*

Departure

> *Kaiser Traffic, Cherokee One Four Six One Tango taking Two One, departing to the east, Kaiser.*

After takeoff

> *Kaiser Traffic, Cherokee Six One Tango clear of the area to the east, Kaiser.*

Touch-and-goes

As with multicom, communicate your intentions before taking off:

> *Kaiser Traffic, Cherokee One Four Six One Tango taking Two One, closed pattern for touch-and-goes, Kaiser.*

On downwind:

Kaiser Traffic, Cherokee Six One Tango turning downwind for Two One, touch-and-go, Kaiser.

Turning base:

Kaiser Traffic, Cherokee Six One Tango turning base for Two One, touch-and-go, Kaiser.

Turning final:

Kaiser Traffic, Cherokee Six One Tango turning final for Two One, touch-and-go, Kaiser.

Landing or departing the pattern after touch-and-goes

When you're finished practicing and want to land, advise the traffic accordingly on downwind, base, and final:

Kaiser Traffic, Cherokee Six One Tango downwind for Two One. Full stop, Kaiser.

And so on.

If departing the pattern, the following will keep other traffic informed:

Kaiser Traffic, Cherokee Six One Tango departing the area to the east, Kaiser.

And a few minutes later:

Kaiser Traffic, Cherokee Six One Tango clear of the area to the east, Kaiser.

Now stay tuned to the Kaiser CTAF until you're 10 miles or so from the airport. This will help alert you to possible inbound traffic that might be in your general line of flight. Also, if you happen to hear an inbound pilot calling Kaiser unicom for a field advisory, don't jump in and volunteer the information, as suggested in the multicom chapter. Let unicom provide the information. However, if unicom doesn't respond after a number of efforts, it's proper to help the other pilot, based on what you knew a few minutes ago:

Aircraft calling Kaiser unicom, Cherokee Six One Tango just departed Kaiser. Favored runway is Two One, winds about one niner at two zero, Kaiser.

A thought for instrument and GPS-equipped aircraft and pilots

This may be old hat for most of you, but a thought struck me recently when I was riding with an experienced pilot in his GPS-equipped Cessna en route to Abilene, Kansas, a small but active unicom-only airport. A couple of attempts to reach the unicom operator for a field advisory failed, so about 15 miles out, the pilot made his initial position report on the unicom frequency. Using the approach-plate fixes, he identified his position as "Over GPS Ezpix for landing Abilene." A few minutes later came his second report at "GPS Adela," the third at "GPS Ikehu," with the added "...landing Three Five, Abilene." And so on until we were on the final approach.

Now there was absolutely nothing out of order with any of the several position reports, except maybe this: If I were a local or transient VFR pilot with no instrument rating, no GPS knowledge or equipment, would I have any clue as to the position of this reporting aircraft? I think not. Instead, I'd be searching the skies trying to locate the traffic. I might be reasonably sure it was south of the field, but that could only increase my concern if I, too, were in that area and planning to land on Runway 35.

Visualizing this scenario, it seemed only logical to conclude that position reports based on approach plate fixes should always include the caller's altitude, geographic location of the fix relative to the airport (as north, south, southeast, or whatever), and miles out from the airport. For example: After contacting unicom for winds and favored runway, the initial call to local traffic would go something like this:

> *Abilene traffic, Cessna 1234 Alpha over GPS Ezpix, fifteen southeast at three thousand five hundred, landing Three Five Abilene.*

Then a few minutes later:

> *Abilene traffic, Cessna 34 Alpha at GPS Adela, ten south at two thousand for straight-in approach Three Five Abilene.*

And so on.

As I've said enough times, effective communication is sharing information so that there is a commonality of understanding. As the listener to position reports, I shouldn't have to wonder where my traffic is. I should know. I should have no doubts.

I do wonder, though, how often VFR pilots have been left in doubt at both controlled and uncontrolled airports when position reports from other aircraft have been basically limited to just the approach plate-reporting fixes. Such identifications would be meaningful to the listening IFR pilot (if he or she were familiar with the area or had the applicable plate at hand), but what about all the other pilots who are also nearing the field or are in the pattern or about ready to take off? Has the communication been complete, or do the listening pilots only know that there's an airplane out there over such and such fix? But where is that fix? How far is it from the airport? What is the pilot's altitude? What are his or her intentions?

A lot of people might have a lot of questions. If so, if there are uncertainties, we'd have to say that the originating caller hasn't really communicated...and that failure could conceivably be the birth of an unnecessary, perhaps serious, situation.

25

Multicom airport radio communications

A few years ago, I was driving with a friend down a well-traveled street in Riyadh, Saudi Arabia. Many side roads intersected this main street, but walls or buildings blocked the driver's view, making it impossible to see any cross traffic that might be approaching. Now driving in Saudi Arabia is a thrill in itself, but when there are few stop signs or traffic lights and you can't see cars that might jump out at you from the left or the right, extreme caution is the only alternative for your continued physical well-being.

In this case, my driver friend slowed down at every blind intersection and honked his horn. At night, he honked the horn while blinking his lights. These signals alerted others that he was there. They were his nonverbal communication signifying his presence as well as his intentions.

In a more sophisticated sense, multicom is akin to the horn and lights of our Saudi driver. At Class G airports, which have no ground-based traffic control or advisory service, no "red or green lights," no "stop signs," multicom provides the aural communications that reveal your presence and your intentions. Just like the Saudi driver, you are transmitting in the blind to anyone who is tuned to the standard multicom frequency. You don't know if anyone is really listening or is even in the immediate vicinity, but like the driver, you take that added step—just in case.

In its simplest terms, multicom is nothing more than communication between two aircraft, whether in the traffic pattern or flying at altitude along the same route. While it can be comforting and perhaps important to talk to another pilot during a cross-country flight to exchange weather information and informal pilot reports (PIREPs), the real value of multicom comes to the fore around uncontrolled, nonunicom airports. That's when you need to know who is there, where they are, and what they intend to do—just as they need to know the same about you. Multicom provides the vehicle for that exchange of information over a common frequency.

The key here is a common frequency. The FAA has thus established what it calls Common Traffic Advisory Frequencies (CTAFs). The CTAFs may be for airports that have no ground radio facilities at all, and are thus referred to as "multicom" airports, those that have only field-advisory radio services provided by the local Fixed Base Operator (unicom airports), Flight Service Stations, or tower-controlled airports that operate only part time.

Two sources tell you what the CTAF is for a given airport. One is the *Airport/Facility Directory (A/FD)*, as illustrated in Fig. 25-1. The other is the sectional chart (Fig. 25-2). If you refer to the sectional chart, you'll note the small circle with the letter "C," which indicates the CTAF for that airport. Immediately to the left of the circle is the frequency itself, in this case, 122.9. CTAFs that are printed in slanted, or italic, numbers and are magenta in color identify non-tower-controlled airports.

Keep in mind these points about CTAFS: (1) "Common" doesn't mean one universal frequency for all airports but rather the common frequency that all pilots should use for a given airport; (2) Despite what I just said, multicom—and only multicom—airports do have one common frequency nationwide—and that frequency, again, is 122.9. Whatever the case, however, reference to the sectional chart or the A/FD is essential to determine the type of airport you are planning to enter and its correct radio frequency.

But back to operating in a multicom airport environment. Despite the importance of radio communications in such an environment,

---

WAMEGO MUNI   (69K)   3 E   UTC−6(−5DT)   N39°11.83′ W96°15.53′   KANSAS CITY
      966   B   FUEL 100LL   L−6H
   RWY 17−35: H3170X30 (ASPH)   RWY LGTS (NSTD)
      RWY 17: Thld dsplcd 170′. P-line.   **RWY 35:** Trees.
   AIRPORT REMARKS: Unattended. Parachute Jumping. Ultralight activity on and in vicinity of arpt. 8′ ditch across apch
      end Rwy 17. For fuel call Wamego Police Dept 913−456−9553. Rwy 17−35 NSTD LIRL. lgts placed 25′ from rwy
      edge. ACTIVATE LIRL Rwy 17−35—122.9.
   COMMUNICATIONS: CTAF 122.9 ◄——
      WICHITA FSS (ICT) TF 1−800−WX−BRIEF. NOTAM FILE ICT.
   RADIO AIDS TO NAVIGATION: NOTAM FILE MHK.
      MANHATTAN (T) VORW/DME 110.2   MHK   Chan 39   N39°08.73′ W96°40.12′   075° 19.4 NM to fld. 1060/6E.
      HIWAS.

Fig. 25-1. *The Airport/Facility Directory (A/FD) identifies the Wamego CTAF as 122.9.*

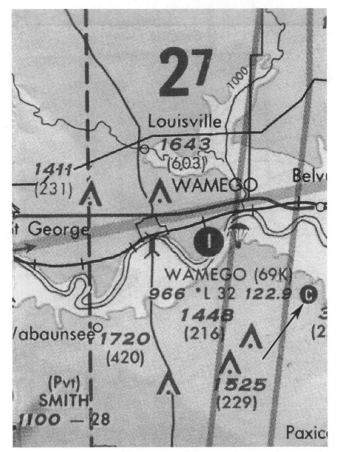

Fig. 25-2. *The sectional chart also identifies the Wamego airport and its 122.9 multicom frequency.*

let's not be naive. No matter how clearly and explicitly you transmit your intentions, not all aircraft have radios. Even if they do, their pilots may not be tuned to the 122.9 frequency. Indeed, they might not have their radios on at all. ("Why bother about such things at Pea-patch Municipal?")

The alternative is obvious. Flying around an uncontrolled airport demands a swivel neck and sharp eyes. To rely solely on blind transmissions is to invite a few thrills or unexpected encounters of the worst kind. Keep in mind, too, that your transmitter might suddenly go on vacation. You think you're broadcasting to all and sundry, but nothing is passing beyond the mouthpiece of your mike. Such failures have happened—and they could happen to you anytime, anywhere. Open eyes and constant head-turning are your best defenses against near hits or close misses.

Why use multicom?

Why use multicom? The answer is already evident: safety. If you use it, other aircraft in the area might hear you and use it too. Then everyone will be informed about who is doing what, where, and when.

At the same time, don't be dumb. Be willing to back off when judgment so dictates. You've done a good job of advising others of your actions and intentions, but just as you're about to turn on final approach, you see some guy coming from nowhere on a long straight-in approach. Decision time—do you assume he knows you're there? Do you assume he'll give way because you're apparently number one to land?

Discretion says give way. In this case, it's better to be number two than arrive at the same point in a dead heat. Piggybacking might be the way to get a space shuttle to Cape Kennedy, but it's not a very comfortable way for a Cherokee and a Cessna to make a landing. That's a "dead heat" in a very literal sense.

Using multicom

This is repetitious, but it should be said again: as with any transmission, follow the Navy's admonition about report writing and correspondence: KISS—Keep It Simple, Stupid. Know what you're going to say, say it in plain English, say it clearly, and keep it short. If you've practiced your radio technique as suggested, you know how you sound. You should have learned to speak distinctly and slowly enough to be understood and to convey your message in an organized sequence. If you speak with the listener in mind, you'll communicate more effectively with fewer words and less monopolizing of the airwaves.

Approach to the field

Tune to 122.9 at least 10 miles out. If there's any activity, you might get an idea of the traffic volume and pick up the favored runway and wind direction. Assuming this to be the case, start the self-announcing process by identifying your aircraft, position, altitude, and intentions. As there is no control agency of any sort on the field, your call is designed to advise other aircraft of your presence in the area. Consequently, open the transmission with the name of the airport, followed by "Traffic." Let's assume that you're going into Wamego (Kansas) Municipal Airport (Figs. 25-1 and 25-2), with its elevation of 966 feet and a 17–35 runway. Let's further assume that you've heard at least one other aircraft report its position in the traffic pattern and have learned that Runway 17 is the runway in use. The call, then, would go like this:

Wamego Traffic, Cherokee One Four Six One Tango is ten north at five thousand five hundred. Will enter left downwind for full stop One Seven Wamego.

Note that the name of the airport is repeated at the end of the transmission. This is to make certain that there is no confusion on the part of other aircraft as to the airport to which you are going or at which you are operating. This positive identification is especially important when two or three other airports are in the general vicinity, each with several aircraft in the air, and all transmitting on 122.9. Identifying the intended airport twice minimizes the potential for confusion.

Going back to the arrival example, if you hear nothing after tuning to 122.9, don't be lulled into the belief that the skies are clear. Give yourself every safety edge you can. Announce your intentions to those who might be listening:

Wamego Traffic, Cherokee One Four Six One Tango is ten north at five thousand five hundred. Will cross midfield at two thousand five hundred for wind tee check and landing Wamego.

Over the field

Assuming that you've heard no other traffic, you're over the field and see that the wind tee or sock favors Runway 17. Again, announce your position and intentions to the seen or unseen audience:

Wamego Traffic, Cherokee Six One Tango over the field at two thousand five hundred. Will enter left downwind for Runway One Seven, full stop, Wamego.

Note the omission of two digits when stating the aircraft's N-number in this second call: Instead of "Cherokee One Four Six One Tango" the identification has been shortened to "Cherokee Six One Tango." After initially transmitting the full identification, succeeding calls can be reduced to aircraft type plus the last two numbers and the assigned phonetic alphabet letter (or the third number, if the aircraft has no assigned letter, as is sometimes the case).

The only time this call-sign shortening should be avoided is when there could be confusion with another plane in the area or pattern that has a somewhat similar call sign. For example, using the abbreviated "Cherokee Six One Tango" as the base for comparison purposes, you might find closely related shortened alphanumerics such as "Cessna Six One Tango," "Cherokee Six Two Tango," "Mooney Six One Papa," or another Cherokee with the five-character call sign of "Cherokee Eight Three Six One Tango." In that admittedly rare coincidence, it would be just a little confusing to other aircraft in the vicinity to have two "Six One Tangos," both Cherokees, flying around in the same general area. Once duplication of the three-character designation is discovered, both aircraft should revert to their full character.

Barring such N-number similarities or sources of possible confusion with another aircraft, this sort of verbal shorthand, after your initial full identification and position call, is very much in order. It shortens succeeding calls while conserving air time. Those within the sound of your voice will appreciate your thoughtfulness.

Entry to downwind

You're entering the downwind leg at pattern altitude — approximately 800 feet agl. Get on the air again:

Wamego Traffic, Cherokee Six One Tango entering left downwind for Runway One Seven, full stop, Wamego.

Turning base

Going into the turn from downwind, make your next call:

Wamego Traffic, Cherokee Six One Tango turning left base for Runway One Seven, full stop, Wamego.

At any uncontrolled airport, make this call when turning onto the base leg. It's a lot easier for other aircraft to see you when you're in a bank as opposed to straight-and-level flight. Also, a fairly wide pattern with a distinct base leg is preferable to a hotshot U-turn. This gives you time to scan the area for other aircraft on the same leg that haven't announced their presence. It also allows you to check the final approach course for someone who might be making a straight-in approach. This is the altar on which many in-flight marriages have been consummated—unwanted but nevertheless eternal marriages.

Turning final

Once again, announce what you're doing while in the turn to final approach:

> *Wamego Traffic, Cherokee Six One Tango turning final for Runway One Seven, full stop, Wamego.*

Clear of the active

On the ground, get off the runway as quickly but safely as possible. Somebody you never heard of might be on your tail. When clear, don't keep it a secret:

> *Wamego Traffic, Cherokee Six One Tango clear of One Seven, Wamego.*

A lot of talk? Yes, but at least you've fulfilled your responsibilities. You've kept others informed, and you've made the air just that much safer. Now if only somebody as considerate is listening....

You've gassed up, coffeed up, and are ready to go again. Once the engine has fired and the radio is on, listen for a moment or two while you're still on the ramp to see if any traffic has developed. As in the prelanding, your actions can be planned according to what you hear—or don't hear. Regardless, don't assume! Announce your intentions. At most one-strip fields, a taxiway is an unknown luxury. Back-taxiing on the active runway is the only way to get into takeoff position. But let the other guy, if he's out there, know what you're going to do.

Taxi and back-taxi

Assuming that there is no taxi strip and that you'll have to taxi back (or "back-taxi") on the active runway for takeoff, don't make your initial call when you're still on the ramp, which may be some distance from the runway itself. Instead, move out to the hold line and set the brakes. Then, if there's a run-up area at the departure end of the runway, proceed with your call thusly:

> *Wamego traffic, Cherokee One Four Six One Tango, back-taxiing Runway One Seven for takeoff Wamego.*

On the other hand, if there is no run-up area at the end of the runway, taxi to the hold line and go through the complete pretakeoff checklist and engine run-up. When you're finally ready to go, make this initial call:

> *Wamego traffic, Cherokee One Four Six One Tango back-taxiing Runway One Seven for takeoff, departing to the (direction).*

Or, if you're staying in the pattern for touch-and-goes:

> *Wamego traffic, Cherokee One Four Six One Tango back-taxiing Runway One Seven, closed pattern, Wamego [or "closed traffic," or "for touch-and-goes"; "closed pattern" is the correct phraseology, however].*

If you do have to back-taxi on the active runway, first scan the approaches for both Runways 17 and 35 before venturing on to the runway itself. It's possible that some unheard-from individual is landing downwind or is indeed on the final approach for Runway 17 but hasn't bothered to tell anybody. But, if the air is clear, apply power and get to the end as quickly and safely as possible. Otherwise, an approaching aircraft might have to go around. Or worse yet, its pilot might not see you in time to abort the landing—and that could develop into a rather messy situation. Then, if you've completed the pretakeoff check, do a 180-degree turn, apply the power, and get going. This is no place to dillydally.

Preflight run-up

Assuming, on the other hand, that there is a run-up area, and if that area is so structured, back-taxi to the end of the runway and then turn left so that you can park on the departure runway's right side. The reason for this positioning is that after you've completed the engine run-up, magneto checks, and all, and are ready to go, you can turn the plane so that it is at a 90-degree angle to the departure runway. Now, sitting in the left seat, you'll have a clear view of the final approach and any unannounced aircraft that might be about ready to land. If you were on the other side of the runway, and in the left seat, scanning the final approach area would be much more difficult, particularly in a high-wing plane.

This might be a small point, but I've had more than one pilot pull out on the runway when I was on final approach. They simply hadn't seen me, and a go-around was the only alternative. Were they using multicom? That's a silly question.

Taking the active runway

While still in this 90-degree position, look upwind as well as downwind (some pilots, at this point, even swing a complete 360 degrees so they can scan the whole traffic pattern). Then, when you're sure everything is clear, make your departure call:

Wamego traffic, Cherokee One Four Six One Tango taking Runway One Seven, departing to the east. Wamego.

On the matter of someone landing downwind, four conditions could bring about such a landing:

1. Very Light wind
2. The pilot's complete disregard of the wind tee or sock
3. The pilot's failure to monitor or use multicom
4. A genuine emergency.

With observant eyes and the radio tuned to 122.9, there is no excuse for aircraft landing in opposite directions at an uncontrolled airport. And yet, because of pilot ignorance, lack of radio equipment, or failure to use the equipment, this sort of thing happens too frequently.

To wit: I recently saw an individual barrel-in downwind, narrowly missing a landing plane that was doing everything correctly. He discharged his passenger and then roared off from the taxiway-runway intersection, which left him about 2500 feet of a 4000-foot strip. This time, however, he went into the wind. Did this hotshot in his sleek Bonanza use the sophisticated radio equipment that the plethora of antennas implied? Not once. He was apparently above such trivialities.

Departure

Although you've already stated your flight intentions, it doesn't hurt to repeat them, once airborne, as a courtesy to those still in the pattern:

> *Wamego Traffic, Cherokee Six One Tango off Runway One Seven, departing the pattern to the east, Wamego.*

Until you're about 10 miles out, stay tuned to 122.9 to pick up any traffic that might be inbound, in your line of flight, or at your altitude. Also, you could help an arriving pilot who has just made an initial call by giving him or her the wind and runway information. For example:

> *Aircraft calling Wamego Traffic, Cherokee Six One Tango just departed Wamego. Favored runway is One Seven, winds about two two zero at six.*

There's no set pattern for such a call, so help the other pilot in your own words. Use the word "favored," however, when giving runway information. "Active" implies that a particular runway must be used—which, at an uncontrolled airport, is not the case.

Touch-and-goes

Let's suppose that instead of departing the pattern, you want to make a few touch-and-goes. If you're parked at the ramp, the calls are the same up to the time you're ready to take off. Then get on the air:

> *Wamego Traffic, Cherokee One Four Six One Tango taking Runway One Seven, closed pattern for touch-and-goes, Wamego.*

On the downwind, turning base, and turning final approach, repeat your intentions, just as you did for the full-stop landing:

> *Wamego Traffic, Cherokee Six One Tango turning downwind for Runway One Seven, touch-and-go, Wamego.*

> *Wamego Traffic, Cherokee Six One Tango turning base for Runway One Seven, touch-and-go, Wamego.*

> *Wamego Traffic, Cherokee Six One Tango turning final for Runway One Seven, touch-and-go, Wamego.*

Yes, that's a total of four messages for one touch-and-go, but you never know who has just entered the pattern. Your last message could be the first he has received. You can't be sure—so be safe.

When you've had enough for the day, you're going to land or leave the pattern. In either case, keep the other traffic informed. If it's the final landing, make the downwind, base, and final calls similar to those cited above, substituting "full stop" for "touch-and-go:"

Wamego Traffic, Cherokee Six One Tango turning downwind for Runway One Seven, full stop, Wamego.

If you're leaving the pattern, make the call after the last takeoff and when you have the aircraft safely under control:

Wamego Traffic, Cherokee Six One Tango departing the pattern to the east, Wamego.

26

Flight service stations

The FAA states that three major components comprise the air traffic control system: en-route air traffic control, terminal air traffic control, and Flight Service Stations. If you are considering the basic flight-related components, the FAA is correct. Literally, though, it's incorrect. Flight Service Stations are an essential element of the air traffic system, but they do not "control" traffic in any way. If the FSS role could be reduced to three principal functions, they would probably be to inform, advise, and search.

To inform

A given FSS has a wealth of data immediately available to help the IFR or VFR pilot intelligently plan a flight. Everything is there: local, en route, and destination weather, winds and temperatures aloft, icing, cloud levels, visibilities, forecasted conditions, outlooks, NOTAMs, military operations, etc.

To advise

Here is where the FSS is not a controlling agency. It can neither authorize nor deny any pilot the right to take off, or a VFR pilot to operate in IFR conditions or enter an active military operations area. It can suggest a course of action, such as, "VFR is not recommended," but that's all. It's the pilot's decision to go or not go. With the experience the FSS specialist has, however, coupled with the data available, it behooves a pilot to heed the specialist's counsel. There are enough pilots who didn't and who aren't around today to regret their indiscretion.

To search

This is a role with which many of us are perhaps not thoroughly familiar. I'll review the sequence of events later, but let's just say for now that when a pilot on a VFR flight plan fails to close out a flight plan within 30 minutes of the estimated arrival, the FSS takes the first steps to locate the aircraft. No, no one from a Flight Service Station literally goes out and searches for the missing plane, but its people do initiate and participate in what could be an exhaustive as well as expensive operation.

To put a little meat on these somewhat bare bones, then, the services that make up the FSS's three functions could be summarized in this way:

- Accepting and closing flight plans
- Conducting preflight weather briefings
- Communicating with VFR pilots en route
- Assisting pilots in distress
- Disseminating weather information
- Publicizing NOTAMs
- Working with search-and-rescue units in locating missing aircraft
- Monitoring air navigation radio aids

FSS consolidation

Back in 1985, there were 294 FSSs across the country. In 1988, the number had dropped to around 200, while currently (1998) there are only 61 to serve the 50 states. The cutback over the years was the result of just one step in an extensive equipment modernization program designed to upgrade all the ATC facilities. In the case of the FSSs, the plan was to consolidate the nearly 300 stations into 61 highly sophisticated *Automated Flight Service Stations* (AFSSs). The plan was realized by the middle 1990s, and the locations of most of the current AFSSs are depicted in Fig. 26-1. (A few additional installations have been added since this charting was issued by the FAA, but I have not been able to locate an updated version.)

This move, of course, originally met with considerable opposition from the general-aviation population. Alaskan pilots were particularly vocal, along with those in the lower 48 states, where local weather characteristics were such that on-site, not remote, FSSs were considered essential. As the result of heavy lobbying by, among others, the Aircraft Owners and Pilots Association (AOPA), the FAA reexamined several of the proposed closings and concluded that 31 should remain open because they were located in "significant

FSS SITE LOCATIONS

Fig. 26-1. *Locations of the 61 FSSs.*

weather areas." These were called *Auxiliary Flight Service Stations,* or XFSSs. According to the FAA's Air Traffic Operations Office in Washington, D.C., however, no XFSSs remain, except for about 13 in Alaska. Meanwhile, the AFSS count remains at 61 (barring the temporary closing of the St. Louis facility because of a damaging flood a couple of years ago).

Consolidation does have some drawbacks, most notably the loss of in-person briefings, except at those airports where an AFSS is physically on the property. (Let's refer to it simply as an FSS from here on.) Nothing can replace the advantages of seeing for oneself the radar screens, charts, tables, and so on and discussing conditions face-to-face with a specialist.

On the opposite side, however, the pluses are many:

- Greatly improved automation and computer capabilities
- Faster flight plan filing
- More complete and more real-time weather data
- With the Model 1 or Model 2 full-capacity (M1FC) computer, instant weather reports of conditions 25 miles either side of an extended cross-country route, such as between New York and St. Louis

- Color radar graphics, similar to but more detailed than those used in TV weather broadcasts
- Automatic radar warnings of approaching significant weather
- Fewer FSS specialists, with resultant lowered personnel costs

The minuses are thus few and the pluses many. Still, although greatly diminishing in volume, there have been complaints. With the telephone the only means for most of us to talk to an FSS, busy signals and waiting times have been sources of pilot irritation; of course, the wait time depends upon the weather conditions and the pilots requesting a briefing. When everything is socked in, all the briefers are naturally busy. But then in good weather or bad, there are those pilots who offer disorganized information and/or aren't prepared for the sequence of data they will be given. These folks obviously tie up the lines and cause excessive, as well as unnecessary, delays to other callers.

All in all, according to most pilots, the scales tip in favor of the pluses. Once you are in contact with a specialist, the briefings are faster and more complete, flight plan filing is almost instantaneous, and many more automated services are available through pilot menu accessing.

A typical FSS: Positions and staffing

The Columbia, Missouri, FSS is typical of the majority of consolidated stations. Some are smaller, others larger, but this facility ranks ninth in activity and is equipped with all the automated hardware available today. To get an idea of its layout, which, again, is representative, a floor plan (Fig. 26-2) and a photograph (Fig. 26-3) might help. The positions available for staffing are

13 Briefing

3 Flight watch

3 In-flight

1 NOTAM update

2 DATA (to correctly phrase arriving data for computer input)

1 Broadcast [*Hazardous In-Flight Weather Advisory Service* (HIWAS) and *Transcribed Information Briefing Service* (TIBS)]. TIBS is the acronym replacing *Pilot Automatic Telephone Weather Answering Service* (PATWAS).

1 Supervisory

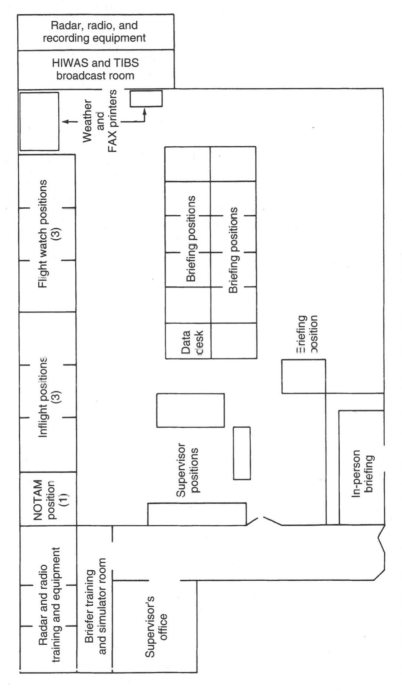

Fig. 26-2. *The functions and positions in the Columbia, Missouri, FSS.*

Fig. 26-3. *Some of the typical briefing positions.*

The current personnel headcount at Columbia is 60:

- 44 Specialists
- 6 Supervisors
- 1 Quality assurance specialist
- 1 Area manager
- 1 Chief
- 1 Administrative assistant
- 1 Secretary

The typical daytime staffing, depending, of course, on need, is one in-flight specialist, five to nine briefers, and one to two flight watch specialists, or roughly, about 10 specialists, plus staff, on duty at the same time. That doesn't sound like a lot of people, especially considering that the FSS is currently responsible for all of Missouri and one very aviation-active county in Kansas. However, I've been in a number of older FSSs that have only a relatively small area of responsibility, and a staffing of 7 to 10 specialists and supervisors was not unusual. Automation, plus sophisticated computerization, makes the difference.

Roaming through the FSS

With the general layout in mind, let's walk through the FSS, again remembering that what is seen is typical of almost all FSSs. Also, during this tour, suggestions will be offered for pilots using the various services.

NOTAM desk

The specialist at this desk is responsible for recording and updating all NOTAMs and entering them in the computer so that they are available to the briefers. While it is a momentary digression from the subject at hand, a few words about NOTAMs are pertinent at this point because of their importance to the pilot.

The *Aeronautical Information Manual* defines a NOTAM as "time-critical aeronautical information that is of either a temporary nature or not sufficiently known in advance to permit publication on aeronautical charts or in other operational publications." AIM then goes on to say that the "...aeronautical information...could affect a pilot's decision to make a flight. It (the NOTAM) includes such information as airport or primary runway closures, changes in the status of navigational aids, ILSs, radar service availability, and other information essential to planned en-route, terminal, or landing operations."

Accordingly, and depending on their source and importance, NOTAMs fall into one of three categories: *Flight Data Center* (*FDC*) *NOTAMs, NOTAM Ds* (distant), or *NOTAM Ls* (local).

An important point to keep in mind is that conditions reported in an FDC or NOTAM D that are expected to exist for an extended period are included in the next biweekly issue of the *Notices to Airmen* publication (NTAP). Once published, those NOTAMs are not volunteered during weather briefings. Consequently, if a pilot has not had access to the latest NTAP, it's his or her responsibility to ask if any 2-week-old or longer NOTAMs exist that would affect the planned flight.

FDC NOTAMs

The National Flight Data Center (FDC) in Washington issues FDC NOTAMs, which are regulatory in nature. Most amend standard instrument approaches, designate restricted airspace, or alter the airway structure. They are transmitted nationwide to all FSSs, and the FSSs maintain a file of current, unpublished NOTAMs that affect the airspace within 400 nautical miles of their facility. FDC NOTAMs more than 400 miles from the FSS or NOTAMs that have already been published are given to the pilot only on request.

NOTAM Ds

All navigational facilities that are part of the National Airspace System receive NOTAM Ds, which communicate changes to the published status of airspace obstructions or components of the airspace system, such as radio or TV tower lighting outages, runway closures and commissionings, navigational aids out of service, or other system operational deviations. NOTAM Ds generally have potential concern to a wide pilot population and are thus given distant dissemination. Current (nonpublished) NOTAM Ds are sent by the weather computer network, are compiled and retransmitted every hour, and are appended to the weather of the affected station until published in the NTAP.

To illustrate how a NOTAM D is initiated, assume that an airport runway is to be closed for resurfacing. The airport manager so advises its area FSS, at which point the NOTAM desk prepares the NOTAM and enters it in the computer. The information then goes to the U.S. NOTAM Service office in Washington, which transmits the final, possibly slightly edited NOTAM to the Weather Message Switching Center (WMSC) in Atlanta. At this point, the WMSC appends the NOTAM information to each hourly weather report for dissemination to all FSSs, public-use airports, seaplane bases, and heliports that are listed in the *Airport/Facilities Directory*.

NOTAM Ls

Local NOTAMs are usually of interest to relatively few pilots and are thus reported to and retained by each FSS for only the facilities within its own geographic area of responsibility. The Ls include such information as taxiway closures, repair crews near or crossing a runway, rotating beacon outages, or airport lighting aid outages that don't affect instrument approaches.

While this is only a brief summary, its intent, in part, is to alert pilots to the importance of asking the briefer if any NOTAMs exist that might affect a planned flight. NOTAMs that have not yet been published and are still valid will be volunteered. Otherwise, as I said, the responsibility falls on the pilot to initiate the request, assuming the pilot hasn't had access to the biweekly publication. It's at the NOTAM desk, then, where the recording, updating, and necessary editing occur and where the NOTAMs are entered in the computer for briefing purposes.

In-flight

You reach the *in-flight specialist* when you contact the FSS by radio to open, amend, or close a flight plan; request a weather update; file a flight plan while en route (not a recommended procedure); or make a position report.

Figure 26-4 shows a series of lights, 48 in all. While there are some duplications, each light represents a radio frequency on which you can call the in-flight specialist. Including the few duplications (those where two radio outlets use the same frequency but are geographically miles apart), this panel could handle up to 48 frequencies.

The main issue here is that when you call in, the light representing the frequency you're using blinks, but it blinks only while you're talking. The instant you release the microphone button, the blinking stops. Perhaps the specialist heard your call, but if otherwise occupied and unable to see the blinking, he or she has no way of knowing which frequency to reply on. So you get no response. The point is this, a point that specialists stressed to me time after time: When you call, state the frequency you're using and your actual, or approximate, geographic location.

Fig. 26-4. *This is the bank of 48 frequency lights that the in-flight position must monitor.*

For example, you're at or over St. Joseph, Missouri, about 140 miles northwest of Columbia. The St. Joseph remote communications outlet (RCO) frequency to the FSS is 122.3, but Sedalia, Missouri, 45 miles west of Columbia, also has a 122.3 RCO. If you call in and merely say, "Columbia Radio, Cherokee Eight Five One Five November," the specialist, who might be talking to another aircraft, is unlikely to spot the blinking light and will have no way of knowing what frequency you're on.

Next time you call, "Columbia Radio, Cherokee Eight Five One Five November, on 122.3." That's more helpful, but are you at St. Joseph or Sedalia? The specialist still can't tell, if the blinking light went unnoticed.

One more time you call, "Columbia Radio, Cherokee Eight Five One Five November, on 122.3 at Saint Joe." Now, even if the light was missed, the specialist knows which transmitter to activate to acknowledge your call.

The format, then, is "(Blank) Radio" (always "Radio," not "Flight Service," in these calls), aircraft type, N number, frequency that you're calling on (which might be a VOR, so mention the VOR by name as well), and location. Now you'll get a response.

I said a moment ago that you could file a flight plan while en route by contacting the in-flight position. That's true, but again, it's not a recommended procedure. It takes up airtime and possibly blocks out others who are trying to get a call through.

As an example, while observing the one Columbia in-flight specialist on duty at the time, I noted at least 10 minutes was spent briefing an airborne pilot who then wanted to file an IFR flight plan. Both requests were thoughtless and most likely unnecessary. These were details the pilot should have handled on the ground. Meanwhile, the specialist wasn't available to respond to other calls that were coming in.

Another point worth noting: You're on a VFR cross-country with or without a filed flight plan. And you have or haven't been in contact with a Center for en-route traffic advisories. Whatever the combination, you'd like to be sure that someone down there knows where you are and has an immediately retrievable record of your call. So you get on the air to the nearest FSS and make a brief position report upon acknowledgment of your call: "Jonesboro Radio, Cherokee Eight Five

One Five November position report. Over Walnut Ridge at 1510, VFR to Tupelo."

Now, if a problem arises and you go down before having time to dial 7700 or contact another facility, the search for you that the FSS initiates will be limited to the territory between Walnut Ridge and Tupelo. That one position report could be the lifesaver, should you end up in Farmer Brown's cow pasture.

Being in contact with a Center as you cruise along is an excellent practice, but Center exists primarily for IFR operations and is not designed to keep an en-route progress check of VFR aircraft. The FSS does, however, maintain flight-following records, if you periodically call in. Like insurance, you might never need it, but that one call is a small price to pay for the additional "insurance coverage" that it provides.

Flight watch

The next positions are *flight watch,* often referred to as EFAS, the En route Flight Advisory Service. You're cruising down an airway and would like an updated, detailed report on the weather ahead. If you've been getting advisories from Center, contact the controller with whom you've been talking and advise that you're temporarily leaving the assigned frequency to call flight watch (Fig. 26-5). (That, by the way, is a procedure always to keep in mind. Never leave one facility to go to another without informing the first facility, which might have a need to reach you.)

With Center advised, tune to 122.0, the common low-altitude flight watch frequency across the country and address the call to the ARTCC responsible for the area in which you are flying, not the FSS. Because you're transmitting on the 122.0 frequency, you'll still be contacting the FSS facility, despite the oral address. Following your aircraft identification number, also include the name of the VOR closest to your present position. This the specialist needs to know in order to select the most appropriate transmitter/receiver outlet for communications coverage. For example, to reach the Macon (Georgia) flight watch, with the ARTCC located in Atlanta: "Atlanta Flight Watch, Cherokee Eight One Five One November, Albany VOR."

If you're not sure which FSS or Center is responsible for the area you're in, just address the call to "Flight Watch," again indicating the VOR station you are near, such as "Flight watch, Cherokee Eight Five

Fig. 26-5. *The flight watch position is equipped with charts, video images, and related data to help the specialist provide in-flight pilots with the most up-to-date weather reports.*

One Five November, Durango VOR." Once contact has been established, go ahead with your message.

Not all FSSs provide EFAS. Those that do are shown on the inside back cover of the *A/FD*. In the North Central *A/FD*, for example, although there are several other FSSs, only Columbia, Missouri (COU), and Princeton, Minnesota (PNM), offer the service through their various remote outlets (Fig. 26-6).

Keep in mind that EFAS means *en-route* flight advisories. It is thus designed to communicate weather conditions after departure climb-out and before the descent for landing, although immediate destination conditions will be provided on request. Also, you should be flying between 5000 feet AGL and 17,500 feet MSL. Between 18,000 and 45,000 feet MSL, EFAS can be contacted on discrete frequencies established for each ARTCC area.

Perhaps a question has already come to mind. If FSS has an in-flight position, why call flight watch, or vice versa? In-flight has responsibilities other than just reporting weather, and the weather information available is not as complete or detailed as flight watch's. Weather is a

ENROUTE FLIGHT ADVISORY SERVICE (EFAS)
Radio Call: Flight Watch-Freq. 122.0

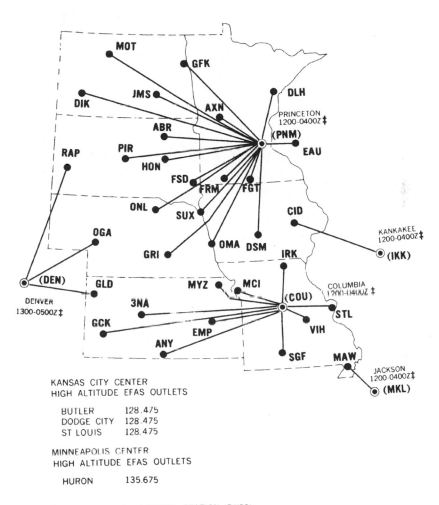

KANSAS CITY CENTER
HIGH ALTITUDE EFAS OUTLETS

BUTLER	128.475
DODGE CITY	128.475
ST LOUIS	128.475

MINNEAPOLIS CENTER
HIGH ALTITUDE EFAS OUTLETS

| HURON | 135.675 |

◉ FLIGHT WATCH CONTROL STATION (FWCS)

● COMMUNICATION OUTLETS

Fig. 26-6. *The inside back cover of the appropriate A/FD indicates the FSSs that provide flight watch service.*

flight watch specialist's sole responsibility, with no responsibility for flight plans, position reports, or the like. Flight watch is thus much more likely to be available for the specific weather-related information you want. Also, the specialist is more thoroughly trained in meteorology than the in-flight positions and is thus better equipped to read, interpret, and determine potential weather conditions. In-flight can certainly be of help, but flight watch is the FSS's primary source for short-range and real-time weather.

Pilot reports (PIREPs)

Something else is important: While flight watch gives weather information, it also wants to know what is actually happening right now where you are. If all is calm and routine, that's one thing; but if conditions, forecast or not, are such that they might pose problems for other pilots, the EFAS specialist wants to know.

What I'm talking about, of course, are pilot reports (PIREPs). Charts, radar graphics, maps, and so forth are essential, but nothing can replace first-hand descriptions of actual now-conditions that a pilot is experiencing. Just as flight watch will give us PIREPs it has received, it asks that pilots volunteer them, when necessary. One of this specialist's principal duties is to serve as a collection point for the exchange of PIREPs with other en-route aircraft. Of course, a PIREP can be given to any ground facility with which you may be communicating: Center, Approach, tower, or the FSS in-flight position; however, en route, flight watch is the primary contact.

Once received, and depending upon the nature of the conditions reported, PIREPs are used by

- Air traffic control towers to expedite the flow of traffic in the airport vicinity and for the avoidance of hazardous weather
- Air Route Traffic Control Centers to expedite en-route traffic flow, for favorable altitude determination, and to communicate hazardous weather information to en route aircraft
- FSSs for briefings, in-flight advisories, and weather avoidance information to en-route aircraft
- The National Weather Service to amend or verify aviation forecasts; to issue, when necessary, advisories; and for meteorological studies and research

A PIREP is thus not just a casual communication. Depending upon what is reported, it could be critical to the well-being of others.

Of primary concern to ground facilities are reports of these weather phenomena:

- Icing: trace, light, moderate, severe
- Turbulence: light, moderate, severe, extreme
- Clear-air turbulence: light, moderate, severe, extreme
- Thunderstorms
- Wind shear: loss or gain of altitude and/or airspeed experienced
- Clouds: bases, tops, layers
- Flight visibility or restrictions: haze, smoke, dust
- Precipitation
- Actual wind direction and speed, and temperature aloft

Be sure to report the conditions, particularly icing and turbulence, in accordance with *AIM's* definitions as summarized in Tables 26-1 and 26-2. For example, turbulence in a Cessna 152 might seem severe but, by definition, is only moderate or light. Misuse of terms because of unfamiliarity with them could mislead others to whom the PIREP is relayed.

When you submit a PIREP, of course it depends on what conditions you are reporting. If it's a routine PIREP reporting moderate turbulence (per the Table 26-1 description) in basically clear air, you'd go through the first six elements of the PIREP element code chart (Table 26-3), perhaps also element 7, if pertinent, and then skip to element 11, turbulence. As an example, after you establish contact with the Princeton, Minnesota, EFAS, the call would go like this:

> *PIREP. About 60 west of Grand Forks VOR on Victor 430 at 1845 UTC. Eight thousand five hundred. Cherokee 180. Scattered clouds. Continuous moderate turbulence last 40 miles. Cherokee One Five November.*

If any portion of the PIREP needs clarification, flight watch or the ground facility you have contacted will ask the necessary questions. Once your report is in, it will be organized and coded for transmission to other weather circuits.

Table 26-1.   The various intensities of turbulence and how they are described.

Turbulence Reporting Criteria Table

Intensity	Aircraft Reaction	Reaction Inside Aircraft	Reporting Term–Definition
Light	Turbulence that momentarily causes slight, erratic changes in altitude and/or attitude (pitch, roll, yaw). Report as Light Turbulence; [1] or Turbulence that causes slight, rapid and somewhat rhythmic bumpiness without appreciable changes in altitude or attitude. Report as Light Chop.	Occupants may feel a slight strain against seat belts or shoulder straps. Unsecured objects may be displaced slightly. Food service may be conducted and little or no difficulty is encountered in walking.	Occasional–Less than $1/3$ of the time. Intermittent–$1/3$ to $2/3$. Continuous–More than $2/3$.
Moderate	Turbulence that is similar to Light Turbulence but of greater intensity. Changes in altitude and/or attitude occur but the aircraft remains in positive control at all times. It usually causes variations in indicated airspeed. Report as Moderate Turbulence; [1] or Turbulence that is similar to Light Chop but of greater intensity. It causes rapid bumps or jolts without appreciable changes in aircraft altitude or attitude. Report as Moderate Chop. [1]	Occupants feel definite strains against seat belts or shoulder straps. Unsecured objects are dislodged. Food service and walking are difficult.	NOTE 1. Pilots should report location(s), time (UTC), intensity, whether in or near clouds, altitude, type of aircraft and, when applicable, duration of turbulence. 2. Duration may be based on time between two locations or over a single location. All locations should be readily identifiable.
Severe	Turbulence that causes large, abrupt changes in altitude and/or attitude. It usually causes large variations in indicated airspeed. Aircraft may be momentarily out of control. Report as Severe Turbulence. [1]	Occupants are forced violently against seat belts or shoulder straps. Unsecured objects are tossed about. Food service and walking are impossible.	EXAMPLES: a. Over Omaha. 1232Z, Moderate Turbulence, in cloud, Flight Level 310, B707.
Extreme	Turbulence in which the aircraft is violently tossed about and is practically impossible to control. It may cause structural damage. Report as Extreme Turbulence. [1]		b. From 50 miles south of Albuquerque to 30 miles north of Phoenix, 1210Z to 1250Z, occasional Moderate Chop, Flight Level 330, DC8.

[1] High level turbulence (normally above 15,000 feet ASL) not associated with cumuliform cloudiness, including thunderstorms, should be reported as CAT (clean air turbulence) preceded by the appropriate intensity, or light or moderate chop.

Briefing and the flight plan

The next part of our tour looks at briefing. The briefing positions are located in the center of the room (Fig. 26-3). Here are the people you talk to when you telephone the FSS. Each position is equipped with a computer screen, keyboard, and radar display (Fig. 26-7), which puts a wealth of data at the specialist's fingertips, including local conditions where you are, en-route and destination weather, ceilings, visibility, winds and temperatures aloft, icing information, forecasts, PIREPs, NOTAMs—just about anything you could want to plan a flight.

At this point, a few words are in order regarding preparation before you call the FSS and the sequence of information you can expect to receive. First, there is the matter of preparation. Many specialists have commented on the lack of information pilots initially supply and/or the disorganization of the information. The briefer, or specialist, will give you what you ask for but can't read your mind. If you're in Albuquerque and want the weather in Salt Lake City, that's what you'll get, but no more, unless you request it.

Table 26-2. The icing conditions that describe the four degrees of accumulation.

1. *Trace:* Ice becomes perceptible. Rate of accumulation is slightly greater than the rate of sublimation. It is not hazardous even though deicing/anti-icing equipment is not utilized unless encountered for an extended period of time (over 1 hour).

2. *Light:* The rate of accumulation may create a problem if flight is prolonged in this environment (over 1 hour). Occasional use of deicing/anti-icing equipment removes/prevents accumulation. It does not present a problem if the deicing/anti-icing equipment is used.

3. *Moderate:* The rate of accumulation is such that even short encounters become potentially hazardous and use of deicing/anti-icing equipment or flight diversion is necessary.

4. *Severe:* The rate of accumulation is such that deicing/anti-icing equipment fails to reduce or control the hazard. Immediate flight diversion is necessary.

EXAMPLE–
PILOT REPORT: GIVE AIRCRAFT IDENTIFICATION, LOCATION, TIME (UTC), INTENSITY OF TYPE, ALTITUDE/FL, AIRCRAFT TYPE, INDICATED AIR SPEED (IAS), AND OUTSIDE AIR TEMPERATURE (OAT).

NOTE–
☐ RIME ICE: ROUGH, MILKY, OPAQUE ICE FORMED BY THE INSTANTANEOUS FREEZING OF SMALL SUPERCOOLED WATER

Table 26-3.

The sequence in which a PIREP should be submitted and the contents of the report, as applicable to the current conditions.

PIREP ELEMENT CODE CHART

	PIREP ELEMENT	PIREP CODE	CONTENTS
1.	3–letter station identifier	XXX	Nearest weather reporting location to the reported phenomenon.
2.	Report type	UA or UUA	Routine or Urgent PIREP
3.	Location	/OV	In relation to a VOR
4.	Time	/TM	Coordinated Universal Time
5.	Altitude	/FL	Essential for turbulence and icing reports
6.	Type Aircraft	/TP	Essential for turbulence and icing reports
7.	Sky cover	/SK	Cloud height and coverage (sky clear, few, scattered, broken, or overcast)
8.	Weather	/WX	Flight visibility, precipitation, restrictions to visibility, etc.
9.	Temperature	/TA	Degrees Celsius
10.	Wind	/WV	Direction in degrees true north and speed in knots
11.	Turbulence	/TB	See AIM paragraph 7–1–20
12.	Icing	/IC	See AIM paragraph 7–1–19
13.	Remarks	/RM	For reporting elements not included or to clarify previously reported items

Fig. 26-7. *Another view of a typical briefing position.*

Form Approved: OMB No. 2120-0026								
U.S. DEPARTMENT OF TRANSPORTATION FEDERAL AVIATION ADMINISTRATION **FLIGHT PLAN**		(FAA USE ONLY) ☐ PILOT BRIEFING ☐ VNR ☐ STOPOVER				TIME STARTED	SPECIALIST INITIALS	
1. TYPE VFR IFR DVFR	2. AIRCRAFT IDENTIFICATION	3. AIRCRAFT TYPE/ SPECIAL EQUIPMENT	4. TRUE AIRSPEED KTS	5. DEPARTURE POINT	6. DEPARTURE TIME PROPOSED (Z) | ACTUAL (Z)		7. CRUISING ALTITUDE	
8. ROUTE OF FLIGHT								
9. DESTINATION (Name of airport and city)		10. EST. TIME ENROUTE HOURS | MINUTES		11. REMARKS				
12. FUEL ON BOARD HOURS | MINUTES		13. ALTERNATE AIRPORT(S)	14. PILOT'S NAME, ADDRESS & TELEPHONE NUMBER & AIRCRAFT HOME BASE				15. NUMBER ABOARD	
			17. DESTINATION CONTACT/TELEPHONE (OPTIONAL)					
16. COLOR OF AIRCRAFT		CIVIL AIRCRAFT PILOTS. FAR Part 91 requires you file an IFR flight plan to operate under instrument flight rules in controlled airspace. Failure to file could result in a civil penalty not to exceed $1,000 for each violation (Section 901 of the Federal Aviation Act of 1958, as amended). Filing of a VFR flight plan is recommended as a good operating practice. See also Part 99 for requirements concerning DVFR flight plans.						

FAA Form 7233-1 (8-82)   **CLOSE VFR FLIGHT PLAN WITH_____ FSS ON ARRIVAL**

Fig. 26-8. *The information to give the briefer at the outset of the call is simple—boxes 1 through 10 on the FAA flight plan form—then stop. The briefing will begin at this point.*

Quite obviously, then, the first thing to do before you pick up the phone is to plot the flight and then complete the FAA flight plan form (Fig. 26-8). Once you have the briefer on the line, state your qualifications (as student, private, commercial) and whether you are instrument-rated. From that point on, read off the information you have entered in boxes 1 through 10 on the FAA flight plan form:

1. Type of flight—VFR or IFR
2. Aircraft N number
3. Aircraft type and special equipment, which means the alphabetical identification of the avionics you have on board, the most common of which are the following (if you have GPS or flight management systems on board, see AIM for better identification):
   A DME, transponder with altitude encoding
   B DME, transponder with no altitude encoding
   C RNAV, transponder with no altitude encoding
   D DME, no transponder
   R RNAV, transponder with altitude encoding
   T Transponder with no altitude encoding
   U Transponder with altitude encoding
   W RNAV, no transponder
   X No transponder

Example: If a Piper Cherokee 180 had a DME and a transponder with altitude encoding, the entry in box 3 would be PA-180/A.

4. True airspeed
5. Departure point
6. Proposed departure time [UTC (Zulu) time, not local time]
7. Intended cruising altitude
8. Route of flight
9. Destination, or first stopping point if the intended ground time will exceed 1 hour
10. Estimated time en route

After box 10, tell the specialist what type of briefing you want: *standard, abbreviated,* or *outlook.* A standard briefing is just what it says. The specialist will give you all the data pertinent to your route of flight and proposed altitude. If you've had a standard briefing and a couple of hours later want an update, ask for the abbreviated. The outlook should be requested when your departure is 6 hours or more from the briefing; however, in this case, be sure to call back, tell the briefer you've received the outlook, and then request the full standard.

Assuming you've asked for a standard briefing, the specialist will give you the information in this sequence:

1. Adverse weather conditions.

2. A synopsis of existing fronts and other weather systems along your line of flight. If conditions are obviously unfavorable, based on 1 and 2, it is at this point that the briefer will probably state, "VFR is not recommended." Otherwise, or if you still want more information, the briefing will continue.

3. A summary of current weather along the route of flight, based on surface observations, radar observations, and PIREPs.

4. A summary of forecast en-route weather.

5. Detailed forecast destination weather.

6. A summary of forecast winds at your proposed altitude range and, if requested, temperature information.

7. Unpublished NOTAMs that could affect your flight, if requested.

8. Additional information you request, such as active MOAs or MTRs in your line of flight, or published NOTAMs.

As a suggestion, before you call an FSS, put these main headings on a piece of paper, leaving space between each for notes:

1. Adverse weather

2. Synopsis (along route of flight)

3. Current en-route weather

4. Forecast en-route weather

5. Forecast destination weather

6. Winds and temperature (at proposed altitude)

7. NOTAMs

8. Other: MOAs, MTRs, published NOTAMs, and the like

Now you'll know what's coming and in what order—you just fill in the blanks.

Filing the flight plan

Before the days of FSS consolidation, the typical practice was to get the briefing, hang up the phone, replot the flight as necessary based on the briefing, and then call back to file the flight plan. There was, and perhaps still is, an element of logic to that in the nonautomated FSSs.

It's a different story, however, today. As you're giving the briefer the data you have entered in boxes 1 through 10 on the flight plan form, the briefer is entering those data into the computer. After box 10, the weather is reported in the sequence I've just outlined. Once any questions you might have had have been answered, the briefer will ask for the balance of the flight plan form information, boxes 11 through 17.

At this point, assuming that the weather will have no material effect on your route of flight or will require extensive replanning, ask the briefer to file the flight plan, which, in effect, has already been done because it's in the computer.

Here is something to keep in mind: While you provide the data on the flight plan form, speak slowly enough and clearly enough that, as one FSS manager put it, "the specialist typing the information can absorb and digest it." Doing as the manager suggests saves time, especially if the specialist doesn't have to ask you to repeat data that could not be understood or that you transmitted too rapidly.

Afterward, and based on the briefing, do whatever replotting is necessary. If your en-route or estimated arrival times are going to change to any measurable degree, just correct the filed times when you radio the FSS to open the flight plan. Thus you save another telephone call and possibly a wait, and you won't be duplicating information that the original briefer had already entered in the computer.

Another means of filing a flight plan offered by the FSSs is *fast file*. When you make the initial telephone contact, a recording will read off the menus of service available, including fast file, and the subsequent number to dial. When you do so, all that's necessary is to give the data in the exact sequence of the flight plan form. Everything is recorded and stored in the FSS's computer, ready for reference when you make the radio call opening the flight plan.

A suggestion made by the McAlester, Oklahoma, FSS is worth noting. In essence, the comment was that fast file is heavily used in areas with large volumes of traffic and when the FSS is very busy. In areas of less traffic or on less busy days, the preferred method is to establish contact with a specialist who copies the flight plan directly, asks questions that might need to be asked, and determines where the caller can be reached over the next few minutes if a discrepancy of some sort is discovered.

To go back to what I said earlier, the FSS is not a controlling agency. If the briefer observes that "VFR is not recommended," the decision to go or not go is yours; however, the fact that the recommendation was made is recorded as part of the briefing, just in case....The briefer has made the recommendation; but if you want to fly in the face of knowledgeable advice, well, it's your neck.

As somewhat of an aside, when you or I receive a briefing or file a flight plan, we're not competing with the air carriers. Almost all airlines have their own weather sources and dispatchers, and their flight plans are filed directly with the originating Center. Even more, flight plans for daily scheduled flights are stored in Center's computer and retrieved prior to the scheduled flight departure. In essence, then, the service I've been discussing is for general-aviation IFR and VFR operations; so let's use it.

Closing the flight plan

Failure to close a flight plan after landing is one of the more common and costly pilot oversights. Admittedly, it's an easy thing to forget. You're tired after a few hours in the air; friends, family, or business associates are waiting for you; you're in a hurry to get one of the last rental cars; you're a student and your instructor is anxious to hear how things went on your first cross-country; you just want to get the airplane tied down or hangared so you can get home. Whatever the case, you forget that final responsibility. The last line on the flight plan form says, in capital letters, CLOSE VFR FLIGHT PLAN WITH _____ FSS ON ARRIVAL.

To minimize the chance for oversight, immediately radio the nearest FSS, if it's possible to reach it on the ground, when you come to a halt on the ramp. Otherwise, make a mental note to pick up a phone as soon as you're in the terminal or FBO's shop, and simply dial the nationwide number: 1-800-WX BRIEF. Don't let outside interferences distract you. The results could be costly.

What happens when a flight plan is not closed

The following is the sequence of events when you file a flight plan but don't close it out:

1. Upon departure, you radio the appropriate FSS to open your flight plan; let's call the first FSS "Alpha."

2. The in-flight specialist sends the flight plan to the FSS that is responsible for the area of your destination airport; let's call the second FSS "Bravo." All that is transmitted is your aircraft type, N number, destination, and ETA.

3. You land at your destination, or some other airport, and fail to close out the flight plan with any FSS.

4. Thirty minutes after your ETA, the computer in FSS Bravo flashes the flight plan data sent by Alpha, indicating you are overdue and haven't closed the flight plan.

5. FSS Bravo sends a query to FSS Alpha to determine if you actually departed or were delayed.

6. FSS Alpha calls your departure airport to verify your actual departure.

7. Because you did depart, FSS Alpha so notifies Bravo and sends to Bravo the balance of your flight plan information: route of flight, fuel on board, number of passengers, your name and address.

8. Bravo calls the tower or FBO and your intended destination to see if you landed

9. If so, and your aircraft is located, that is the end of the search. Let's assume, however, that you landed at a different airport than originally intended, or that neither you nor your plane can be located at your planned destination. The search then goes on.

10. One hour after your ETA, FSS Bravo initiates an *information requested* (INREQ). This goes to all FAA facilities along your route of flight, including Centers and towers to determine if any facility has heard from you. If one has, the search can be focused on the territory between your last reporting point and destination. A copy of the INREQ is also usually sent to the applicable search-and-rescue unit, alerting it to the possibility of a downed aircraft.

11. If there has been no recorded or immediately available evidence of any en-route contact, all FSSs along the route of flight begin telephoning every airport 50 miles either side of the route.

12. If these actions draw a blank, 11/2 hours after your ETA, an *alert notice* (ALNOT) is sent to every FAA facility along your route.

13. Each facility does further checking, such as playing back tapes to find any record of radio contact with you. The search-and-rescue unit also alerts the Civil Air Patrol. The Air Force might enter the investigation, contacting family or business associates to determine if they have heard from you or if you had indicated any possible route deviation before departing. All FSSs in the area broadcast over the VORs each hour that an aircraft has been lost and asks pilots to monitor 121.5 (the emergency frequency) for an activated emergency locator transmitter (ELT).

14. The physical search begins.

See what problems can be caused by a simple oversight? In one case, hardly exceptional, an FSS made more than 100 telephone calls to airports along a pilot's route before the pilot was located safely at an airport other than the planned destination. The pilot's only excuse? He forgot!

All this only emphasizes the value of making periodic position reports to an FSS. If you should have a problem, the search can then be narrowed to the area between the last reporting point and your destination.

Today's aircraft are dependable, but things do go wrong. How much better to keep people down there informed of your progress than to fly merrily along and perhaps flounder for hours in an obscure cornfield while others are searching thousands of square miles for a small speck on the ground. And don't disregard the expense of a search—justified or not—in both time and money.

Filing to the first stop

You've got a 6-hour journey ahead of you, but you plan to make a pit stop about midway. Should you file a flight plan to the final destination or to the midpoint airport? The answer is rather apparent: If you're going to be on the ground 1 hour or more at that airport, file to the next point of landing, no matter how many legs the trip might have.

Say you leave Point A at 1200 local, planning to stop briefly at Point B, but you file to your destination, Point C, with an ETA of 1800 local. Thirty minutes out, the prop grinds down to windmilling status, you hit a furrow in the forced landing, and you end up on your back. It's now 1230, but it will be 1830 before the FSS at Point C even starts asking questions about you. The more intensive wheels of inquiry won't begin to roll until 1900 local. That might mean 6½ hours trapped upside down in a bent airplane before a search even starts to get serious.

If you had filed to the midway point, Point B, with a 1500 ETA, the search would have started at 1530, a full 3 hours earlier. Also, the search efforts would be limited to the area between Points A and B, further enhancing the chances of rescue and survival. Caution and wisdom do contribute to pilot longevity.

Transcribed services

The next stop on the tour of the FSS is a small room where two transcribed services are recorded. One is *Hazardous In-flight Weather Advisory Service* (HIWAS), and the other is *Transcribed Information Briefing Service* (TIBS).

HIWAS

This service is a continuous broadcast over those VOR navigation aids so indicated by a small square in the lower right corner of the VOR identification box (Fig. 26-9). Included in the broadcasts, as pertinent, are *severe weather alerts* (AWWs), *airman's meteorological information* (AIRMETs), *significant meteorological information* (SIGMETs), *convective sigmets* (WSTs), *urgent pilot reports* (UUAs), *area forecasts* (FAs), and *Center weather advisories* (CWAs). If no weather advisories have been issued, an hourly statement to that effect is recorded and broadcast over the selected VORs.

The HIWAS is provided 24 hours per day, and the FSS completes the recording of the applicable data within 15 minutes after receipt. The actual broadcast message includes

- Statement of introduction
- Summary of the reported conditions (AWW, SIGMET, AIRMET)
- Recommendations to contact the FSS for further details
- Request for PIREPs

Fig. 26-9. *The small square in the lower right corner identifies the VORs that transmit HIWAS.*

Example:

> *HIWAS FOR JACKSONVILLE CENTER AREA AND PORTIONS OF ATLANTA AND MIAMI CENTER AREAS RECORDED AT ONE FIVE THREE ZERO ZULU. CONVECTIVE SIGMET ONE SEVEN ECHO. FROM FIVE ZERO SOUTH OF ST. PETERSBURG TO THREE ZERO SOUTH OF COLUMBUS, LINE OF THUNDERSTORMS THREE FIVE MILES WIDE MOVING EAST AT ONE FIVE KNOTS. MAXIMUM TOPS FOUR SEVEN THOUSAND. ISOLATED THUNDERSTORMS OBSERVED ON WEATHER RADAR VICINITY OF CRESTVIEW. IFR CONDITIONS ARE REPORTED AT VALDOSTA AND JACKSONVILLE. CONTACT FLIGHT WATCH OR FLIGHT SERVICE FOR ADDITIONAL DETAILS. PILOT REPORTS ARE REQUESTED.*

In addition, Centers and airport traffic control towers alert pilots to the existence of a HIWAS condition on all except the emergency frequency, 121.5. A typical Center alert is shorter:

> *ATTENTION ALL AIRCRAFT: MONITOR HIWAS OR CONTACT A FLIGHT SERVICE STATION ON FREQUENCY MEGAHERTZ OR _____ MEGAHERTZ FOR NEW CONVECTIVE SIGMET (NUMBER) INFORMATION.*

TIBS

This FSS service is similar to what has been known as *Pilots' Automatic Telephone Weather Answering Service* (PATWAS), but covers many more geographic areas.

To be more specific, once you have reached the FSS and listened to the recorded menu of services available, dial 201, which is the universal number to access TIBS. Another recording then tells you what subsequent 200-series number to dial for the desired information. Those numbers range from 202 to 224, although not all numbers might be currently in use. You'll then hear a transcript of the meteorological conditions within a 50-mile radius of the location you have selected. For example, at the Columbia FSS, by dialing the following numbers, except 202, 203, 211, 214, and 215, the weather data at the selected cities, locations will be summarized:

202Weather synopsis

203Thunderstorm activity

206Springfield

207Joplin

204Columbia

208Cape Girardeau

205Kansas City

209St. Louis

210Kirksville

211Winds aloft

214ATC delays

215How to get a quality briefing

Under the PATWAS system, the caller was limited to the conditions within a 50-nautical-mile radius of the FSS location. TIBS, however, offers a much wider range of locations, plus other pertinent data, simply by dialing the applicable numbers.

Despite the benefits of TIBS, it should not be a substitute for a standard briefing. Everything might be fine at departure and destination points, but what about the in-between weather? TIBS won't really tell you that. What about VORs that might be out, runway construction at the en-route airports, or active MOAs on the route of flight? TIBS provides area weather information and can be helpful in arriving at a go/no-go decision. If things look good, fine. But get the other details from a briefing specialist.

In-person briefing

The final stop on the FSS tour is the *in-person briefing* position (Fig. 26-10). Here the pilot can see firsthand the weather radar displays, the computerized forecasts, winds aloft printouts, and all the

Fig. 26-10. *The in-person briefing position.*

information the telephone briefer provides. This type of briefing can be more instructive, particularly when conditions are marginal or questionable. It's sometimes difficult to visualize things over the telephone. Unfortunately, this advantage is being lost with consolidation, except for those operating where an FSS is located. Here is something to consider: When you plot a cross-country flight with overnight stops, and if it is practical to do so, perhaps you should plan to land at an airport with an FSS. You'd then be able to enjoy the benefits of an in-person briefing and flight plan filing prior to departure the next day.

Insurance and the FSS

If one word could sum up the value of Flight Service to the VFR pilot, it would probably be *insurance*. Indeed, the services available are optional. You don't have to get a briefing, file a flight plan, make periodic position reports, get weather updates from flight watch, offer PIREPs, and—except for FAA test purposes—know anything about Flight Service, what it does, or what it offers.

But that seems more than a little shortsighted. Here's a facility with a wealth of information at its disposal and one that could literally be a lifesaver in an emergency. There are those among us, however, who don't know what services are available or how to use them. And there are those who consider themselves above the need for briefings, filing a flight plan, and the rest. Perhaps, just perhaps, this chapter might persuade the nonusers that taking advantage of Flight Service is simply common sense.

27

When an emergency occurs

Understanding the roles that ground facilities play during an emergency is important, but it's equally critical to be well versed in our own pilot responsibilities when things are tight and help is needed. Radio failure is a good place to start, followed by an examination of other in-flight emergencies.

Radio failure

A potential radio failure is pretty hard to predict, although a thorough preflight check could give some clues. A loose or broken antenna is an obvious signal; as is a popped circuit breaker; or a loose, worn, or frayed alternator belt, if you can see or test the belt. Another clue is a radio that tends to slip in and out of its rack. And still another is garbled reception on the ground or controller comments that your transmission is weak or scratchy.

If any of these symptoms exist on the ground, either the radio's sick right now or it could die on you at an inappropriate time. The only solution: Get it fixed or the potential cause corrected before you venture forth.

A failure is suspected

Let's assume, though, that everything checks out on the ground, so off you go. After a while, you become conscious of the fact that there hasn't been much chatter over the air for several minutes, which causes you to wonder if....These are some things you can do when a problem is suspected:

- Adjust the squelch or turn up the volume. If you hear the typical static, the set's probably OK; there's just been a period of unusual radio silence.

- Push the radio in a little. Vibration or turbulence might have caused it to slip slightly from its rack.

- Wiggle and push all microphone and headset plugs to verify that they are firmly in place. (A spare microphone is inexpensive insurance.)

- Check the circuit breakers. If one has popped, let it cool for a couple of minutes and then reset it. You might get your radio back, but a popped breaker is symptomatic of a problem, so have a mechanic investigate the cause when you're back on the ground. If the breaker pops out again, leave it alone; never try to force a breaker to remain reset.

- Check the ammeter. If it shows no charge, test it by turning on the landing light. If the needle doesn't move, you can be sure the alternator has died or its belt has broken, in which case the only electric power will come from the battery. The engine won't quit, because the ignition system is independent of the alternator-battery system; but without an alternator, the life of the battery is only about 2 hours. After that, you'll have no electric power at all. Consequently, turn off all nonessential electrical equipment, except one radio, and head for home or the nearest airport. Enough battery power might be left to make whatever radio contacts are necessary before you land.

The failure is real

It's no longer a matter of suspicion: The radio is dead or rapidly dying. Now, what are your options? Five of the most likely answers depend largely on the situation.

1. You're flying locally or on a short cross-country trip away from Class B, C, or D airspaces and intend to terminate the flight back at your own uncontrolled airport. Once you realize the radio is gone, the best thing to do is head for home or some other uncontrolled field where you can get the radio repaired. There is no need to squawk the 7600 RF (radio failure) code in this environment, but do use extra caution when landing. You're coming in unannounced with presumably no knowledge of who or where anyone else is in the pattern.

2. You're going into a Class B or Class C primary airport. To make it simple, let's say that you had enough battery juice left to monitor the ATIS and to be cleared into the Class B or to

establish contact with the Class C Approach Control. Once you are inside the Class B or C, though, the radio dies completely, and you've lost all transmitting and receiving contact with Approach.

At this point, squawk the 7600 code and continue through B or C airspace to the airport. Approach knows your intentions and, seeing the RF code on the radarscope, will protect you as well as advise the tower of your predicament.

When you're within the tower's 5-mile area of control, watch the tower for light signals, as summarized in Fig. 27-1, that will clear you to land or tell you to keep circling. Once on the ground, pull off on a taxiway, and continue to watch the tower for the green light that authorizes you to taxi to the ramp.

This same scenario applies if you're transiting a Class B or Class C airspace or if you're already in one of the airspaces but intend to land at a satellite airport. The main point is to keep right on going in accordance with your announced intentions. Don't wander around in those high-density areas. If you deviate from what you've told the controller, the controller won't know what to expect, which could cause confusion in his or her efforts to maintain an orderly flow of traffic.

3. You want to land at a Class D airport that has no Approach Control and is miles from a Class B or C. Some distance out, the radio dies, so you squawk the RF transponder code. The nearest Center or Approach, seeing the code on its radar, will

Meaning			
Color and Type of Signal	Movement of Vehicles Equipment and Personnel	Aircraft on the Ground	Aircraft in Flight
Steady green	Cleared to cross, proceed or go	Cleared for takeoff	Cleared to land
Flashing green	Not applicable	Cleared for taxi	Return for landing (to be followed by steady green at the proper time)
Steady red	STOP	STOP	Give way to other aircraft and continue circling
Flashing red	Clear the taxiway/runway	Taxi clear of the runway in use	Airport unsafe, do not land
Flashing white	Return to starting point on airport	Return to starting point on airport	Not applicable
Alternating red and green	Exercise extreme caution	Exercise extreme caution	Exercise extreme caution

Fig. 27-1. *The various colors and meanings of the tower light gun signals.*

track your flight route and if it appears that you're headed for the controlled airport, will notify the tower of your probable landing intentions.

As you near the field, keep a sharp eye out for the light gun signal, especially a red signal that tells you, in essence, not to land. If you see no signal, cross over the field about 500 feet above traffic pattern altitude, note the flow of traffic, fly upwind over the active runway, and watch for the green "cleared to land" signal. Keep an eye on the tower, though, even after the clearance signal. For a variety of possible reasons, you might get a red light on the final approach. Do a go-around then for another approach, as long as you get the green light again.

4. As opposed to the second situation cited above, this time you have not been using Center or been in contact with Approach, but you want to land at the primary airport in a B or C airspace, and the radio has failed. Your alternative? Only one: Land outside the area and request entry approval from Approach by telephone. You probably won't get it, though, unless the traffic is very light and you're close to the B or C airspace. Even then, the odds are against approval of any aircraft without an operating two-way radio.

5. You're on the ground at an uncontrolled field. The radio is dead, but you want to fly to a nearby Class D airport to have it repaired. Your only option is to telephone the airport tower, explain the situation, and ask for approval to enter the area.

Depending on the probable traffic at your estimated arrival time, approval might or might not be forthcoming. You're on the ground and there is no emergency, so airport conditions will largely determine what the Class D tower says. In most cases, though, tower supervisors want to help and, traffic permitting, will attempt to accommodate your request.

Know what to do—and have alternatives

A VFR radio failure, when viewed objectively, is really more of a nuisance than a true emergency and should not be the cause of cockpit panic. The best advice, unless you're actually in a Class B, C, or D airspace when the failure occurs, is to land at the most convenient uncontrolled airport.

On the other hand, if the set dies while you're in one of those controlled airspaces, stay with your already announced intentions and proceed directly to the field. With the RF code on the radar screen, the controller will protect you from other aircraft until you're in the tower's traffic area. From that point on, the tower controller will clear you for landing with the help of the light gun.

Quite apparently, you and I have to know what those on the ground expect of us when we're faced with a radio failure. Otherwise, we could be guilty of causing a lot of confusion and perhaps creating hazardous conditions for others and ourselves. The only answer is to be prepared for the problem and have a clear set of alternative actions in mind if the problem should ever become a reality.

In-flight emergencies

The FAA makes it very clear that when an emergency develops in flight, only one person has the final responsibility for the operation of the aircraft. That person is the pilot. The FAA further makes it clear that rules and regulations are pretty much tossed aside when action is required to meet the emergency. In the process, however, the FAA stresses that the pilot should request immediate help through radio contact with a tower, a Center, or a Flight Service Station.

Emergency classifications

Emergencies are classified as *distress* or *urgency*. A distress condition is one of fire, engine failure, or structural failure. In other words, the situation is dire, immediate, and life-threatening. An urgency is not necessarily immediately perilous but could be potentially catastrophic, such as being lost, a low fuel supply, a seriously malfunctioning engine, pilot illness, weather, or any other condition that could affect flight safety. When any of these conditions arise, the pilot should ask for help now. Don't wait until the urgency becomes a distress.

While *AIM* outlines the appropriate emergency procedures in some detail, the following summarizes the essential elements for the VFR land pilot.

Transponder operation

When either an urgency or a distress situation develops, immediately enter the 7700 emergency squawk in the transponder. That code then appears on the screens of all radar-equipped facilities within radar range and, by sound as well as the flashing blip, attracts the controllers' attention. As controllers put it, "Lights light and bells ring." Ground help is, of course, hard to offer if you haven't been in contact with a Center or any other facility, but at least they know that there's an aircraft in trouble out there and know its location.

Radio communications

If you have been in routine contact with a ground facility, such as a Center, and a distress situation suddenly arises, the first thing to do is to dial the 7700 transponder emergency code and communicate as quickly as possible with the Center controller. Once you are in contact with him or her, and assuming you have initially volunteered little information, the controller is likely to ask as many questions about you as the situation permits, such as the number of people aboard, color of your aircraft, and the like.

How much actual help the controller can offer at this juncture depends on how long you can remain airborne. Given time, he or she might be able to guide you to another airport, if one is near your present position, or alert you to ground obstructions that could pose a problem, or perhaps lead you to a major highway that could serve as a landing strip. At the worst, in distress situations, the controller may be limited to just making sure that rescue forces and equipment have been alerted and kept advised of where you finally put down.

Let's say, though, that you haven't been in contact with any ground facility and the emergency—distress or urgency—develops rapidly. What do you do in this case? For one, you probably wouldn't have time to search for the frequency of a tower, a Center, or whatever; so you quickly dial the 7700 emergency transponder code, tune to 121.5, the universal emergency-only frequency that is guarded by direction-finding stations, civil aircraft, Centers, military towers, Approach Control facilities, and FSSs, and make your call. If the emergency is of the distress nature, start the call with "Mayday," repeated three times. This is the universal term asking for assistance and, to refresh your memory, comes from the French word "*M'aidez*," pronounced "Mayday," meaning "Help me." Distress messages have priority over all others, and the word Mayday commands silence on the frequency in use.

If the situation is of an urgent nature, begin the call with "Pan-," also repeated three times. Urgency messages have priority over all others except distress and warn others not to interfere with the various transmissions.

Then, recognizing the problem of time, particularly in distress situations, the FAA suggests that you communicate in your initial message as much of the following as possible, preferably in this sequence:

1. "Mayday, Mayday, Mayday" or "Pan-pan, Pan-pan, Pan-pan"
2. Name of facility addressed
3. Aircraft identification and type
4. The nature of the distress or urgency
5. Weather
6. Pilot's intentions and request
7. Present position and heading, or if lost, last known position, time, and heading since that position
8. Altitude or flight level
9. Hours and minutes of fuel remaining
10. Any other useful information, such as visible landmarks, aircraft color, emergency equipment on board, number of people on board
11. Activate the emergency locator transmitter (ELT) if possible

Pilot responsibilities after radio contact

Once in contact with a ground facility, you have certain responsibilities:

- Maintain control of the aircraft.
- Comply with advice and instructions, if at all possible.
- Cooperate.
- Ask questions or clarify instructions not understood or with which you cannot comply.
- Assist the ground facility in controlling communications on the frequency. Silence interfering stations.
- Don't change frequencies or change to another ground facility unless absolutely necessary.
- If you do change frequencies, always advise the ground facility of the new frequency and station before making the change.
- If two-way communication with the new frequency can't be established, return immediately to the frequency where communication last existed.

- Remember the four Cs:
  - *Confess* the predicament to any ground station.
  - *Communicate* as much of the distress or urgency message as possible.
  - *Comply* with instructions and advice.
  - *Climb*, if possible, for better radar detection and radio contact.

The emergency locator transmitter (ELT)

Emergency locator transmitters, designed to assist in locating downed aircraft, are required by FAR 91 for most general-aviation aircraft, although certain exceptions are allowed. An *ELT* is a battery-operated transmitter that when subjected to crash-generated forces, transmits a distinctive and continuous audio signal on 121.5 and 243.0. The life of a transmitter is supposed to be 48 hours over a wide range of temperatures.

Depending on its location in the aircraft, some ELTs can be activated by the pilot while airborne. In other installations, the ELT is secured elsewhere in the fuselage and cannot be accessed in flight. These are activated only upon impact or by the pilot when on the ground and out of the airplane.

Because of their importance in search-and-rescue (SAR) operations, ELT batteries are legal for 50 percent of their manufacturer-established shelf life, after which they must be replaced. Periodic ground checks of an ELT should be made, but only in accordance with the procedures outlined in *AIM*.

Anticipation and alternatives

In any discussion of emergencies, one pilot responsibility hopefully stands out: the responsibility to be prepared mentally and physically for the unexpected, the nonroutine. If you fly long enough, sooner or later you're going to encounter an emergency situation of some nature. Perhaps it will be very minor and easily correctable; perhaps it will require every bit of knowledge and expertise you have amassed. Whatever the case, the odds of something going sour sometime are close to 100 percent.

If that's a reasonable bet, preparation (and all that preparation implies) is absolutely essential. Beginning with every pilot's very first flight, he or she should be asking two questions: What could go wrong? Then, if what could go wrong did go wrong, what would I do? This is only the logical process of potential problem analysis (PPA).

A fire in flight; the engine bucks and coughs for no apparent reason; the engine quits entirely; a bird strike smashes the windscreen; a passenger has an apparent heart attack; you're lost; the fuel gauges are showing close to empty; electric power is interrupted: What would you do? It's a matter of anticipating these as well as other potential problems and then having alternative plans of action firmly in mind in case a potential problem ever became a reality. (As always, though, and above all else, fly the aircraft first, then follow through with your planned and alternative actions.)

While transponder and radio procedures are perhaps only small elements in handling a distress or urgency situation, they could be major factors in helping you get that airplane down safely and in one piece. Our entire ATC system is designed to maximize safety, and the folks on the ground have had drilled into them what to do when a pilot calls Mayday or Pan-pan or squawks the emergency or RF code. But controllers can only help pilots to the extent that pilots can help controllers. That's where personal preparation and the skill with which pilots handle the emergency come into play.

When was the last time you asked, "What could go wrong?"

Part Four

Integrate

Armed with a bag of tools: an airplane, a map, and "street smarts," you're ready to fly on your own...

28

A cross-country flight: Putting it all together

Now let's put it all together with a mythical cross-country trip. Hopefully, the "flight" will serve as a reasonable model for the real-life excursions you might make. Naturally, your itineraries will differ, as will the frequencies, but the principles illustrated should not vary because of that.

The flight will take you from the Kansas City Downtown Airport to Omaha's Eppley Field, where you'll pick up two friends. From there, you'll go to the Minneapolis International Airport. After completing your business in the Twin Cities, you'll drop off one of your friends at Mason City, Iowa, then the other friend at Newton, Iowa. From Newton, it's nonstop back to Kansas City.

The entire flight presupposes that you have a Mode C transponder, two navcoms, distance measuring equipment (DME), and no loran or GPS. If you don't have the luxury of multiple navcoms, the principles are the same, but frequency changing is a little less convenient.

The flight route

Without loran, you decide to fly the airways whenever possible, even though doing so will add a few miles to the trip. Accordingly, the route of flight, VORS, course headings, and point-to-point nautical mileage resemble Fig. 28-1.

As 900 statute miles is obviously too much territory to cover in one day, with stops and business en route, you plan to stay overnight in Minneapolis. You'll also refuel at each stop, except Mason City, and use the appropriate Air Route Traffic Control Centers for VFR advisories.

The reason for this itinerary is simply to review the communication procedures involved in the various situations, airports, and controlling agencies we discussed in the previous chapters. To wit:

657

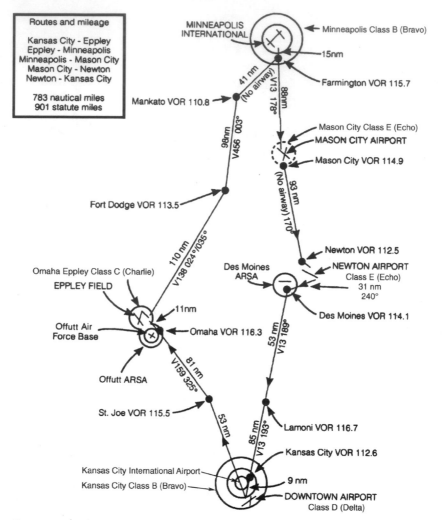

Fig. 28-1. *The route of the simulated flight.*

- Kansas City Downtown Airport, a Class D airport, lies under a Class B airspace.
- Omaha Eppley is a Class C airspace.
- Minneapolis International Airport is a Class B airport.
- Mason City is a unicom Class E airport with a weather observer, but no tower or FSS is on the field.
- Newton is an uncontrolled field with only unicom.

Let's assume that all preliminaries have been completed, including weight-and-balance calculations and filing the flight plan.

Recording the frequencies

As part of the preflight planning, write down in sequence the known or probable frequencies you will use. Some, particularly Center's, might differ from what you expect, but at least you'll be ready for the majority that come into play.

I suggest you do not record all the frequencies on one piece of paper for a flight like this with five different legs. Enter the frequencies to Omaha on one sheet, those from Omaha to Minneapolis on another, and so on (Figs. 28-2 through 28-6). Then number each sheet in sequence. As you leave one frequency and progress to the next, draw a line through (but don't obliterate) the one you have just left. This preliminary recording and progressive deleting provides cockpit organization and minimizes some of the confusion that is often the bane of the private pilot.

Two other suggestions: On each segment page, provide space for the ATIS information at the destination airport. When you're on the ground before departure, this isn't so important because you can listen to the local information as many times as necessary. In the air, however, it's another matter. Center has handed you off to Approach, but before contacting Approach, you should have moni tored the ATIS—which means that you don't have a lot of time to absorb the data being transmitted. Approach is expecting to hear from you rather promptly after the handoff.

Consequently, to expedite matters, line out a box on the flight segment page so that you can record the crucial information:

1. The phonetic alphabet
2. The sky or cciling
3. Visibility
4. Temperature
5. Dewpoint
6. Altimeter setting
7. Runway in use
8. Other important information that might be included.

Now, when you call Approach, you can advise the controller that you "have Charlie," or whatever, and be sure that you have it accurately. Memories do fail us.

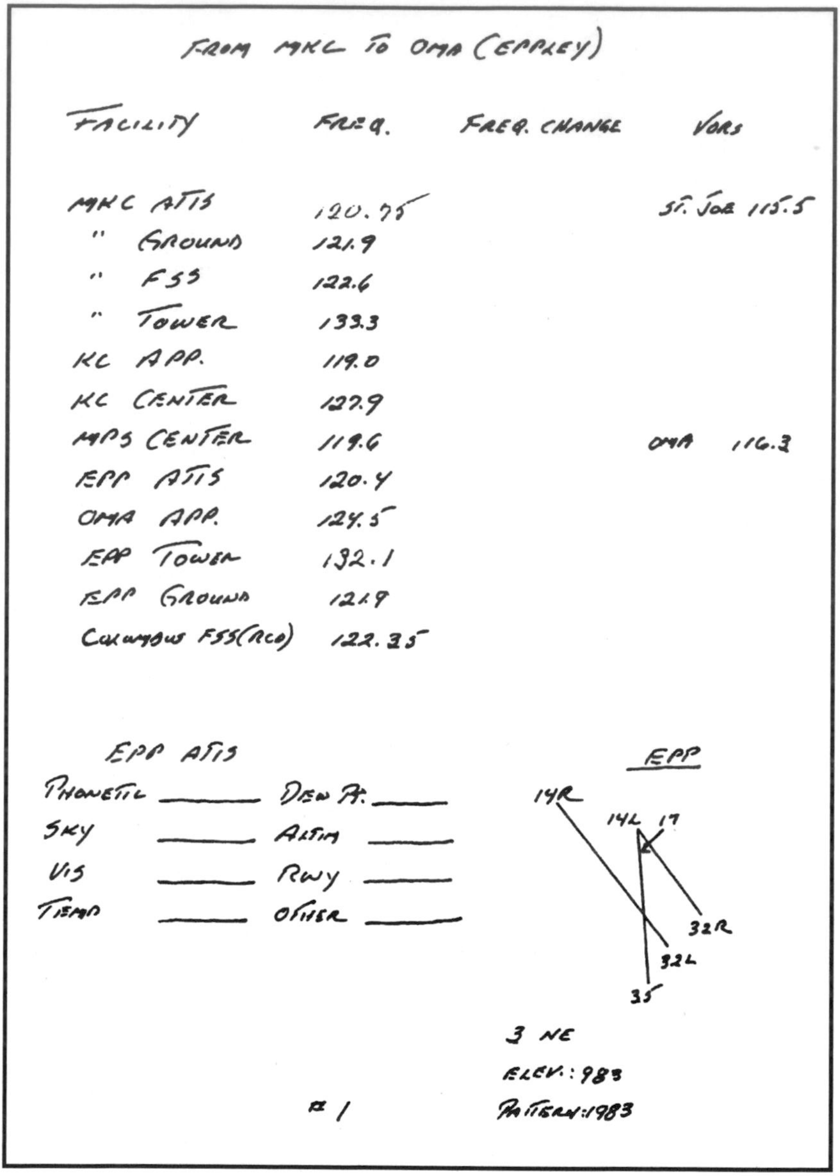

From MKC to OMA (Eppley)

Facility   Freq.   Freq. Change   Vors

MKC ATIS   120.75   St. Joe 115.5
   " Ground   121.9
   " FSS   122.6
   " Tower   133.3
KC APP.   119.0
KC Center   127.9
MPS Center   119.6   OMA 116.3
EPP ATIS   120.4
OMA APP.   124.5
EPP Tower   132.1
EPP Ground   121.9
Columbus FSS(RCO)   122.35

EPP ATIS   EPP

Phonetic _____ Dew Pt. _____   14R
Sky   _____ Altim   _____   14L 17
Vis   _____ Rwy   _____
Temp   _____ Other   _____
                                                   32R
                                                  32L
                                                 35

                                          3 NE
                                          Elev.: 983
                          # 1   Pattern: 1983

Fig. 28-2. *The flight planning notes from MKC to OMA (Eppley).*

Second, sketch in on the same segment page a rough diagram of the destination airport runways, including the distance and direction from town, field elevation, and pattern altitude. If you want, you can add the taxiways and building locations. *AOPA's Aviation USA,* published by the Aircraft Owners and Pilots Association, provides dia-

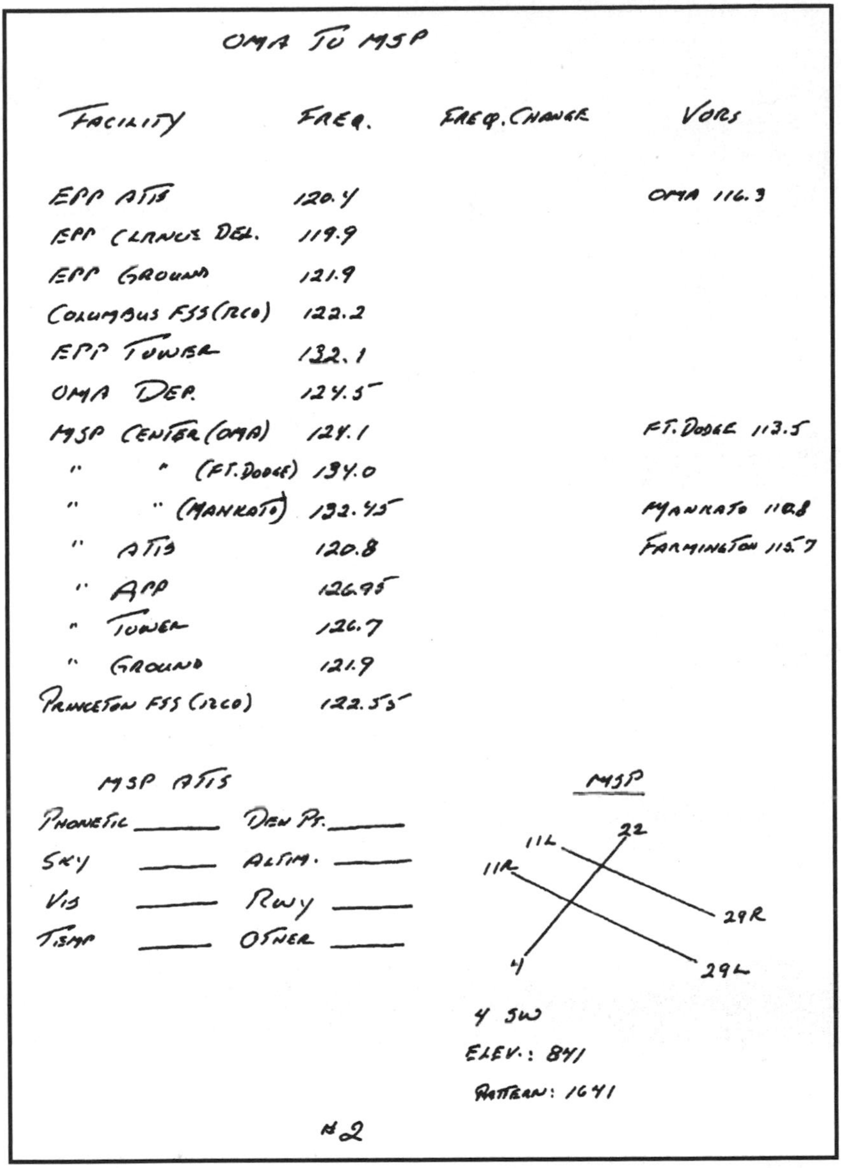

OMA TO MSP

FACILITY	FREQ.	FREQ. CHANGE	VORS
EPP ATIS	120.4		OMA 116.3
EPP CLRNCE DEL.	119.9		
EPP GROUND	121.9		
COLUMBUS FSS (RCO)	122.2		
EPP TOWER	132.1		
OMA DEP.	124.5		
MSP CENTER (OMA)	124.1		FT. DODGE 113.5
" " (FT. DODGE)	134.0		
" " (MANKATO)	132.45		MANKATO 110.8
" ATIS	120.8		FARMINGTON 115.7
" APP	126.95		
" TOWER	126.7		
" GROUND	121.9		
PRINCETON FSS (RCO)	122.55		

MSP ATIS

PHONETIC ____ DEW PT. ____
SKY ____ ALTIM. ____
VIS ____ RWY ____
TEMP ____ OTHER ____

4 SW
ELEV.: 841
PATTERN: 1641

#2

Fig. 28-3. *The flight planning notes from OMA to MSP.*

grams and data of more than 13,000 airports in the United States and its possessions. It's an excellent source for determining the layout and runway data of whatever airport you have in mind. Similar diagrams are found on state aeronautical charts and on instrument approach charts.

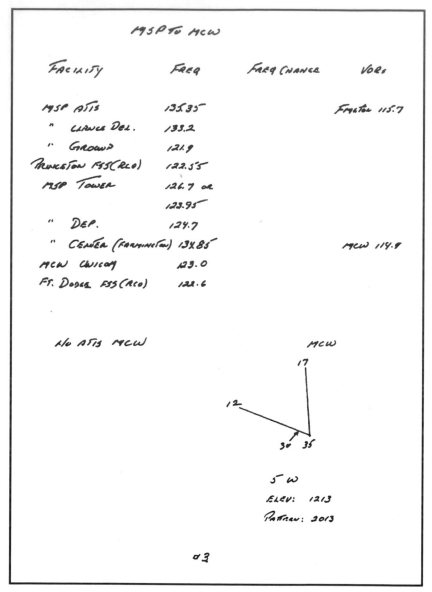

Fig. 28-4. *The flight planning notes from MSP to MCW.*

The purpose of the sketch is apparent: it minimizes mental or spatial confusion when going into a strange airport for the first time. Just as important, it can reduce the number of questions or inquiries you might have to make of Approach, the Tower, or Ground Control.

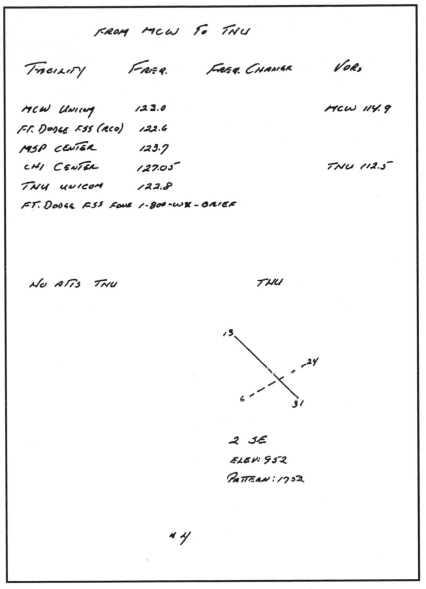

Fig. 28-5. *The flight planning notes from MCW to TNU.*

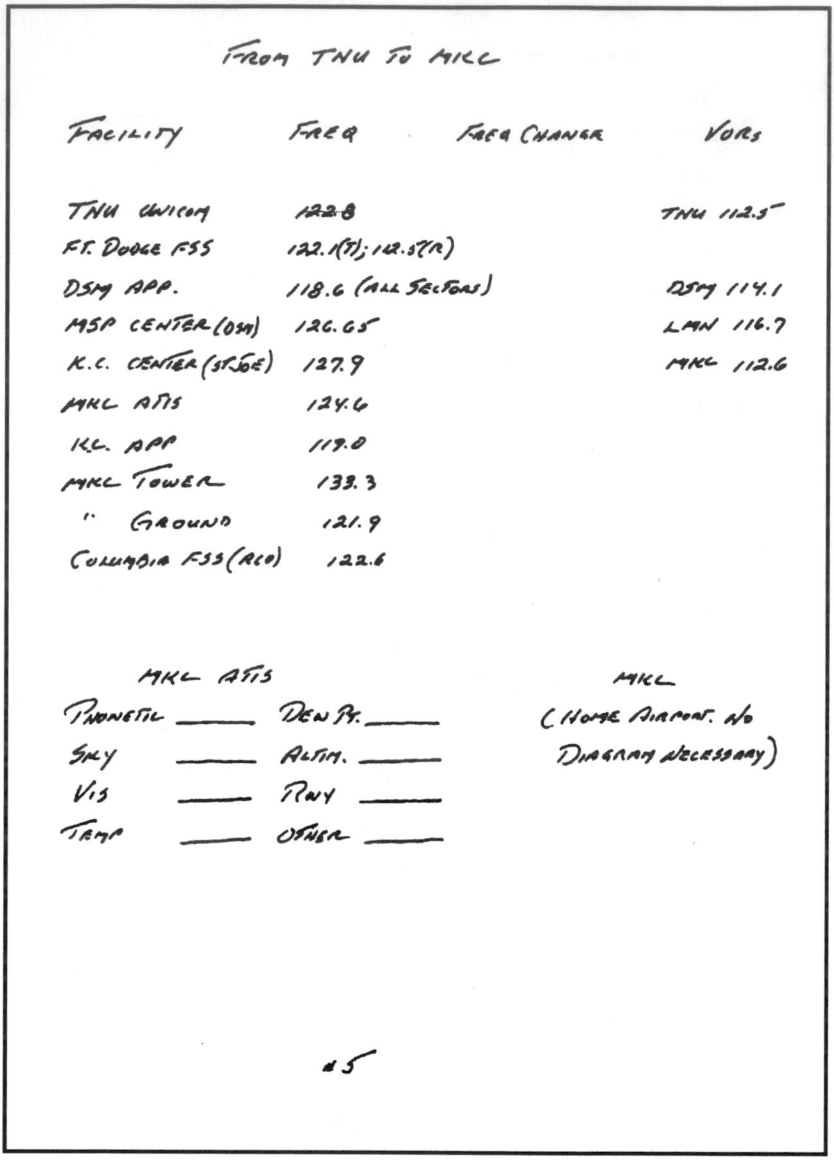

From TNU to MKC

Facility	Freq	Freq Change	VORs
TNU Unicom	122.8		TNU 112.5
Ft. Dodge FSS	122.1(T); 112.5(R)		
DSM APP.	118.6 (all sectors)		DSM 114.1
MSP CENTER (OSM)	126.65		LMN 116.7
K.C. CENTER (ST. JOE)	127.9		MKC 112.6
MKC ATIS	124.6		
KC. APP	119.0		
MKC TOWER	133.3		
" GROUND	121.9		
Columbia FSS (RCO)	122.6		

MKC ATIS

PHONETIC _____   DEW PT. _____
SKY _____   ALTIM. _____
Vis _____   RWY _____
TEMP _____   OTHER _____

MKC
(HOME AIRPORT. NO
DIAGRAM NECESSARY)

45

Fig. 28-6. *The flight planning notes from TNU to MKC.*

The examples (Figs. 28-2 through 28-6) aren't very fancy, but that's not the point. Practicality is the objective. (Just don't rely on the frequencies cited as being current. They do change.)

The flight and radio contacts

Equipped with the necessary charts—sectional charts as well as the En-Route Low Altitude charts (ELAC)—the frequencies recorded, the flight planned, and the flight plan filed, you're ready to depart.

Kansas City to Omaha

With the engine started, the first order of business is to tune to the Downtown Airport ATIS on 120.75. Put this in the first radio and set up Ground Control, 121.9, on the second.

> **ATIS:** *This is Kansas City Downtown Airport Information Delta. One six four five Zulu weather. Five thousand scattered, measured ceiling ten thousand broken, visibility eight. Temperature seven eight, dewpoint five five, wind one seven zero at ten, altimeter two niner niner eight. ILS Runway One Niner in use, land and depart Runway One Niner. Advise you have Delta.*

Now change the first radio to the FSS frequency of 122.6, which is the frequency that Flight Service gave you to open your flight plan. Put the transponder on STANDBY and call Ground Control on the second radio:

> **You:** *Downtown Ground, Cherokee One Four Six One Tango at Hangar 6, VFR Omaha with Information Delta.*

> **GC:** *Cherokee One Four Six One Tango, taxi to Runway One Niner.*

> **You:** *Roger. Taxi to One Niner, Cherokee Six One Tango.*

Stay on the Ground frequency. Clearance to taxi doesn't mean that the controller might not have subsequent instructions for you:

> **GC:** *Cherokee Six One Tango, give way to the Baron taxiing south.*

> **You:** *Wilco, Cherokee Six One Tango.*

After completing the pretakeoff check, call Ground and advise them that you're leaving the frequency momentarily to go to Flight Service. Now change radio 2 to the tower frequency of 133.3, switch to radio 1, which is already tuned to Flight Service, and open the flight plan:

> *You:* *Columbia Radio, Cherokee One Four Six One Tango on one two two point six, Kansas City.*
>
> *FSS:* *Cherokee One Four Six One Tango, Columbia Radio.*
>
> *You:* *Would you please open my VFR flight plan to Omaha Eppley at one three three five UTC?*
>
> *FSS:* *Cherokee Six One Tango, Roger We'll open your flight plan at one three three five UTC.*
>
> *You:* *Roger, thank you. Cherokee Six One Tango.*

Two points to remember:

1. Be sure to add five to ten minutes to your expected takeoff time in case your departure is delayed.
2. Opening a flight plan while still on the ground is possible only when there is an FSS, RCO, or VOR voice facility on the field.

As your course to Omaha is northwesterly, the most direct routing is through the Kansas City Class B airspace, so change radio 1 to Approach Control on 119.0. (Remember that you're under the Class B airspace and must contact Approach to enter it.) With radio 1 set up, the next call is to the tower on radio 2:

> *You:* *Downtown Tower, Cherokee One Four Six One Tango ready for takeoff with north departure.*
>
> *Twr:* *Cherokee One Four Six One Tango, hold short. Landing traffic.*
>
> *You:* *Cherokee Six One Tango holding short.*
>
> *Twr:* *Cherokee Six One Tango, cleared for takeoff. Left turn for north departure approved. Remain clear of the final approach course. Contact Approach when airborne.*
>
> *You:* *Will do. Cherokee Six One Tango. [Now switch the transponder to ALT.]*

Even if the tower has advised you to contact Approach after takeoff, it doesn't hurt to request the frequency-change approval or to inform the tower that you're about to make the change. The controller might have reasons for wanting you to stay with him for a few minutes. The next call then would be:

> **You:** *Tower, Cherokee Six One Tango requests frequency change [or "going to Approach"].*
>
> **Twr:** *Cherokee Six One Tango. Frequency change approved.*
>
> **You:** *Cherokee Six One Tango. Good day.*

Go now to radio 1 and call Approach on 119.0:

> **You:** *Kansas City Approach, Cherokee One Four Six One Tango.*
>
> **App:** *Cherokee One Four Six One Tango, Kansas City Approach, go ahead.*
>
> **You:** *Cherokee One Four Six One Tango is off Downtown at two thousand, requesting six thousand five hundred to Omaha, and would like clearance through the Bravo airspace.*
>
> **App:** *Cherokee Six One Tango, squawk zero two five two and ident. Remain clear of the Bravo airspace.*
>
> **You:** *Cherokee Six One Tango squawking zero two five two. Remaining clear.*

Remember that you have not yet been cleared into the airspace, so stay below this Class B's 2,400-foot floor until you hear the next message:

> **App:** *Cherokee Six One Tango, radar contact. Cleared to enter the Class B airspace. Fly heading three four zero and maintain four thousand five hundred.*
>
> **You:** *Roger. Cleared to enter the Class B airspace, heading three four zero, leaving two thousand for four thousand five hundred. Cherokee Six One Tango.*

As you start your turn and begin the climb, you might hear this before actually entering the Bravo airspace.

> ***App:*** *Cherokee Six One Tango, traffic at twelve o'clock, three miles, southbound. Unverified altitude two thousand niner hundred.*

> ***You:*** *Cherokee Six One Tango looking.*

And then:

> ***App:*** *Cherokee Six One Tango, traffic no longer a factor.*

> ***You:*** *Roger. Cherokee Six One Tango.*

> ***You:*** *Cherokee Six One Tango, level at four thousand five hundred.*

> ***App:*** *Cherokee Six One Tango, Roger. Climb and maintain six thousand five hundred.*

> ***You:*** *Out of four thousand five hundred for six thousand five hundred. Cherokee Six One Tango.*

> ***You:*** *Cherokee Six One Tango level at six thousand five hundred.*

> ***App:*** *Cherokee Six One Tango, Roger. Turn left heading three one zero. Proceed direct St. Joe when able.*

> ***You:*** *Roger. Three one zero on the heading. Direct St. Joe when able. Cherokee Six One Tango.*

The St. Joe VOR lies about 50 nm north of Kansas City. Approach is saying that you are to tune the nav receiver to the VOR frequency of 115.5. When you have the altitude and distance to get a steady needle reading, you're cleared to proceed directly on course to St. Joe.

Other instructions or traffic alerts might follow. While you have the time, however, you should be setting up radio 2 (until now still on the tower frequency) to Kansas City Center, which you expect to be 127.9. Approach might give you a different frequency, but if not, you're ready to contact Center without delay.

As you work your way northward toward St. Joe, you reach the Class B limits:

> ***App:*** *Cherokee Six One Tango, position 30 miles northwest of International, departing the Bravo airspace. Radar service terminated. Squawk one two zero zero. Frequency change approved.*

You: Cherokee Six One Tango. Can you hand us off to Center?

App: Unable at this time, Six One Tango. Contact Center on one two seven point niner. Good day.

You: One two zero zero and one two seven point niner. Roger. Cherokee Six One Tango.

There is no handoff, so the call to Center requires the full IPAI/DS:

You: Kansas City Center, Cherokee One Four Six One Tango.

Ctr: Cherokee One Four Six One Tango, Kansas City Center, go ahead.

You: Cherokee One Four Six One Tango is two zero south of St. Joe VOR, level at six thousand five hundred en route Omaha Eppley. Squawking one two zero zero. Request VFR advisories, if possible.

Ctr: Cherokee Six One Tango, squawk four one five zero and ident.

You: Cherokee Six One Tango squawking four one five zero.

Ctr: Cherokee Six One Tango. Radar contact. St. Joe altimeter two niner niner eight.

You: Roger. Cherokee Six One Tango.

What comes over the air now depends primarily on the traffic along your line of flight. You might be told to assume a different heading; you might be alerted to the proximity of other aircraft; you might be alerted to military training flights. Or you might hear nothing.

Regardless of the communiqués, or lack thereof, you have some navigating to do, along with keeping an ear out for your call sign. You're still on course to the St. Joe VOR, with the VOR head needle centered and the DME recording the distance to the VOR, the time to the station, and your current ground speed.

As soon as you pass over St. Joe and get a "FROM" reading on the VOR, turn to a heading of 325 degrees. Once the needle has centered itself, indicating that you're now on V159, stay tuned to that station for another 35 or 40 miles. If you have a second navcom, you can tune in the Omaha VOR on 116.3 at any time, but at your altitude you probably won't get a very reliable "TO" indication until you're within a 60- or 70-mile range of the station. So maintain the outbound course from St. Joe with the "FROM" reading on one VOR, and then when the other VOR needle settles down to a steady centered position with a "TO" reading, rely on it to lead you to the Omaha station.

Meanwhile, don't get too enthralled with VOR-to-VOR navigating alone. Something might go wrong, so it's always wise to check your progress against the route you've laid out on the sectional chart. If you should lose the nav portion of the radio, or if the VOR suddenly had a mechanical failure, it could be rather important to know where you were along the route. Things electronic are great, but they're not infallible.

Eventually, as you move down V159, Kansas City Center comes on the air:

> *Ctr:* *Cherokee Six One Tango, contact Minneapolis Center now on one one niner point six. Good day.*

> *You:* *One one niner point six. Thank you for your help. Cherokee Six One Tango.*

No comment here about radar service being terminated, so this is a handoff by Kansas City to Minneapolis. If you've already changed radio 1 from Kansas City Approach to Minneapolis, all you have to do now is go back to radio 1 and introduce yourself:

> *You:* *Minneapolis Center, Cherokee One Four Six One Tango with you, level at six thousand five hundred.*

> *Ctr:* *Cherokee Six One Tango, Roger. Altimeter three zero zero two.*

Unless you are told otherwise, maintain your present heading and altitude. Again, there might or might not be instructions or advice for you. Eventually, however, as you near the Omaha Class C Charlie airspace, Center will come back on:

> *Ctr:* *Cherokee Six One Tango, position 30 miles south of Eppley. Contact Omaha Approach on one two four point five. Good day.*

> *You:* *Roger. One two four point five. Cherokee Six One Tango. Good day.*

This, too, is a handoff, so tune in to Approach. Before making the call, however, if you haven't done so already, get the ATIS on 120.4 for the current Eppley information. Now you're ready to contact Approach:

> *You: Omaha Approach, Cherokee One Four Six One Tango is with you, level at six thousand five hundred with Information Echo.*

> *App: Cherokee Six One Tango, Roger. Maintain present heading and altitude.*

As you enter the Class C airspace and draw close to the field, you're likely to be given a variety of instructions that will sequence you into the existing traffic flow. Whatever the messages, be sure to acknowledge and repeat them in an abbreviated form:

> *App: Cherokee Six One Tango, turn right heading three five zero. Descend and maintain four thousand.*

> *You: Right to three five zero. Leaving six thousand five hundred for four thousand. Cherokee Six One Tango.*

In another few minutes, you'll hear from Approach again:

> *App: Cherokee Six One Tango, Eppley is at twelve o'clock, six miles. Contact Eppley Tower on one two seven point six.*

> *You: Roger. One two seven point six— and we have the field in sight. Cherokee Six One Tango. Good day.*

Another handoff:

> *You: Eppley Tower, Cherokee One Four Six One Tango with you, level at four thousand.*

> *Twr: Cherokee Six One Tango, enter left downwind for Runway One Four left. Sequence later.*

Two points here:

1. "Sequence later" simply means the tower will advise you in due time whether you're cleared to land, are "number two following a Cessna on downwind" "number four following the Duchess," or whatever. Just remember to tell the tower that you "have the Cessna" or "negative contact on the Duchess" or "have the traffic." Always keep the tower informed—don't leave them in the dark.

2. You'll note that you've used the same squawk from Kansas City Center through Minneapolis Center, Omaha Approach, and Eppley Tower. No controlling agency has asked you to change. This is not always the case. Any one of them could have requested a different squawk and an ident. Leave the transponder on the current squawk until directed otherwise.

With a little time available now, dial in Eppley's Ground frequency, 121.9, on radio 1. Then, in due course, you'll hear on radio 2:

Twr: *Cherokee Six One Tango, cleared to land.*

You: *Roger, cleared to land. Cherokee Six One Tango.*

During the landing rollout, the tower makes its final contact with you:

Twr: *Cherokee Six One Tango, contact Ground point niner clear of the runway.*

You: *Cherokee Six One Tango, wilco.*

Suppose, however, that you're not sure whether to make a left or right turnoff. You want to go to a Texaco dealer but don't know where one is located. Despite the uncertainty, don't tie up the tower frequency by asking the controller for directions. Let Ground do this for you, even if it means being cleared back across the active runway because you turned right instead of left, or vice versa. Merely acknowledge the tower's instructions and go to Ground's frequency:

You: *Eppley Ground, Cherokee One Four Six One Tango Clear of One Four Left. Request progressive taxi to the Texaco dealer.*

GC: *Cherokee One Four Six One Tango, turn right next taxiway. Taxi to One Four Right and hold for departing traffic.*

You: *Roger, right and hold at One Four. Six One Tango.*

After a minute or two:

GC: *Cherokee Six One Tango, Clear to cross One Four Right. Texaco is to your right at the Sky Harbor FBO.*

You: *Roger, Ground. Clear to cross One Four—and we have the FBO. Six One Tango.*

You're at the ramp, with the engine cut. Now don't forget to call Flight Service by phone or radio to close out your flight plan.

Omaha to Minneapolis

After an hour on the ground for refueling, a bite to eat, and a call to the Columbus, Nebraska, FSS for a weather briefing and filing the flight plan, you're ready to go again, with friends and baggage on board. The only new element is the need to call Clearance Delivery for initial VFR instructions within the Charlie airspace. As you've already determined, that frequency is 119.9, and you were told to contact Flight Service on the Omaha RCO frequency of 122.35. That's a change from the 122.2 you had listed on the frequency chart you prepared back in Kansas City (Fig. 28-3). So cross off 122.2 on that chart and enter 122.35 under the "Changes" column.

As usual, the communication chain begins by monitoring ATIS: "This is Eppley Information Gulf. One niner four five Zulu weather. Eight thousand scattered, visibility five, haze and smoke. Temperature eight five, dewpoint six two. Wind one four zero at one five. Altimeter three zero two five. ILS Runway One Four Right in use, land and depart Runway One Four Right. Advise you have Gulf."

Next, the call to Clearance (CD):

You: Eppley Clearance, Cherokee One Four Six One Tango

CD: Cherokee One Four Six One Tango, Eppley Clearance.

You: Cherokee One Four Six One Tango will be departing Eppley, VFR northeast for Minneapolis at seven thousand five hundred.

CD: Cherokee Six One Tango, Roger. Turn left heading zero four five after departure. Climb and maintain three thousand. Squawk two four four zero. Departure frequency one two four point five.

You: Roger. Zero four five on the heading, maintain three thousand, two four four zero, and one two four point five. Cherokee Six One Tango.

CD: Cherokee Six One Tango, readback correct.

You: Cherokee Six One Tango.

Before calling Ground on radio 2, dial out the Clearance frequency in radio 1 and replace it with Flight Service's—122.35. Also, put the transponder on STANDBY.

You: Eppley Ground, Cherokee One Four Six One Tango at Sky Harbor with Information Gulf. Ready to taxi with clearance.

GC: Cherokee Six One Tango, taxi to Runway One Four Right for intersection departure.

You: Ground, Cherokee Six One Tango would like full length.

GC: Cherokee Six One Tango, Roger. Full length approved.

You: Cherokee Six One Tango.

After taxiing to the end of Runway 14, and with the pretakeoff check completed, you next call Flight Service after advising Ground that you were going to change frequencies momentarily.

You: Columbus Radio, Cherokee One Four Six One Tango on one two two point three five, Eppley.

FSS: Cherokee One Four Six One Tango, Columbus Radio.

You: Columbus, Cherokee Six One Tango. Would you open my VFR flight plan to Minneapolis International at this time?

FSS: Cherokee Six One Tango, Roger. We'll open your flight plan at two five.

You: Roger, thank you. Cherokee Six One Tango.

That done, advise Ground Control that you're back with them. Now taxi to the hold and call the tower:

You: Eppley Tower, Cherokee One Four Six One Tango ready for takeoff, northeast departure.

Twr: Cherokee Six One Tango, taxi into position and hold.

You: Position and hold. Cherokee Six One Tango. [Switch the transponder from STANDBY to ALT as you're taxiing to the runway.]

Twr: Cherokee Six One Tango, cleared for takeoff.

You: Roger, cleared for takeoff. Cherokee Six One Tango. Is northeast departure approved?

Twr: *Affirmative, northeast departure approved.*

You: *Roger. Cherokee Six One Tango.*

When airborne, request the frequency change to Departure—if the tower has not already advised you to do so:

You: *Tower, Cherokee Six One Tango requests frequency change to Departure.*

Twr: *Cherokee Six One Tango, frequency change approved.*

You: *Roger. Cherokee Six One Tango. Good day.*

To Departure Control:

You: *Omaha Departure, Cherokee One Four Six One Tango is with you out of one thousand eight hundred for three thousand. Request seven thousand five hundred.*

Dep: *Cherokee Six One Tango, radar contact. Maintain present heading. Report level at three thousand.*

You: *Roger, report three thousand, Cherokee Six One Tango.*

You: *Cherokee Six One Tango level at three thousand.*

Dep: *Cherokee Six One Tango, Roger. Turn left heading zero three zero.*

You: *Left to zero three zero. Cherokee Six One Tango.*

Dep: *Cherokee Six One Tango, climb and maintain seven thousand five hundred.*

You: *Roger. Out of three thousand for seven thousand five hundred. Cherokee Six One Tango.*

You: *Cherokee Six One Tango level at seven thousand five hundred.*

Dep: *Cherokee Six One Tango, Roger. Proceed on course and contact Minneapolis Center on one two four point one. [An unsolicited handoff for advisories, although rare, is possible.]*

You: *One two four point one. Thank you for your help. Cherokee Six One Tango.*

To Center:

> *You:* *Minneapolis Center, Cherokee One Four Six One Tango with you, level at seven thousand five hundred.*
>
> *Ctr:* *Cherokee Six One Tango, Roger. Radar contact. Eppley altimeter Three Zero Two Five.*
>
> *You:* *Roger. Cherokee Six One Tango. Three Zero Two Five.*

Down the road apiece, the summer turbulence at 7500 feet is getting a bit too much for one of your passengers, so you decide to climb to 9500. But first:

> *You:* *Minneapolis Center, Cherokee Six One Tango is out of seven thousand five hundred for niner thousand five hundred due to turbulence.*
>
> *Ctr:* *Cherokee Six One Tango, Roger. Report level at niner thousand five hundred.*
>
> *You:* *Wilco, Cherokee Six One Tango.*
>
> *You:* *Cherokee Six One Tango level at niner thousand five hundred.*
>
> *Ctr:* *Cherokee Six One Tango, Roger.*

Just remember that on a VFR flight plan outside of a terminal airspace, you're free to deviate from existing headings and altitudes. But since you're asking Center for en-route advisories, don't make changes without advising the controller of your intentions. Although he or she can spot heading changes on the screen (and altitude changes, if you're equipped with a Mode C transponder), be sure to warn the controller in advance.

As you proceed toward Fort Dodge on V138, Center comes on again:

> *Ctr:* *Cherokee Six One Tango, contact Minneapolis Center now on one three four point zero. [This is the Fort Dodge remote outlet.]*
>
> *You:* *Roger. One three four point zero. Cherokee Six One Tango. Good day.*

Change the frequency accordingly and reestablish yourself with Center:

You: Minneapolis Center, Cherokee One Four Six One Tango with you level at niner thousand five hundred.

Ctr: Cherokee Six One Tango, Roger. Fort Dodge altimeter three zero one six. [Plus any instructions or traffic advisories the controller might have for you.]

Crossing the Fort Dodge VOR, you make a time check against your flight plan and find that you're running about 20 minutes behind schedule, which is quite a difference for the 110-mile flight from Eppley. Either the forecast winds have changed or those at the 9500-foot altitude are from a different direction or velocity. Whatever the case, and because you intend to stay at the same altitude and generally northeast direction, it's fair to assume that your arrival in Minneapolis will be later than planned. With another 150 miles to go and at the present ground speed, you could easily be 45 minutes to an hour later than the flight plan ETA.

Before taking any arbitrary action, you decide that more information about the winds is in order, thus a call to Flight Service. First, however, advise Center if you're temporarily going to leave the frequency:

You: Center, Cherokee Six One Tango is leaving you temporarily to go to Flight Service.

Ctr: Cherokee Six One Tango, Roger. Advise when you're back with me.

You: Will do. Cherokee Six One Tango.

The FSS call is made on 122.3, the Fort Dodge transmit/receive frequency: (Note: the Fort Dodge FSS does not have Flight Watch service—which is why you make the call on 122.3 rather than 122.0.)

You: Fort Dodge Radio, Cherokee One Four Six One Tango on one two two point three.

FSS: Cherokee One Four Six One Tango, Fort Dodge Radio, go ahead.

You: Fort Dodge, Cherokee Six One Tango is just north of the Fort Dodge VOR at niner thousand five hundred. Request winds at niner thousand feet.

FSS: Cherokee Six One Tango, Roger. Stand by.

> **FSS:** *Cherokee Six One Tango, winds at niner thousand are three five zero at four zero. Fort Dodge altimeter two niner two five.*

> **You:** *Three five zero at four zero and two niner two five. Thank you. Cherokee Six One Tango.*

Without going through the mechanics of how you recompute your ground speed and ETA based on this information, let's just say that you determine that your arrival will be 48 minutes later than your flight plan forecast. This conclusion warrants another call to Flight Service:

> **You:** *Fort Dodge Radio, Cherokee One Four Six One Tango on one two two point three.*

> **FSS:** *Cherokee One Four Six One Tango, Fort Dodge Radio, go ahead.*

> **You:** *Cherokee One Four Six One Tango is one zero north of the Fort Dodge VOR on VFR flight plan to Minneapolis International with a one six four five local ETA. Would like to extend the ETA to one seven three zero.*

> **FSS:** *Cherokee Six One Tango, Roger. Will extend your ETA to one seven three zero local. Fort Dodge altimeter two niner two five.*

> **You:** *Roger, thank you. Cherokee Six One Tango.*

The next move is to go back to Center and reestablish contact:

> **You:** *Center, Cherokee Six One Tango is back with you.*

> **Ctr:** *Cherokee Six One Tango, Roger.*

With that taken care of, you can rest more easily. You have until 1800 before Flight Service will start asking questions as to your whereabouts. You've also allowed yourself an additional 27 minutes as an extra cushion.

Moving northward along V456, you'll be approaching another Center change point—this time to the remote Mankato site. Somewhere near Mankato, you get this call:

> **Ctr:** *Cherokee Six One Tango, contact Minneapolis Center now on one three two point four five.*

You: One three two point four five. Cherokee Six One Tango. Good day.

You: Minneapolis Center, Cherokee One Four Six One Tango with you, level at niner thousand five hundred.

Ctr: Cherokee Six One Tango, Roger. Mankato altimeter two niner two three.

You: Roger, two niner two three. Cherokee Six One Tango.

Immediately after passing the Mankato VOR, you leave the airway and turn to about 30 degrees in the direction of the Farmington VOR. Because you're only 50 miles from the airport at Mankato, however, it's almost certain that Center will turn you over to Approach and Approach will vector you to the airport area. That means you might not come anywhere near Farmington. Meanwhile, the closer you edge toward the Minneapolis Class B airspace, the greater the likelihood that Center will offer traffic advisories and possible heading changes. There might be none, but moving into a busy terminal environment increases the possibility.

About 20 miles out from the airspace limit, Center will conclude its radar surveillance. Perhaps the controller will hand you off to Approach. If so, the usual "with you," plus your altitude, is all that's necessary. If there is no handoff, though, you'll have to give the IPAI/DS—which means, among other things, knowing your position or approximate distance from the airport.

In this case; let's assume that Center is getting busy and doesn't have time to contact Approach. The controller comes on with:

Ctr: Cherokee One Four Six One Tango, radar service terminated. Squawk one two zero zero. Contact Minneapolis Approach on one two five point zero. Descend your discretion.

You: One two six point niner five. Will do. Cherokee Six One Tango.

"Descend your discretion" simply means what it says: You are clear to begin losing altitude whenever you wish and at whatever rate you wish. With that approval, you throttle back so that you'll be down to about 4000 feet by the time you near the 30-mile Bravo airspace veil.

There's one more thing: Before calling Approach, dial in 120.8 for the current ATIS. Then go to 125.0, which is the Approach frequency for aircraft arriving from the south and west below 4500 feet, and make the introductory call:

> **You:** *Minneapolis Approach, Cherokee One Four Six One Tango.*

> **App:** *Cherokee One Four Six One Tango, Minneapolis Approach.*

> **You:** *Approach, Cherokee Six One Tango is thirty-five southwest over Le Center at five thousand three hundred descending for landing International and squawking one two zero zero with Delta.*

> **App:** *Cherokee Six One Tango, squawk four one two two and ident. Remain clear of the Bravo airspace.*

> **You:** *Roger, four one two two, and remaining clear. Cherokee Six One Tango.*

> **App:** *Cherokee One Four Six One Tango, radar contact thirty-two miles southwest. Cleared into the Minneapolis Bravo airspace. Turn left to zero two zero, descend and maintain four thousand five hundred.*

> **You:** *Roger, Cherokee Six One Tango cleared into Bravo airspace. Left to zero two zero and descending to four thousand five hundred.*

> **You:** *Approach, Cherokee Six One Tango level at four thousand five hundred.*

> **App:** *Roger Six One Tango. Maintain heading and altitude.*

> **You:** *Will do. Six One Tango.*

A short time later:

> **App:** *Cherokee One Four Six One Tango, turn left to zero one zero and descend to two thousand five hundred.*

> **You:** *Roger, left to zero one zero and down to two thousand five hundred, Cherokee Six One Tango.*

You: Approach, Cherokee Six One Tango level at two thousand five hundred.

App: Roger, Six One Tango. Contact Minneapolis Tower on one two six point seven.

You: One two six point seven. Will do, and thank you. Cherokee Six One Tango.

You: Minneapolis Tower, Cherokee One Four Six One Tango is with you, level at two thousand five hundred.

Twr: Roger, Cherokee One Four Six One Tango. Descend to one thousand eight hundred for straight-in approach, landing Runway Four. Sequence later.

You: Roger, descending to one thousand eight hundred and Runway Four. Cherokee Six One Tango.

You: Tower, Cherokee Six One Tango level at one thousand eight hundred.

Twr: Roger, Six One Tango, you will be number two to land behind the Baron. Advise when you have traffic in sight.

You: Number two, and will advise. Six One Tango.

You: Tower, Cherokee Six One Tango has the Baron.

Twr: Roger Six One Tango. Thank you.

Twr: Cherokee Six One Tango, cleared to land Runway Four.

You: Cherokee Six One Tango, cleared to land.

When you're down:

Twr: Cherokee Six One Tango, contact Ground point niner.

You: Will do, tower. Six One Tango.

When you're clear of the active and at a full stop:

You: Minneapolis Ground, Cherokee One Four Six One Tango clear of Four. Taxi to Signature.

GC: *Roger. Cherokee One Four Six One Tango, taxi to Signature.*

Once again, when you're parked or in the pilot's lounge, close out the flight plan with Flight Service—which, in this case, is located in Princeton, Minnesota.

Minneapolis to Mason City

It's the next morning and you're ready to set out for Mason City, Newton, and then back to Kansas City. First the normal routines: checking the weather, filing the flight plan, and determining the FSS frequency to open the flight plan. With the engine started, monitor the Minneapolis departure ATIS on 135.35. Remember that you're in a Class B airspace, so the first call goes to Clearance Delivery on 133.2:

You: *Minneapolis Clearance, Cherokee One Four Six One Tango.*

CD: *Cherokee One Four Six One Tango, Clearance.*

You: *Cherokee One Four Six One Tango will be departing VFR to Mason City. Request seven thousand five hundred.*

CD: *Cherokee Six One Tango, cleared into the Bravo airspace. Turn right heading one seven five after departure. Climb and maintain three thousand. Squawk zero four two zero. Departure frequency one two four point seven.*

You: *Cherokee Six One Tango, cleared into the Bravo airspace, right heading one seven five after departure, maintain three thousand, zero four two zero, and one two four point seven.*

CD: *Cherokee Six One Tango, readback correct.*

You: *Roger. Cherokee Six One Tango.*

Now to Ground Control on 121.9:

You: *Minneapolis Ground, Cherokee One Four Six One Tango at Airmotive with Information Echo and clearance.*

GC: *Cherokee Six One Tango, taxi to Runway One One Right.*

You: *Roger, One One Right, Cherokee Six One Tango.*

The pretakeoff check has been completed. Before taxiing to the hold line, ask Ground for permission to switch frequencies, then call Flight Service to open the flight plan.

> *You:* *Princeton Radio, Cherokee One Four Six One Tango on one two two point five five [the RCO frequency FSS gave you when filing], Minneapolis.*

> *FSS:* *Cherokee One Four Six Tango, Princeton Radio.*

> *You:* *Roger, would you please open my flight plan to Mason City at 1410 UTC?*

(Note: It is now 0900 Central Daylight Time, or 1400 UTC Greenwich Time. You have added 10 minutes to your departure time in case of any delay.)

> *FSS:* *Cherokee Six One Tango, Roger. Will open your flight plan at one zero.*

> *You:* *Thank you. Cherokee Six One Tango.*

As you move toward the hold line, you see that two aircraft are ahead of you awaiting takeoff permission. Regardless, you pull behind the second plane and call the tower on 126.7:

> *You:* *Minneapolis Tower, Cherokee One Four Six One Tango ready for takeoff, number three in sequence, south departure.*

> *Twr:* *Cherokee One Four Six One Tango, taxi around the Mooney and Skymaster. Cleared for takeoff, south departure approved.*

> *You:* *Roger, cleared for takeoff, Cherokee Six One Tango.*

At this point, switch the transponder from SBY to ALT. When airborne, turn to your assigned heading of 175 degrees. The tower will probably authorize the frequency change to Departure, but if it doesn't, request the change:

> *You:* *Tower, Cherokee Six One Tango requests frequency change to Departure.*

> *Twr:* *Cherokee Six One Tango, frequency change approved. Good day.*

> *You:* *Roger. Cherokee Six One Tango. Good day.*

Switch to your other radio, already dialed in to 124.7:

You: *Minneapolis Departure, Cherokee One Four Six One Tango with you out of one thousand seven hundred for three thousand.*

Dep: *Cherokee One Four Six One Tango, report level at three thousand.*

You: *Wilco, Cherokee Six One Tango.*

You: *Cherokee Six One Tango level at three thousand.*

Dep: *Cherokee Six One Tango, Roger.*

Dep: *Cherokee Six One Tango, climb and maintain four thousand five hundred.*

You: *Roger. Out of three thousand for four thousand five hundred. Cherokee Six One Tango.*

You: *Cherokee Six One Tango level at four thousand five hundred.*

Dep: *Cherokee Six One Tango, Roger. Cleared direct Farmington VOR.*

You: *Roger, cleared direct Farmington. Cherokee Six One Tango.*

Dep: *Cherokee Six One Tango, climb and maintain seven thousand five hundred.*

You: *Roger. Out of four thousand five hundred for seven thousand five hundred. Cherokee Six One Tango.*

You: *Cherokee Six One Tango level at seven thousand five hundred.*

Dep: *Cherokee Six One Tango. Roger. Report Farmington VOR.*

You: *Roger, report Farmington. Cherokee Six One Tango.*

It isn't long before the Course Direction Indicator (CDI) swings erratically and the VOR flag changes from TO to FROM. You're over the Farmington VOR. You report in and are cleared to turn to the 178-

degree heading, which establishes you on Victor 13. As you're still in the Bravo airspace, other instructions might be forthcoming from Departure. In a very few minutes, you'll hear something like this:

Dep: *Cherokee Six One Tango, position fifteen miles south of Farmington VOR, departing the Bravo airspace. Radar service terminated. Squawk one two Zero Zero. Frequency change approved. Good day.*

You: *Departure, Cherokee Six One Tango. Can you hand us off to Center?*

Dep: *Cherokee Six One Tango, stand by. [Pause] Cherokee Six One Tango, unable at this time. Squawk one two zero zero. Suggest you contact Minneapolis Center on one three four point eight five.*

You: *Roger. One two zero zero and one three four point eight five. Cherokee Six One Tango.*

You: *Minneapolis Center, Cherokee One Four Six One Tango.*

Ctr: *Cherokee One Four Six One Tango, Minneapolis Center.*

You: *Cherokee One Four Six One Tango is fifteen south of the Farmington VOR at seven thousand five hundred, VFR to Mason City, squawking one two zero zero. Request VFR advisories, if possible.*

This time, the press of traffic in and out of the Minneapolis area is of such density that Center can't accept your request:

Ctr: *Cherokee One Four Six One Tango, unable at this time. Suggest you monitor this frequency.*

You: *Roger, Center. Understand. Cherokee Six One Tango.*

With only 80 miles or so to go, this isn't much of a problem. However, it does mean that the need for constant sky-scanning is more important than ever. If Center is too busy to give you advisories, you can be reasonably certain that there's a fair amount of activity in the surrounding area. Maximum alertness is thus in order.

As you've already noted in your flight planning, Mason City is a non-tower-controlled Class E airport with only unicom on 123.0 available for airport advisories. The Fort Dodge FSS can be reached on the 122.6 RCO frequency for opening and closing flight plans. First things first, though, so you begin letting down about 15 miles north of the airport and call unicom on its CTAF:

> *You:* *Mason City Unicom, Cherokee One Four Six One Tango.*
>
> *Uni:* *Cherokee One Four Six One Tango, Mason City Unicom.*
>
> *You:* *Unicom, Cherokee One Four Six One Tango is fifteen north at five thousand three hundred descending for landing. Request field advisory.*
>
> *Uni:* *Six One Tango, wind is two zero zero at one zero, altimeter two niner four five. Favored runway is One Seven. No reported traffic.*
>
> *You:* *Roger, thank you. Six One Tango.*

From this point on, remember to address all flight operation reports to "Mason City Traffic," not unicom. If, however, your remaining passenger wants a cab or a telephone call made, or you have a request of a nonoperational nature, the message is addressed to "Mason City Unicom."

Even though there is no "reported" traffic and a straight-in approach to Runway 17 would be the fastest way to get on the ground, you're aware that non-tower-controlled airport pattern procedures call for standard downwind and base legs. Consequently, your next call to Mason City traffic goes like this:

> *You:* *Mason City Traffic, Cherokee One Four Six One Tango is 10 miles north at three thousand. Will cross over the airport to the southeast at three thousand for one eighty and entry to left downwind, landing Mason City.*

A few minutes later, as you complete the 45-degree entry to the downwind, you report your position:

> *You:* *Mason City Traffic, Cherokee Six One Tango entering left downwind at midfield for landing Runway One Seven, Mason City.*

Then, as you're turning to base and final:

You: *Mason City Traffic, Cherokee Six One Tango turning left base for landing One Seven, Mason City.*

You: *Mason City Traffic, Cherokee Six One Tango turning final, landing Mason City.*

Finally, when down and clear of the runway:

You: *Mason City Traffic, Cherokee Six One Tango clear of One Seven, taxiing to the ramp, Mason City.*

At the ramp you decide to close the flight plan by radio rather than by telephone. Using the RCO you call the FSS:

You: *Fort Dodge Radio, Cherokee One Four Six One Tango on one two two point six, Mason City.*

FSS: *Cherokee One Four Six One Tango, Fort Dodge Radio, go ahead.*

You: *Cherokee Six One Tango is on the ramp at Mason City. Would you close out my VFR flight plan from Minneapolis at this time?*

FSS: *Cherokee Six One Tango, Roger. Will close out your flight plan at five five.*

You: *Thank you. Cherokee Six One Tango.*

Mason City to Newton

The flight plan to Newton, 90 miles away, has been filed, you know the weather, and it's departure time again. Tune to 123.0 and announce your initial intentions:

You: *Mason City Traffic, Cherokee One Four Six One Tango at the terminal, taxiing to Runway One Seven, Mason City.*

After the pretakeoff check, with the second radio already turned to 122.6, open the flight plan:

You: *Fort Dodge Radio, Cherokee One Four Six One Tango on one two two point six, Mason City.*

FSS: *Cherokee One Four Six One Tango, Fort Dodge Radio, go ahead.*

> *You:* Fort Dodge, would you please open my flight plan to Newton at this time?
>
> *FSS:* Cherokee Six One Tango, Roger. Will open your flight plan to Newton at two zero.
>
> *You:* Roger. Thank you. Cherokee Six One Tango.

Now go back to 123.0:

> *You:* Mason City Traffic, Cherokee Six One Tango taking One Seven, straight-out departure, Mason City.

When clear of the Class E (Echo) surface area:

> *You:* Mason City Traffic, Cherokee Six One Tango is clear of the area to the south, Mason City.

Following this call, you request Minneapolis Center to give you VFR advisories. Center in this location is remoted to Mason City on 127.3:

> *You:* Minneapolis Center, Cherokee One Four Six One Tango.
>
> *Ctr:* Cherokee One Four Six One Tango, Minneapolis Center, go ahead.
>
> *You:* Center Cherokee One Four Six One Tango is off Mason City at four thousand five hundred, climbing to seven thousand five hundred, en route Newton, and squawking one two zero zero. Request VFR advisories.
>
> *Ctr:* Cherokee Six One Tango, squawk two five two five and ident.
>
> *You:* Cherokee Six One Tango, two five two five.
>
> *Ctr:* Cherokee Six One Tango, radar contact. Traffic at ten o'clock, four miles, southbound. Altitude unknown.
>
> *You:* Cherokee Six One Tango is looking.

A minute or so later, you spot the target a little above you at the eleven o'clock position:

> *You:* Cherokee Six One Tango has the traffic.
>
> *Ctr:* Cherokee Six One Tango, Roger.

When at your altitude:

> **You:** *Center, Cherokee Six One Tango level at seven thousand five hundred.*

> **Ctr:** *Cherokee Six One Tango, Roger.*

Very shortly, according to the En-route Low Altitude Chart (ELAC), you'll be leaving the airspace of Minneapolis Center and entering that controlled by Chicago. You can't be sure, but you'll probably be asked to change to the Des Moines remote outlet on 127.05. Assuming that will be the frequency, dial it in so that you'll be prepared. After a few more minutes, Center comes on:

> **Ctr:** *Cherokee Six One Tango, contact Chicago Center now on one two seven point zero five. Good day.*

> **You:** *Roger, one two seven point zero five. Thank you for your help. Cherokee Six One Tango.*

> **You:** *Chicago Center, Cherokee One Four Six One Tango is with you, level at seven thousand five hundred.*

> **Ctr:** *Cherokee Six One Tango, radar contact. Des Moines altimeter two niner niner eight.*

Nearing Newton, you'll hear something like this:

> **Ctr:** *Cherokee Six One Tango, position one five miles north of the Newton VOR. Radar service terminated. Squawk one two zero zero. Frequency change approved. Good day.*

> **You:** *Roger. One two zero zero. Thank you for your help. Cherokee Six One Tango.*

Newton is another non-tower-controlled Class E Airport, with only unicom on 122.8. About 10 miles out, you announce your presence:

> **You:** *Newton Unicom, Cherokee One Four Six One Tango is five north of Newton VOR. Request airport advisory.*

> **Unicom:** *Cherokee One Four Six One Tango, Newton Unicom. Wind is two one zero at one five. Altimeter three zero one five. Runway One Three in use. Three Cessnas reported in the pattern.*

> **You:** *Roger. Cherokee Six One Tango.*

This is another situation where your position and the runway-in-use invite a straight-in approach and landing. Again, though, you intend to go by the rules and adhere to the standard traffic pattern procedures. Hence this call:

> *You: Newton Traffic, Cherokee One Four Six One Tango is eight miles northwest at four thousand, descending. Will cross over the airport at two thousand five hundred for standard entry to downwind for landing One Three Newton.*

While in the process of turning to the downwind leg from the 45-degree entry, your next calls follow the routine non-tower-controlled Class E or G format:

> *You: Newton Traffic, Cherokee Six One Tango entering left downwind at midfield for landing One Three Newton.*

> *You: Newton Traffic, Cherokee Six One Tango turning left base for landing One Three Newton.*

> *You: Newton Traffic, Cherokee Six One Tango turning final for landing One Three Newton.*

And when down and clear:

> *You: Newton Traffic, Cherokee Six One Tango clear of One Three. Taxiing to the terminal, Newton.*

When in the terminal, don't forget to cancel the flight plan. At Newton, this has to be done by phone to the FSS in Fort Dodge on 1-800-WX-BRIEF.

Newton to Kansas City

With the remaining passenger dropped off and full fuel tanks, you're ready for the last leg back to Kansas City. But first comes another call to Flight Service for a weather check and flight plan filing. Local conditions are determined from the unicom operator (or review the winds, etc., yourself if he's out gassing an airplane).

The radio contacts will first be to local traffic and then to Flight Service, which you won't be able to reach until you have some altitude. Heading southwest to the Des Moines VOR, you'll be well above the 5000 foot msl ceiling of the Des Moines Class C, so calling Approach isn't necessary. From the VOR southbound on V13, however, you'll be in the Class C airspace's outer area, which rises to approximately 12,000 feet msl, for a few minutes. Even though the weather is fine, you conclude that monitoring Approach would be a good idea, and, if the volume of traffic indicates, asking for advisories.

While you're still on the ramp at Newton:

> ***You:*** *Newton Traffic, Cherokee One Four Six One Tango at the terminal, taxiing to Runway One Three, Newton.*

After engine run-up:

> ***You:*** *Newton Traffic, Cherokee One Four Six One Tango taking One Three, southwest departure, Newton.*

Off the ground and at 300 feet or so, you contact Fort Dodge Flight Service over the Newton VOR. In this instance, you transmit on 122.1 and receive on the VOR frequency of 112.5.

> ***You:*** *Fort Dodge Radio, Cherokee One Four Six One Tango listening Newton VOR.*

> ***FSS:*** *Cherokee One Four Six One Tango, Fort Dodge Radio, go ahead.*

> ***You:*** *Fort Dodge, Cherokee Six One Tango was off Newton at three five past the hour. Would you please open my flight plan to Kansas City Downtown?*

> ***FSS:*** *Cherokee Six One Tango, Roger. We show you off Newton at three five and will activate your flight plan to Kansas City. Des Moines altimeter three zero zero one.*

> ***You:*** *Roger. Thank you. Cherokee Six One Tango.*

Finally, one more local call is in order:

> ***You:*** *Newton Traffic, Cherokee Six One Tango now clear of the area to the southwest, Newton.*

Monitoring Des Moines Approach as you pass through the Class C airspace's outer area, you call Minneapolis Center when clear of the area.

> ***You:*** *Minneapolis Center, Cherokee One Four Six One Tango.*

Ctr: *Cherokee One Four Six One Tango, Minneapolis Center, go ahead.*

You: *Center, Cherokee One Four Six One Tango is fifteen southwest of the Newton VOR at five thousand two hundred, climbing to six thousand five hundred, VFR to Kansas City Downtown via Victor One Three. Squawking one two zero zero. Request VFR advisories, workload permitting.*

Ctr: *Cherokee Six One Tango, squawk zero five two three and ident.*

You: *Cherokee Six One Tango squawking zero five two three.*

Ctr: *Cherokee Six One Tango, radar contact. Report level at six thousand five hundred.*

Ctr: *Cherokee Six One Tango, Roger.*

There might or might not be further advisories from Center, depending on traffic. Whichever the case, you're soon over the Des Moines VOR and heading outbound on Victor 13.

After passing the Lamoni VOR, 53 miles out of Des Moines, you leave Minneapolis Center and enter the Kansas City Center area. As you cross the line:

Ctr: *Cherokee Six One Tango. Contact Kansas City Center now on one two seven point niner. Good day.*

You: *Roger. One two seven point niner. Cherokee Six One Tango. Good day.*

You: *Kansas City Center, Cherokee One Four Six One Tango is with you at six thousand five hundred.*

Ctr: *Cherokee Six One Tango, Roger. St. Joe altimeter two niner niner six.*

You: *Roger, Cherokee Six One Tango.*

About now is the time to tune the second radio to the Kansas City VOR on 112.6. With one VOR head tracking you outbound from Lamoni and the other inbound to Kansas City, you should be smack in the middle of V13.

When you're approximately due east of St. Joseph, Missouri, you spot some lightning not too far distant and just to the right of your course. This can be an omen of bad stuff, so you decide to check with Center.

> ***You:*** *Center, Cherokee Six One Tango. Request.*
>
> ***Ctr:*** *Cherokee Six One Tango, go ahead.*
>
> ***You:*** *Cherokee Six One Tango has lightning at one o'clock. Will my present course keep me clear of the storms?*
>
> ***Ctr:*** *Cherokee Six One Tango, affirmative. Scattered thunderstorms are moving northeast, but you should be past the area at your present ground speed.*
>
> ***You:*** *Roger. Thank you. Cherokee Six One Tango.*

On the other hand, if a storm encounter seems likely, Center might offer this advice: "Cherokee Six One Tango, the storm activity is moving due east. Suggest right heading of two seven zero past St. Joe to Topeka and come in behind the weather." Keep in mind that such a suggestion is not a command. You're VFR, and have freedom as well as options.

Assuming the weather is not going to be a factor, Center will call you as you near the Kansas City TCA:

> ***Ctr:*** *Cherokee Six One Tango, position one five miles north of the Kansas City Bravo airspace. Contact Kansas City Approach on one one niner point zero.*
>
> ***You:*** *Roger, one one niner point zero. Thank you for your help. Cherokee Six One Tango.*

Before contacting Approach, be sure you have monitored the Kansas City Downtown ATIS. Then, with the information clearly in mind, call Approach:

> ***You:*** *Kansas City Approach, Cherokee One Four Six One Tango is with you, level at six thousand five hundred with Foxtrot.*
>
> ***App:*** *Cherokee One Four Six One Tango, cleared into the Bravo airspace, direct Kansas City VOR. Descend and maintain three thousand. Remain VFR at all times.*

You: *Cherokee Six One Tango, Roger. Cleared into the Bravo airspace, direct Kansas City VOR, down to three thousand and remain VFR.*

You: *Approach, Cherokee Six One Tango level at three thousand.*

App: *Roger, Six One Tango.*

Approach sees you nearing the VOR:

App: *Cherokee Six One Tango, descend to two thousand five hundred.*

You: *Cherokee Six One Tango out of three for two thousand five hundred.*

You: *Approach, Cherokee Six One Tango level at two thousand five hundred.*

App: *Roger, Six One Tango. Maintain heading and altitude.*

You: *Will do. Six One Tango.*

As you cross the VOR, only 9 miles from the airport:

App: *Cherokee Six One Tango, Downtown is niner miles at twelve o'clock. Advise when you have the airport in sight.*

You: *Approach, Six One Tango has the airport.*

App: *Roger, Six One Tango. Contact Downtown Tower on one three three point three.*

You: *Will do. Cherokee Six One Tango.*

You: *Downtown Tower, Cherokee One Four Six One Tango is with you, level at two thousand five hundred.*

Twr: *Cherokee One Four Six One Tango, continue straight in for Runway One Niner. Sequence later.*

You: *Roger, straight in for One Niner, Cherokee Six One Tango.*

Twr: *Cherokee Six One Tango, you'll be number two to land following a Citation on base.*

You: Roger. *Cherokee Six One Tango has the Citation.*

Twr: Cherokee Six One Tango, Roger. Caution wake turbulence landing Citation. Cleared to land Runway One Niner.

You: Cleared to land, Cherokee Six One Tango.

You watch the Citation touch down. To plan your landing because
of the wake turbulence, you'd like a current wind reading.

You: Tower, Cherokee Six One Tango. Wind check.

Twr: Cherokee Six One Tango, wind two one zero at one five.

You: Cherokee Six One Tango.

Because you're coming in on Runway 19, these winds should blow
the wake to the left of the runway, so you decide to land on the
right, or upwind, side. You do so without difficulty and complete
the rollout.

Twr: Cherokee Six One Tango, contact Ground point niner
when clear.

You: Cherokee Six One Tango.

You: Downtown Ground, Cherokee One Four Six One
Tango clear of One Niner. Taxi to Hangar Six.

GC: Cherokee Six One Tango, taxi to Hangar Six.

By radio when parked, or by phone, you close out the flight plan—
and the trip concludes without incident.

Home at last

The point of this cross-country was to illustrate the typical radio procedures when using Clearance Delivery Contacting and Ground
Control, Tower, Approach/Departure, Center, Flight Service, when
operating within Class B, C, D, and E airspaces, and at unicom airports. The trip tried to encapsulate the more common phrases and
phraseologies discussed in the various previous chapters.

Yes, there were several instances of what you might have considered
needless repetition, but a cross-country involves repetition. Many of
the same things are said to different agencies. Besides, repetition has

a way of cementing habits in our minds and preventing errors and misunderstandings. Not every possible contact was included (as, for example, a call to Flight Watch), but many of the most common dialogues were recreated.

And, yes, some of the dialogues might seem a little stilted. As earlier stated, however, the examples are based on those illustrated in the *Aeronautical Information Manual* (*AIM*) and established as policy in the controller's *Air Traffic Control* manual, 7110.65G.

If you fly enough, you'll occasionally hear minor variations or local adaptations, particularly on the part of pilots. It's probably inevitable that a little slang or jargon, neither of which necessarily distorts the message, creeps in. Such liberties, however, don't conform to FAA procedures.

In the process, the use of the radio and the services available to every pilot are made just a bit clearer. If such is the case, perhaps some of the communicating concerns experienced by so many VFR pilots—both new and experienced—have been allayed, at least a little.

29

Practical cross-country flights

A safe cross-country flight consists of two distinct phases: preflight planning and the flight itself. If your preplanning is thorough, you will be able to fly your trip with greater confidence and probably enjoy it a lot more. The more planning you do on the ground, the less busy you will be in the air. On your first few cross-country flights, you will have plenty of things to do to occupy your time and mind without trying to find that frequency, heading, or some other important detail that you should have already written down and put in a safe place.

The first things you need to know before you go on your cross-country are where you are going, the time of day you will be flying, and the charts that will be necessary to safely complete the trip (Fig. 29-1). You also have to plan whether the trip will be made by VOR, pilotage, or a combination of the two. Most VFR flights are made by using a combination of VOR and pilotage. However, with the advent of the FAA's changes to the FARs of 1997, pilotage is the method you must use to plan and execute your cross-country flight for your private pilot checkride.

Pilotage

Today's pilot can utilize U.S. government-supplied sectional charts that depict the landmarks in very minute detail. In fact, there is so much information on these sectional charts that in a heavily populated area there is sometimes almost too much information. For instance, these charts display all towers, tell you how high they are above both sea level and the ground, as well as almost any other identifiable object you might imagine. A short list of items displayed on these charts would contain drive-in movie theatres, rivers, lakes, towns, towers, roads, airports, hills, mountains, cliffs, valleys, power

Fig. 29-1. *Check the course and distance as part of your initial preparation.*

plants, race tracks, railroad tracks, and just about any other solid object you can imagine. If you could see it, hit it, or use it, it will be displayed on the government sectional charts.

I have a small but significant piece of advice for you new or would-be pilots concerning the proper method of using the sectional chart. As is the case with all maps I am aware of, sectional charts are oriented to north. That is the top of the map, when held so it is easily readable, is north. Okay, so that's not earthshaking of itself. But human nature is such that we all like to be able to read things easily, and so it is with sectional charts. We like to hold the map so that the map is always right side up. It's just easier to read that way.

Trouble is, our flights aren't necessarily always going from south to north. Sometimes we are going to fly southwest or whatever. When this happens, if you continue to hold the map so that it's right side up, the objects appear out the window of the aircraft out of place. They aren't in their proper perspective. Say a lake shows itself to your right, or west, of your course on the sectional. If you are flying south while holding the map upright, it's easy to believe the lake should appear out the left window of your aircraft. Of course this is not the case, but when you hold the map in what amounts to be an upside down position, funny things can happen to your navigation. Ask Wrong-Way Corrigan.

The cure for this malady is to always hold the map with the course you are following pointed out the front windshield of your aircraft. Turn the map so the departure point is nearest your body and the destination is at the point farthest away from you. It's only when landmarks appear out the window in the same relative positions in which they appear on the charts that the potential problem of reverse orientation is most likely avoided.

VOR

Without getting too complicated, let's review the functions and normal use of the very high frequency omnidirectional radio range (VOR). The VOR is called the *omni*. Omni is Latin for "in all directions." It can be a very good friend or a frightening enemy, depending on your understanding of its use and purpose. Well understood, it is of invaluable assistance in navigation. If you try to fly the VOR without fully understanding its use, you might actually wind up going in the wrong direction.

From its ground base, the VOR transmits signals in all directions. Each VOR has its own frequency, and you tune it in on the navigation side of your radio. The following describe the features of your VOR receiver.

Frequency selector

The *frequency selector* is manually rotated to select any of the frequencies in the VOR range of 108.0 to 117.95 MHz.

Course selector

By turning the OBS (omni bearing selector), the desired course is selected. This usually appears in a window or under an index on the VOR receiver head on your instrument panel.

Course deviation indicator

(CDI) The *course deviation indicator* is composed of a dial and a needle. The needle centers when you are on the selected course or its reciprocal, regardless of your heading (Fig. 29-2). Full needle deflection

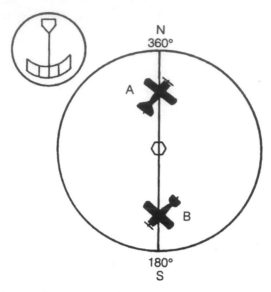

Fig. 29-2. *Course deviation indicator.*

from the center position to either side of the dial indicates the aircraft is 10 degrees or more off course (assuming normal needle sensitivity).

TO/FROM indicator

The *TO/FROM indicator* is also called sense indicator or ambiguity indicator. The TO/FROM indicator shows whether the selected course will take the aircraft to or from the station. It does not indicate whether the aircraft is heading to or from the station (Fig. 29-3).

Flags

Flags can be labeled as signal strength indicators. The device to indicate whether a signal is usable or an unreliable signal is called an *OFF/VOR* flag. This flag retracts from view or says OFF when signal strength is sufficient for reliable instrument indications. When the VOR signal is strong enough to give reliable navigation guidance, the flag switches to VOR. Insufficient signal strength might also be indicated by a blank TO/FROM window.

There are a couple of very important points to remember when using the VOR. The VOR transmits 360 possible magnetic courses to and from the station. These courses are called *radials*. They are oriented from the station (Fig. 29-4). For example, the aircraft at A,

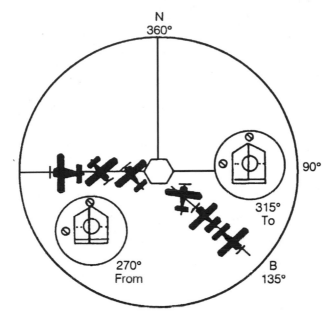

Fig. 29-3. *TO/FROM indicator.*

heading 180 degrees, is flying to the station on the 360 radial. After crossing the station, the aircraft is flying from the station on the 180 radial at 2A. Aircraft B is shown crossing the 225 radial. Similarly, at any point around the station, an aircraft can be located somewhere on a VOR radial. The important point is if you want to know what radial you are on, turn the OBS until the CDI centers and the TO/FROM indicator reads FROM. That is the radial you are on.

To properly utilize your VOR, you must first know where you are (what radial you are on), where you are going (what radial you have to intercept), and how to track the radial once you get there. This process utilizes three steps: orientation, interception of the radial, and tracking.

The following demonstration can be used to intercept a predetermined inbound or outbound track. The first three steps may be omitted if you turn directly to intercept a course without initially turning to parallel the desired course.

Turn to a heading to parallel the desired course. Turn in the same direction as the course to be flown.

- Determine the difference between the radial to be
  intercepted and the radial on which you are located.

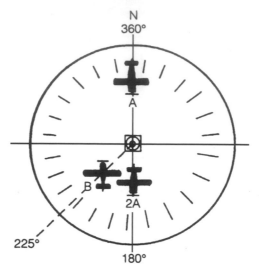

Fig. 29-4. *VOR courses called radials are oriented FROM the station.*

- Double the difference to determine the intercept angle not less than 20 degrees or more than 60 degrees.
- Rotate the OBS to the desired radial or inbound course.
- Turn to the interception beading (magnetic).
- Hold this magnetic heading until the CDI centers, indicating the aircraft is crossing the desired course.
- Turn to the magnetic heading corresponding to the selected course and track inbound or outbound on the radial.

VOR tracking also involves drift correction sufficient to maintain a direct course to or from a station. The course selected for tracking inbound is the course shown on the course index with the TO/FROM indicator showing TO. If you are off course to the left, the CDI is deflected right; if you are off course to the right, the CDI is deflected to the left. Turning toward the needle returns the aircraft to the course centerline and centers the needle.

To track inbound with the wind unknown, proceed using the following steps (Fig. 29-5). Outbound tracking is the same.

- With the CDI centered, maintain the heading corresponding to the selected course.
- As you hold the heading, note the CDI for deflection to the left or right. The direction of CDI deflection from centerline

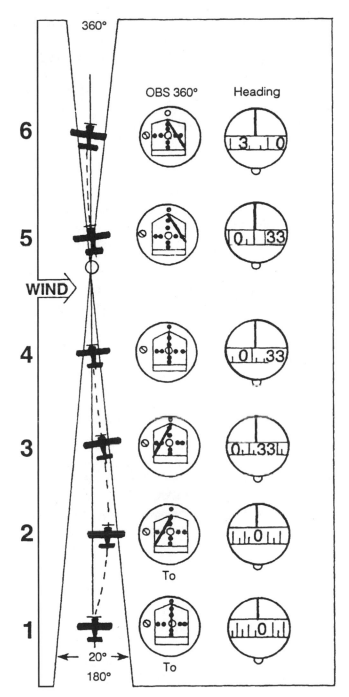

Fig. 29-5. *Tracking inbound with the wind unknown.*

shows you the direction of the crosswind. Figure 29-5 shows a left deflection, therefore a left crosswind.

- Turn 20 degrees toward the needle and hold the heading until the needle centers.

Reduce drift correction to 10 degrees left of the course setting. Note whether this drift-correction angle keeps the CDI centered. Subsequent left or right needle deflection indicates an excessive or insufficient drift-correction angle—either add or remove some correction (heading). With the proper drift correction established, the CDI will remain centered until the aircraft is close to the station. Approach to the station is shown by a flickering of the TO/FROM indicator and CDI as the aircraft flies into the no-signal area (almost directly over the station). Station passage is indicated by a complete reversal of the TO/FROM indicator.

Following station passage and TO/FROM reversal, course correction is still toward the needle to maintain course centerline. The only difference is that now you are tracking *away* from the station instead of to the station. In the previously listed steps, you were tracking inbound on the 180 radial. After station passage, although your heading hasn't changed, you are tracking outbound on the 360 radial.

Loran

Loran is an acronym for *LOng RAnge Navigation.* An old form of navigation, Loran has gained unqualified acceptance with pilots. The advent of better and smaller computers has made the units more size and price competitive. Also, the FAA, until quite recently, was constructing more transmitting stations within the inner United States. Originally, Loran was utilized during World War II to assist with ocean navigation. Later, the Coast Guard further developed its usefulness with the newer Loran-C, which we now call Loran.

The operating principle of Loran is the measurement of time between two electronic impulses received from a chain of transmitters. The transmitters are spaced many hundreds of miles apart and are divided into master and secondary stations. Each station, master and secondary, transmits an impulse at precisely the same instant. The Loran receiver measures the difference in time that it takes for these impulses to travel to the receiver and computes its position based on these slight time differences. These position cross-checks are accurate to within one-quarter of a mile or better, and you'd better believe a position check from several hundred miles away that is that accurate is a welcome companion to any pilot. (Fig. 29-6.)

As computer technology advanced and the weight and size of receivers reduced, Loran receivers became more readily accessible to aircraft. The newer models are approximately the size of any other communication receiver. This sizing, along with more affordable pricing, has made the Loran a viable option to many pilots.

The Loran's small computer gives pilots an enormous amount of navigation information, so much so that some are intimidated by the Loran. Witness the information, ready at the pilot's fingertips, from a modern Loran:

- Present position
- Bearing and range to the station
- Bearing and distance from current position to origination point
- Estimated time en route
- Ground speed
- Distance from desired track and physical depiction of correction needed

Additionally, the Loran can store in its memory thousands of bits of information concerning U.S. navigation aids, altitude restrictions, wind correction information, airports of the world, and much more. Is it any wonder this form of navigation has caught on with pilots in all walks of aviation?

The Loran does suffer a couple of points of detraction. It is very sensitive to precipitation, and proper antenna installation is critical to its

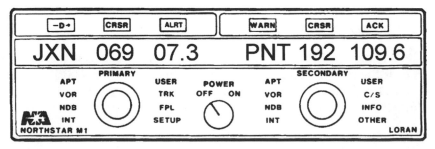

Fig. 29-6. *The LORAN can give a pilot a great deal of information.*

accuracy. Also, the FAA is talking about reducing funding for Loran stations and going with GPS (Global Positioning Systems) in the near future. But, personally, I doubt this will happen any time soon. The popularity of the Loran and the multiple tens of thousands of them in daily use make me believe the Loran will be with us for quite a few years. Should the FAA make any serious attempt to withdraw these inexpensive and user friendly little navigation tools from use, the outcry from pilots would be resounding. However, the FAA is giving quite a bit of priority to the GPS since it is much more accurate and has more possibilities for use than does the Loran.

The Loran chain

The 27 U.S. Loran transmitters that provide signal coverage for the continental United States and the southern half of Alaska are distributed from Caribou, ME, to Attu Island in the Aleutians. Station operations are organized into subgroups of four to six stations called "chains." One station in the chain is designated the "Master" and the others are "secondary" stations.

The Loran Navigation signal is a carefully structured sequence of brief radio frequency pulses centered at 100 kilohertz. The sequence of signal transmissions consists of a pulse group from the Master (M) station followed at precise time intervals by groups from the secondary stations, which are designated by the U.S. Coast Guard with the letters V. W. X. Y. and Z. All secondary stations radiate pulses in groups of eight, but the Master signal for identification has as additional ninth pulse.

The time interval between the recurrence of the Master pulse group is the Group Repetition Interval (GRI). The GRI is the same for all stations in a chain and each Loran chain has a unique GRI. Since all stations in a particular chain operate on the same radio frequency, the GRI is the key by which a Loran receiver can identify and isolate signal groups from a specific chain.

The line between the Master and each secondary station is the "baseline" for a pair of stations. Typical baselines are from 600 to 1,000 nautical miles in length. The continuation of the baseline in either direction is a "baseline extension."

Loran transmitter stations have time and control equipment, a transmitter, auxiliary power equipment, a building about 100 by 30 feet in size and an antenna that is about 700 feet tall. A station generally requires approximately 100 or more acres of land to accommodate guy lines that keep the antenna in position. Each Loran station transmits from 400 to 1600 kilowatts of signal power.

The USCG operates 27 stations, comprising eight chains, in the United States. Four control stations, which monitor chain performance, have personnel on duty full time. The Canadian east and west coast chains also provide signal coverage over small areas of the NAS.

When a control station detects a signal problem that could affect navigation accuracy, an alert signal called "Blink" is activated. Blink is a distinctive change in the group of eight pulses that can be recognized automatically by a receiver so the user is notified instantly that the Loran system should not be used for navigation. In addition, other problems can cause signal transmissions from a station to be halted.

The Loran Receiver

Before a Loran receiver can provide navigation information for a pilot, it must successfully receive, or "acquire," signals form three or more stations in a chain. Acquisition involves the time synchronization of the receiver with the chain GRI, identification of the Master station signals from among those checked, identification of secondary station signals, and the proper selection of the point in each signal at which measurements should be made.

Signal reception at any site will require a pilot to provide location information such as approximate latitude and longitude, or the GRI to be used, to the receiver. Once activated, most receivers will store present location information for later use. The basic measurements made by Loran receivers are the difference in time-of-arrival between the Master signal and the signals from each of the secondary stations of a chain.

An aircraft's Loran receiver must recognize three signal conditions:

- Usable signals
- Absence of signals
- Signal blink

The most critical phase of flight is during the approach to landing at an airport. During the approach phase the receiver must detect a lost signal, or a signal Blink, within 10 seconds of the occurrence and warn the pilot of the event.

Loran signals operate in the low-frequency band around (100 kHz) that has been reserved for Loran use. Adjacent to the band, however, are numerous low-frequency communications transmitters. Nearby signals that can distort the Loran receivers have selective internal filters. These filters, commonly known as "notch filters" reduce the effect of interfering signals.

Careful installation of antennas, good metal-to-metal electrical bonding, and provisions for precipitation noise discharge on the aircraft are essential for the successful operation of Loran receivers. A Loran antenna should be installed in accordance with the manufacturer's instructions. Corroded bonding straps should not be used, and static discharge devices installed at points indicated should be by the aircraft manufacturer.

Loran Navigation

An airborne Loran receiver has four major parts:

- Signal processor
- Navigation computer
- Control/display
- Antenna

The signal processor acquires Loran signals and measures the difference between the time-of-arrival of each secondary station pulse group and the Master station pulse group. The measured TDs depend on the location of the receiver in relation to the three or more transmitters.

The first TD will locate an aircraft somewhere on a line-of-position (LOP) on which the receiver will measure the same TD (time-distance) value. A second LOP is defined by a TD measurement between the Master station signal and the signal from another secondary station. The intersection of the measured LOPs is the position of the aircraft.

The navigation computer converts TD values to corresponding latitude and longitude. Once the time and position of the aircraft is established at two points, distance to destination, cross-track error, ground speed, estimated time of arrival, etc., can be determined. Cross-track error can be displayed as the vertical needle of a course deviation indicator, or digitally, as decimal parts of a mile left or right of course. During a non-precision approach, course guidance must be displayed to the pilot with a full-scale deviation of 0.30 nautical miles or greater.

Loran navigation for nonprecision approaches requires accurate and reliable information. During an approach the occurrence of signal Blink or loss of signal must be detected within 10 seconds and the pilot must be notified. Loran signal accuracy for approaches is 0.25 nautical miles, well within the required accuracy of 0.30 nautical miles.

Flying a Loran nonprecision approach is different from flying a VOR approach. A VOR approach is on a radial off the VOR station, with guidance sensitivity increasing as the aircraft nears the airport. The Loran system provides a linear grid, so there is constant guidance sensitivity everywhere in the approach procedure. Consequently, inaccuracies and ambiguities that occur during operations in close proximity to VORs (station passage, for example) do not occur in Loran approaches.

The navigation computer also provides storage for data entered by pilot or provided by the receiver manufacturer. The receiver's database is updated at local maintenance facilities every 60 days to include all changes made by the FAA.

Global Positioning System (GPS)

GPS is a U.S. satellite based radio navigational, positioning, and time-transfer system operated by the Department of Defense (DOD). The system provides highly accurate position, velocity, and precise time information on a continuous global basis to an unlimited number of properly equipped users. Unlike LORAN, the GPS system is unaffected by weather and provides a worldwide common grid reference system based on an earth-fixed coordinate system.

GPS provides two levels of service: Standard Positioning Service (SPS) and Precise Positioning Service (PPS). SPS provides, to all users, horizontal positioning accuracy of 10 meters, or less, with a probability of 95 percent and 300 meters with probability of 99.99 percent. PPS is more accurate than SPS: However, this is limited to authorized U.S. civil and military users who meet specific requirements.

GPS operation is based on the concept of ranging and triangulation from a group of satellites in space that act as precise reference points. A GPS receiver measures distance from a satellite using the travel time of a radio signal. Each satellite transmits a specific code, called a course/acquisition (CA) code, which contains information on the satellite's position, the GPS system time, and the health and accuracy of the transmitted data. Knowing the speed at which the signal traveled, the speed of light, (approximately 186,000 miles per second), and the exact broadcast time, the distance traveled by the signal can be computed from the arrival time.

The GPS receiver matches each satellite's CA code with an identical copy of the code contained in the receiver's database. By shifting its copy of the satellite's code in a matching process, and by comparing this shift with its internal clock, the receiver can calculate how long it took the signal to travel from the satellite to the receiver. The distance derived from this method of computing distance is called a pseudorange because it is not a direct measurement of distance, but a measurement based on time.

In addition to knowing the distance to a satellite, a receiver needs to know the satellite's exact position in space; this is known as its ephemeris. Each satellite transmits information about its exact orbital location. The GPS receiver uses this information to precisely establish the position of the satellite.

Using the calculated pseudorange and position information supplied by the satellite, the GPS receiver mathematically determines its position by triangulation. The GPS receiver needs at least four satellites to yield a three-dimensional position (latitude, longitude, and altitude) and time solution. The GPS receiver computes navigational values such as distance and bearing to a waypoint, ground speed, etc., by using the aircraft's known latitude/longitude and referencing these to data built into the receiver.

The GPS constellation of 24 satellites is designed so that a minimum of five is always observable by a user anywhere on earth. The receiver uses data from a minimum of four satellites above the mask angle (the lowest angle above the horizon at which it can use a satellite).

The GPS receiver verifies the integrity of the signals received from the GPS constellation through receiver autonomous integrity monitoring (RAIM) to determine if a satellite is providing corrupted information. At least one satellite, in addition to those required for navigation, must be in view for the receiver to perform the RAIM function; thus, RAIM needs a minimum of 5 satellites in view, or 4 satellites and a barometric altimeter (baro-aiding) to detect an integrity anomaly. For receivers capable of doing so, RAIM needs 6 satellites in view (or 5 satellites with baro-aiding) to isolate the corrupt satellite signal and remove it from the navigation solution. Baro-aiding is a method of augmenting the GPS integrity solution by using a nonsatellite input source. GPS-derived altitude should not be relied upon to determine aircraft altitude since the vertical error can be quite large. To ensure that baro-aiding is available, the current altimeter setting must be entered into the receiver as described in the operation manual.

The Department of Defense declared initial operational capability of the U.S. GPS on December 8, 1993. The Federal Aviation Administration (FAA) has granted approval for U.S. civil operators to use properly certified CPS equipment as a primary means of navigation in oceanic airspace and certain remote areas. Properly certified GPS equipment may be used as supplemental means of IFR navigation for domestic en route, terminal operations, and certain instrument approach procedures. This approval permits the use of GPS in a manner that is consistent with current navigation requirements as well as approved air carrier operations specification.

Civilian pilots are authorized to conduct any GPS operation under IFR provided that the GPS navigation equipment used is approved in accordance with the requirements specified in TSO C-129, or equivalent, and the installation must be done in accordance with Notice 8110.47 or 8110.48, or the equivalent.

Procedures must be established for use in the event that the loss of RAIM capability is predicted to occur. In situations where this is encountered, the flight must rely on other approved navigation equipment (VOR, ADF, etc.), delay departure, or cancel the flight. (Fig. 29-7.)

The GPS operation must be conducted in accordance with the FAA-approved aircraft flight manual or flight manual supplement. Flight crewmembers must be thoroughly familiar with the particular GPS

Fig. 29-7. *The GPS is the FAA's chosen method of navigation now and into the 21st century.*

equipment installed in the aircraft, the receiver operation manual, and the AFM or flight manual supplement. Unlike ILS and VOR, the basic operation, receiver presentation to the pilot, and some capabilities of the equipment can vary greatly. Due to these differences, operation of different brands, or even models of the same brand, of GPS receiver under IFR should not be attempted without thorough study of the operation of that particular receiver and installation. Most receivers have a built-in simulator mode that will allow the pilot to become familiar with operation prior to attempting operation in the aircraft. I would suggest using the equipment in flight under VFR conditions prior to attempting IFR operation.

Aircraft navigation by IFR-approved GPS are considered to be RNAV aircraft and have special equipment suffixes as related to ATC flight plans. You must file the appropriate equipment suffix on the ATC flight plan. If GPS avionics become inoperative, the pilot should advise ATC and amend the equipment suffix.

GPS position orientation

As with most RNAV systems, pilots should pay close attention to position orientation while using GPS. Distance and track information are provided to the next active waypoint, not to a fixed navigation aid. Receivers may sequence when a pilot is not flying along an active route, such as when being vectored or deviating due to weather, due to the proximity to another waypoint in the route. This can be prevented by placing the receiver in a nonsequencing mode. When the receiver is in the nonsequencing mode, bearing and distance are provided to the selected waypoint and the receiver will not sequence to the next waypoint in the route until placed back in the auto sequence mode or the pilot selects a different waypoint. On an overlay approach, the pilot may have to compute the track distance to step-down fixes and other points due to the receiver showing track distance to the next waypoint rather than DME to the VOR or ILS ground station.

GPS versus conventional navigation

You may find slight differences between the heading information shown on the navigational charts and the GPS navigation display when flying on an overlay approach or along an airway. All magnetic tracks defined by a VOR radial are determined by the application of magnetic variation at the VOR. However, GPS operations may use an algorithm to apply the magnetic variation at the current position, which may produce small differences in the displayed course. Both operations should produce the same desired ground track. Due to the use of great circle courses, and the variations in magnetic variations, the bearing to the next waypoint and the course from the last waypoint may not be exactly 180 degrees apart when long distances are involved.

Variations in distances will occur since GPS distance-to-waypoint values are along track (straight-line) distances computed to the next waypoint and the DME values published on underlying procedures are the slant range distances measured to the station. The difference increases with aircraft altitude and proximity to the NAVAID.

Waypoints

GPS approaches make use of both flyover and fly-by waypoints. Fly-by waypoints are used when an aircraft should begin a turn to the next course prior to reaching the waypoint separating the two route segments. This is known as turn anticipation and is compensated for in the airspace and terrain clearances. Approach waypoints, except for the missed approach waypoint (MAWP), and the missed approach holding waypoint (MAHWP), are normally fly-by waypoints Flyover waypoints are used when the aircraft must fly over the point prior to starting a turn. New approach charts depict flyover waypoints as a circled waypoint symbol. Overlay approach charts and some early stand-alone GPS approach charts may not reflect this convention.

On overlay approaches (titled "or GPS"), if no pronounceable five-character name is published for an approach waypoint or fix, it may be given an ARINC database identifier consisting of letters and numbers. These points will appear in the list of waypoints in the approach procedure database, but may not appear on the approach chart. Procedures without a final approach fix (FAF), for instance, have a sensor final approach waypoint (FAWP) added to the database at least 4 NM prior to the MAWP and MAHWP to allow the receiver to transition to the approach mode. Some approaches also contain an additional waypoint in the holding pattern when the MAWP and MAHWP are colocated. Arc and radial approaches have an additional waypoint that is used for turn anticipation computation when the arc joins the final approach course. These coded names will not be used by ATC.

Unnamed waypoints in the database will be uniquely identified for each airport but may be repeated for another airport (e.g., RW36 will be used at each airport with a runway 36 but will be at the same location for all approaches at a given airport).

The runway threshold waypoint, which is normally the MAWP, may have a five-letter identifier (e.g., SNEEZ) or be coded as RW## (e.g., RW36, RW36L). Those thresholds which are coded as five-letter identifiers are being changed to the RW## designation. This may cause the approach chart and database to differ until all changes are complete. The runway threshold waypoint is also used as the center of the MSA on most GPS approaches. MAWPs not located at the threshold will have a five-letter identifier.

Charting the course

Let's say you are going from Lawrenceville, Illinois, to Evansville, Indiana. For this trip, the St. Louis Sectional fits the bill, as the trip is well within the boundaries of this particular chart.

Open the St. Louis Sectional and locate the two airports. Using a straightedge, draw a line from the center of the departure airport to the center of the destination airport. You will want to circle prominent checkpoints along your route of flight. Select ones that can be easily seen and recognized from the air. Then, use your plotter and obtain the true course for the cross-country. In this case, the true course is 175 degrees. You will also want to use your personal computer to access weather information or call the FSS to find out what type of weather to expect along your route as well as obtain the forecast winds aloft.

Using your flight computer, find your true airspeed, true heading, ground speed, and magnetic and compass headings. The easy way to remember the sequence for this computation are the initials TVMDC. *TVMDC* stands for true, variation, magnetic, deviation, and compass.

Use both the St. Louis Sectional and the *Aeronautical Information Manual* as well to obtain radio frequencies, airport data, NOTAMs, new and closed airports, and any other information relevant to your proposed flight. Be sure to write this information down so you can find it quickly. Many pilots use a flight log that can be found on the back of FAA Flight Plans. Some merely use a piece of paper. In any case, don't try to commit it to memory-write it down.

When you have gathered all the pertinent data, fill out and file a VFR flight plan with the nearest FSS. As I said at the beginning of this book, this filing is the cheapest insurance you can have, so don't neglect it.

As you taxi out, set your OBS to 175 degrees and tune in the Lawrenceville VOR on 108.8. Arrange all your flight plans and notes and fold your Sectional Chart so only the portion you are using shows. As I said before, turn your Sectional Chart so that it corresponds with your route of flight. Place it on your lap with your destination airport furthest away from your body. When you look outside, everything will then be in the proper perspective. A town that should be off to your right will be off to your right. You will learn to read upside down very quickly. Keep the sectional turned in the direction of your course.

I have advised you to set your VOR and keep your sectional in place at the same time. Then, you can fly the VOR and double-check your route with the sectional, or vice versa. A strong word of caution: If you ever decide to fly either VOR or pilotage exclusively, fly pilotage (the map). Radios have a very peculiar habit of malfunctioning at the most inopportune times. Know where you are on the sectional. I've known a couple of pilots who liked to fly the VOR with the sectional somewhere in the back seat, out of sight, and then actually flew right over their intended destination and never even knew it. I guess it's all right if you're trying to build flight time because you will certainly have to turn around and hunt for your destination.

The flight

Once airborne, make sure you get off to a good start on your cross-country by being on course from the very beginning. In order to assure this, some pilots take off, circle the airport, and fly directly over the departure airport so they are absolutely sure they are right on target when they part. Not a bad idea, at first.

Headings

If your plan calls for flying both the VOR and pilotage, you should take up a heading that allows you to intercept the 175-degree radial from the Lawrenceville VOR. At the same time, you should begin to pick up your visual checkpoints. In this case, the first good one is when you cross the four-lane highway just south of the Lawrenceville airport. From this position, the city of Vincennes should be visible off to your left and slightly ahead. The city of Lawrenceville will be just about under your right wingtip. You are off to a good start, and it is only a matter of keeping close track of the other checkpoints as they come up. You can crosscheck your position by reference to the CDI indication. If you are directly on course, it will read FROM the 175-degree radial of the Lawrenceville VOR, and the needle will be in the center.

Visual checkpoints

As you continue on towards Evansville (Fig. 29-8), the next set of visual checkpoints will be about 5 miles south of the four-lane highway, where you should see the bridge that crosses the Wabash River at St. Francisville about a mile to the right of your course. On a fairly clear day, you should be able to see the Mt. Carmel airport a little further to the west. You're still on course. It's a good feeling when the checkpoints come up at the proper time and in the proper place.

If you want to do a crosscheck with the VOR to really confirm your exact position, tune in the Evansville VOR shown in the bottom left corner of the picture of the St. Louis Sectional in Fig. 29-8 on the frequency shown in the box just under the northern most edge of the VOR compass rose. This frequency is 113.3. Let's say you are crossing over the city of Princeton. With the frequency set in, rotate the OBS until the needle centers on a FROM indication. (Remember, you find the radial you are on with a FROM indication.) It should read 017 degrees FROM as you pass over Princeton. Now, you know exactly where you are. This method can be utilized anywhere, at any time, as long as you are receiving a strong enough signal to know the VOR is reliable. You know you are receiving a usable signal when the TO/FROM indicator is firmly on either TO or FROM, not bouncing from one to the other, and certainly never when the flag box reads OFF.

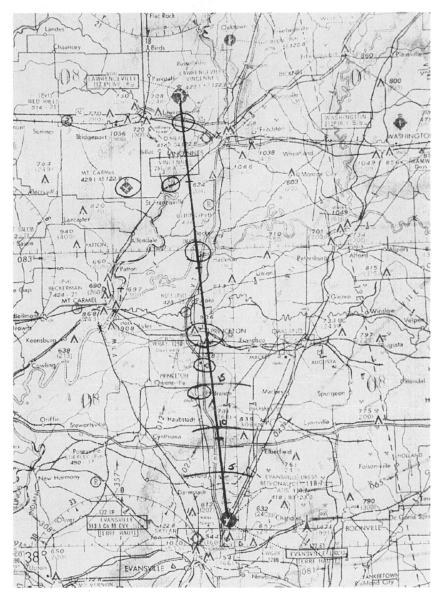

Fig. 29-8. *This sample cross-country, shown on the St. Louis Sectional, is from Lawrenceville, Illinois to Evansville, Indiana.*

Appendix A

NTSB accident synopses

An introduction to the case studies

The probable causes that follow each synopsis are those listed in the NTSB report, while the comments after the causes are basically, but not exclusively, my additions or opinions about what happened and why—the purpose being to raise issues that might not have been apparent in the synopsis or the statement of causes.

As you read these (most are relatively short), you might think about them in terms of your own personal experiences and of what lessons, if any, a given case illustrates. In other words, if you had been the pilot in the case, what would you have done? Or what would you have done to prevent this situation from developing in the first place? Or what mistakes did the pilot make that good judgment would have avoided?

But, should these cases not raise meaningful issues for you (and that's always a possibility), bring up NTSB on the internet (http://www.ntsb.gov), click on Aviation, then Accident Synopses, and finally "Select from monthly list of accidents." A month-by-month calendar from 1983 to the present will then appear. Once you start opening the many files, you'll find enough learning examples to keep you busy for many a TV-less evening.

The most common causes—in summary

The Aircraft Owners and Pilots Association's 1996 "Nall Report" indicates, once all the facts are established, that 70 to 80 percent of the general-aviation accidents can be traced to human factors, while only about 8 percent are the result of mechanical or maintenance failures. "Other" or "unknown" make up the rest. If you can accept the Nall Report and then have the opportunity to review the National Transportation Safety Board (NTSB) accident summaries compiled over the past 15 years, the role that radio communications has played becomes strikingly apparent. Sometimes the role is major, sometimes it's incidental, but it's there, nonetheless. And, typically associated with so many of the NTSB accident studies are comments such as these:

- "Air/ground communications not attained."
- "Communications/information delayed."
- "Air/ground communications not used."
- "Improper interpretation of instructions."
- "Communication inattention."
- "Communications not understood."
- "Improper use of radio equipment."
- "Radio communications not maintained."
- "Failure to communicate on the Common Traffic Advisory Frequency (CTAF)."
- "Communications inadequate."

Although other factors may also have contributed to a given accident, the frequency with which some form of communicating failure is mentioned leaves at least one major impression: We need to sharpen up on our radio sending, receiving, and listening skills!

It's the human element that needs the attention, the training, the re-training, and, in tune with the subject at hand, a continuing emphasis on radio communicating skills. As pilots, one human characteristic we all share is the ability to communicate. We may not be very good communicators, but we at least are physically able to speak and to hear, and to convey and receive messages. If deficient in either of those abilities, it's most unlikely that you or I would even possess a current FAA medical certificate.

Being capable of speech and hearing, however, does not mean that, as pilots, we use these capabilities skillfully or effectively. Should you tend to question that, just listen carefully to some of the chatter that fills the air on a sunny Sunday—or any other good flying day. You'll hear all sorts of pilot-originated messages that reflect a dire need for subject organization, on-the-air brevity, clarity of message, and, in effect, plain and simple communications training.

A few case histories

To illustrate the need for training, or at least for better radio communications, let's look briefly at a few NTSB-recorded accidents, the cause(s) of same, as concluded by the NTSB, and the extent to which radio communications played a part—major or minor—in what happened. As implied earlier, these cases have been taken directly from files provided by the National Transportation Safety Board in Washington.

The NTSB has recorded, by month, over 37,000 accidents that have occurred since January 1983, up to the present. With the help of its Aviation Accident Data Specialist Analysis and Data Division, accidents in which radio communications, or lack thereof, played a part were isolated for me. Only a few (13, to be specific) are summarized here simply because of space limitations.

To protect the guilty or avoid possible embarrassment to any party, some of the details included in the NTSB printouts, such as aircraft N-numbers, dates of the incidents, or pilots' names, are not included. "Probable Causes" and "Contributing Factors," cited at the end of each case, are summaries of the NTSB's conclusions as to the cause(s) of the accident. Then, where it seems appropriate, I have added a further comment or a question. To make it easier to visualize the situations, I've also included skeletal diagrams of the various airport runway systems, based on AOPA's (Aircraft Owners and Pilots Association) 1997 Airport Directory, all rights reserved.

The case studies: Synopsis, causes, comments

Case 1

Location:California

Date: 05/22/87

Aircraft: Cessna 206; T-38

Aircraft Damage: Both destroyed

Fatalities: 4

A U.S. Air Force T-38 aircraft and a Cessna 206 collided in midair during visual flight operations in a military operations area (MOA). Radar data indicated that the overtaking T-38 collided with the Cessna from the right. The Cessna was on a local government contract terrain photographic mission and had almost completed a left 360° turn. The T-38 was transiting the MOA in connection with a military mission.

Both aircraft were operating under VFR and were not in contact with the MOA controllers. Neither pilot was required to establish or maintain positive radio or radar contact with the air traffic control facility responsible for the area. However, the *Aeronautical Information Manual (AIM)* states, "Pilots operating under VFR should exercise extreme caution while flying within an MOA...prior to entering an active MOA, pilots should contact the controlling agency for traffic advisories." Radar and radio assistance was available to either pilot. The T-38 was using a transponder with Mode C readout. The collision occurred at about 8700 feet MSL.

Probable cause(s)

Visual lookout inadequate—both pilots; procedures or directives not followed; communications not used; radar assistance to VFR aircraft not used.

Comments

Said another way, both pilots must have had their heads in the cockpit and not scanning the skies around them. Neither was in contact with a traffic controlling facility in a frequently busy military airspace. This was apparently a simple but tragic example of inattention in flight, lack of radio communications, and failure to establish contact with the responsible ATC facility.

Case 2

Date: 08/31/86
Location: Cerritos, California, Los Angeles International Airport
Aircraft: Piper PA-28; McDonnell-Douglas DC-9
Aircraft Damage: Both destroyed
Fatalities: 82 Injuries: 8

At 1140 PDT, a Piper PA-28 departed Torrance, California, on a VFR trip to Big Bear, California. After takeoff, the pilot turned east toward the Paradise VORTAC with his transponder squawking 1200. At that time, Aeromexico Flight 498 was on arrival, receiving northbound vectors from LAX Approach Control for an ILS approach to the LAX International Airport. At 1151:04, the controller asked Flight 498 to reduce speed to 190 knots and to descend from 7000 to 6000 feet. During this time, the controller was handling other traffic and providing radar advisories but didn't see a display for the PA-28 on his scope. At 1152:09, the Piper and Flight 498 converged and collided at approximately 6560 feet, then fell to the ground. An investigation revealed the PA-28 had inadvertently entered the LAX Terminal Control Area (a Class B airspace, but called a *TCA* in those days) and wasn't in radio contact with ATC. And LAX TRACON (Terminal Radar Approach Control) wasn't equipped with an automatic conflict alert system, and the analog beacon responder from the Piper's transponder wasn't displayed due to equipment configuration. The Piper's position was displayed by an alphanumeric triangle, but the primary target wasn't displayed due to atmospheric inversion.

Probable cause(s)

The principal causes included inadequate visual lookout, both pilots; no identification of aircraft on radar; procedures or directives not followed by the PA-28 pilot; inadequate FAA procedures.

Comments

Several factors contributed to this accident, but the one that started it all was the PA-28's violation of the LAX Class B airspace (TCA), later compounded by lack of radio contact with LAX Approach Control, followed by transponder equipment difficulties, and finally the fact that neither pilot saw the other aircraft—or at least not in time to take evasive action. This was a tragic and unnecessary accident that gave birth to stricter equipment and operating requirements relative to the then-TCAs and today's Class B airspaces. Think about it, though: Could it have all been avoided if those in both cockpits had been more alert to what was going on in the immediate world outside them?

Case 3

Location: Grand Canyon, Arizona
Date: 10/05/96
Aircraft: Cessna 172RG
Aircraft Damage: Destroyed
Fatalities: 4

The recently certified private pilot (he had logged 141 hours) ordered the aircraft to be fueled to capacity, which resulted in its being about 90 pounds over its maximum certificated gross takeoff weight. The Grand Canyon airport elevation is 6606 feet above sea level, but at his departure time, the density altitude was 8100 feet. The aircraft took off from the mountain airport and climbed toward gradually rising terrain. In response to a concerned call from the tower, the pilot stated that he was unable to climb and might have to land on a road. The pilot of another aircraft observed the Cessna flying approximately 5 miles at treetop level with the wings rocking and that it was in a nose-high attitude before impacting treetops and crashing. The aircraft had climbed approximately 400 feet from the runway to the accident site. No evidence of a mechanical discrepancy was found.

Probable cause(s)

Inadequate preflight preparation, resulting in inability to outclimb rising and wooded terrain; failure of the pilot to ensure that the aircraft's gross takeoff weight was not exceeded; high-density altitude takeoff conditions; the pilot's lack of experience in high-density altitude operations.

Comments

Watch out for hot days at any airport, but especially when taking off or landing at airports in mountainous areas where the air is naturally thinner than at sea level or lower altitudes. Then add heat to the mixture, and you increase even more the thinness of the air, resulting in decreased engine and propeller efficiency and lift. Under such conditions, exceeding the maximum gross takeoff weight produces a most dangerous operating formula.

Case 4

Location: Oregon
Date: 05/24/97
Aircraft: Piper PA-32-301T
Aircraft Damage: Destroyed
Fatalities: 2

The last radio communications from the pilot indicated that they were in instrument meteorological conditions (IMC), experiencing updrafts and picking up ice on the structure. Widely scattered thunderstorms and rain were reported in the area. Light to occasional moderate rime ice and mixed icing in the precipitation were reported from the freezing level to 12,000 feet. Moderate turbulence was reported for the entire state. Witnesses reported that a "pretty nasty-looking cloud" with the sounds of thunder was in the area when they heard a sound similar to a sonic boom. After this sound, the aircraft's engine was no longer heard. Evidence at the accident site indicated that the aircraft broke up in flight. The left wing separated at the root; the right wing separated about 109 inches from the root. Both the horizontal stabilizer and the vertical stabilizer separated from the empennage. Metallurgical examination of the fractured components was typical of overstressed separations. Both the wings and the horizontal stabilizer failed in a downward loading condition.

Probable cause(s)

Overload of the airframe structure. Thunderstorm and turbulence were factors.

Comments

On the surface, this is a difficult accident to explain, in part because both the pilot and her passenger held private pilot certificates for single-engine-land aircraft and both were instrument-rated. The pilot-in-command had accumulated about 660 hours in "all make and model" of aircraft and approximately 220 hours in the accident aircraft, while her pilot-rated passenger had 892 hours and 220 hours in the same accident aircraft. The results of all postmortem medical and pathological examinations proved negative. Although the weather was not good throughout the area, the reported meteorological information cited in the NTSB summary did not indicate the likelihood of the severe weather that the aircraft encountered in the area. The pilot had filed an instrument flight plan, however, for her cross-country flight to Redmond, Oregon, and first initiated contact with Seattle Center at 1106. Between 1133 and 1218, nine dialogues took place in efforts to assist the pilot in response to her requests for various altitude and heading changes. After 1219, there was no further word from the pilot. The plane crashed at approximately 1220 Pacific daylight time. From the written materials available, it would appear that the Center personnel did everything possible to accommodate the pilot and help guide her to a safe landing.

With the given weather conditions, this would appear to be one of those situations in which qualified pilots found themselves and from which escape was impossible. Fighting the combination of icing and turbulence, the pilot may have become disoriented and lost control of the aircraft; or the turbulence could have reached the level of severity that caused the aircraft to break up in the air. Wreckage and impact information leaves no question about a breakup, as one wing was found about 800 feet from the fuselage, other parts 300 to 500 feet from the crash site, and wreckage scattered over an area of 1400 feet by 800 feet. Furthermore, metallurgist reports stated that the various fractures were typical of overstress separations.

With their combined experience, it isn't logical that the pilots would have intentionally flown into weather as extreme as what they encountered. Something to remember, though: Embedded thunderstorms are not always easy to spot, and venturing into one of these storms could be an experience full of unpleasant surprises. Also, hail, turbulence, and strong winds are not necessarily confined just to the immediate area of a thunderstorm itself, but can be experienced 20 or 30 miles from the storm. The moral, then, is to keep lots of distance between you and any of those black and threatening creatures of nature. Go around, go back, but don't go through. You'll never regret the few extra miles or minutes that storm-avoidance deviations might cost you!

Case 5

Date: 08/31/86
Location: New Jersey
Aircraft: Cessna 152
Aircraft Damage: Destroyed
Fatalities: 1

The student pilot, on a supervised solo cross-country trip, diverted from his planned route and was observed flying low over a residential area, performing maneuvers at or below treetop levels. The pilot apparently lost control of the aircraft. The aircraft impacted the ground in a nose-down attitude. The pilot received fatal injuries.

Probable causes

Intentional buzzing, violations of Federal Aviation Regulations (FARs), disregard of written and/or verbal instructions, loss of control of aircraft.

Comments

All causes listed in the NTSB findings relate directly to things the pilot-in-command did or didn't do. This is unfortunately a not-uncommon example of what too many beginning pilots tend to do after they have soloed or have a few hours in their logbooks. They can't resist the tendency to buzz the old neighborhood or the girlfriend's house to show off their newly acquired piloting "skills." Dumb as that is, doing so also violates FARs about flights over populated areas and minimum altitudes for so-called aerobatics. Even if this type of pilot survives the buzzing experience, he or she may well face the disciplinary action of the FAA. Smart pilots don't do dumb things—and buzzing can be very dumb.

Case 6

Date: 02/03/97

Location: Oklahoma

Aircraft: Piper PA-28-200

Aircraft Damage: Destroyed

Fatalities: 3

The pilot met his two passengers at an airport where he purchased 10 gallons of fuel. The airport attendant estimated that there was about 1 inch of fuel in each wing tank before he added 5 gallons to each tank. The pilot commented to the attendant, "Let me see if I can't scare the guys to death." Witnesses observed the airplane flying over a lodge area. They said the plane was "tilting at 90° angle on its wings." Also they reported the airplane went "inverted" and then nosed down and disappeared from view. It then crashed in a lake and was destroyed by impact. One witness stated that "the motor was making a noise, as if it was getting gas and then not getting gas." Another witness observed the airplane earlier that day doing "tricks and stunts." The airplane was restricted from inverted flight and was not approved for aerobatic maneuvers. During examination of the wreckage, flight control continuity was established, and no anomalies were found that would have resulted in a loss of power. Toxicology tests of the pilot's blood showed 0.008 microgram per milliliter (mg/mL) tetrahydrocannabinol (marijuana), 0.014 mg/mL tetrahydrocannabinol carboxylic acid (metabolite of marijuana), and 0.052 mg/mL alprazolam (similar to Valium, a minor tranquilizer). Tests of his urine showed 0.695 mg/mL tetrahydrocannabinol carboxylic acid (metabolite of marijuana), 0.071 mg/mL dihydrocodeine (derivative of codeine), 0.155 mg/mL hydrocodone (derivative of codeine), 0.338 mg/mL hydromorphone (derivative of morphine), 4.8 mg/mL acetaminophen (Tylenol), and an undetermined amount of alprazolam and alpha-hydroxalprazolam (tranquilizers). According to the FAA's southwest regional flight surgeon, "the combined effects of these drugs could have caused impairment in the cockpit."

Probable cause(s)

The pilot's impairment of judgment and performance due to drugs, his resultant improper planning and decision, and his failure to maintain sufficient altitude (clearance above a lake) while performing acrobatic flight. Factors relating to the accident were the pilot's use of an aircraft that was not certified for aerobatic flight and possible distraction (diverted attention) when the engine momentarily lost power due to fuel starvation during inverted flight.

Comments

The NTSB summary, quoted almost verbatim above, provides little personal data about the pilot. A separate narrative report on the accident, however, states that he had a private pilot certificate, obtained in 1995, and a total of approximately 350 hours. Earlier in 1995, still as a student at the Spartan School of Aeronautics, he videotaped aerobatic maneuvers he performed during a solo training flight and gave the tape to his flight instructor, telling the instructor he thought the instructor would be interested in it. The school convened a board of inquiry to review the incident, and on July 17, 1995, the pilot was put on academic probation. In January 1996, he was observed performing acrobatic maneuvers in a Cessna 152. The FAA suspended his certificate for 90 days for performing acrobatic maneuvers contrary to the operating limitations specified in the aircraft's flight manual and for performing acrobatic flight within 4 nautical miles of a federal airway's centerline.

His flying history suggests a pattern of irresponsibility and lack of judgment. This, coupled with the fact that he consumed the drugs found in his blood and urine, was bad enough, but then to try to fly when under the influence of marijuana, codeine, and morphine made almost certain the tragic ending of this brief flight. Distorted awareness and bravado ("Let me see if I can't scare the guys to death") won out over any semblance of good sense, and if the "guys" weren't already scared to death, the eventual plunge into the lake achieved the pilot's ultimate "objective."

The distorting effect of drugs and alcohol, individually or in combination, is drastic enough when one is driving a car. In the relatively rarefied atmosphere of flight, the distortion is even more dramatic and debilitating. If you tend to use alcohol or drugs, that's your business, but don't mix either with flying. You're playing with your life, the lives of any others with you, and the lives of innocent people on the ground who fall victim to an out-of-control aircraft. Also, the FAA comes down very hard on violators of its substance abuse regulations. To show how far it goes, if you're convicted of driving a car under the influence, you must report the violation to the FAA Civil Aviation Security Division in Oklahoma City not later than 60 days after the motor vehicle action [FAR 61.15(e)]. What the FAA then decides may be harmful to your flying aspirations. Intelligence and good judgment say, Lay off the stuff. Be smart and don't disregard what that small inner voice of reason tells you—or should be telling you.

Case 7

Date: 08/31/86
Location: Kansas
Aircraft: Piper PA-28-140
Aircraft Damage: Destroyed
Fatality: 1

The student pilot intended to fly from his home airport to a nearby airport where he was receiving flight instruction. No record of a weather briefing was found. Local authorities said there were low clouds and fog in the area at the time of the accident. No one saw the aircraft take off, but the pilot was seen in a local restaurant a short time before the accident. The aircraft crashed on the municipal golf course about 100 feet from two residences. An occupant of one of the residences saw the aircraft crash and said it was descending nearly vertical just before impact. He also said the engine noise increased just before impact.

Probable cause(s)

Preflight planning and preparation inadequate; lack of instrument time; VFR flight into IMC weather; airplane control not maintained; spatial disorientation. All causes are directly attributable to the pilot in command.

Comments

There's not much more to say about this case. Disregarding VFR weather minimum regulations, the noninstrument pilot lost control of his airplane once low ceilings and fog obscured any reference to the ground. Until you have that instrument ticket, don't lose sight of the ground and don't get caught on top of an overcast (you're going to have to come down through it, which is both illegal and dangerous).

Case 8

Date: 09/07/95
Location: Midland, Texas
Aircraft: Grumman American AA-5A
Aircraft Damage: Destroyed
Fatalities: 2

The recently certified pilot received his private pilot certificate on August 8, 1995, and had only 8 hours of flying time in the accident aircraft. Nonetheless, he and the passenger/owner of the airplane were en route to Alaska on a 3-week hunting and fishing trip. After topping off at their first fueling stop at the Midland, Texas, Airpark, the pilot taxied to runway 29[(290°). This synopsis heading has to be in error. The comparable runways at Midland Airpark are 25 (250°) and 7 (70°).] and held while another plane landed into the wind on runway 7. The Grumman plane took off on runway 29 (?) with an 18-knot tailwind, gusting to 24 knots. After a "long" takeoff run, as the airplane came within 500 feet of the departure end of the 4380-foot runway, a "very steep" angle of climb was established and the airplane cleared the ground. After clearing commercial buildings off the airport, the pilot lost control of the airplane and impacted a single-story residence. The winds were reported from 070 degrees at 18 knots, gusting to 24 knots, the temperature 92°F, and the density altitude was calculated as 5300 feet. (The Midland Airpark field elevation is 2802 feet MSL.) At the time of takeoff, the airplane was calculated to be 118 pounds over its maximum gross weight.

Probable cause(s)

The pilot's selection of a downwind runway for takeoff. Contributing factors were the tailwind, the density altitude, and the overloading of the airplane.

Comments

Yes, the principal cause was the downwind takeoff, but what prompted that takeoff decision? Didn't the pilot know the wind direction and its velocity? If not, why not? Didn't he understand the effect a downwind takeoff would have on lift and takeoff run? Was he aware that the density altitude was equivalent to that of an airport at 5300 feet? Or did he know these things but disregard them in an aura of personal invincibility? Also, did he consider the risks to which he was exposing other aircraft that might have been on or turning to the final approach for landing, as he flew directly toward them in his futile effort to get airborne? Finally, did he know the aircraft was 118 pounds over its maximum gross weight? If he did, he was taking a chance, even if he had taken off into the wind. One other question: Did he even know how to compute the weight and balance of the aircraft? These are unanswered but inevitable questions when one sees or reads about such unnecessary accidents and lost lives.

Case 9

Date: 02/01/97
Location: Alaska
Aircraft: Piper PA-22
Aircraft Damage: Destroyed
Fatalities: 3

The non-instrument-rated private pilot received a weather briefing from the Kenai Flight Service Station at 1113 on February 1, 1997, for a flight from Port Alsworth to Anchorage's Merrill Field. At the same time, he also filed a VFR flight plan for a 1200 departure, with an en-route time of 21U2 hours to his destination and 3 hours of fuel on board. The weather briefing indicated low fog and stratus clouds over Cook Inlet and the Kenai Peninsula, with pockets of IFR conditions. The forecast was for improving weather, with VFR conditions along the route of flight after 1200. The pilot and his two family member passengers took off at 1200, as scheduled.

At about 1350, the pilot was on top of an overcast at 10,000 feet MSL and contacted the Anchorage Approach Control facility for assistance in determining his position. He told the controller that he thought he was a few miles from Anchorage, but with the help of Approach controllers and Flight Service Station specialists, at **1415** his position was determined to be about 124 miles northwest of Anchorage. Meanwhile, the pilot had said that his only electronic navigation instrument aboard, a loran, was not reliable and that there was a large disparity between the readings of his wet (magnetic) compass and his directional gyro. At 1413 the controller asked if he could see the ground at all, to which the pilot responded, "Ah no, that's a negative." At 1414, he added, "I can't see anything now. I'm not exactly sure where we are, but we need to do something quick." A minute later he said he was at 10,100 feet at the cloud tops and going through some "right now." When a Flight Service Station specialist offered a heading to the west side of the Alaska range near McGrath, where there were VFR conditions, the pilot said he didn't have enough fuel to fly over the range.

Radio contact was lost at 1424. At 1521, an emergency locator transmitter (ELT) was received from the accident airplane. Rescue workers, however, couldn't reach the site until February 5. The airplane was discovered crashed in a near-vertical position on a glacier. Postaccident inspection disclosed no mechanical anomalies with the airplane and a functional loran. About 5 to 6 gallons of fuel was remaining in the wing fuel tank.

Probable cause(s)

The pilot's continued VFR flight into instrument meteorological conditions and subsequent failure to maintain control of the airplane. Factors associated with the accident are the pilot's inadequate weather evaluation, his becoming lost or disoriented, and spatial disorientation.

Comments

What else is there to say? Even with a preflight weather briefing, the pilot got caught above the overcast, was lost and low on fuel, and then became totally disoriented as he flew into IFR conditions—conditions with which he was unable to cope. When he saw the overcast ahead of him, instead of trying to climb above it to maintain VFR, why didn't he turn 180° and get out of there, especially when he had to keep going ever higher to stay out of the overcast? We'll never know the answers to many such questions, but this is just another case with at least a couple of meaningful messages: For one, until you're IFR-qualified, *never* allow yourself to get caught above an overcast. You might have to come down through it, and that could spell much trouble. Another message: Centers, Approach Controls, and Flight Service Stations are there to help you, but their potential services are greatly limited if your fuel gauge needles are bouncing on empty when you first contact them. *Don't* wait until you're really in trouble before you call one of the facilities. By then, it may be too late.

As for the pilot himself, he was 27 years old and held a private pilot certificate with an airplane single-engine-land-rating. His pilot and medical certificates indicated he was prohibited from night flying. It is estimated that his total time was approximately 120 hours. An autopsy simply noted that cause of death was "blunt impact injuries."

Case 10

This case is not from the NTSB records but appeared in an issue of *Flyer,* a biweekly newspaper published in Tacoma, Washington. It has a happier ending than the others, but as it is still in the various hearing stages, I'm omitting all names, aircraft N numbers, and potentially identifying data. I've also capsulized the *Flyer* article in the interests of space and brevity.

The incident began at 9:30 in the morning when a non-instrument-rated pilot of a Bonanza A-35 requested flight following from a Center while over the Sierra mountains. All seemed routine until the pilot climbed without approval to 20,000 feet to get above the clouds that were building up ahead of him. As the *Flyer* article stated, the controller contacted the Bonanza and learned that the pilot was not instrument-qualified and had no oxygen on board.

Seeing the potential of a problem on the horizon, another controller, who was also an instrument-rated commercial pilot, took over the controlling responsibilities. As he did, other Center personnel tried to find breaks in the overcast through which the Bonanza might descend. With no autopilot on board, those involved in the efforts to help feared that the VFR-only pilot would lose control of the plane if he had to descend through a thick overcast. There were no breaks, though, and after about an hour in the rarefied 20,000-foot altitude, the pilot's speech was becoming slurred because of hypoxia (the lack of oxygen). And, of course, with hypoxia, judgment, problem-solving abilities, and normal motor skills begin to fail. Then, unless followed by counteraction of some nature, unconsciousness and ultimately death are inevitable. With the first hypoxia symptoms showing, it was clear to the controllers that the Bonanza would have to start down through the overcast—now.

After turning him toward the airport, the instrument-rated controller began a continuous communication with the pilot, directing him to reduce power, keep his wings level, and begin a descent to 12,000 feet. Taking him through the overcast, the pilot descended until he broke into the clear at 14,500 feet. The drama, intense for several minutes, was over.

Shortly afterward, the Bonanza landed safely, with the pilot left with nothing more than a headache for his time in the relatively rarefied atmosphere. Following engine shutdown, he phoned the Center to express his appreciation for the help the controllers gave him.

The *Flyer* article closes with this paragraph: "How long those good feelings for the FAA will last is unknown. An FAA spokesman said the pilot is now under investigation and possibly facing sanctions for departing an assigned altitude without ATC clearance and for operating at altitude without supplemental oxygen."

Comments

The case speaks for itself. Apart from the sanctions for the violations cited, other questions might be asked: Did the pilot contact a Flight Service Station for a preflight briefing? Was he aware of the cloud cover, its floor, its ceiling, and the conditions within the overcast, such as icing, turbulence, rain, lightning? Did he file a flight plan? Why did he allow himself to get caught on top of the cloud cover? Why didn't he make a 180° turn and get back to where the weather had been clear?

These several questions should be answered. At the same time, though, the Center controllers deserve hearty kudos for what they did to escort the pilot through a 6000-foot overcast into clear air and an eventual incident-free landing. This is one example of what ATC can do for you when you get into trouble and of why VFR pilots should learn to use and then take advantage of the FAA services at their disposal. It does seem fair to mention, though, that this pilot, with no instrument rating, had to have done a good job to descend safely through almost 6000 feet of overcast. Not many untrained instrument pilots would have been as fortunate, as skilled, or perhaps as lucky.

Case 11

Location: Martinsburg, WV
Aircraft: Cessna 150
Aircraft damage: Substantial
Injuries: One, minor

The pilot left the airport without contacting the tower. He returned 30 minutes later and made several missed approaches, still without contacting the tower. Tower personnel tried to communicate with the pilot and used light signals while directing other traffic. The pilot stated (later) that he saw a C-130 on the taxiway as well as several other light civil aircraft in the area. He then stated he became disoriented and lost control. Witnesses reported that the aircraft entered a steep circling maneuver which terminated when the left wing hit the ground, and the aircraft finally came to rest between runway 8/26 and the parallel taxiway. The 75-year old pilot had no previous experience in the Cessna 150, had not flown in the previous 90 days, and had never received a biennial flight review.

Probable cause/contributing factors

Poor airplane handling, no radio communications, poor judgment, lack of experience in aircraft type, and several other causes, all pilot-directed.

Comment

Radio communications might not have prevented this fiasco, but at least the tower would have been in a position to help guide the pilot and prevent disorientation. It also could have saved much disruption of traffic in the airport vicinity and unnecessary directions to other pilots.

Case 12

Location: Glendale, AZ
Aircraft: Starduster II
Aircraft damage: Destroyed
Injuries: Two, serious

The pilot crossed the airport at midfield and made a left turn for a close-in run downwind. Witnesses estimated he made a 180-degree descending turn to final approach. While on the final descent, the pilot observed another aircraft that was already on final slightly ahead and slightly below his aircraft. The pilot performed a hard pull-up to avoid the other aircraft and allowed his aircraft to stall without sufficient altitude to effect recovery before ground impact occurred.

Probable causes/contributing factors

Pull-up excessive, airspeed not maintained; visual lookout inadequate; radio communications not maintained—pilot of other aircraft.

Comment

The NTSB summary doesn't indicate whether the Starduster was radio equipped. If it was, whether hand-held or otherwise, this pilot was as much to blame for the near-midair collision as the pilot of "the other aircraft." If the Starduster had no radio, then the NTSB's last comment about the "pilot of other aircraft" really isn't pertinent. One can only assume, then, that the Starduster was radio-equipped and that its pilot either didn't have the set on or wasn't paying attention to advisories from other aircraft in the area. Whatever the case, radio communications could have prevented the near-miss and the eventual crash from ever happening.

Case 13

Location: Temple, TX
Aircraft: Cessna 150 and Cessna 152
Aircraft damage: C-150, substantial; C-152, destroyed
Injuries: One fatality; two serious

A midair collision occurred between a Cessna 150 in the traffic pattern and a Cessna 152 on a long, low final approach to the same runway. (Note: The runway number was not identified in the NTSB
report.) The C-152's radio was mistuned and the C-150 had no radio
installed. Consequently, neither could have heard warnings broadcast by another aircraft. At the time of the accident, the C-152 came
between the C-150 and the sun during the last minute of the flight.
The C-150 was hidden from the C-152's pilot's view by the left wing.
No obstruction was found to account for either aircraft not seeing or
avoiding the other prior to those positions.

Combined probable causes/contributing factors

Communications/navigation equipment—lack of; radio not tuned to
local frequency; visual lookout inadequate; pattern procedures not
followed; light conditions (sunglare).

Comment

Could two-way radios in both aircraft, tuned to the Temple Common
Traffic Advisory Frequency, and properly used by the pilots have
prevented this accident? We'll never know, but logic says "yes."

Case 14

Location: West Palm Beach, FL
Aircraft: Cessna 403
Aircraft damage: None
Injuries: None

A Learjet 35 was cleared onto Runway 09L and held. The Learjet
crew confirmed with the tower personnel that the transmission was
intended for them and taxied into position and held. The tower
then cleared a Cessna 403 into "position and hold" on Runway 13,
and cleared the Learjet for departure. The Learjet acknowledged the
call and the clearance. Both aircraft then began to take off and
missed each other by about 10 feet after becoming airborne. The
Cessna pilot stated later that he thought the take-off clearance was
for him.

Probable cause and comment

The Cessna pilot did not acknowledge tower instructions and obviously did not listen to instructions to the other aircraft, the Learjet. (In later chapters, I'll emphasize, a lot more than once, the importance of terse but complete acknowledgments of ATC [Air Traffic Control] instructions or information.) This was a pure case of radio communications failures on the part of the Cessna, with only 10 feet saving the combined passenger count of 11 occupants from almost certain injury or death.

Case15

Location: Ruston, LA

Aircraft: Cessna 310; Cessna 152

Aircraft damage: Cessna 310, minor; Cessna 152, substantial

Injuries: Three uninjured

Both airplanes were in the traffic pattern. The C-152 made radio calls in the blind; the C-310 did not. Both planes were on final at the same time, with the C-152 in the lead and the C-310 lower. The C-310 passed the C-152 just before touchdown and was on the ground as the C-152 rounded out right behind it. The C-152 pilot then saw the C-310, added full power, and pulled up. The left horizontal stabilizer of the C-152 struck the C-310's vertical stablizer. The C-310 stopped and the C-152 landed in front of the C-310 on the same runway.

Probable causes/contributing factors

Standard traffic pattern procedures not followed, inadequate visual scanning of the pattern area, and lack of radio communications by the Cessna 310.

Comment

The accident report does not state whether the C-310 had its radio on. If it did, the pilot(s) was obviously not paying attention or he (she) would have heard the C-152's various calls. This again appears to be an incident primarily caused by radio communications failures, compounded by visual inattention and nonadherence to traffic pattern procedures.

Case 16

Location: Zephyrhills, FL

Aircraft: Cessna 152

Aircraft damage: Substantial
Injuries: One uninjured

The pilot stated that when he contacted unicom, he heard Runway 32 was active, but since it did not exist, he interpreted the active as Runway 22. After touchdown with a right quartering tail wind, the aircraft bounced, drifted off the runway, flew over a ditch, stalled, then touched down in sand and nosed over. The unicom operator stated that he heard the pilot call for a runway advisory, but before he could issue an advisory, the pilot responded, "Understand Runway 22 Zephyrhills." The operator then advised that the winds were favoring Runway 36, but there was no response. The pilot also stated that he did not fly over the field to observe the windsock.

Probable causes and comments

The student pilot incorrectly interpreted the favored runway information and then either did not hear or disregarded the advisory about the winds and the favoring of Runway 36. After landing on the wrong runway, the damage to the aircraft was primarily the result of improperly recovering from a bounced landing, the tail wind conditions, and the pilot's lack of experience.

Case 17

Location: Cove Neck, NY
Aircraft: Boeing 707-321B (Avianca Airlines)
Aircraft damage: Destroyed
Injuries: 73 fatal, 81 serious, one minor

On January 25, 1990, at approximately 2134 EST, Avianca Airlines Flight 052 (AVA052) crashed in a wooded residential area in Cove Neck, Long Island, New York. The flight was a scheduled international passenger flight from Bogota, Colombia, to John F. Kennedy International Airport, New York, with an intermediate stop at Jose Maria Cordova Airport, near Medellin, Colombia. Because of poor weather conditions in the northeastern part of the United States, the flight was placed in holding three times by ATC, for a total of one hour and 17 minutes. During the third period of holding, the flight crew reported that the aircraft could not hold longer than five minutes, that it was running out of fuel, and that it could not reach its alternate airport, Logan International, Boston. While trying to return to the airport (JFK), the aircraft experienced a loss of power to all four engines and crashed approximately 16 miles from the airport.

Probable cause

The failure of the flight crew to manage the plane's fuel load adequately and their failure to communicate an emergency fuel situation to ATC before fuel exhaustion occurred. Contributing to the accident was the flight crew's failure to use an airline operational control dispatch system to assist them during the international flight into a high-density airport in poor weather. Also contributing was inadequate traffic flow management by the FAA and the lack of standardized, understandable terminology for pilots and controllers for minimum and emergency fuel states. The NTSB also determined that wind-shear, crew fatigue, and stress were factors that led to the unsuccessful completion of the first approach (at JFK) and thus contributed to the accident.

Comments

Not clearly stated in the above NTSB report were factors such as the possibility of a language barrier problem and the failure, for whatever reason, of the AVA 052 crew to declare clearly and forcefully the existence of a bona fide emergency. It has been reported that had ATC been aware of the dire fuel shortage situation, the aircraft would have been given landing priority. Other factors contributed to the accident, but it all might have been avoided had radio communications, particularly from the aircraft, been clearer, more forceful, and more descriptive of the true emergency the flight was facing. This is a commercial airline example of communicating difficulties, but it's included here to show that not even the professionals are immune from falling prey to confusing, conflicting, or misleading messages.

Case 18

Location: Statesboro, GA
Aircraft: Taylorcraft; Cessna 150
Aircraft damage: Both substantial
Injuries: Three uninjured

The Taylorcraft entered a standard traffic pattern for Runway 14 at about the same time the Cessna 150 was maneuvered for a straight-in landing on Runway 5. Subsequently, the two aircraft converged and collided at the intersection of the two runways. The Taylorcraft pilot said he didn't see the Cessna before impact. The C-150 pilot obtained an airport advisory from unicom and was told of work on the end of Runway 14. He said he was also told that Runway 5 was the active runway and that there was no other reported traffic in the area. Also, the C-150 pilot said that he made another advisory call on short final approach, but did not hear any advisory calls from the Taylorcraft. The C-150 pilot saw the Taylorcraft just before impact and attempted to avoid a collision by initiating a go-around. The Cessna, however, did not gain enough altitude and it struck the Taylorcraft. The *AIM* states that standard traffic pattern entry at uncontrolled airports is from a 45-degree entry to the downwind leg. *AIM* also recommends inbound traffic advisories be made 10 miles out and entering downwind, base, and final approach.

Probable cause

Failure of both pilots to perform a visual lookout sufficiently adequate to avoid a collision at an uncontrolled airport. A related factor was the failure of the Taylorcraft to communicate on the Common Traffic Advisory (CTAF) and to provide inbound advisories.

Comment

Each pilot was right and wrong in this case. The Taylorcraft made the proper entry into the traffic pattern, but failed, for whatever reason, to communicate his or her position and intentions. The Cessna pilot made the proper final approach calls but violated the recommended pattern entry and pattern procedures. Fundamentally, however, if the two aircraft had been in radio communication, or had known where each other was and the intentions of each, the accident probably would never have occurred.

Case 19

Location: Old Bridge, NJ
Aircraft: Cessna 172; Piper PA-28-180
Aircraft damage: Both, substantial
Injuries: Five uninjured

Runway 06 has a displaced landing threshold because of tall trees at the approach end. The Piper was landing as the Cessna was starting its takeoff roll. The two planes collided on the runway. Neither pilot heard radio position reports from the other pilot. This is an uncontrolled airport.

Probable cause

The failure of both pilots to maintain proper visual lookout. Related to the accident was inadequate radio communications by the pilots.

Comment

Even more at fault than the lack of visual lookout, in my opinion, was the apparent absence of any radio communications. Had both aircraft had their radios on, tuned to the proper CTAF, and both pilots communicated their positions and intentions, this is another accident that probably wouldn't have happened.

Case 20

Location: Pittsfield, MA
Aircraft: Piper PA-28-180; Piper PA-38-112
Aircraft damage: Both, substantial
Injuries: One minor, four uninjured

Both aircraft were on final approach to Runway 26 when they collided about one-quarter of a mile from the runway threshold and about 200 feet above the ground. Both pilots stated that they reported their positions on unicom. Several witnesses, however, who were monitoring the radio, stated that they only heard the PA-38 pilot report her position. The radios of the PA-28 were found tuned to the unicom frequency of its departure airport.

Probable cause

The failure of the PA-28 pilot to tune his radios to the proper frequency and his failure to see and avoid the PA-38.

Comment

This seems to be pretty cut and dry: The radio is a great instrument, but it's next to worthless when it's not tuned to the right frequency. Just another example of an accident that probably should never have happened.

Case 21

Location: Conroe, TX, (Montgomery County)
Aircraft: Bellanca 17-30A; Piper PA-32-300
Aircraft damage: Bellanca, destroyed; Piper, substantial
Injuries: One, serious; one, uninjured

The right main landing gear of the Bellanca landing on Runway 19 struck the windshield and engine cowling of the Piper on a takeoff roll on Runway 14. Both planes were equipped with two-way radios and both pilots reported that they had announced their intentions on the manned unicom on 122.95. The VHF radio on the departing

Piper, however, was found tuned to 122.7. The unicom operator did not recall hearing either pilot announce his intentions prior to the accident. The winds at the time of the accident were reported from 120 degrees at 8 knots, and Runway 14 was in use at the time.

Probable cause

The failure of both pilots to maintain visual lookout. Contributing factors were inadequate radio communications by both pilots and the failure of both to follow procedures and directives.

Comment

More than anything, this appears to be primarily a case of radio communications. Yes, the favored runway was 14, and the Bellanca landing on Runway 19 created the potentially dangerous cross-traffic situation. The incorrect tuning of the Piper's radio to 122.7, however, made it impossible for any transmission to be heard locally. By the same token, the Piper, not being tuned to 122.95, couldn't receive any transmission the Bellanca might have originated. Operating thus in a communications vacuum, a cross-traffic meeting of aircraft was invited. Coupled, in this case, with apparent inattention to the possibility of other traffic in the area, the collision was almost fore-ordained.

Case 22

Location: Quincy, IL
Aircraft: Beech 1900C; Beech King Air
Aircraft damage: Both destroyed
Injuries: 14 fatal

A United Express Beechcraft 1900C, Flight 5925, collided with a Beechcraft King Air A90 at Quincy Municipal Airport, near Quincy, IL. The United Express flight was completing its landing roll on Runway 13 and the King Air was departing on Runway 4. Both pilots and 10 passengers on the 1900C and both pilots on the King Air were killed. The 1900C was on an IFR flight plan and operating under FAR Part 135. The A90 had not filed a flight plan and was operating under FAR Part 91.

Comment

At the time of writing, the NTSB had not completed, or at least published, its final analysis and probable cause(s) of the accident. With no knowledge of the accident other than what has already been printed in the various media, it would nonetheless seem most likely that a radio communications failure played a significant, if not exclusive, causal role. With experienced pilots in the cockpits of both aircraft, it is hard to believe that the King Air would have attempted a cross-traffic takeoff had he been aware of the landing United Express. And it's equally hard to believe that the United pilot would have continued an apparent normal postlanding rollout had she known that the King Air was taking off on a heading that would cross United's path at almost a 90-degree angle. For whatever reason, a breakdown in radio communications, either transmitting or receiving, by one pilot or the other, would appear to have been a major contributing factor to this accident.

Perhaps by now you've had enough of accident summarizing. Unfortunately, though, these are only a handful of literally thousands of incidents, ranging from near-misses up to total aircraft destruction, that have been influenced by pilot misuse or nonuse of radio communications. But just the few examples, picked at random from the NTSB files, would hopefully convince even the most dubious of the role the radio plays in making general aviation a safer source of pleasure, travel, and, yes, for some, a means of livelihood.

If you read the history of most aircraft accidents, one fact will probably hit you: It's a rare accident that is caused simply and solely by a mechanical failure that the pilot could not anticipate or take action to prevent. It's rare that a well-maintained, properly serviced engine just up and quits, and it's even rarer that a plane crashes because some part of the airframe failed or broke off. Yes, engines do quit and airframe failures have happened; but more than anything, accidents, fatal or otherwise, serious or minor, are mostly the result of pilots doing something they shouldn't or not doing something they should. It's almost that simple. The best air traffic control system and the most complete list of regulations will never make aviation accident free. They will go a long way in that direction, but nothing will replace the pilot's exercise of good judgment and self-control as she or he operates within the overall system.

Appendix B

Federal Aviation Regulations (changes to)

The FAA is responsible for the administration of the aviation rules and regulations that are divided into chapters the FAA likes to call Parts. The title of each chapter (Part) gives us a hint of what lies within. For instance, FAR Part 1 is called Definitions and Abbreviations and contains—that's right, definitions and abbreviations. FAR Part 61 is called Certification of Pilots and Instructors and contains the minimum requirements for the licensing of pilots, flight, and ground instructors. FAR Part 91 is called General Operating and Flight Rules and contains—you guessed it—general operating and flight rules that we pilots must abide by when we operate an aircraft.

Every so often, the FAA will make small revisions or update the FARs to conform with new information, safer policies, or merely to adhere to a new political climate. Usually this is done by rescinding a sentence or two and the insertion of new information.

However, FAR Part 61, Certification of Pilots and Instructors, the FAR that governs how pilots are licensed has just recently been given a complete overhaul by the FAA. Because of the importance of both FAR Part 61 and how it affects a soon-to-be private pilot or a pilot planning to upgrade to a Commercial Certificate or Instrument Rating, following is a review of the major changes of the new FAR Part 61.

FAR Part 61 Revisions

The long-awaited revision to regulations affecting pilots, flight, and ground instructors has been released. On April 4, 1997, the FAA announced the final rule, which became effective August 4, 1997. It significantly modified Part 61 of the Federal Aviation Regulations (FARs). The original proposal, which drew thousands of suggestions from pilots around the United States, included numerous changes, including medical self-certifications for some pilots, separate instrument ratings for single and multiengine airplanes, a new balloon flight instructor certificate, and several new aircraft categories and classes.

As a result of the comments received by the FAA, many of the original proposed changes were dropped and others have been implemented in a highly modified form. In the paragraphs below I have outlined some of the more significant changes having widespread effect. Be aware that these are not verbatim FARs; I have paraphrased the actual regulations. In order to remain abreast of the complete changes to the FARs, and to view the format of their composition, it would be wise for you to refer to the actual FARs.

Applicability and definitions (61.1)

This paragraph, which covers the hows and whys of FAR Part 61, has been greatly expanded. Among the new and revised definitions are aeronautical experience, authorized instructor, flight simulator, flight training device, pilot time, knowledge test, practical test, and training time. For instance, we old certified flight instructors are now called authorized flight instructors. I don't know if it pays any more or not, but that's what the FAA wants us to be called. When studying the provisions required for certification contained within Part 61, it will be important to refer to this paragraph.

Definition of cross-country

The definition of cross-country is now contained entirely within paragraph 61.1 (applicability and definitions). For the purposes of meeting the requirements for private, instrument, or commercial certification, any point of landing must be a straight-line distance greater than 50 nautical miles from the original point of departure. Another change is commercial and airline transport pilots may log cross-country time on any flight that includes a straight-line distance of greater than 50 nautical miles from the original point of departure, regardless of whether a landing is made.

Logging of flight time (61.51)

Most of the changes to this section involve clarification of what have been long-standing points of confusion and disagreement. Here are the key items. Recreational, private, commercial, and airline transport pilots may log pilot-in-command (PIC) time whenever they are sole manipulators of the controls of an aircraft for which they are rated. Flight instructors may log PIC time whenever acting as an instructor.

A major change, student pilots may log PIC time when they are the sole occupant of the aircraft, have a current solo endorsement, and are either undergoing training for a certificate or rating or building PIC time for a certificate or rating. Second-in-command time may be logged when more than one pilot is required either under the type certification of the aircraft or by the regulation under which the flight is being conducted. This includes when acting as safety pilot for another pilot under simulated instrument conditions.

Instrument currency (61.57)

The new regulations eliminate the requirement for six hours of instrument time every six months to remain current. Instrument pilots must now accomplish the following within the preceding six months; six instrument approaches, holding procedures, and intercepting and tracking courses through the use of navigation systems. This may be accomplished in flight in either actual or simulated (with a safety pilot) instrument conditions or in an approved simulator or flight training device. The only alternative to these currency requirements would be for the instrument pilot to choose to accomplish an instrument competency check with an authorized instrument instructor in lieu of the approaches and holding.

Complex versus high performance (61.31)

The distinction between complex and high-performance airplanes has been formalized. Separate training and separate instructor endorsements are required prior to acting as pilot-in-command. Complex land airplanes are defined as those having retractable landing gear, flaps, and a controllable pitch propeller. High-performance airplanes are those with an engine of more than 200 horsepower.

Recreational pilot privileges

In a change heavily *lobbied* for by aviation advocacy groups, the requirement to remain within the 50-mile distance limitation for recreational pilots has been lifted. Recreational pilots may now venture far and wide, providing they receive the same cross-country training specified for private pilots. Other existing limitations remain, including no flying after sundown, or where two-way communication with ATC is required (tower controlled airports, etc.). Nor may the recreational pilot carry more than one passenger at a time in an aircraft with no more than four seats or engines greater than 180 horsepower.

Flight instructor renewal (61.197)

While the various methods for CFI renewal remains the same, CFIs will now be able to renew up to 90 days prior to their certificate expiration date with the new expiration date 24 months after the original date. This translation is a welcome relief to the CFIs who may have had trouble renewing on, or by, their expiration dates.

Retesting

For those of us who sometimes fall a little bit short when test time comes around, the FAA has deemed that retesting after the second failure of a written or practical (flight) test no longer requires a 30-day waiting period. This 30-day wait after failing a test for the second time was a good plan that didn't work well. The thinking was that by forcing the 30-day wait, the applicant would seek more dual instruction and therefore be better prepared for the failed exam. Sadly, about the only thing most of these folks did was wait the 30 days and then retake the test. It has been my observation, and I know there are exceptions, that any pilot who fails the same test twice usually won't do much better whether on the fourth or fortieth try.

Hourly requirements for certification
Private pilot (61.109)

Applicants for the Private Pilot Certificate still must have 40 hours of total flight time, of which 20 hours must be dual instruction and 10 hours must be solo flight training specific to the areas of operation listed in FAR 61.107. The remaining 10 hours may be either dual or solo. Up to 2.5 hours of the training may be completed in an FAA-approved simulator or flight training device.

There must be a minimum of 3 hours of daytime dual cross-country instruction. Additionally, there must be 3 hours of night dual instruction that must include one cross-country of over 100 nautical miles total distance and at least 10 takeoffs and landings to a full stop. The option of not conducting night training and receiving a "night flying prohibited" limitation no longer exists. The only exception is for pilots training and residing in Alaska.

The Private Pilot applicant needs a minimum of 3 hours of dual instrument flight training. This represents a significant change; previously there was no specified minimum instrument training. The total instrument time was left to the discretion of the flight instructor who had to declare the student was proficient at the instrument flight maneuvers required for the private license.

The minimum of 3 hours of dual flight instruction in preparation for the practical test, within 60 days prior to that test, remains the same. This regulation is designed so the applicant for the license will have had some recent formal dual instruction prior to the checkride.

The solo cross-country flight time has been reduced to 5 hours total flight time. One solo cross-country must be at least 150 nautical miles total distance, must include one nonstop leg of at least 50 nautical miles, and must include full stop landings at a minimum of three points.

The Private Pilot applicant must also have made at least 3 full-stop landings at an airport with an operating control tower. These full-stop landings may be dual or solo and are designed so that all private pilots will have had at least a minimal exposure to a tower-controlled airport prior to their receiving their license.

Instrument rating (61.65)

The big change in the instrument rating is that the previous requirement for 125 hours of total flight time has been eliminated. This change will allow newly certified private pilots to begin training immediately for an instrument rating. Applicants are still required to have 50 hours of pilot-in-command cross-country (again, see definition above). However, PIC cross-country time obtained as a student pilot will now count along with all PIC cross-country flight time logged after receiving the private license.

Forty hours of actual or simulated instrument time must be logged to include a minimum of 15 hours of dual instruction, with at least 3 hours of that 15 hours coming within 60 days prior to taking the practical test. Up to 20 hours of instrument instruction in an approved simulator or flight training device may be counted.

Instrument cross-country dual instruction must include at least one flight of at least 250 nautical miles along the airways or direct as cleared by ATC. Additionally, there must be one instrument approach at each airport, and these approaches must use three different kinds of navigation systems.

Commercial pilot (61.129)

Total flight time required for commercial pilot applicants remains at 250 hours (of which 50 hours may be in a simulator or flight training device). This must include 100 hours in powered aircraft (including 50 in airplanes), 100 hours PIC time (including 50 in airplanes), and 50 hours of PIC cross-country (see definition above). However, some very significant additional requirements have been adopted.

There must be twenty (20) hours of dual instruction on the areas of operation specific to commercial certification (listed in 61.127), which must include the following:

- 10 hours of instrument instruction
- 10 hours in a complex or turbine-powered aircraft
- One day-VFR cross-country flight of at least two (2) hours and more than 100 nautical miles from the original departure point
- One night-VFR cross-country flight of at least two (2) hours and more than 100 nautical miles from the original departure point

There must also be three (3) hours in preparation for the practical test within 60 days preceding and ten (10) hours of solo flight (this means sole occupant of the aircraft), which must include at least one cross-country flight of at least 300 nautical miles, with landings at a minimum of 3 points, one of which is at least 250 nautical miles from the original departure point. There must also be five (5) hours of night-VFR, which must include at least 10 takeoffs and landings at an airport with an operating control tower.

This sums up the major changes to FAR Part 61. If you are on the path to certification, it would be wise for you to pick up a copy of the current version of the FAR/AIM at your local airport or bookstore. As its title implies, the FAR/AIM contains the major FARs and the total *Aeronautical Information Manual*. This compact $10 or $12 book has more information in it than one could ever absorb, comes out yearly, and is literally the Bible of aviators.

Appendix C

NOTAM and chart contractions

A

AADC approach and departure control
ABV above
A/C approach control
ACFT aircraft
ACLTG accelerating
ACR air carrier
ACTV active
ACTVT activate
ADCUS advise customs
ADF automatic direction finder
ADJ adjacent
ADV advise
ADVY advisory
ADZ advise
A/FD (AFD) *Airport/Facility Directory*
AFDK after dark
AIM *Aeronautical Information Manual*
ALQDS all quadrants
ALS approach light system
ALSF approach light system with sequence flashing lights
ALSTG altimeter setting
ALSEC all sectors
ALT altitude
ALTM altimeter
ALTN alternate
AMDT amendment
AMPLTD amplitude
AOE airport of entry
APCH approach

APL airport lights
APP CON approach control
ARFF airport rescue and fire fighting
ARPT airport
ARR arrive (arrival)
ARSR air route surveillance radar
ASDE airport surface detection equipment
ASOS automated surface observing system
ASPH asphalt
ASR airport surveillance radar
ATCSCC Air Traffic Control System Command Center
ATCT airport traffic control tower
ATIS automatic terminal information service
AUTH authority
AVBL available
AWOS automated weather observing system
AWY airway
AZM azimuth

B

BC back course
BCN beacon
BERM snowbank(s) containing earth/gravel
BFDK before dark
BLO below
BND bound
BNDRY boundary
BRAF braking action fair
BRAG braking action good
BRAN braking action nil
BRAP braking action poor
BRG bearing
BRK break
BTWN between
BYD beyond

C

C circling
CAAS class A airspace
CAT category

CBAS class B airspace
CBSA class B surface area
CCAS class C airspace
CCLKWS counterclockwise
CCSA class C surface area
CD clearance delivery
CDAS class D airspace
CDSA class D surface airspace
CDAS class E airspace
CESA class E surface airspace
CFA controlled firing area
CGAS class G airspace
CHG change
CIG ceiling
CLKWS clockwise
CLNC clearance
CLNC DEL clearance delivery
CLSD closed
CMB climb
CMSND commissioned
CNCL cancel
CNDN Canadian
CNTRLN centerline
CONC concrete
COND condition
CONFDC confidence
CONT continuous
COP changeover point
CRS course
CTAF common traffic advisory frequency
CTC contact
CTGY category
CTL control
CW clockwise

D

DALGT daylight
DCMSND decommissioned
DCT direct
DEGS degrees
DEP depart/departure

DF direction finder
DH decision height
DLAD delayed
DLT delete
DLY daily
DME distance measurement equipment
DMSTN demonstration
DPCR departure procedure
DR dead reckoning
DRCT direct
DSPLCD displaced
DSCT distance
DURC during climb
DURD during descent

E

EBND eastbound
ELEV elevation
ELNGT elongate
ENERN east-northeastern
ENEWD east-northeastward
ENTR entire
ENRT en route
ESERN east-southeastern
ESEWD east-southeastward
EXCP except

F

FA final approach
FAC facility
FAF final approach fix
FIR flight information region
FL flight level
FM fan marker
FPM feet per minute
FRH fly runway heading

G

GC ground control
GCA ground control approach

GP glide path
GPI ground point of interception
GPS global positioning system
GRAD gradient
GRDL gradual
GRVL gravel
GS glide slope

H

HAA height above airport
HAL height above landing
HAT height above touchdown
HAZ hazard
HDG heading
HEL helicopter
HELI heliport
HF high frequency
HIRL high intensity runway lights
HIWAS hazardous in-flight weather advisory service
HOL holiday
HP holding pattern
HR hour

I

IAF initial approach fix
IAP instrument approach procedure
IBND inbound
ICAO International Civil Aviation Organization
ID identification
IDENT identification
IF intermediate fix
ILS instrument landing system
IM inner marker
IMDT immediate
INDEFLY indefinitely
INFO information
INOP inoperative
INST instrument
INT intersection
INTCP intercept
INTL international

INTMT intermittent
INTS intense
INTST intensity
IR ice on runway

L

LAA local airport advisory
LAHSO listing of land and hold short operations
lat-long latitude/longitude
LC local control
LCL local
LCTD located
LDA localizer type directional aid
LDIN lead in lighting system
LGRNG long range
LGT light
LIRL low intensity runway lights
LLWAS low level wind shear alert system
LLWS low level wind shear
LMM compass locator at the middle marker
LN line
LNDG landing
LOC localizer
LOM compass locator at the outer marker
LR lead radial
LRA landing rights airport
LRN loran
LSR loose snow on runway
LT left turn after takeoff
LVL level
LWR lower

M

M minus
M missing
MAG magnetic
MAINT maintenance
MALS medium intensity approach lighting system
MALSF medium intensity approach lighting system with sequence
   flashers

MALSR medium intensity approach lighting system with runway alignment indicator lights

MAP missed approach point

MCA minimum crossing altitude

MDA minimum descent altitude

MEA minimum en-route altitude

MED medium

MIDN midnight

MIN minute

MIRL medium intensity runway lights

MM middle marker

MNM minimum

MOA military operations area

MOCA minimum obstruction clearance altitude

MONTR monitor

MOV move

MRA minimum reception altitude

MRGL marginal

MRNG morning

MS minus

MSA minimum safe altitude

MSTLY mostly

MSAW minimum safe altitude warning

MSL mean sea level

MTR military training route

MU designate a friction value representing runway surface conditions

MULT multiple

MUNI municipal

MXD mixed

N

NA not authorized

NASA National Aeronautics and Space Administration

NAS National Airspace System

NAV navigation

NAVAID navigational aid

NBND northbound

NDB nondirectional radio beacon

NELY northeasterly

NERN northeastern

NGT night

NM nautical mile
NMBR number
NMR nautical mile radius
NMRS numerous
NNERN north-northeastern
NNEWD north-northeastward
NNWRN north-northwestern
NNWWD north-westward
NOPT no procedure turn
NRW narrow
NTAP *Notice to Airmen Publication*
NWLY northwesterly
NWRN northwestern

O

OCNL occasional
ODALS omnidirectional approach light system
OM outer marker
ONSHR on shore
OPER operate
OPN operation
ORIG original
OTS out of service
OTP on top
OROCA off-route obstruction clearance altitudes
OTRW otherwise
OVHD overhead
OVR over

P

P plus
P time proposed departure time
PAEW personnel and equipment working
PAJA parachute jumping
PAPI precision approach path indicator
PAR precision approach radar
PARL parallel
PAT pattern
PBL probable
PCL pilot controlled lighting

PERM permanent
PDMT predominant
PDMT predominate
PLA practice low approach
PLW plow (snow)
PN prior notice required
PPR prior permission required
PREV previous
PRIRA primary radar
PRJMP pressure jump
PROC procedure
PROP propeller
PRSNT present
PS plus
PSG passing
PSGR passenger
PSR packed snow on runway
PT procedure turn
PTCHY patchy
PTLY partly
PTN portion
PVT private

Q

QUAD quadrant

R

R radial
RA radio altimeter setting height
RAIL runway alignment indicator lights
REIL runway end identifier lights
RCAG remote communication facility
RCC rescue coordination center
RCL runway centerline
RCLS runway centerline light system
RCO remote communication outlet
RCV receive
REF reference
REIL runway end identifier lights
RELCTD relocated

RESTR restrict
RMDR remainder
RMRK remark
RNAV area navigation
RPD rapid
RPI runway point of interception
RPLC replace
RPRT report
RQRD required
RRL runway remaining lights
RSG rising
RSVN reservation
RT right turn after takeoff
RTE route
RTR remote transmitter/receiver
RTS return to service
RUF rough
RVR runway visual range
RVRM runway visual range (midpoint)
RVRN RVR not available
RVRR runway visual range (rollout)
RVRT runway visual range (touchdown)
RVV runway visibility value
RVVNO RVV not available
RWY runway
RY runway

S

S straight-in
S snow
SBND southbound
SDF simplified directional facility
SECRA secondary radar
SELY southeasterly
SERN southeastern
SFL sequence flashing lights
SGD solar-geophysical data
SGFNT significant
SHRTLY shortly
SI straight-in approach

SID standard instrument departure
SIMUL simplified short approach light system with sequenced flashers
SIR snow and ice on runway
SKED scheduled
SLD solid
SLGT slight
SLR slush on runway
SND sanded
SNBNK snowbank
SNGL single
SNRS sunrise
SNST sunset
SOIR simultaneous operations on intersecting runways
SPD speed
SR sunrise
SS sunset
SSALF simplified short approach lighting system with sequenced flashers
SSALR simplified short approach lighting system with runway alignment indicator lights
SSALS simplified short approach lighting system
SSERN south-southeastern
SSEWD south-southeastward
SSNO no STARs, no SIDs
SSWRN south-southwestern
SSWWD south-southwestward
STAR standard terminal arrival
STFR stratus fractus
STG strong
STNRY stationary
SVC service
SVRL several
SWLY southwesterly
SWRN southwestern
SXN section
SYS system

T

TA transition altitude
TAC TACAN

TACAN tactical air navigational aid
TCH threshold crossing height
TDWR terminal Doppler weather radar
TDZ touchdown zone
TDZ/CL touchdown zone and runway centerline lighting
TDZE touchdown zone elevation
TDZL touchdown zone lights
TFC traffic
TFR temporary flight restriction
TGL touch-and-go landing
THK thick
THN thin
THR threshold
TKOF takeoff
TMPA traffic management program alert
TMPRY temporary
TRML terminal
TRNG training
TRRN terrain
TRSN transition
TSNT transient
TWD toward
TWR tower
TWRG towering
TWY taxiway

U

UNAVBL unavailable
UNLGTD unlighted
UNMKD unmarked
UNMON unmonitored
UNOFFL unofficial
UNRELBL unreliable
UNSBL unseasonable
UNSTBL unstable
UNSTDY unsteady
UNUSBL unusable
UPDFTS updrafts
UPR upper
UTC coordinated universal time

V

V variable
VASI visual approach slope indicator
VCNTY vicinity
VDP visual descent point
VIA by way of
VICE instead of
VLCTY velocity
VMC visual meteorological conditions
VOL volume
VR veer
VRBL variable
VSBY visibility

W

WDLY widely
WDSPRD widespread
WI within
WK weak
WKDAYS Monday through Friday
WKEND Saturday and Sunday
WNWRN west-northwestern
WNWWD west-northwestward
WPT waypoint
WR wet runway
WRM warm
WRNG warning
WSR wet snow on runway
WSWRN west-southwestern
WSWWD west-southwestward
WTR water

X

XCP except
XPC expect

Z

Z coordinated universal time (UTC)

Appendix D

Aviation professions

Studies show that the majority of people do not like their chosen profession. Further, most of us dream of being something, or someone, else. Farmers want to be teachers and teachers want to be accountants. Nearly every person I know can't wait for Friday night and age 65. Not so with most professional pilots. Chances are, a professional pilot wouldn't change places with anyone. The phrase most often heard when a pilot describes his job is, "I have the best damn job in the world."

Flying holds a certain fascination for almost everyone. In over three decades as a pilot, I have spent more time at school, parties, and family reunions talking about flying than all other subjects combined. And I am rarely the originator of the conversation. It seems that when people find out I fly for a living, that's all they want to talk about. I suppose I have been told by at least 10,000 people, "I've always wanted to learn to fly since I was little. I guess someday I'll get around to it."

Flying intrigues many people. In fact, it probably weaves its way through the daydreams of most people, but not to the extent that they are willing to drop everything and pursue it. In short, that's not what they really want to do.

And then there are the true fliers. For we pilots, there is absolutely nothing we could be doing for a living that would come close to flight. To us, the worst flying job on the planet would be 10 times better than whatever comes in second place.

If you feel you would like a career in aviation, you quite likely feel it deeply. Few people are drawn to flight by a superficial urge. It seems to be a longing—an almost insatiable desire to fly—which is not diminished by neglect. If I have not flown for a period of time,

I feel, as do most other pilots, a sort of inner anguish. And it can only be satisfied by flight.

If you love flying, you'll know it and no one will have to convince you to go get that flying job. In fact, no one will be able to stop you. To a pilot, flying is the ultimate career, and the flying jobs are there! All you have to have is a real desire, a modicum of ability, and the tenacity to stick it out during the early years. For the competent, dedicated pilot, the sky really is the limit!

Airlines

An airline career seems to hold the most fascination to the most people. Without a doubt, this career has the most visibility, good and bad, and has the highest earning potential of all the current legal aviation careers. Additionally, there seems to be little doubt that the airline industry will continue to grow at a steady pace well into the twenty-first century. This, coupled with the natural retirement rate, should guarantee a need for quite a few airline pilots over the coming decade and beyond. Some people have gone so far as to predict the need for 3000 new airline pilots per year well into the next century.

I believe these figures might be a bit optimistic since our present airport and Air Traffic Control system is near its capacity. Until we build more airports, or expand our present ones, I cannot see a dramatic rise in the need for airline pilots. And I haven't a clue as to how ATC can handle many more aircraft than it now handles and maintain the same margin of safety in the traffic flow.

Another factor that will figure into the potential airline job market is the advent of two-man crews. Every airliner built in the United States today is crewed by two pilots instead of the traditional two pilots and a flight engineer used for so many years in the old 707s, DC-8s, etc. This is the price we pay for our technical advances. The sophistication of EFIS (electronic flight information systems) instrument panels has so centralized pilot information that it has eliminated the need for one human. This also cuts by one-third the need for manpower.

Up until the early 80s, if a pilot was fortunate enough to be hired by a major airline and gathered a few years of seniority, he had it made. Then, seemingly in a few short months, it all changed. It changed with the Federal deregulation of the airline industry. For the first time in their history, the airlines were free to do as they pleased.

The federal guidelines were gone, and what followed was a frenzy beyond the wildest dreams of most aviators. There was a wellspring of new carriers, mergers, buyouts, takeovers, fare-wars, restructuring, and ultimately, bankruptcies. The air carriers became convinced they would have to grow or be consumed. They were right.

A new idea sprang onto the scene—commuter airlines. Nearly every city with an airport had a commuter airline. Never mind that it had no passengers and made no money, Smallsville, U.S.A., had an airline. One by one, these commuters were bought, merged with, or simply taken over.

This growth brought on a need for pilots, which triggered a surge of hiring by the airlines, especially the commuters. The problem was that there was a lack of qualified pilots from which to choose. The commuters' solution to this problem was to hire pilots that not only weren't dry behind the ears—they'd never been wet. This caused a great cry of concern from seasoned pilots and much of the flying public. The concern was that safety was being sacrificed for the sake of routes. And several very serious accidents were directly attributed to lack of flight crew experience.

All of these buyouts and mergers left the airline pilots in a state of flux. A senior captain with one airline today could very easily be wearing a different hat with a low seniority number the very next day. Often, pilots with many years of seniority were furloughed indefinitely when their line was merged with another. It all depended on which line you happened to fly for. Some pilots lost everything when their financial future that had looked so solid turned sour. In short, it would have been easier to nail Jell-O to a tree than to predict the future of an airline pilot in the 80s. It was wild.

About 1990, as quickly as it had begun, the hiring boom was quieted. The major airlines began a more methodical, safety-paced approach. Hiring minimums began to take on the look of yesteryear. More experience would be needed to qualify for that coveted seat aboard an airliner once again. The quick trip to the majors was over, at least for now.

Then, about 1995, the airlines began to notice something they had not prepared for. The economy had been spiraling upward for several years, the average person had more disposable income, and people were flying to their vacation spots instead of driving. The airlines, leaner and smarter than before, were beginning to make money. And looking into the twenty-first century, there appears to be no end in sight.

The point I'm attempting to convey with this brief history lesson is that the airline job is not all the sweetness and security that so many would have you believe. It is a tough, often disappointing, career that is not for the timid. If you desire to be an airline pilot, you had better go in with your eyes open, your head on straight, and your proficiency unquestioned. If not, it will eat you alive.

This is not to say that there aren't any great jobs flying for the airlines. For those who do make it into the left seat, and manage to remain high in seniority, it can be a very rewarding career. How many people do you know that work three days and then get four days off? And the pay's not bad, either.

I asked many airline pilots what the most important prerequisite was for landing a job with a major airline. Their answer: experience, experience, experience. Naturally, the airlines would prefer an ex-military pilot who has already been trained and could come to them with several thousand hours of heavy multiengine jet time. If you want a job with the airlines, this is as close as I can come to guaranteeing you a seat. I know of no airline that turns away pilots with an adequate amount of military multiengine jet time.

Interestingly, when I asked several airlines about the greatest asset they sought in a pilot candidate, the resounding answer was attitude. The close work environment of modern airliners demands a cohesiveness and interpersonal ability to work and play well with others that the airlines find absolutely invaluable in today's pilot applicant. Several airline human resource managers told me that attitude had in fact replaced experience as their number-one hiring priority.

Some airlines are not enamored by the fighter pilots. It seems that some fighter pilots have a single-minded, cover-your-own-butt mentality. And who could blame them? It is reasoned that this could interfere with the teamwork necessary in an airline cockpit. While this is certainly not a given, it is a consideration, and most airlines seem to prefer the four-engine types.

For those of you not in the mood to give those years to Uncle Sam in exchange for receiving invaluable training, the next best path to the airline cockpit is to somehow build flight time. The airlines prefer that you have a couple thousand hours of flight time, preferably turbine. Their dream pilot would have a degree in aeronautics, 3,000 hours of flight time, of which about 2,000 hours was in turbine-powered aircraft, and be about twenty-eight years old. That's their dream, but they will settle for a bit less.

The personnel I talked with told me that a college degree is as important as it ever was. In fact, one airline HR person told me flatly that a full 99.9 percent of their new hires had a four-year degree. So naturally, if you have a degree, you will be considered (read that *hired*) before someone with the same qualifications who does not have a college diploma. In reality, a degree is very helpful both at the time of hiring and later during promotion time.

There are many ways to build the flight time desired by the airlines, and no one system works equally well for everyone. (I was told one point, pretty plainly. The airlines do not care if you have 8,000 hours in your Cessna 172; they want turbine time.)

An acquaintance of mine took the following route to his airline seat. He soloed at the age of 16, obtained his commercial when he was 18, and then went to college to earn his bachelor's degree. During his college summers, he ferried new aircraft to dealers for a large aircraft manufacturer and managed to build up another 1,000 flight hours. More importantly, he got paid to fly. Smart!

Upon graduating from college, he managed to land a corporate job flying a King-Air for a power company. He spent eight years working for this corporation, during which time they conveniently added a Sabreliner and a Cessna Citation. During these years, he flew an additional 3,000 flight hours, all turbine.

He was now ready, at the ripe old age of 30, to apply to the airlines. He found a place with a commuter airline and flew with them for two years before making the jump to the majors. He flew five years as a First Officer (copilot) on a DC-9 and currently flies right seat on an A-320 Airbus. He is 37 and should make captain in the next five years.

That's all well and good, you say, but does it pay? You'd better believe it. As I said, if you can land the job and manage to remain with a solvent company for six or seven years, you will be well compensated. Our friend in the above paragraph made $98,110 per year in

1998. And he was a third-year copilot. A 10-year captain flying a 747 made $202,500 for the same year. In fact, this year the lowest-paying flying job with the airline I am quoting will be $59,880 for a second officer on a 727. You could live on this to start, couldn't you? Also, the major airlines offer great benefit packages including life, accident, and health insurance, profit sharing, stock options, annuities, and so on.

The pros and cons depend on to whom you happen to be talking. The obvious negatives are the takeovers, which cause loss of seniority, and the possible layoffs. In this volatile time in the evolution of the airlines, this seems to be the main complaint of all airline pilots. Following this on the negative side are union disputes, long hours (like 16-hour days), and being away from home. The latter was also mentioned as a plus. I guess it depends on your priorities.

On the plus side, money, money, and money were mentioned in first, second, and third place. They were followed by free travel, short work weeks, and excellent retirement plans as solid reasons for working with the airlines.

Speaking of retirement, airline pilots *must* retire from the flight controls at the age of 60. They can remain as flight engineers but cannot man the two front seats any longer. This is a mandatory retirement age as decreed by the feds, and they seem unwilling to discuss it.

The reason for this arbitrary age of 60 seems to be predicated on the FAA's assumption that a pilot of 60 is more likely to die suddenly than a pilot of 25. This is known in aviation circles as *FAA intelligence*. There is a fairly large push among pilots to set the retirement age back to 65, saying that the experience gained by this age is irreplaceable. However, most airlines seem to favor the 60 rule, probably for financial reasons. The pilots who remain in place to age 60 are usually making some serious money. Anyway, an age for mandatory retirement had to be chosen, and for now, it remains 60.

Air traffic control

If you don't want to fly for a living but desire to remain close to aviation, one of the most attractive nonflying vocations could be a career with ATC. ATC, which includes control towers, air route traffic control centers, and flight service stations, is run by the government—the FAA to be precise. The folks in ATC work with all types of aircraft on a daily basis, assisting their every move from clearance delivery before engine start to closing flight plans after shutdown. Given the ever-increasing number of aircraft sharing our seemingly shrinking airspace, job security should be solid for the foreseeable future.

In order to land a position as a controller, you must first be placed on the Federal Register. The register is a list of all prospective employees the government utilizes to fill vacancies as they occur. You are placed on this list by successfully passing a government test designed specifically for ATC hopefuls, called the *ATC Operators Test.* This written test is scheduled for various parts of the country at periodic intervals, or for specific areas when the need for ATC personnel is apparent. You can obtain a current test schedule by contacting your nearest Federal Office of Personnel Management or by calling your nearest FAA Flight Standards District Office.

Once you pass the written test, the FAA will screen you for physical, psychological, and security reasons. If you clear this, you are summoned to Oklahoma City for four months of initial training at the FAA's training facility. The high scores from the Oklahoma City schooling are given the choice of entering either the tower or the enroute (center) phase of the ATC system. Either way, you will serve two years of apprenticeship before being certified as having reached ATC FPL (full performance level) status.

ATC personnel work a standard eight-hour day, 40-hour week. And since the stress of separating numerous high-speed aircraft can be counterproductive, their contract calls for a short break once every two hours.

ATC salaries are figured by several criteria, including the traffic count at the facility worked in comparison to all the others. For instance, ATC control towers are divided into five levels based on traffic count. An entry-level controller based at a level-one facility (light traffic) made about $24,000 in 1998. An FPL controller working a level-five tower (very busy) earned about $58,000. When you add in government health and life insurance benefits to this, you begin to see possibilities of a career that provides a pretty comfortable living.

Additionally, all ATC personnel earn paid leave at the rate of a half day off per pay period at the entry level. This rate grows proportionally until, after 15 years of service, they earn one day of leave per pay period. This figures out to between 13 and 26 paid personal days per year in addition to regular days off and 10 paid federal holidays per year. Controllers also accumulate a half day of sick leave for every pay period, or 13 days per year. This all adds up to a very tidy seven weeks off with pay for a veteran controller. Not bad.

Now, an age-old question: Should an air traffic controller also be a pilot? Should they not know something about what it is pilots do since, after all, pilots are placing their lives in their hands. Some say no; other say absolutely; and some are ambivalent towards the whole subject. One camp says that an ATC person who can relate to the pilot and identify with the problems faced up there in three-dimensional airspace is better prepared to help the pilot overcome obstacles to safe navigation.

Another camp says that this identity with the pilot and his or her plight is the exact reason a controller should *not* be a pilot. They say the controller cannot, and therefore should not, attempt to fly the plane. They maintain that the controller should do the controlling and the pilot should fly the plane.

I believe the best solution, as usual, lies somewhere in between. Given the choice, I will take a controller who flies, and yet knows both his and my limitations. This controller will know enough about the problems I am facing in the aircraft to be of great help to me, thus allowing me to do my work in the most efficient manner. Further, this type of controller can most likely be counted on to help, and perhaps equally important, will not compound an already difficult situation by issuing an impossible instruction.

The comments, pro and con, I heard most concerning a career with ATC had to do with opposite ends of the same thought. The consensus was that the worst part of a controller's job was either stress or boredom. Imagine that! They seem to be either too busy, or not busy enough. I guess pilots had better become more skilled at spacing arrivals so as to spread them more evenly over the day. Come to think of it, that is not such a bad idea.

The consensus was that the best part of the controller's job description was the chance to interact with and aid the pilots. I saw the warm glow of satisfaction in the eyes of more than one controller as they described how they had helped a pilot, who had flown past his fatigue or experience level, culminate his flight successfully. And the pride was all the more evident if they hinted that they had to bend the FAA's stringent rules in order to discreetly suggest a better way to a tired or frightened pilot. That is a kind of job satisfaction few of us will ever have the chance to enjoy.

ATC is a viable career with much growth potential and job satisfaction. Most controllers I talk to are absolutely in love with their profession. If you would like to be around aviation, and yet not fly for a living, this could be the career you have been looking for.

Flight instruction

"What's Al do for a living?" "Oh, he's 'just' a flight instructor. As soon as he gets enough hours logged, he's gonna move up to corporate flying."

I wonder how many times I have heard this pathetic exchange between two otherwise intelligent people. Some believe that flight instruction is the poor step-child of aviation. They think of flight instruction as the sort of flying job that should make you hang your head when someone asks what you do for a living; the sort of job you take to build flying time so you can go get a real job. Sadly, these naive thoughts seem to have become "tradition" in some areas of aviation. But they are only the traditional thoughts of the ill taught and the unknowing. The rest of us realize the absolute undeniable importance of the flight instructor and know that they are vital to all who desire to learn to fly. The official position of the governing body of aviation, the FAA, is that the CFI is the backbone of aviation.

It's true that over the years many pilots have used flight instruction as a method of building flight time. Far too many CFIs have used the position to simply ride along with their students, building time instead of teaching. Those of us who have chosen to make instructing our career hold the aforementioned in the utmost contempt. And some of us who are in a position to hire these joy-riders as flight instructors, don't.

I'll tell you a secret. You will fly more like the instructor who teaches you early flight (Private Pilot) than like *anyone else* you will ever fly with. It's only natural. You talk like the people who taught you to talk and walk like the people who taught you to walk, don't you? And you drive like the people who taught you to drive, don't you? You bet you do. You learned the basics from watching and copying the first role model you ever had. And so it is with flight. I hope you had, or will have, a great primary flight instructor because you will be a clone of his or her ways, good or bad. And if you are contemplating a career as a CFI, remember this and teach accordingly.

In flight instruction, as with all vocations, not all who teach are suited to the job. A flight instructor's duties include the ability to teach, counsel, coach, coddle, trick, exhort, and support the student in one of the most difficult teaching environments imaginable. The cockpit is a tough classroom, and it takes a special personality to maintain self-control as a student repeatedly tries to kill you. This gets even more problematic as you and the student begin to develop the natural bond that comes with many hours shared. It is difficult to critique accurately and subjectively anyone with whom you have developed a fondness.

For this reason, I try to never get too close to the people I train to fly—for their sake. I don't want to be in the position of endangering their lives because I like them so much that I might not want to risk hurting their feelings. I hope I wouldn't do this, but I have seen it happen to others. When my son learned to fly, I sent him to the best instructor I knew rather than risk the tendency to overlook problem areas. My son says he wouldn't have let me train him anyhow. He says I would have been an unmerciful tyrant who would have settled for nothing less than perfection. He's probably right. Either way, I believe it is best not to train people you are close to.

I'll tell you another secret. If you are a pilot of better-than-average abilities, no matter where you work, you will be asked to teach. It is a high honor, indeed, to be asked to teach. Flight schools, colleges, universities, the armed forces, airlines, the FAA, and many corporate flight departments have their own flight instructors. And they usually don't offer the position of training the pilots they will have to depend on to the underqualified or the inept. I offer as an example the U.S. Navy "Top Gun" Aerial Warfare School. These guys are "just" flight instructors, only they happen to work for the Navy.

It is a solemn responsibility to accept the flight instructor task. I consider it the highest form of compliment when people are willing to entrust me with the flight training of their son, daughter, husband, or wife. It is my honor to teach them a craft that few attempt, and at which even fewer excel. It is hard, sometimes frustrating work, but it's always rewarding.

If you would like to become a flight instructor, you will first need to attain your Commercial Pilot Certificate with Instrument Rating. You must then pass two written exams: Fundamentals of Instruction and Flight Instructor Airplane. With the writtens safely in the bag, all that remains is to complete the flight time and pass the oral and flight check. While all this only fills a couple of lines here, it will generally take about two years to complete these requirements as a full-time student starting from the beginning. It can also be done several other ways, such as in your spare time (which could take forever) or at one of those accelerated flight schools. My personal preference would be to opt for either of the first two choices. I am not a fan of forced training.

Now, you're ready to work. But where? The ranges of opportunities for employment as a new flight instructor are quite varied. Some new CFIs will be asked to work for the flight school that trained them. This invitation is usually predicated upon being in the above-average range I talked about previously. Others will have to use their initiative and read the trade magazines to learn where work is available.

If you desire to work in a certain area of the country, be advised that the areas that are described as attractive will often pay less than their counterparts that are perceived as less desirable. What I'm saying is that you will most likely receive a smaller salary in the warmer, tourist-laden climates like Florida or California than you will in upper peninsula of Michigan. The main thing to remember is that there are opportunities in nearly every area. All you have to do is decide where you want to live, and then go for it. Somewhere nearby there is always work for a good CFI.

In order to attain serious earning potential as a CFI, you will have to land a position with one of the large fixed-base flight schools, a university, corporation, or the military. These training institutions usually offer a steady salary, great benefits, and a constant stream of students.

For instance, a CFI I am very familiar with began his career in a small fixed-base operation in Texas. He worked there for a year and a half, flying over 1000 hours as an Air Force ROTC instructor. During this time, he also gained a lot of experience crop dusting, another high-tech, high-proficiency area that would prove invaluable to him in the near future.

One day, an acquaintance told him of a large university flight school that was interested in initiating an aerial application curriculum. He applied and was accepted. This combination of luck, talent, and proficiency culminated in his teaching for 20 years at the university. For him, it was the best flight instructor job he could have asked for. He received a very livable salary, had great medical and life insurance, annuities, etc., and worked only nine months per year. During this 20-year period, he flew approximately 360 to 370 hours per year, or about two hours per day. Oh, yes, he also taught ground classes three afternoons a week for two hours.

The university connections allowed him to work with many other areas of aviation such as the EPA, the FAA, and various state and local officials, and to speak at large gatherings at other universities about his favorite subject: aviation safety. The FAA appointed him an Accident Safety Counselor and a Designated Pilot Examiner, one of only 1700 worldwide. Another attractive bonus of college employment was the opportunity to attend classes without cost. This allowed his two children to obtain a nearly cost-free college education.

When the 20-year mark rolled around, this CFI retired with a hefty teacher's retirement account and a handsome annuity paid for by the university. He then moved to another college, where he went into administration and presently earns about twice as much money as he did at the previous university. This man is just over 50, has flown many thousands of hours, seen sights most men only dream of, was home nearly every night, loved every minute of his career, and is very close to being financially set for life. Who would have thought this possible for a person who began as, and still remains, "only" a flight instructor? Believe me, it can be a wonderful career.

Corporate pilot

"I just got back from 18 days in Europe," a corporate pilot friend said to me one day. "We were scheduled to be over there for six days, but business was better than they expected."

Now, this friend flies a Falcon 900 tri-jet, an extremely fine aircraft, and I imagine if I had to go to Europe, this would be as good a way to go as any. But 18 days for a six-day trip? Not me. He, on the other hand, wouldn't trade places with any other pilot in the world. He loves it!

Depending on your point of view, corporate flying has to be either the greatest thrill of all times, or the worst flying job imaginable. If the idea of rarely knowing where you will be going on a trip, or when you will return, sets you on fire, this is surely the flying job for you. The corporate world of flight carries with it some of the most interesting flying imaginable sometimes coupled with the most irregular hours that one could imagine.

On the other hand, many corporate flight departments are handled on a strict schedule. Crews know where and when they will fly, and for how long. They are run as if they were an airline. Many flight departments employ dispatchers who coordinate the scheduling of company aircraft with airlines, car rentals, etc., so the boss can minimize his travel time. With fewer and fewer "mom and pop" corporate aircraft flying, and profitability factoring heavily into the decision of whether to keep the aircraft or not, utilization has become a very important piece of the corporate picture.

I have many friends who are employed as corporate pilots, and they all, to a man, love their work. They like the fact that unlike airline pilots, they never get bored with a route, because they never know where they will be going next. They also seem to gain some odd sense of pleasure from this lack of regimen. In short, with most corporate flying, you fly to where the business is located. Sometimes this will take you to a certain city repeatedly, and sometimes it will be months between seeing a familiar area. You literally go when and where the boss wants to go and remain until he is ready to go home. You are, after all, carrying the corporation's finest in the back.

This diversity of time and locale seems to heighten the enjoyment of the true corporate pilot. The controlled chaos of juggling a schedule that matches business needs with aircraft schedules, weather delays, crew duty times, and scheduled and unscheduled maintenance is not a job for the fainthearted or the easily distracted. It takes a person with a high degree of flexibility to manage a career as a corporate pilot.

If a career in corporate aviation interests you, begin with your education. A degree in aviation with a minor in business will be the best two friends you could have at interview time. Your next best friends will be a couple thousand hours in your logbook, with about 500 hours multiengine time, and an A&P. Since many corporations like a combination pilot/A&P, an A&P certificate wouldn't hurt, if you can muster the time, energy, and expense in addition to your other schooling.

Whoa, wait a minute! This is getting to be real serious, isn't it? You bet it is. Don't think for one second that a Fortune 500 company is going to throw you into the front office of their Gulfstream IV just because you think it is something you'd like to do. With corporate aircraft in the 20-million-dollar range, and given the very nature of the cargo (executives) in the rear, it takes a very skilled person to break into the corporate ranks. This is some serious flying, and you will have to prepare accordingly. In fact, most corporate flight departments are run to airline standards. They legally operate under FAR Part 91, but boost it to the FAR Part 121 (airline) level for their flight operations.

Don't let these requirements slow you down. Just chart an intelligent course, build some flight time, preferably multiengine, and be prepared to settle for a copilot position for a few years. I know of several pilots who fly King-Airs for corporations who are not even typed in the aircraft. They signed on as copilots, and they wash the aircraft and even sweep up the hangar. They don't care. They know that one day the King-Air will turn into a Falcon 900 and they will be in the left seat. It's called *paying your dues*. After all, you can't be handed the knowledge required to fly a corporate jet to far-off places and arrive safely. You must start at the beginning and learn the trade. That is the primary reason for the corporate pilot's paycheck—the boss has faith he will get to and from the desired destination in the greatest comfort and safety. For that, he wants a pro.

The rate of pay varies from company to company, but a beginning pilot with a good background can start with a fair-sized corporation in the $26,000 to $35,000 range. Most corporate flight jobs peak at approximately $80,000 per year. The standard life, accident, and health insurance is usually a given, and many companies offer a bonus incentive program driven through corporate earnings. When you figure that many corporate pilots fly only three or four days per week, this begins to become pretty attractive. Most corporate pilots have about 120 hours of duty time per month, out of which they will fly about 40 hours.

Before you get too excited, let's look at what some pilots said about the good and bad of corporate flying. The most-often mentioned downside seemed to be the long days. Some pilots told me of days that begin at 4:30 a.m. and last until 7:00 p.m., or longer. The flight time on a day such as this is usually in the five-to-six hour range with the rest of the time spent on preflight and postflight duties or just waiting. Though the corporate pilot usually flies just three or four days per week, the average duty day lasts about 12 hours.

Independence, or the feeling of being self-employed without the financial responsibility, was one of the most used responses for the positive side of the corporate career. Variety in aircraft flown as well as destination was a close second. Good pay, the feeling of being needed, and equal administrative treatment rounded out the pluses.

The one point on which nearly all corporate pilots are united is that their career area will grow steadily more competitive as economics force the elimination of many flight departments in the near future. Quite a few corporate crews will find that they have lost their aircraft, as companies trim to their respective fighting weights for the predicted monetary competition in the next century. This will most likely require the successful applicant to be more skilled as the decade unfolds. There will be jobs, but there will also be competition for those jobs.

If your desire ranges in the corporate area, go for it. It has been my experience that the proficient, prepared pilot is always employed, and sooner or later, that employment is in the desired area.

Military

To many aviators, and would-be aviators, a career in military aviation is the very pinnacle of success. Just stop and think for a moment: How many pilots ever get the opportunity to fly an F-15, F-18, or C-5 Galaxy? If you make the grade as a military aviator, these are but a few of the aircraft you could potentially be assigned to fly. The opportunities are as boundless as the sky. You could catapult from the deck of a carrier in an F-14, pilot a C-5 to drop a load of cargo into some far-off country about which you have only read, or fly a low-level bomb run in an FB-111.

Since the map to the pilot's seat for all American military operations are nearly identical, I have arbitrarily chosen the Air Force to use as an example here. While the basic steps will usually remain unchanged, sometimes the requirements can change quickly. In order to get up-to-the-minute information for the military branch that interests you the most, I suggest you see your local recruiter. For instance, the Navy has only recently relaxed their vision requirements from the rigid 20-20 or better, which has been required for years.

In order to be considered for Air Force pilot training, you must be a college graduate, not less than $20^{1}/_{2}$ years of age nor more than $26^{1}/_{2}$ when you enter pilot training. These are the absolute musts.

The first step in earning your Air Force wings is to be commissioned as an Air Force officer. There are three paths to arrive at this goal:

1. You can enlist in the Air Force ROTC (Reserve Officer Training Corps) while attending one of the 600 participating colleges and universities across the United States. You simply enroll in the aerospace studies course at the time you register for your other freshman courses. Then, during your college years, the Air Force ROTC will prepare you for duty in the Air Force after graduation;

2. You can be chosen to attend the Air Force Academy. The Academy is a fully accredited four-year college, and upon graduation you are automatically qualified for your commission;

3. You can be commissioned by successfully graduating from the Officers Training School (OTS), a 12-week course offered by the Air Force at Lackland AFB, Texas. In OTS, people who already have earned their college degree are training to become Air Force officers.

Prior to entering the undergraduate pilot program, all prospective Air Force pilots have to have had flight training in ROTC or at the Academy, or possess a civilian private pilot license. Those entering training through OTS will go through a Flight Screening Program (FSP), during which time they must advance to solo in a T-41, a military version of the Cessna 172.

Upon graduation from ROTC, OTS, or the Academy, you will enter Air Force undergraduate pilot training. This is where the fun really begins. During these 49 weeks, you will receive 176 hours of intensive flight training divided between the T-37 and the T-38 jet trainers. The initial 4 $^{1}/_{2}$ months consist of training in the T-37, a subsonic jet trainer manufactured by Cessna. In this phase, you will be taught aerobatics, instrument and formation flying, and navigation.

After completing the T-37 phase, you train for an additional six months and 101 flight hours in the supersonic, twin-engine, T-38 "Talon." In this phase of training, you will delve more deeply into aerobatics, navigation, and instrument and formation flight while flying this advanced trainer, sometimes at altitudes as high as 55,000 feet.

After completing this final phase of undergraduate pilot training, you will receive your Air Force silver wings. You are now ready for advance training in the operational aircraft of the Air Force flight line. You will be selected to fly a type of aircraft based, for the most part, on the present needs of the Air Force. However, some latitude is given in allowing graduates to choose their desired aircraft. The training period for the operational aircraft is usually an additional three to six months.

The rate of pay in the Air Force, while not up to par with most civilian industry, is pretty good. It's good enough that a fair amount of Air Force pilots choose to make a career out of military aviation. The Air Force brochure says that it is "a unique way of life with an excellent compensation system to support it."

During training, a single, second lieutenant receives a base pay of $1862 each month and pays income taxes on this amount. Upon graduation, as a captain, the base pay jumps to $3062 per month. Additional bonuses are free housing in Junior Officer's Quarters and a subsistence (food) allowance. A major with 12 years service earns a base pay of $3149 per month, $706 for housing, and $127 for subsistence.

And, oh yes, flight pay. Flight pay ranges from $125 per month for the first two years through a scale that peaks during years seven through 17 with a monthly flight pay of $650. This is in addition to all other salaries.

Additionally, you receive nearly free medical and dental care for you and your family, 30 days of vacation annually, discount shopping at the base exchange and commissary, and base housing with free utilities.

Well, there you are. A college education, OTS, and about a year and a half of additional training buy you a career that begins with an eight-year commitment to the Air Force. That's really not bad considering the government invests over a million dollars in your training—and you get paid while you train.

There is one more consideration I want you to remember before you run awestruck for your nearest recruiter: In real life they shoot real bullets; so do you—and the bullets go both ways. That's the down side of a military career. If you can handle that, go get it; it can be a great way of life!

Other possibilities

The career areas I have discussed are merely the tips of a vast aviation complex we have to choose from. The uses for aircraft and pilots seem to grow with each passing day. The majority of the areas I talked about are for fixed-wing pilots, and I didn't touch on the career possibilities for pilots of helicopters, balloons, and blimps. I also didn't mention such nonflying areas such as A&P mechanic, flight dispatcher, steward, airport personnel, airport management, and support areas such as FBOs (fixed base operators), fuel retailers, etc. These areas make up a vast portion of aviation, and we couldn't get along without any of them, but I really wanted to deal with the area I know best—the cockpit and the opportunities that lie in the front of the fixed-wing aircraft.

There are many other flight positions that you could consider in your aviation career search. A few that come quickly to mind are fish spotting, fire bombing, crop dusting, sightseeing tours, aerial photography, pipeline patrol, powerline patrol, and air taxi and charter. I'm certain I left out a few, but I believe you can see from this the great variety of potential aviation careers.

A career in aviation is also, no matter what or where you fly, the absolute best way to make a living that exists on the face of this earth. At least to us pilots, it is. It is my sincere hope that you find your place in aviation; that you serve it well; and that you become as safe and competent a pilot as is humanly possible.

A good friend of mine perhaps said it best when he said to me, "Wouldn't it be horrible if we had to punch a clock like so many other poor mortals do? God, what a life we live. I can't imagine not flying for a living." Nor could I, my friend. Nor could I.

So, if flight is your dream, go get it. Don't let anyone talk you out of it for any reason. If you desire it, it can happen. Good luck and God bless you.

Abbreviations and acronyms

AAS	Advanced Automation System, or Airport Advisory Service
ACF	Area Control Facility
ADF	Automatic Direction Finding
ADIZ	Air Defense Identification Zone
AERA	Automated en Route Air Traffic Control
A/FD	Airport/Facility Directory
AFSS	Automated Flight Service Station
A/G	Air-ground communications
AGL	Above ground level
AIM	Aeronautical Information Manual
AIRMET	Airman's Meteorological Information
ALNOT	Alert Notice
AOPA	Aircraft Owners and Pilots Association
App	Approach Control
ARF	Airport Reservation Function
ARSA	Airport Radar Service Area
ARSR	Air Route Surveillance Radar
ARTCC	Air Route Traffic Control Center
ARTS	Automated Radar Terminal System
ARU	Airborne Radar Unit
ASDE	Automated Surface Detection Equipment
ASOS	Automated Surface Observation System
ASR	Airport Surveillance Radar
AT	Air Traffic
ATA	Airport Traffic Area
ATC	Air Traffic Control
ATCAA	Air Traffic Control Assigned Airspace
ATCRBS	Air Traffic Control Radar Beacon System
ATCS	Air Traffic Control Specialist
ATCT	Air Traffic Control Tower
ATIS	Automatic Terminal Information Service
ATS	Air Traffic Service
AWOS	Automated Weather Observation System
AWW	Severe Weather Alert
BRITE	Bright Radar Indicator Tower Equipment
CA	Conflict Alert
CAA	Civil Aeronautics Authority

CAB	Civil Aeronautics Board
CAMI	Civil Aeromedical Institute (FAA in Oklahoma City)
CA/MSAW	Conflict Alert/Minimum Safe Altitude Warning
CARF	Central Altitude Reservation Function
CD	Clearance Delivery
Center	Air Route Traffic Control Center
CERAP	Combined Center Radar Approach Control
CIP	Capital Investment Plan
CL	Clearance (Delivery)
CONUS	Continental, Contiguous, or Conterminous United States
CRT	Cathode-Ray Tube
CT	Control Tower
CTAF	Common Traffic Advisory Frequency
CWA	Center Weather Advisory
CWSU	Center Weather Service Unit
CZ	Control Zone
D	Developmental controller or specialist
Dep	Departure Control
DF	Direction Finder
DLP	Data Link Processor
DME	Distance-Measuring Equipment
DME/P	Precision Distance-Measuring Equipment
DOD	Department of Defense
DOT	Department of Transportation
DUAT	Direct User Access Terminal
DVFR	Defense Visual Flight Rules
DVOR	Doppler Very High-Frequency Omnidirectional Range
EARTS	En Route Automated Radar Tracking System
EFAS	En Route Flight Advisory Service
ELAC	En Route Low-Altitude Chart
ELT	Emergency Locator Transmitter
EPA	Environmental Protection Agency
ESP	En Route Spacing Program
ETA	Estimated Time of Arrival
ETD	Estimated Time of Departure
ETE	Estimated Time en Route
FA	Area Forecast
FAA	Federal Aviation Administration
FARs	Federal Aviation Regulations
FBO	Fixed-Base Operator
FDC	Flight Data Center

FDEP	Flight Data Entry and Printout
FPL	Full-Performance Level controller or specialist
FSAS	Flight Service Automation System
FSP	Flight Strip Printer
FSS	Flight Service Station
GC	Ground Control
GCA	Ground Control Approach
GOES	Geostationary Operational Environmental Satellite
GPS	Global Positioning System
HIWAS	Hazardous In-Flight Weather Advisory Service
HVAC	Heating, Ventilating, and Air Conditioning
IATA	International Air Transport Association
ICAN	International Convention for Air Navigation
ICAO	International Civil Aviation Organization
IFAPA	International Federation of Airline Pilots Association
IFR	Instrument Flight Rules
ILS	Instrument Landing System
IMC	Instrument Meteorological Conditions
INREQ	Request for Information
IPAI/D	Identification-Position-Altitude-Intentions (or) Destination
IR	Military Instrument Flight Training Route
ISSS	Initial Sector Suite Subsystem
IVRS	Interim Voice Response System
LLWAS	Low-Level Wind Shear Alert System
LORAN	Long Range Navigation
MARSA	Military Assumes Responsibility for Separation of Aircraft
MLS	Microwave Landing System
MOA	Military Operations Area
Mode 3/A	Standard transponder without altitude-reporting capability
Mode C	Standard transponder with altitude-reporting capability
Mode S	Selectively addressable transponder with data link
MRU	Military Radar Unit
MSAW	Minimum Safe Altitude Warning
MSL	Mean Sea Level
MTR	Military Training Route
MULTICOM	Nongovernment air-air radio communications frequency
NAR	National Airspace Review
NAS	National Airspace System

NASP	National Airspace System Plan
Navaid	Navigational Aid
NDB	Nondirectional Beacon
NEXRAD	Next Generation Weather Radar
NFCT	Nonfederal Control Tower
NM	Nautical Miles
NOAA	National Oceanic and Atmospheric Administration
NOS	National Ocean Service
NOTAM	Notice To Airmen
NPRM	Notice of Proposed Rule-Making
NTAP	Notice To Airmen Publication
NWS	National Weather Service
PAR	Precision Approach Radar
PATWAS	Pilots' Automatic Telephone Weather Answering Service
PCA	Positive Control Area
PIREP	Pilot Report
PVD	Plan Visual Display
RAPCON	Radar Approach Control (military)
RCAG	Remote Center Air-Ground Communications Facility
RCF	Remote Communications Facility
RCO	Remote Communications Outlet
RDT&E	Research, Development, Testing, and Evaluation
RF	Radio Failure
RML	Radar Microwave Link
RMM	Remote Maintenance Monitoring
RNAV	Area Navigation
RTR	Remote Transmitter-Receiver
RVR	Runway Visual Range
SAR	Search And Rescue
SFAR	Special Federal Aviation Regulation
SIGMET	Significant Meteorological Information
SM	Statute Miles
Squawk	Activate specific number code in the transponder
SR	Sunrise
SS	Sunset
SUA	Special Use Airspace
SVFR	Special Visual Flight Rules
TA	Transition Area
TAC	Terminal Area Chart
TACAN	Tactical Air Navigation
TCA	Terminal Control Area

TCAS	Traffic Alert and Collision Avoidance System
TDWR	Terminal Doppler Weather Radar
TIBS	Transcribed Information Briefing Service
TML	Television Microwave Link
TMS	Traffic Management System
TMU	Traffic Management Unit
TPX	Military Beacon System
TRACAB	Terminal Radar Approach Control in the Tower Cab
TRACON	Terminal Radar Approach Control
TRSA	Terminal Radar Service Area
TWEB	Transcribed Weather Broadcast
UA	Pilot Report
UHF	Ultra High Frequency
UUA	Urgent Pilot Report
UNICOM	Nongovernment air/ground radio communications facility
UTC	Coordinated Universal Time
VFR	Visual Flight Rules
VHF	Very High Frequency
VOR	VHF Omnidirectional Range
VOR/DME	VOR also equipped with DME
VORTAC	VOR collocated with TACAN
VR	Military VFR Training Route
VRS	Voice Response System
VSCS	Voice Switching Communications System
WMSC	Weather Message Service Center
WST	Convective SIGMET
XFSS	Auxiliary Flight Service Station
ZULU	Sometimes used to mean UTC time, e.g., "1500 zulu"

Glossary

above ground level (agl) Height, usually in feet, above the surface of the earth.

above sea level (ASL) *See* mean sea level.

administrator The Federal Aviation Administrator or any person to whom he has delegated his authority in the matter concerned.

air defense identification zone (ADIZ) The area of airspace over land or water, within which the ready identification, location, and the control of aircraft are required in the interest of national security for the United States.

air traffic clearance An authorization by air traffic control for the purpose of preventing collision between known aircraft and for an aircraft to proceed under specified traffic conditions within controlled airspace.

air traffic control A service operated by appropriate authority to promote the safe, orderly, and expeditious flow of air traffic.

air traffic Aircraft operating in the air or on an airport surface, exclusive of loading ramps and parking areas.

aircraft engine An engine that is used or intended to be used for propelling aircraft. It includes turbosuperchargers and accessories necessary for its functioning, but does not include propellers.

aircraft A device that is used or intended to be used for flight in the air.

airframe The fuselage, booms, nacelles, cowlings, fairings, airfoil surfaces (including rotors but excluding propellers and rotating airfoils of engines), and landing gear of an aircraft and their accessories and controls.

airplane A heavier-than-air, engine-driven, fixed-wing aircraft that is supported in flight by the dynamic reaction of the air against its wings.

airport traffic area (Unless otherwise specifically designated in FAR Part 93) that airspace within a horizontal radius of five statute miles from the geographical center of any airport at which a control tower is operating, extending from the surface up to, but not including, an altitude of 3000 feet above the elevation of the airport.

airport An area of land or water that is used or intended to be used for the landing and takeoff of aircraft. Includes its buildings and facilities, if any.

airship An engine-driven, lighter-than-air aircraft that can be steered.

alternate airport An airport at which an aircraft may land if a landing at the intended airport becomes inadvisable.

altitude engine A reciprocating aircraft engine having a rated take-off power that is producible from sea level to an established higher altitude.

approved (Unless used with reference to another thing) approved by the administrator.

area navigation (RNAV) A method of navigation that permits aircraft operation on any desired course within the coverage of station-referenced navigation signals or within the limits of a self-contained system capability. Types include VORTAC, OMEGA/VLF, inertial, loran, and GPS.

area navigation high route An area navigation route within the airspace extending upward from and including 18,000 feet MSL to flight level 450.

area navigation low route An area navigation route within the airspace extending upward from 1200 feet above the surface of the earth to, but not including, 18,000 feet MSL.

ARFF *See* certificated airport.

automated weather observing system (AWOS) A computerized system that measures some or all of the following: sky conditions, visibility, precipitation, temperature, dewpoint, wind, and altimeter setting.

balloon A lighter-than-air aircraft that is not engine driven.

Bernoulli effect The venturi effect of terrain that causes a decrease in air pressure, resulting in altimeter error.

brake horsepower The power delivered at the propeller shaft (main drive or main output) of an aircraft engine.

calibrated airspeed Indicated airspeed of an aircraft, corrected for position and instrument error. Calibrated airspeed is equal to true airspeed in standard atmosphere at sea level.

category (1) As used with respect to the certification, rating, privileges, and limitations of airmen, means a broad classification of aircraft. Examples include: airplane; rotorcraft; glider; and lighter-than-air; and (2) as used with respect to the certification of aircraft, means a grouping of aircraft based upon intended use or operating limitation. Examples include: transport; normal; utility; acrobatic; limited; restricted; and provisional.

catenary On aeronautical charts, a cable, power line, cable car, or similar structure suspended between peaks, a peak and valley below, or across a canyon or pass.

ceiling The height above the earth's surface of the lowest layer of clouds or obscuring phenomena that is reported as *broken, overcast,* or *obscuration,* and not classified as *thin* or *partial.*

certificated airport (FAR 139) An airport certified for commercial air carriers under FAR Part 139, which relates to the requirement for crash, fire, and rescue equipment (ARFF).

changeover point (COP) The charted point where a pilot changes from one navigational facility to the next for course guidance.

CHUM *Chart Updating Manual* used to supplement Defense Mapping Agency charts.

civil aircraft Aircraft other than public aircraft.

class (1)As used with respect to the certification, rating, privileges, and limitation of airmen, means a classification of aircraft within a category having similar operating characteristics. Examples include: single-engine; multiengine; land; water; gyroplane; helicopter; airship; and free balloon; and (2) as used with respect to the certification of aircraft, means a broad grouping of aircraft having similar characteristics of propulsion, flight, or landing. Examples include: airplane; rotorcraft; glider; balloon; landplane; and seaplane.

climb gradient A rate of climb in feet per nautical mile, usually associated with a requirement for a departure procedure.

commercial operator A person who, for compensation or hire, engages in the carriage of aircraft in air commerce of persons or property, other than as an air carrier or foreign air carrier or under the authority of Part 375 of this Title. Where it is doubtful that an operation is for *compensation* or *hire,* the test applied is whether the carriage by air is merely incidental to the person's other business or is, in itself, a major enterprise for profit.

common traffic advisory frequency (CTAF) A frequency designed for the purpose of airport advisory practices (pilot's position and intentions during takeoff and landing) at uncontrolled airports.

Consol/CONSOLAN A long-range radio aid to navigation, the emissions of which, by means of their radio frequency modulation characteristics, enable bearings to be determined.

controlled airspace Airspace designated as a continental control area, control area, control zone, terminal control area, or transition area, within which some or all aircraft may be subject to air traffic control.

coordinated universal time (UTC) Formerly Greenwich mean time (GMT), also known as Z or zulu time, UTC is the international time standard.

crewmember A person assigned to perform duty in an aircraft during flight time.

critical altitude The maximum altitude at which, in standard atmosphere, it is possible to maintain, at a specified rotational speed, a specified power or a specified manifold pressure. Unless otherwise stated, the critical altitude is the maximum altitude at which it is possible to maintain, at the maximum continuous rotational speed, one of the following: (1) the maximum continuous power, in the case of engines for which this power rating is the same at sea level and at the rated altitude; and (2) the maximum continuous rated manifold pressure, in the case of engines, the maximum continuous power of which is governed by a constant manifold pressure.

critical elevation The highest elevation in any group of related and more or less similar relief formations.

critical engine The engine whose failure would most adversely affect the performance of handling qualities of an aircraft.

culture Features of the terrain that have been constructed by man, including roads, buildings, canals, and boundary lines.

datum Any quantity, or set of quantities, that might serve as a reference or base for other quantities.

direct user access terminal (DUAT) A computer terminal where pilots can directly access meteorological and aeronautical information without the assistance of an FSS.

drainage patterns Drainage patterns are the overall appearance of features associated with water, such as shorelines, rivers, lakes, and marshes, or any similar feature.

DUAT *See* direct user access terminal.

equivalent airspeed The calibrated airspeed of an aircraft corrected for adiabatic compressible flow for the particular altitude. Equivalent airspeed is equal to calibrated airspeed in standard atmosphere at sea level.

fixed-base operator (FBO) A private vendor of airport services, such as fuel, repairs, and tiedown facilities.

flame resistant Not susceptible to combustion to the point of propagating a flame, beyond safe limits, after the ignition source is removed.

flammable With respect to fluid or gas, susceptible to igniting readily or exploding.

flap extended speed The highest speed permissible with wing flaps in a prescribed extended position.

flash resistant Not susceptible to burning violently when ignited.

flight crewmember A pilot, flight engineer, or flight navigator assigned to duty in an aircraft during flight time.

flight level (FL) Altitude with the altimeter set to standard pressure (29.92 inches or 1013.2 millibars), pressure altitude. Flight level is the altitude reference for high-altitude flights, usually above 18,000 feet.

flight plan Specified information relating to the intended flight of an aircraft that is filed orally or in writing with air traffic control.

flight time The time from the moment the aircraft first moves under its own power for the purpose of flight until the moment it comes to rest at the next point of landing— *block-to-block time.*

flight visibility The average forward horizontal distance from the cockpit of an aircraft in flight, at which prominent unlighted objects can be seen and identified by day and prominent lighted objects can be seen and identified by night.

FLIP *Flight Information Publication* supplements Defense Mapping Agency charts.

flumes Water channels used to carry water to a source of power, such as a waterwheel.

glider A heavier-than-air aircraft that is supported in flight by the dynamic reaction of the air against its lifting surfaces and whose free flight does not depend principally on an engine.

global positioning system (GPS) A navigational system based on low-earth orbiting satellites.

GPS Global Positioning System is a form of navigation using triangulation of signals from satellite transmitters to receivers in an aircraft.

great circle A circle on the surface of the earth, the geometric plane of which passes through the center of the earth.

Greenwich meridian The meridian through Greenwich, England, serves as the reference for Greenwich time (now coordinated universal time, UTC). It is accepted almost universally as the prime meridian, or the origin of measurements of longitude.

ground visibility Prevailing horizontal visibility near the earth's surface as reported by the U.S. National Weather Service or an accredited observer.

gyrodine A rotorcraft whose rotors are normally engine-driven for taking off, hovering, and landing as well as for forward flight through part of its speed range. Means of propulsion, consisting usually of conventional propellers, are independent of the rotor system.

gyroplane A rotorcraft whose rotors are not engine-driven except for initial starting, but are made to rotate by action of the air when the rotorcraft is moving; and whose means of propulsion, consisting usually of conventional propellers, is independent of the rotor system.

hachures A method of representing relief upon a map or chart by shading in short disconnected lines drawn in the direction of the slopes.

hazardous inflight weather advisory service (HIWAS) A continuous broadcast of hazardous weather conditions over selected NAVAIDs.

helicopter A rotorcraft that for its horizontal motion depends principally on its engine-driven rotors.

heliport An area of land, water, or structure used or intended to be used for the landing and taking off of helicopters.

HIWAS *See* hazardous inflight weather advisory service.

horizontal datum A geodetic reference point that is the basis for horizontal control surveys, where latitude and longitude are known.

hummocks A wooded tract of land that rises above an adjacent marsh or swamp.

hydrography The science that deals with the measurements and description of the physical features of the oceans, seas, lakes, rivers, and their adjoining coastal areas, with particular reference to their use for navigational purposes.

hypsography The science or art of describing elevations of land surfaces with reference to a datum, usually sea level.

hypsometric tints A method of showing relief on maps and charts by coloring, in different shades, those parts that lie between selected levels.

ICAO *See* International Civil Aviation Organization.

idle thrust The jet thrust obtained with the engine power control lever set at the stop for the least-thrust position at which it can be placed.

IFR conditions Weather conditions below the minimum for flight under visual flight rules.

IFR over-the-top With respect to the operation of aircraft, the operation of an aircraft over-the-top on an IFR flight plan when cleared by air traffic control to maintain *VFR conditions* or *VFR conditions on top.*

indicated airspeed The speed of an aircraft as shown on its pitot static airspeed indicator calibrated to reflect standard atmosphere adiabatic compressible flow at sea level uncorrected for airspeed system errors.

instrument A device using an internal mechanism to show visually or aurally the attitude, altitude, or operation of an aircraft or aircraft part. It includes electronic devices for automatically controlling an aircraft in flight.

International Civil Aviation Organization (ICAO) A specialized agency of the United Nations whose objective is to develop the principles and techniques of international air navigation and foster planning and development of international civil air transport.

isogonic lines Lines of equal magnetic declination for a given time. The difference between true and magnetic north.

Julian date The date based on the Julian calendar. Days of the year are numbered consecutively from 001 for January 1; the year precedes the three-digit day group (91244 means September 1, 1991).

landing gear extended speed The maximum speed at which an aircraft can be safely flown with the landing gear extended.

landing gear operating speed The maximum speed at which the landing gear can be safely extended or retracted.

large aircraft Aircraft of more than 12,500 pounds, maximum certificated takeoff weight.

latitude A linear or angular distance that is measured north or south of the equator.

lighter-than-air aircraft Aircraft that can rise and remain suspended by using contained gas weighing less than the air that is displaced by the gas.

load factor The ratio of a specified load to the total weight of the aircraft. The specified load is expressed in terms of any of the following: aerodynamic forces, inertia forces, or ground or water reactions.

location identifier Consisting of three to five alphanumeric characters, location identifiers are contractions used to identify geographical locations, navigational aids, and airway intersections.

longitude A linear or angular distance that is measured east or west from a reference meridian, usually the Greenwich meridian.

loran A *lo*ng range *ra*dio *n*avigation position fixing system using the time difference of reception of pulse type transmissions from two or more fixed stations.

loran TD correction A correction factor, due to seasonal variations in loran signals, that must be set into loran receivers before beginning a loran RNAV approach.

mach number The ratio of true airspeed to the speed of sound.

magenta A purplish-red color used on aeronautical charts to distinguish different features.

mangrove Any of a number of evergreen shrubs and trees growing in marshy and coastal tropical areas. A nipa is a palm tree indigenous to these areas.

manifold pressure Absolute pressure as measured at the appropriate point in the induction system and usually expressed by inches of mercury.

maximum authorized altitude (MAA) The highest altitude, for which an MEA is designated, where adequate NAVAID signal coverage is assured.

maximum elevation figure (MEF) The MEF represents the highest elevation, including terrain or other vertical obstacles, bounded by the ticked lines of the latitude/longitude grid on a chart.

mean sea level (MSL) Altitude above mean or average sea level. This is the reference altitude for most charted items. In Canada, it is called above sea level (ASL).

medical certificate Acceptable evidence of physical fitness on a form prescribed by the Administrator.

meridian A north-south reference line, particularly a great circle through the geographical poles of the Earth, from which lines of longitude are determined.

minimum reception altitude (MRA) The lowest altitude required to receive adequate NAVAID signals to determine specific fixes.

NAVAID Any type of radio aid to navigation.

navigable airspace Airspace at and above the minimum flight altitudes prescribed by regulations, including airspace needed for safe takeoff and landing.

night The time between the end of evening civil twilight and the beginning of morning civil twilight, as published in the American Air Almanac, converted to local time.

nipa *See* mangrove.

nonperennial *See* perennial.

North American datum The horizontal datum for the United States revised in 1983, referred to as the North American Datum 1983 (NAD 83).

NOTAM file As used in a flight supplement, the associated weather or facility identifier (OAK) where notices to airmen for the associated facility will be located.

OMEGA A long-range hyperbolic navigation system designed to provide worldwide coverage for navigation.

operate With respect to aircraft, means use, cause to use, or authorize to use aircraft for the purpose (except as provided in FAR 91) of air navigation including the piloting of aircraft, with or without the fight of legal control (as owner, lessee, or otherwise).

operational control With respect to a flight, the exercise of authority over initiating, conducting, or terminating a flight.

parallel A circle on the surface of the earth, parallel to the plane of the equator and connecting all points of equal latitude.

penstock *See* flume.

perennial A feature, such as a lake or stream that contains water year round, as opposed to nonperennial, a feature that is intermittently dry.

person An individual, firm, partnership, corporation, company, association, joint-stock association, or governmental entity. It includes a trustee, receiver, assignee, or similar representative of any of them.

pilot in command The pilot responsible for the operation and safety of an aircraft during flight time.

pilotage Navigation by visual reference to landmarks.

pitch setting The propeller blade setting as determined by the blade angle measured in a manner, and at a radius, specified by the instruction manual for the propeller.

planimetry Planimetry is the depiction of man-made and natural features, such as woods and water, but does not include relief.

positive control Control of all air traffic, within designated airspace, by air traffic control.

preventive maintenance Simple or minor preservation operations and the replacement of small standard parts not involving complex assembly operation.

prime meridian *See* Greenwich meridian.

prohibited area Designated airspace within which the flight or aircraft is prohibited.

projection A systematic drawing of lines on a plan surface to represent the parallels of latitude and the meridians of longitude of the earth.

propeller A device for propelling an aircraft that has blades on an engine-driven shaft and that, when rotated, produces by its action on the air a thrust approximately perpendicular to its plane of rotation. It includes control components normally supplied by its manufacturer, but does not include main and auxiliary rotors or rotating airfoils of engines.

public aircraft Aircraft used only in the service of a government or a political subdivision. It does not include any government-owned aircraft engaged in carrying persons or property for commercial purposes.

QNE This is the altitude shown on the altimeter with the altimeter set to 29.92 inches. See pressure altitude.

QNH This is the altitude above mean sea level displayed on the altimeter when the altimeter-setting window is set to the local altimeter setting.

rated takeoff power (With respect to reciprocating, turbopropeller, and turboshaft engine type certification) the approved brake horse-power that is developed statically under standard sea level conditions, within the engine operating limitations established under Part 33, and limited in use to periods of not over five minutes for takeoff operation.

rating A statement that, as a part of a certificate, sets forth special conditions, privileges, or limitation.

relief Relief is the inequalities of elevation and the configuration of land features on the surface of the earth.

reporting point A geographical location in relation to which the position of an aircraft is reported.

restricted area Airspace designated within which the flight of aircraft, while not wholly prohibited, is subject to restriction.

rhumb line A line on the surface of the earth cutting all meridians at the same angle.

RNAV waypoint (W/P) A predetermined geographical position used for route or instrument approach defined for progress-reporting purposes that is defined relative to a VORTAC station position.

rotorcraft A heavier-than-air aircraft that depends principally for its support in flight on the lift generated by one or more rotors.

route segment A part of a route. Each end of that part is identified by (1) a continental or insular geographical location; or (2) a point at which a definite radio fix can be established.

runway visual range (RVR) The horizontal distance a pilot will be able to see high-intensity lights down the runway from the approach end.

scale The ratio or fraction between the distance on a chart and the corresponding distance on the surface of the Earth.

SCATANA *See* security control of air traffic and navigation aids.

sea-level engine A reciprocating aircraft engine having a rated take-off power that is producible only at sea level.

second in command A pilot who is designated to be second in command of an aircraft during flight time.

security control of air traffic and navigation aids (SCATANA) A plan for the security control of civil and military air traffic and NAVAIDs under various conditions.

shaded relief A method of shading areas on a map or chart so that they would appear in shadow if illuminated from the northwest.

show (Unless the context otherwise requires) to show to the satisfaction of the Administrator.

small aircraft Aircraft of 12,500 pounds or less maximum certification takeoff weight.

special-use airspace (SUA) Airspace where activities must be confined because of their nature, such as military operations or national security. Restrictions to flight might be imposed, or they might alert pilots to hazards of concentrated, high-speed, or acrobatic military flying.

spot elevation A point on a chart where elevation is noted, usually the highest point on a ridge or mountain range.

standard atmosphere The atmosphere defined in *U.S. Standard Atmosphere,* 1962 (Geopotential altitude tables).

standard service volume (SSV) The distances and altitude that a particular NAVAID can be relied upon for accurate navigational guidance.

stopway An area beyond the takeoff runway, no less wide than the runway and centered upon the extended centerline of the runway, able to support the airplane during an aborted takeoff without causing structural damage to the airplane and designated by the airport authorities for use in decelerating the airplane during an aborted takeoff.

takeoff power (1) With respect to reciprocating engines, the brake horsepower that is developed under standard sea-level conditions, and under the maximum conditions of crankshaft rotational speed and engine manifold pressure approved for the normal takeoff and limited in continuous use to the period of time shown in the approved engine specification; and (2) with respect to turbine engines, the brake horsepower that is developed under static conditions at a specified altitude and atmospheric temperature, and under the maximum conditions of rotorshaft rotational speed and gas temperature approved for the normal takeoff, and limited in continuous use to the period of time shown in the approved engine specification.

topography The configuration of the surface of the earth, including its relief, the position of its streams, roads, cities, and the like.

touchdown zone The first 3000 feet of the runway.

touchdown zone elevation (TDZE) The highest elevation in the first 3,000 feet of the landing surface.

traffic pattern The traffic flow that is prescribed for aircraft landing at, taxiing on, or taking off from an airport.

true airspeed The airspeed of an aircraft relative to undisturbed air.

tundra A rolling, treeless, often marshy plain, usually associated with arctic regions.

type (1) As used with respect to the certification, rating, privileges, and limitations of airmen, a specific make and basic model of aircraft, including modifications thereto that do not change its handling or flight characteristics. Examples include: DC-10, 1049, and F-14; (2)

as used with respect to the certification of aircraft, those aircraft that are similar in design. Examples include: DC-7 and DC-7C; 1049G and 1049H; and F-27 and F-27F; and (3) as used with respect to the certification of aircraft engines, those engines that are similar in design. For example, JT8D and JT8D-7 are engines of the same type and JT9D-3A and JT9D-7 are engines of the same type.

unicom A nongovernment communications facility that can provide airport advisory information.

United States (In a geographical sense) (1) the states, the District of Columbia, Puerto Rico, and the possessions, including the territorial waters; and (2) the airspace of those areas.

UTC *See* coordinated universal time.

V speeds In general, the reference-indicated airspeeds used to limit an aircraft's performance.

V_a Design maneuvering speed. The maximum speed at which the pilot may use full, abrupt control travel without causing structural damage to the aircraft.

V_{fe} Maximum flap extension speed. Shown at the top (highest speed) of the white arc on the airspeed indicator.

vignette A gradual reduction in density so that a line appears to fade in one direction, often used to distinguish airspace boundaries.

visual descent point (VDP) The point on a nonprecision straight-in approach where normal descent from the minimum descent altitude to the runway touchdown point can begin.

V_{le} Maximum landing gear extension speed.

V_{lo} Maximum speed to operate the landing gear up or down.

V_{lof} Liftoff speed.

V_{mc} The minimum airspeed at which it's possible to maintain directional control of the aircraft within 20 degrees of heading and, thereafter, maintain straight flight with not more than 5 degrees of bank if one engine fails suddenly.

V_{ne} The maximum speed at which an aircraft may be operated under any circumstances.

V_{no} Maximum velocity for normal operations found at the top of the green arc on the airspeed indicator.

V_r Rotation airspeed.

V_{s1} Stall speed in the clean configuration with landing gear and flaps up and power at idle.

V_{so} Stall speed in the landing configuration with landing gear and flaps down and power at idle. Is shown on the airspeed indicator as the lowest number on the white arc.

V_{sse} Minimum speed at which to purposely fail an engine during flight.

V_x Best angle of climb airspeed. This airspeed gives the aircraft the greatest gain of altitude in a given distance.

V_{xse} Best angle of climb—single-engine. This airspeed gives the aircraft the greatest gain of altitude in a given distance when operating on a single engine.

V_y Best rate of climb airspeed. This airspeed gives the aircraft the greatest gain of altitude in a given period of time.

V_{yse} Best rate of climb airspeed. This airspeed gives the aircraft the greatest gain of altitude in a given period of time when operating on a single engine.

zulu (Z) *See* coordinated universal time.

Index

Note: **Boldface** numbers indicate illustrations; *italic t* indicates a table.

About the Author

Lewis Bjork, a veteran pilot and aviation writer, is the author of *Piloting for Maximum Performance* and *Piloting at Night.* He is also co-editor of the upcoming *Van Sickle's Modern Airmanship*, Eighth Edition. He lives in Taylorsville, Utah.